IBN AL-HAYTHAM AND ANALYTICAL MATHEMATICS

This volume provides a unique primary source on the history and philosophy of mathematics and the exact sciences in the medieval Arab world. The second of five comprehensive volumes, this book offers a detailed exploration of Arabic mathematics in the eleventh century as embodied in the legacy of the celebrated polymath al-Hasan ibn al-Haytham.

Extensive analyses and annotations from the eminent scholar, Roshdi Rashed, support a number of key Arabic texts from Ibn al-Haytham's treatises in infinitesimal mathematics, translated here into English for the first time. Rashed shows how Ibn al-Haytham's works demonstrate a remarkable mathematical competence in mathematical subjects like the quadrature of the circle and of lunes, the calculation of the volumes of paraboloids, the problem of isoperimetric plane figures and solid figures with equal surface areas, along with the extraction of square and cubic roots, etc.

The present text is complemented by the first volume of *A History of Arabic Sciences and Mathematics*, which focused on founding figures and commentators in the ninth- and tenth-century Archimedean–Apollonian mathematical 'School of Baghdad'. This constellation of works illustrates the historical and epistemological development of 'infinitesimal mathematics' as it became clearly articulated in the oeuvre of Ibn al-Haytham.

Contributing to a more informed and balanced understanding of the internal currents of the history of mathematics and the exact sciences in Islam, and of its adaptive interpretation and assimilation in the European context, this fundamental text will appeal to historians of ideas, epistemologists and mathematicians at the most advanced levels of research.

Roshdi Rashed is one of the most eminent authorities on Arabic mathematics and the exact sciences. A historian and philosopher of mathematics and science and a highly celebrated epistemologist, he is currently Emeritus Research Director (distinguished class) at the Centre National de la Recherche Scientifique (CNRS) in Paris, and is the former Director of the Centre for History of Arabic and Medieval Science and Philosophy at the University of Paris (Denis Diderot, Paris VII). He also holds an Honorary Professorship at the University of Tokyo and an Emeritus Professorship at the University of Mansourah in Egypt.

THE FORMATION OF HANBALISM
Piety into power
Nimrod Hurvitz

ARABIC LITERATURE
An overview
Pierre Cachia

STRUCTURE AND MEANING IN MEDIEVAL ARABIC AND PERSIAN LYRIC POETRY
Orient pearls
Julie Scott Meisami

MUSLIMS AND CHRISTIANS IN NORMAN SICILY
Arabic-speakers and the end of Islam
Alexander Metcalfe

MODERN ARAB HISTORIOGRAPHY
Historical discourse and the nation-state
Youssef Choueiri

THE PHILOSOPHICAL POETICS OF ALFARABI, AVICENNA AND AVERROES
The Aristotelian reception
Salim Kemal

PUBLISHED BY ROUTLEDGE

1. THE EPISTEMOLOGY OF IBN KHALDUN
Zaid Ahmad

2. THE HANBALI SCHOOL OF LAW AND IBN TAYMIYYAH
Conflict or concilation
Abdul Hakim I Al-Matroudi

3. ARABIC RHETORIC
A pragmatic analysis
Hussein Abdul-Raof

4. ARAB REPRESENTATIONS OF THE OCCIDENT
East–West encounters in Arabic fiction
Rasheed El-Enany

5. GOD AND HUMANS IN ISLAMIC THOUGHT
Abd al-Jabbār, Ibn Sīnā and al-Ghazālī
Maha Elkaisy-Friemuth

6. ORIGINAL ISLAM
Malik and the Madhhab of Madina
Yasin Dutton

7. AL-GHAZALI AND THE QUR'AN
One book, many meanings
Martin Whittingham

8. BIRTH OF THE PROPHET MUHAMMAD
Devotional piety in Sunni Islam
Marion Holmes Katz

9. SPACE AND MUSLIM URBAN LIFE
At the limits of the labyrinth of Fez
Simon O'Meara

10. ISLAM AND SCIENCE
The intellectual career of Nizam al-Din al-Nizaburi
Robert G. Morrison

11. IBN 'ARABÎ – TIME AND COSMOLOGY
Mohamed Haj Yousef

12. THE STATUS OF WOMEN IN ISLAMIC LAW AND SOCIETY
Annotated translation of al-Ṭāhir al-Ḥaddād's *Imra'tunā fi 'l-sharīʿa wa 'l-mujtamaʿ*, with an introduction
Ronak Husni and Daniel L. Newman

13. ISLAM AND THE BAHA'I FAITH
A comparative study of Muhammad 'Abduh and 'Abdul-Baha 'Abbas
Oliver Scharbrodt

14. COMTE DE GOBINEAU AND ORIENTALISM
Selected Eastern writings
Translated by Daniel O'Donoghue Edited by Geoffrey Nash

15. EARLY ISLAMIC SPAIN
The history of Ibn al-Qūṭīya
David James

16. GERMAN ORIENTALISM
The study of the Middle East and Islam from 1800 to 1945
Ursula Wokoeck

17. MULLĀ ṢADRĀ AND METAPHYSICS
Modulation of being
Sajjad H. Rizvi

18. SCHOOLS OF QUR'ANIC EXEGESIS
Genesis and development
Hussein Abdul-Raof

19. AL-GHAZALI, AVERROES AND THE INTERPRETATION OF THE QUR'AN
Common sense and philosophy in Islam
Avital Wohlman, translated by David Burrell

20. EASTERN CHRISTIANITY IN THE MODERN MIDDLE EAST
Edited by Anthony O'Mahony and Emma Loosley

21. ISLAMIC REFORM AND ARAB NATIONALISM
Expanding the crescent from the Mediterranean to the Indian Ocean (1880s–1930s)
Amal N. Ghazal

22. ISLAMIC ETHICS
Divine command theory in Arabo-Islamic thought
Mariam al-Attar

23. MUSLIM FORTRESSES IN THE LEVANT
Between Crusaders and Mongols
Kate Raphael

24. BEING HUMAN IN ISLAM
The impact of the evolutionary worldview
Damian Howard

25. THE UAE AND FOREIGN POLICY
Foreign aid, identities and interests
Khalid S. Almezaini

26. A HISTORY OF EARLY AL-ANDALUS
The *Akhbār majmūʿa*
David James

27. INSPIRED KNOWLEDGE IN ISLAMIC THOUGHT
Al-Ghazali's theory of mystical cognition and its Avicennian foundation
Alexander Treiger

28. SHI'I THEOLOGY IN IRAN
The challenge of religious experience
Ori Goldberg

29. FOUNDING FIGURES AND COMMENTATORS IN ARABIC MATHEMATICS
A history of Arabic sciences and mathematics, Volume 1
Roshdi Rashed (translated by Roger Wareham, with Chris Allen and Michael Barany, under the supervision of Nader El-Bizri)

30. THE MUSLIM CONQUEST OF IBERIA
Medieval Arabic narratives
Nicola Clarke

31. ANGELS IN ISLAM
Jalal al-Din al-Suyuti's *al-Haba'ik fi akhbar al-mala'ik*
Stephen Burge

32. THEOLOGICAL APPROACHES TO QUR'ANIC EXEGESIS
A practical comparative–contrastive analysis
Hussein Abdul-Raof

33. IBN AL-HAYTHAM AND ANALYTICAL MATHEMATICS
A history of Arabic sciences and mathematics, Volume 2
Roshdi Rashed, translated by Susan Glynn and Roger Wareham

IBN AL-HAYTHAM AND ANALYTICAL MATHEMATICS

A history of Arabic sciences and mathematics

Volume 2

Roshdi Rashed

Translated by Susan Glynn and Roger Wareham

Routledge
Taylor & Francis Group
LONDON AND NEW YORK

مركز دراسات الوحدة العربية
CENTRE FOR ARAB UNITY STUDIES

First published 2013 by Routledge

2 Park Square, Milton Park, Abingdon, Oxon OX14 4RN
711 Third Avenue, New York, NY 10017, USA

Routledge is an imprint of the Taylor & Francis Group, an informa business

First issued in paperback 2017

British Library Cataloguing in Publication Data
A catalogue record for this book is available from the British Library

Library of Congress Cataloging in Publication Data
The Library of Congress has cataloged volume 1 of this title under the LCCN: 2011016464

ISBN 978-0-415-58218-6 (hbk)
ISBN 978-1-138-10913-1 (pbk)

Typeset in Times New Roman
by Cenveo Publisher Services

This book was prepared from press-ready files supplied by the editor.

CONTENTS

FOREWORD

This book is a translation of *Les Mathématiques infinitésimales du IXe au XIe siècle*, vol. II: *Ibn al-Haytham*. The French version, published in London in 1993, also included critical editions of the Arabic texts of all the writings by Ibn al-Haytham that were translated into French and were the subject of analysis and commentary in the volume.

The commentary has been translated by Ms Susan Glynn. Ibn al-Haytham's texts have been translated by Roger Wareham, from the French translation, and Aline Auger, CNRS, and I have revised the translation using the original Arabic. In the course of this revision, we benefited from the help of Dr N. El-Bizri, to whom I wish to express my thanks. I also express my deep thankfulness to Mr Joe Whiting, to Ms Anna Calander and to the proofreader at Routledge. Mme Aline Auger prepared the camera-ready text and compiled the indexes. I take this occasion to offer her my thanks.

Roshdi Rashed
Bourg-la-Reine, June 2012

PREFACE

It falls to the lot of very few thinkers to be counted among that elite group who have succeeded in perfecting a scientific tradition by recasting its very meaning. Each follows their own road when setting out on this adventure, but all ultimately lead to the new idea that is slowly maturing among the old, and each forges the most appropriate tools in order to advance that idea. While the verb 'to perfect' in this sense certainly involves a search through the past to reveal the many potentialities lying there, it also implies the profound thought processes needed to identify new possibilities, and to build the tools to realize them. A scientific tradition can only be accomplished in the wake of victorious science, and through the work of researchers who gain strength on a daily basis from their illustrious forebears. Ibn al-Haytham is an outstanding example of this class and, like those who emulated him, he was unfailingly inventive in the mathematical sciences of his time. His work ranged widely, from his reform of optics, and physics in general, his criticism of the Ptolemaic model in astronomy and the basis he laid down for future research, to his work in the various branches of mathematics, especially infinitesimal mathematics in which finite division is used in the pursuit of infinite refinement. If we have done no more, it is this that we have sought to demonstrate in this volume.

In the first volume of this series, we examined the early research in this field from the first half of the ninth century through to the end of the tenth. Each of the edited[1] and translated texts, and the commentaries that accompany them, are important mileposts in the development and transformation of Archimedean research in the years before Ibn al-Haytham. The road is a long one, and in our journey along it we have met the Banū Mūsā, Thābit ibn Qurra, his grandson Ibrāhīm ibn Sinān, al-Khāzin, Ibn Sahl and al-Qūhī. These names, and the renown in which they are held, confirm the pre-eminent position enjoyed by the Archimedeans for over a century and a half. The work of these mathematicians was not confined to infinitesimal mathematics; they also developed research into geometrical transformations and projections. This was the inheritance that

[1] The *princeps* edition of all these treatises, with their translation into French, have been published in *Les Mathématiques infinitésimales du* IXe *au* XIe *siècle*. Vol. II: *Ibn al-Haytham*, London, al-Furqān Islamic Heritage Foundation, 1993.

came down to Ibn al-Haytham, and it was this rich and living tradition that he worked to renew.

This volume is intended to supply a reconstitution of this historical fact, and it is therefore entirely dedicated to the works of Ibn al-Haytham on infinitesimal mathematics. It was he, as we shall see, who radically overhauled the study of lunes, making him a worthy successor to Hippocrates of Chios, but one who was infinitely closer to Euler. It was he who developed the ancient methods of integration, taking them further in the infinitesimal direction, and achieving results that were later rediscovered by Kepler and Cavalieri. And it was he who truly founded the study of solid angles as part of his work on isoperimetrics and isepiphanics, combining projections with infinitesimal determinations for the first time, to my knowledge. All these points will be developed fully in this volume describing the history of this work. Here, we offer the reader the translation of eight of the nine treatises in this field that have survived to the present day, together with their detailed analyses. The texts and their commentaries are preceded by an introduction, in which the acts and writings of Ibn al-Haytham are treated as rigorously as possible with the aim of dispelling a possible confusion of which we consider him to have been the victim. This volume, the second in this series, may also be considered to be the first volume of the mathematical works of Ibn al-Haytham and of his predecessors and contemporaries.

Roshdi Rashed
Bourg-la-Reine, October 1992

INTRODUCTION

IBN AL-HAYTHAM AND HIS WORKS ON INFINITESIMAL MATHEMATICS

1. IBN AL-HAYTHAM: FROM BASRA TO CAIRO

Arab mathematicians who are as famous as Ibn al-Haytham can be counted on the fingers of one hand. This mathematician who was also a physicist, rapidly gained a celebrated reputation in the world for his work on optics, astronomy and mathematics, which appeared first in Arabic, both in the East and the West, then in Latin under the familiar name of Alhazen (from his first name al-Ḥasan), as well as in Hebrew and Italian.

But such fame, though justified by the importance of his contributions and his scientific reforms, contrasts markedly with the all too obvious lack of information we have on the man himself, on his mentors and on his own scientific milieu. This universally recognized scientist, who, even in the twelfth century, was honoured with the evocative title of 'Ptolemy the Second', was to be further shrouded in the mists of legend by the sheer weight of his work. In fact, our sources are reduced to stories told by eleventh- and twelfth-century biobibliographers, wherein legend becomes confused with historical accounts, stories which modern biobibliographers still reproduce today in their entirety or in part.[1] And yet even the briefest

[1] There is an abundance of biographical and bibliographical material on Ibn al-Haytham, which in the main follows the accounts of Ibn Abī Uṣaybiʿa. Here are some examples:

H. Suter, *Die Mathematiker und Astronomen der Araber und ihre Werke*, Leipzig, 1900; Johnson Reprint, New York, 1972, pp. 91–5.

C. Brockelmann, *Geschichte der arabischen Literatur*, 2nd ed., I, Leiden, 1943, pp. 617–19; 1st ed., pp. 469–70; Suppl. I, Leiden, 1937, pp. 851–4; Suppl. III, Leiden, 1942, p. 1240.

F. Sezgin, *Geschichte des arabischen Schrifttums*, V, Leiden, 1974, pp. 358–74. Cf. also vol. VI, Leiden, 1978, pp. 251–61.

G. Nebbia, 'Ibn al-Haytham, nel millesimo anniversario della nascita', *Physis* IX, 2, 1967, pp. 165–214.

of glances would reveal the contradictions contained in these narratives and highlight the uncertainties which surround Ibn al-Haytham, his life, some of his writings and even his very name. But then contradictions and uncertainties form an inevitable part of the biobibliographies of thinkers and philosophers.

It is well known that historical disciplines have played an important role in Islamic civilization; there was in this classical era an unprecedented growth in annals and chronicles, bibliographical dictionaries of jurists, grammarians and scholars, etc. The latter sources listed authors, giving both biographical notes and bibliographies, sometimes accompanied by accounts given by contemporaries and successors, in order to prove the level of importance of the works. However, in sharp contrast to their professional colleagues whom they studied, these Islamic biobibliographers were not always rigorous in their application of the critical method; and when it came to factual evidence, they were prone to use instead contradictory romanticized and picturesque stories, as a means of attracting attention to their own works. They were keen to create a portrait of the ideal man of the day; a tolerant and objective scholar who had devoted his entire life to the quest for the truth, but whose learning would not lead him to deny religious Revelation. In short, these flights of fancy are a trademark of these biobibliographies and it should be borne in mind that such sources require a meticulous critical study.

The case of Ibn al-Haytham offers a prime example of such biobibliographic practice: the temptation to use the imagination was all the greater as he was not only a distinguished scholar, but also a subject under the Caliph al-Ḥākim, a character who was, to say the very least, unusual. Described by some as a megalomaniac, unpredictable, excitable and violent, quite simply idolized by others, this caliph could not have failed to ignite the romantic imagination of chroniclers and historians. Just one meeting between the two, the scholar and the caliph, would have been enough to

---

A. Sabra: the article concerning Ibn al-Haytham in *Dictionary of Scientific Biography*, ed. C. Gillispie, vol. VI, New York, 1972, pp. 204–8; *The Optics of Ibn al-Haytham, Books I–III, On Direct Vision*, 2 vols, London, 1989, vol. II, pp. XIX–XXXIV.

To these must be added the studies of:

E. Wiedemann, 'Ibn al-Haiṯam, ein arabischer Gelehrter', *Festschrift J. Rosenthal ... gewidmet*, Leipzig, 1906, pp. 149–78.

M. Naẓīf, *Al-Ḥasan ibn al-Haytham, buḥūthuhu wa-kushūfuhu al-baṣariyya*, 2 vols, Cairo, 1942–1943, especially vol. I, pp. 10–29.

M. Schramm, *Ibn al-Haythams Weg zur Physik*, Wiesbaden, 1963, especially pp. 274–85.

enliven a story which was in danger of being extremely ordinary, even dull: the tale of a great scholar born in the second half of the tenth century, who after an intensely hard working life, as witnessed by his writings, died some time after 1040.

This study of Ibn al-Haytham begins by highlighting the particularly uncertain style of biobibliography at this time, not merely as a warning, but in order to pose the following questions:

How, armed with so little information, is it possible to arrive at the truth? How to untangle legend from fact? It will be necessary to examine documents and writings to sift the true from the likely, the unconfirmed and the uncorroborated from the purely fictional imaginings and the window-dressing. The aim is to clearly formulate those questions which remain unanswered to this day and to outline certain problems which, it must be acknowledged, may well remain unsolved for some time to come.

Five biobibliographical sources have been used – not all of equal importance and not all independent of each other. The oldest, and also the shortest, is the book by Ṣā'id al-Andalūsī (420/1029–462/1070): *Ṭabaqāt al-umam*.[2] Then comes *Tatimmat ṣiwān al-ḥikma*,[3] written by al-Bayhaqī (499/1105–6–565/1169–70), a Muslim from the Eastern province, in the region of Nayshabur in Khurāsān. The next – and most important – is from al-Qifṭī (568/1172–646/1248): *Ta'rīkh al-ḥukamā'*.[4] After this comes a text recently discovered in a manuscript in Lahore,[5] written in 556/1161, which includes various titles of Ibn al-Haytham's writings. Ibn Abī Uṣaybi'a (596/1200–668/1270)[6] incorporated the source of the Lahore manuscript and the writings of al-Qifṭī into his more complete version. And finally, there is a catalogue of the works of al-Ḥasan ibn al-Haytham in a manuscript discovered in the town of Kuibychev, Siberia, which differs only

[2] R. Blachère, *Kitāb Ṭabaḳāt al-Umam* (*Livre des Catégories des Nations*), Paris, 1935; ed. H. Bū'alwān, Beirut, 1985.

[3] Ed. M. Kurd 'Alī entitled *Tārīkh ḥukamā' al-Islām*, Arab Language Academy of Damascus, 1st ed., 1946; 2nd ed., 1976. See M. J. Hermosilla, 'Aproximación a la *Taṭimmat ṣiwān al-ḥikma* de Al-Bayhakī', in *Actas de las II Jornadas de Cultura Arabe e Islámica*, Instituto Hispano-Arabe de Cultura, Madrid, 1980, pp. 263–72. Cf. D. M. Dunlop, 'al-Bayhaḳī', EI^2, vol. I, pp. 1165–6.

[4] Al-Qifṭī, Jamāl al-Dīn 'Alī ibn Yūsuf, *Ta'rīkh al-ḥukamā'*, ed. J. Lippert, Leipzig, 1903. Cf. also *corrigenda and addenda* to this edition by H. Suter, as appeared in *Bibliotheca Mathematica*, 3rd series, 4, 1903, see pp. 295–6.

[5] Cf. note 31.

[6] Ibn Abī Uṣaybi'a, *'Uyūn al-anbā' fī ṭabaqāt al-aṭibbā'*, ed. A. Müller, 2 vols, Cairo/Königsberg, 1882–1884, vol. II, pp. 90–8.

slightly from the one given by Ibn Abī Uṣaybiʿa.[7] From this, we can see that easily the most important sources are: Ṣāʿid, al-Bayhaqī, al-Qifṭī, and Ibn Abī Uṣaybiʿa. We will go through them one by one.

Al-Bayhaqī tells, in his *Ptolemy II: Abū 'Alī ibn al-Haytham*,[8] of the arrival of Ibn al-Haytham in Egypt, and his meeting with the Caliph al-Ḥākim, where he presented his scheme to control of the flow of the Nile. The story continues with the project being bluntly refused by the caliph, followed by Ibn al-Haytham's flight into Syria and then a few finishing anecdotal touches, to complete the portrait of Ibn al-Haytham as the ideal scholar of his time. The whole biography ends with al-Bayhaqī's description of a fatal illness and a deathbed scene, complete with last words.

The following passage, taken from this same biography, illustrates the most important part of the story:

> He had written a book on the mechanics in which he proposed an ingenious procedure of controlling the Nile when it flooded the fields. He took this book and went to Cairo. He stopped at an inn. As soon as he sat down, he was summoned: 'The Ruler of Egypt, named al-Ḥākim, is at the door and is asking for you.' Abū ʿAlī went out with his book. Abū ʿAlī was a small man and had to stand on a banquette by the door of the inn to reach up to the Ruler of Egypt who was perched atop an Egyptian donkey, and dressed in full chain mail armour. When the Ruler of Egypt had examined his book he said: 'You are mistaken. The cost of this procedure is greater than the advantages it will bring to the cultivation of the land.' And he ordered that the banquette should be destroyed, and rode away. Abū ʿAlī, much in fear of his life, escaped into the night.[9]

Nothing in this story is precise enough to be checked: the picture of al-Ḥākim perched up on the donkey, violent and raging, is a stereotype that has been passed on by chroniclers before al-Bayhaqī's time; he must have borrowed it from them,[10] as others would do from him. Hence, this writer

[7] B. A. Rozenfeld, 'The list of physico-mathematical works of Ibn al-Haytham written by himself', *Historia Mathematica* 3, 1976, pp. 75–6.

[8] Al-Bayhaqī, *Tārīkh ḥukamā' al-Islām*, pp. 85–8.

[9] *Ibid.*, pp. 85–6.

[10] Al-Qalānsī, *Dhayl Tārīkh Dimashq*, Beirut, 1908, pp. 59–80. Abū al-Maḥāsin ibn Taghrī Birdī, *al-Nujūm al-zāhira fī mulūk Miṣr wa-al-Qāhira*, 4 vols, Cairo, 1933, vol. IV, pp. 176–247. It should be remembered that al-Ḥākim encouraged the study of science and founded in Cairo the 'House of Science' (*Dār al-'ilm*), as described by al-Maqrīzī, *Kitāb al-mawā'iẓ wa-al-i'tibār bi-dhikr al-khiṭaṭ wa-al-āthār*, ed. Būlāq, 2 vols, Cairo, n.d., vol. I, pp. 458–9, recently reissued, undated in Cairo. See article by

embellished his tale with picturesque details found in stories more than a century after the death of Ibn al-Haytham. As for the flight of Ibn al-Haytham into Syria, and his living there, this is mentioned only by al-Bayhaqī and contradicts what we almost definitely know about the mathematician in his lifetime. It would therefore seem best to dismiss all fanciful descriptions as pure invention and to believe only that which can be proved – namely the following:

Ibn al-Haytham was not originally from Egypt, but was there during the reign of al-Ḥākim, and had amongst his papers a hydraulic scheme which might have been of interest to the State; he wrote a book on Ethics, another on Astronomy, promoting alternatives to Ptolemy's models of the movement of heavenly bodies. Although al-Bayhaqī refers to these two books in somewhat vague terms, they are easily identifiable.[11]

Al-Bayhaqī's biography became widely known and was taken up again later in a much talked-about work by al-Shahrazūrī,[12] which was brought out at the same time as al-Qifṭī's biography.

Al-Qifṭī's biography is by far the most important of all: written almost a century after al-Bayhaqī's version, and two centuries after the death of Ibn al-Haytham, it is rather more of a biobibliography, and clearly not based on the earlier al-Bayhaqī's biography. Al-Qifṭī, who was born in Egypt and then lived in Syria,[13] was much better informed of the scientific traditions of the time than al-Bayhaqī, who was born and raised in Khurāsān. Al-Qifṭī's biobibliography was, as mentioned previously, taken up by al-Shahrazūrī, then by Ibn Abī Uṣaybiʿa, Ibn al-ʿIbrī,[14] and then again by later

---

M. Canard about al-Ḥākim, ʿal-Ḥākim bi-Amr Allāh', *EI*², vol. III, pp. 79–84, for an example of the prevailing stereotypes.

[11] This is most probably to do with the similarity of *Doubts on Ptolemy* and *Treatise on Ethics*; both appearing on the list of al-Qifṭī as well as on the list of Ibn Abī Uṣaybiʿa.

[12] Al-Shahrazūrī, *Nuzhat al-arwāḥ wa-rawḍat al-afrāḥ fī tārīkh al-ḥukamā' wa-al-falāsifa*, Osmānia Oriental Publications Bureau, Hyderabad, 1976, vol. II, pp. 29–33.

[13] Al-Qifṭī born in Qifṭ, Upper Egypt, began his education in Cairo, before going at the age of 15 with his father to Jerusalem. He later went to Aleppo. About his life and the history of his book, see especially: A. Müller, 'Über das sogenannte *Ta'rīkh al-ḥukamā'* des al-Qifṭī', in *Actes du VIII*ᵉ *Congrès International des Orientalistes tenu en 1889 à Stockholm et à Christiana*, Sect. I, Leiden, 1891, pp. 15–36; the preface by J. Lippert to the edition of *Ta'rīkh al-ḥukamā'*, pp. 5–10; C. Nallino, *Arabian Astronomy, its History during the Medieval Times*, conferences pronounced in Arabic at the University of Cairo, Rome, 1911, pp. 50–64.

[14] Ibn al-ʿIbrī, *Tārīkh mukhtaṣar al-duwal*, ed. O. P. A. Ṣāliḥānī, 1st ed., Beirut, 1890; repr. 1958, pp. 182–3.

biobibliographers. The following translation of its main part will serve to illustrate just how important the text is:

> Al-Ḥākim, Alawite Ruler of Egypt, who had a liking for philosophy, received information about him (Ibn al-Haytham) and his excellence in this domain. He therefore wanted to see him. Al-Ḥākim was then told what he had said: 'If I was in Egypt, I would have controlled the Nile so that it might be possible to profit from it in any state, in flood or not. I have heard tell the river starts from a high point at the furthest end of Egypt'. Al-Ḥākim became even more anxious to see him, and secretly sent to him a large amount of money, which persuaded him to come. Ibn al-Haytham departed for Egypt and al-Ḥākim went out to meet him on his arrival, in a Mu'izite village at the gates of Cairo, known as al-Khandaq;[15] orders were given to receive him generously. Ibn al-Haytham took some little time to rest and al-Ḥākim then asked him to start his Nile project. He went out with a group of artisans who would help with the construction and architecture in the geometric exercises he imagined. But once he had travelled the length of the country and had seen the monuments of these ancient peoples – which had been constructed using the best and most perfect of geometric art – as heavenly figures, full of marvellous geometry and design, he was convinced that his own project was no longer possible. The knowledge of his forebears was most certainly equal to his own, and if what he imagined was possible, they would already have done it. His enthusiasm therefore was diminished and his project was halted. When he came to an elevated area known as the Cataracts, south of the town of Aswan, from where the waters of the Nile flow, he observed, and he carried out tests on both the ebb and flow of the river; he found that the results did not correspond to those he sought and became convinced that the things he had promised had been based on errors. He came back ashamed and defeated and presented his excuses in such a way that al-Ḥākim seemed to accept them and agreed. Then al-Ḥākim gave him responsibility for certain administrative duties, which he accepted more out of fear than desire. He was sure that it was a mistake to accept, because al-Ḥākim was an unpredictable man and shed blood without cause for the very slightest reason he could think of. Ibn al-Haytham was in great torment for a long time, thinking of ways to be rid of this responsibility, and yet the only solution he could think of was to feign madness and insanity.[16]

[15] Ibn Duqmāq, *Kitāb al-intiṣār li-wāsiṭat 'aqd al-amṣār*, ed. Būlāq, Cairo, n.d., 2nd part, p. 43. The author pinpoints this village. For more see al-Maqrīzī, *Kitāb al-mawā'iẓ wa-al-i'tibār bi-dhikr al-khiṭaṭ wa-al-āthār*, vol. II, pp. 136–7.

[16] Al-Qifṭī, *Ta'rīkh al-ḥukamā'*, pp. 166–7.

Al-Qifṭī then tells us that after the death of al-Ḥākim (411/1020), Ibn al-Haytham stopped being 'mad' and took up his research work again, as well as copying manuscripts – such as works by Euclid, Ptolemy's *Almagest*, and *The Intermediate Books* (*al-Mutawassiṭāt*) to earn a living.[17] He cites as evidence a certain doctor Yūsuf al-Fāsī al-Isrā'īlī,[18] who also confirms that Ibn al-Haytham died in Cairo around the year 431/1039.[19] And finally, he produces a list of about sixty of Ibn al-Haytham's titles, which we will return to later.

Al-Qifṭī's biography differs from the version of al-Bayhaqī on two counts: it is written by an author who was undeniably better informed of the life and work of this learned man of Egypt – that much is certain. However his biobibliography is surprising in its wealth of detail, scattered in great profusion to describe the meeting between Ibn al-Haytham and al-Ḥākim as well as the travels of the mathematician to Upper Egypt, his state of mind and his closest thoughts. Such detail almost two centuries after the event could only have been provided by an autobiography; but al-Qifṭī did not have such a document at his disposal, or he would have said so – as he had done in his article on Avicenna. And so therefore it might be rather incautious to set much store by all these descriptions.

Let us now proceed to define the elements which are common to both biographies of Ibn al-Haytham; although, as we have already stressed, both stand alone and the later version does not derive at all from the earlier one.

[17] *Ibid.*, p. 167.

[18] Instead of *al-Fāsī* J. Lippert read *al-Nāshī*, but mentions the first name as per Ibn Abī Uṣaybi'a. This copyist error should not lead us to believe that this person was unknown to al-Qifṭī and would need a long explanation which would be superfluous to Ibn al-Haytham's biography. In fact, this person was a personal friend of al-Qifṭī, as he himself wrote (p. 393) in an article entirely dedicated to him. There, al-Qifṭī gives the full name 'Yūsuf ibn Isḥāq al-Sabtī al-Maghribī, Abū al-Ḥajjāj, residing in Aleppo, native of Fās' (p. 392). According to al-Qifṭī, he died in the first ten days of Dhū al-ḥijja, 623H, i.e. end of November 1226. Al-Qifṭī reports other sightings of him in Baghdad, for example, in a crowd being harangued by Ibn al-Maristāniyya railing against science (p. 229). He is also mentioned by Ibn Abī Uṣaybi'a (vol. II, p. 213) and by Ibn al-'Ibrī (pp. 242–3). See also M. Munk, 'Notice sur Joseph ben-Iehouda ou Aboul'hadjâdj Yousouf ben-Ya'hya al-Sabti al-Maghrebi, disciple de Maïmonide', *Journal Asiatique*, 3rd series, 14, 1842, pp. 5–70. As he himself says, his evidence is hearsay.

[19] This is what al-Qifṭī records from Yūsuf al-Isrā'īlī: 'I heard that Ibn al-Haytham in the course of one year copied three books in his chosen specialist area: that is, Euclid, the *Intermediate Books*, the *Almagest.* When he began to transcribe them, somebody gave him 150 Egyptian dinars. This became a regular fee, which he lived off during the year. He carried on like this up to his death in Cairo around 430 or soon after' (*Ta'rīkh al-ḥukamā'*, p. 167).

The two authors both confirm that Ibn al-Haytham arrived in Egypt, met al-Ḥākim, and showed him a hydraulic project which was turned down: at least, this is what we see if each of the texts is divested of its more blatant romantic elements and embellishments.

On Ibn al-Haytham's country of origin, al-Bayhaqī is silent, whereas al-Qifṭī tells us it is Baṣra, in Iraq.[20] A manuscript of the *Book on Optics* by Ibn al-Haytham copied by his own son-in-law, Aḥmad ibn Muḥammad ibn Jaʿfar al-ʿAskarī, provides corroboration of al-Qifṭī's statement. For it is in Baṣra that this copy had been made, immediately after the death of his father-in-law. It might be possible to counter this evidence with a quotation from Ṣāʿid al-Andalūsī who describes the mathematician as 'The Egyptian Ibn al-Haytham'.[21] However, a single description by Ṣāʿid does not carry the same weight as the Baṣra connection. And it was in fact common at the time to find people named either after the place where they were born, or, equally, after the place that they had come to regard as home.[22] Moreover, there is only one letter different between al-Miṣrī (the Egyptian) and al-Baṣrī (from Baṣra), and this might easily have been confused in the Maghrebi script used by Ṣāʿid.

It is therefore highly likely that Ibn al-Haytham came to Egypt from Baṣra in the time of al-Ḥākim, that is to say toward the end of the tenth century or in the first few years of the following century. Al-Ḥākim was in fact born in 375/985; his reign began in 386/996 and ended with his assassination in 411/1020. Other sources confirm that Ibn al-Haytham was in Cairo in the following decades, one such source being the judge Abū Zayd ʿAbd al-Raḥmān ibn ʿĪsā ibn Muḥammad ibn ʿAbd al-Raḥmān,[23] exactly as Ṣāʿid[24] had maintained. There is further confirmation within Ibn al-Haytham's own writings, which indicates at the very least a familiarity with

[20] He writes 'Al-Ḥasan ibn al-Ḥasan ibn al-Haytham Abū ʿAlī the geometrician from Baṣra, living in Egypt…'.

[21] Ṣāʿid al-Andalūsī, *Ṭabaqāt al-umam*, p. 150.

[22] Note also that al-Khāzinī, in his *Mīzān al-ḥikma*, also calls him 'Ibn al-Haytham al-Miṣrī', 'Ibn al-Haytham, the Egyptian', Osmānia Oriental Publications Bureau, Hyderabad, 1940–1941, p. 16.

[23] This is what Blachère wrote based on Ibn Bashkuwāl no 725 'Born in Cordoba, was Cadi of Toledo, of Tortosa, then Denia, under the Emir al-Ma'mūn ibn Dhī al-Nūn, protector of Ṣāʿid. He died in 473/1080' (p. 116, note 4). Note that Ibn Bashkuwāl (*Kitāb al-Ṣila*, ed. Sayyid ʿIzzat al-ʿAṭṭār al-Ḥusaynī, Cairo, 1955, no. 728) quotes him under the name of 'Abū Zayd ʿAbd al-Raḥmān ibn Muḥammad ibn ʿĪsā ibn ʿAbd al-Raḥmān'; note the inversion of ibn Muḥammad'.

[24] 'He informed me that he met him (Ibn al-Haytham) in Egypt in the year 430 [1038–39]' (Ṣāʿid al-Andalūsī, *Ṭabaqāt al-umam*, p. 150).

the Egyptian milieu of the time. And finally, there is also the fact that Ibn Riḍwān, the well-known doctor from Cairo and a contemporary of Ibn al-Haytham's, wrote a book entitled *On Discussions which Took Place with Ibn al-Haytham on the Milky Way and the Place*.[25]

But had Ibn al-Haytham, on his arrival in Egypt, really met al-Ḥākim to present his hydraulic project to him? On this point, we are reduced to relying on the statements of al-Bayhaqī and al-Qifṭī. But their retelling of the event (the place, the scene itself, the consequences) is not consistent and the noticeable discrepancies between the two versions suggest that this is indeed a distant echo of a scene that each of the two biographers had struggled to imagine and one which they had reanimated in a flurry of detail. The only sustainable argument is presented by al-Bayhaqī. He produces as evidence of the existence of the hydraulic scheme, a book which had been written by Ibn al-Haytham some time before, and which dealt with 'engineering procedures'[26] or 'Mechanics'; but there is no record of this book and we do not know if it ever existed, as al-Bayhaqī is the only one to refer to it. Yet even if we do question the details given by these biographers, we need not necessarily dismiss as pure invention the 'distant echo' or embellished memories referred to above. Ibn al-Haytham – mathematician and physician – was also an engineer, as will be shown in some of his other works; and it was natural, according to the custom of the age, that a scholar should be received by the Caliph.[27] Briefly then, it is certain that Ibn al-Haytham arrived in Egypt at the end of the tenth century or soon afterwards, most probably from Baṣra, and probably bringing with him a hydraulic project which he was later to present to al-Ḥākim. It is

[25] It is this same Ibn Riḍwān who copied the treatise of Ibn al-Haytham on *The Light of the Moon* which work was finished on the Friday midway through Shaʿbān 422, which is Friday 7 August 1031. Cf. al-Qifṭī, *Ta'rīkh al-ḥukamā'*, p. 444; Ibn Abī Uṣaybiʿa, *'Uyūn al-anbā'*, II, p. 104. Cf. also J. Schacht and M. Meyerhof, *The Medico-Philosophical Controversy between Ibn Buṭlān of Baghdad and Ibn Riḍwān of Cairo*, Cairo, 1937, p. 46.

[26] Perhaps this is the book *'Uqūd al-abniya* (*On Architecture*), quoted by al-Qalqashandī in *Ṣubḥ al-a'shā*, ed. Būlāq, Cairo, n.d.; repr. 1963, vol. 7, p. 476. See also Tashkupri-Zadeh, *Miftāḥ al-Sa'āda*, ed. Kamil Bakry and Abdel-Wahhab Abū' L-Nur, Cairo, 1968, vol. I, p. 375 (see p. 538).

[27] We know from al-Maqrīzī (*Kitāb al-mawā'iẓ wa-al-i'tibār bi-dhikr al-khiṭaṭ wa-al-āthār*, vol. I, p. 459) that in fact al-Ḥākim granted audiences to learned men and joined in discussions with them. It is therefore possible to read 'In 403 (1012–1013) a group of arithmetic and logic specialists from the House of Science, a group of legal experts – among them ʿAbd al-Ghanī ibn Saʿīd – and a group of doctors were invited to an audience with al-Ḥākim. Each group came alone to hold discussions in his presence. He later gave them all payment.'

equally probable that, if al-Qifṭī is to be believed, Ibn al-Haytham lived in the neighbourhood of al-Azhar Mosque.[28]

We know nothing of Ibn al-Haytham's life in Cairo.[29] Al-Qifṭī's reports are rather unsubstantial, especially the episode concerning Ibn al-Haytham's feigning madness up until the death of al-Ḥākim. His death, however, is much better documented; and we see that it was *after* 432 (September 1040). Actually the first report of his death is one to which we have already referred – that of al-Isrā'īlī. He stated that Ibn al-Haytham died around 430 (1038) 'approximately in the year 430, or soon afterwards'. But since we have already seen that the Andalusian Judge Abū Zayd met Ibn al-Haytham in Egypt in 430, he must therefore have died some time after this date. Al-Qifṭī himself wrote, quoting al-Isrā'īlī, 'I saw in his own (Ibn al-Haytham's) handwriting a volume on geometry which he wrote in the year 432', i.e. 1040–41.[30]

Of the two biographies written by al-Bayhaqī and al-Qifṭī, the latter is definitely the worthier of the two. However, even this work later became entangled in another tradition, the result of an unfortunate 'confusion', which lasted until the twelfth century and which had in some way been fuelled by Ibn Abī Uṣaybi'a. This 'confusion' is the subject of the investigation which follows next.

[28] This is confirmed by al-Qifṭī, *Ta'rīkh al-ḥukamā'*, p. 167.

[29] Ibn Abī Uṣaybi'a, still confusing names, talks of two disciples of Ibn al-Haytham. These disciples, although not up to the same high standard as their master, were an Emir and a doctor. The first is Emir Abū al-Wafā' al-Mubashshir ibn Fātik who is not known in the world of mathematical sciences. The second is the doctor Isḥāq ibn Yūnus, who had noted the *Observations* of Ibn al-Haytham on Diophantus' *Art of Algebra* – or *Arithmetic*. Cf. Ibn Abī Uṣaybi'a, *'Uyūn al-anbā'*, vol. II, pp. 98–9. And it may well be that it is Ibn al-Fātik who is the dedicatee in Ibn al-Haytham's *Treatise on the Compasses of Great Circles* to 'His Excellency the Emir, may God increase his power' (ms. India Office, Loth 734, fols 116^v–118^r).

[30] Al-Qifṭī, *Ta'rīkh al-ḥukamā'*, p. 167. Note also that according to the doctor Ibn Buṭlān, as recorded by Ibn Abī Uṣaybi'a (*'Uyūn al-anbā'*, I, pp. 242–3), Ibn al-Haytham as well as other learned men, philosophers, lawyers, men of letters and poets had fallen prey to epidemic illnesses and they all died in the same decade. Among them, al-Sharīf al-Murtaḍā, who died in 449/1044, and Abū al-Ḥusayn al-Baṣrī, who died in 436/1044. But also from the same group Abū al-'Alā' al-Ma'arrī, died 449/1058; the poet Mihyār al-Daylamī himself died in 428/1037. This list also includes the philosopher Ibn al-Samḥ who died in 1027, the doctor and philosopher Abū al-Faraj ibn al-Ṭayyib who died in 1043. In fact the deaths ran over two and not just one decade. However, the majority of the group died in the fourth decade of the eleventh century.

2. AL-ḤASAN IBN AL-ḤASAN AND MUḤAMMAD IBN AL-ḤASAN: MATHEMATICIAN AND PHILOSOPHER

In order of merit, the most important biobibliographies of Ibn al-Haytham are: the one written by al-Qifṭī and the one written by Ibn Abī Uṣaybiʿa. The article devoted to Ibn al-Haytham included in Ibn Abī Uṣaybiʿa's *'Uyūn al-anbā'* is the most thorough and is the one most often referred to by modern biobibliographers. But its importance has as much to do with the fact that Ibn Abī Uṣaybiʿa collects – albeit in a disorderly fashion – information from several different sources. He uses statements from his contemporaries as well as information from the earlier biography by al-Qifṭī. He also uses a text containing the autobiography of Muḥammad ibn al-Ḥasan, and also a catalogue of the writings of al-Ḥasan ibn al-Ḥasan up to the end of 429/October 1038. This text and this catalogue were taken by Ibn Abī Uṣaybiʿa from a previous text compiled before 556/1161 and this is also the source of the Lahore manuscript which was copied at the same time.[31] Essentially, Ibn Abī Uṣaybiʿa considered that both Muḥammad and al-Ḥasan were one and the same person, and this opinion has survived until the present day. But we should ask ourselves whether this is a considered opinion or whether it is the result of simple confusion: this question is all the more serious since it challenges the very authenticity of certain works by al-Ḥasan ibn al-Haytham.

Let us first of all examine the article by Ibn Abī Uṣaybiʿa on Ibn al-Haytham. It is a composite of several fragments, not considered coherent by the author's own standards nor by anyone else later. It begins with a preamble, then goes on to comment on propositions of his contemporary, the geometrician ʿAlam al-Dīn; it continues by quoting *in extenso* al-Qifṭī's biography, and finishes by reproducing the autobiography and catalogue of works of Muḥammad ibn al-Ḥasan up to October 1038. In fact, it is no more than a collection of fragments taken from a wide variety of sources, with a preamble which only serves to emphasize the rhapsodic quality of the whole work. It is also remarkable that Ibn Abī Uṣaybiʿa seems totally unaware of the glaring contradictions which appear as a result of his various

[31] This is the manuscript which belonged to the Nabī Khān family in Lahore. M. Anton Heinen, who discovered the existence of this manuscript, authenticated the two texts: the autobiography of Muḥammad and the list of al-Ḥasan by identifying the two separate names. Cf. 'Ibn al-Haiṯams Autobiographie in einer Handschrift aus dem Jahr 556 H/1161 A. D.', in U. Haarmann and P. Bachmann (eds), *Die islamische Welt zwischen Mittelalter und Neuzeit*, Beiruter Texte und Studien 22, Beirut, 1979, pp. 254–77.

borrowings from different versions, at least where it concerns the name of his subject. In the paragraph taken from al-Qifṭī's biography, the scholar appears under the name 'Abū ʿAlī al-Ḥasan ibn al-Ḥasan ibn al-Haytham'. Yet we see at the end of the biography that Ibn Abī Uṣaybiʿa quotes a catalogue of works as belonging to 'al-Ḥasan ibn al-Ḥasan ibn al-Haytham'. Between this paragraph and this catalogue, Ibn Abī Uṣaybiʿa inserts the autobiography of 'Muḥammad ibn al-Ḥasan', as well as two lists of his writings and works, without the slightest explanation. Perhaps this contradiction is the reason why he felt it necessary, consciously or not, to put together the composite name which is at the top of the preamble to the article: 'Abū ʿAlī Muḥammad ibn al-Ḥasan ibn al-Haytham'.[32] This is what he writes: 'Ibn al-Haytham: Abū ʿAlī Muḥammad ibn al-Ḥasan ibn al-Haytham, native of Baṣra, who travelled to Egypt and lived there until the end of his life.'[33] He goes on to describe his moral and intellectual qualities and adds the following comments: 'he produced reports and commentaries on *many* of Aristotle's books; similarly, he produced commentaries on *many* of Galen's medical books. He was an expert in the principles of the art, laws and rules of medicine generally, but he was not a practitioner and was not trained in the use of medicines.'[34] In a nutshell, Ibn Abī Uṣaybiʿa presents us with a philosopher in the Greek tradition, a medical theoretician, with a good knowledge of the works of Galen, but nothing like a mathematician of great note. We will see that this is an exact portrait of Muḥammad and not of al-Ḥasan, insofar as he is portrayed in the works available to us.

Ibn Abī Uṣaybiʿa delves deep into the propositions of his contemporary, the geometrician ʿAlam al-Dīn ibn Abī al-Qāsim al-Ḥanafī[35] (1178/9–1251). However, as they are drawn from the geometrician's own reading of al-Qifṭī's biography, it really adds nothing new. He goes on to say that Ibn al-Haytham lived at first in Baṣra and the surrounding area, that he was appointed as a minister, that he wanted to devote himself to science, since he was attracted to medieval *vertu* and to wisdom, that he then feigned madness to divest himself of his ministerial responsibilities, and that he finally left for Cairo and settled in the neighbourhood of al-Azhar Mosque. This

[32] In the article on al-Mubashshir ibn Fātik, which directly follows the one devoted to Ibn al-Haytham, Ibn Abī Uṣaybiʿa always writes the name of the latter as 'Abū ʿAlī Muḥammad ibn al-Ḥasan ibn al-Haytham', *'Uyūn al-anbā'*, vol. II, p. 99.

[33] *Ibid.*, vol. II, p. 90.

[34] *Ibid.* Our italics.

[35] Just like al-Qifṭī, this mathematician was born in Upper Egypt in 574/1178–79, emigrated to Syria and died there in Damascus in 649/1251. Cf. H. Suter, *Die Mathematiker und Astronomen der Araber*, p. 243, and C. Brockelmann, *Geschichte der arabischen Literatur*, I, p. 625 [474]; Supp. I, p. 867; Supp. III, p. 1241.

version is as close as it possibly could be to al-Qifṭī's version, with the exception that, most probably as a result of an unreliable memory, ʿAlam al-Dīn transposes the Baṣra years and what al-Qifṭī says are the Cairo years, and on top of that makes Ibn al-Haytham into a minister.

Following on from ʿAlam al-Dīn's evidence, Ibn Abī Uṣaybiʿa then reproduces al-Qifṭī's text, but without bringing out these discrepancies. Next, he quotes from the autobiography of Muḥammad ibn al-Ḥasan, a work which is very much in the tradition of Galen's *Libris Propriis*.[36] Muḥammad describes his *curriculum*, his intellectual views and his writings up to around 417/1026, by which time he had reached the age of 63 lunar years – which would place the time of his birth at around 354/965. His *curriculum* is definitely that of a philosopher, who at 63, had already written twenty-five dissertations on mathematics and astronomy, forty-four on logic, metaphysics and medicine as well as a dissertation to show how worldly matters and religious matters come into being as a result of philosophical disciplines; finally 'other writings where the original documents are no longer with him, but which are with the people of Baṣra and al-Ahwāz'.[37] In this first list, we find for example: 'the answer to seven mathematical problems about which I had been asked in Baghdad and to which I replied' and also 'reply to a problem raised by Ibn al-Samḥ al-Baghdādī',[38] and yet another 'review of the response to a problem raised by some Muʿtazilite in Baṣra'.[39]

There then comes a second list, also in the handwriting of Muḥammad ibn al-Ḥasan, which encompasses works from 417/1026 to 419/1028: fourteen dissertations on philosophy, three on astronomy, one on geometry, two on optics and one on medicine. Amongst these dissertations, we also find for example, 'a geometrical problem which had been addressed to him in Baghdad during the months of the year four hundred and eighteen (1027–8)',[40] a letter addressed to Abū al-Faraj ʿAbdallāh ibn al-Ṭayyib al-

[36] On the relationship between the autobiography of Muḥammad ibn al-Haytham and the model proposed by Galen in the *Libris Propriis*, cf. F. Rosenthal, 'Die arabische Autobiographie', *Studia Arabica: Analecta Orientalia* 14, 1937, pp. 3–40.

[37] Ibn Abī Uṣaybiʿa, *'Uyūn al-anbā'*, vol. II, p. 96.

[38] Philosopher of the school of Baghdad, died in 1027. Cf. S. M. Stern, 'Ibn al-Samḥ', *Journal of the Royal Asiatic Society*, 1956; repr. in S. M. Stern, *Medieval Arabic and Hebrew Thought*, ed. F. W. Zimmermann, London, 1983.

[39] Ibn Abī Uṣaybiʿa, *'Uyūn al-anbā'*, vol. II, p. 95.

[40] *Ibid.*, vol. II, p. 97

Baghdādī, who was a philosopher and medical doctor in Baghdad,[41] on 'some notions on physical sciences and theology' and a treatise in response to the same Abū al-Faraj, criticizing his views on natural forces in the human body, which differed from those of Galen.

Ibn Abī Uṣaybi'a wrote at the bottom of this second list: 'I confirm: this is the end of what I have found in the hand of Muḥammad ibn al-Ḥasan ibn al-Haytham, the author – may God have mercy on him' and goes on immediately 'This is also a catalogue *(fihrist)* that I have found of the books of Ibn al-Haytham up to the end of the year 429 (2 October 1038).'[42] But, in order to shed more light on the behaviour of Ibn Abī Uṣaybi'a and these two statements (in particular the latter, where there is no mention made of a first name), let us now consider the Lahore manuscript, which had access to the same source.

This manuscript contains a collection of treatises from several mathematicians, including some by Ibn al-Haytham, together with various lists of works. It is here, between page 174 and the middle of page 184, that we find the autobiography of Muḥammad ibn al-Ḥasan, including two lists of his works: and this is definitely the text quoted by Ibn Abī Uṣaybi'a. The text is not followed by the list of al-Ḥasan's works as it is in Ibn Abī Uṣaybi'a's version, but is followed by a list of the works of the philosopher al-Fārābī; this fills the other half of page 182 and page 183.[43] It is not until page 184 that we find the 'list of books by al-Ḥasan ibn al-Ḥasan ibn al-Haytham up to the end of the year 429'.[44] This is a shortened list, but you only have to compare this with the list of al-Ḥasan's works copied by Ibn Abī Uṣaybi'a to see that the two lists originate from the same source. It is possible to see from the order of the lists, the insertion of al-Fārābī's list between the autobiography of Muḥammad ibn al-Ḥasan and al-Ḥasan ibn

[41] On Abū al-Faraj 'Abdallāh ibn al-Ṭayyib, died in 1043, see G. Graf, *Geschichte der christlichen arabischen Literatur*, Rome, 1947, vol. II, pp. 160–76; Ibn Abī Uṣaybi'a, *'Uyūn al-anbā'*, vol. II, p. 97.

[42] Ibn Abī Uṣaybi'a, *'Uyūn al-anbā'*, vol. II, p. 97.

[43] We read: 'the list of works by Abū Naṣr Muḥammad ibn Muḥammad ibn Turkhān al-Fārābī based on those copied in the hand of Ibn al-Murakhkhim'. The latter was a judge in Baghdad between 541 and 555/1146–1160 and was a man of philosophy and science. He also copied a work on optics by Ibn Sahl (see R. Rashed, *Geometry and Dioptrics*, pp. 13–14). The copyist of the Lahore manuscript, part of the Niẓāmiyya school, is both a contemporary of Ibn al-Murakhkhim and his fellow citizen.

[44] We read in the manuscript: 'Full and final list of al-Ḥasan ibn al-Ḥasan ibn al-Haytham *(ilā ākhirihi)*.' However this description does not make sense, as it is clear that there is something missing; this can be corrected with the help of Ibn Abī Uṣaybi'a. It should read *ilā ākhiri <sanat 429>* ('until the end of the year 429').

al-Ḥasan's list, that the copyist of the Lahore manuscript, and *a fortiori* his source, did not go as far as identifying Muḥammad and al-Ḥasan as the same person – unlike Ibn Abī Uṣaybiʿa. Whereas on Ibn Abī Uṣaybiʿa's list works were attributed to 'Ibn al-Haytham' without first names, here, the same works are explicitly attributed right from the beginning to 'al-Ḥasan ibn al-Ḥasan ibn al-Haytham'.[45] Moreover, the title of the list referred to above was replaced by Ibn Abī Uṣaybiʿa, with the telling phrase 'this is also a catalogue I found of the works of Ibn al-Haytham' (*wa-hādhā ayḍan fihrist wajadtuhu li-kutub Ibn al-Haytham*),[46] which was meant to establish some sense of continuity between Muḥammad's autobiography and al-Ḥasan's list. If there is any confusion, then, as far as we can tell from our present sources of information, the fault lies with Ibn Abī Uṣaybiʿa.[47]

We seem therefore to have discovered two separate persons: Muḥammad, linked to Baghdad (where he was to be found in 1027) and to ancient Southern Iraq, and al-Ḥasan, who lived in Cairo well before 1020. The following will prove this assertion:

I. Al-Ḥasan always used the name of *al-Ḥasan ibn al-Ḥasan ibn al-Haytham,* never Muḥammad ibn al-Ḥasan. In an Arabic translation of Apollonius' *Conics*,[48] copied by Ibn al-Haytham, the colophon to the third book reads:

[45] This is further confirmed by the list of Ibn al-Haytham's works found in the Kuibychev manuscript. Almost identical to the one Ibn Abī Uṣaybiʿa gives under the name of Ibn al-Haytham, it should really be credited to al-Ḥasan ibn al-Ḥasan ibn al-Haytham.

[46] Ibn Abī Uṣaybiʿa, *'Uyūn al-anbā'*, vol. II, p. 97.

[47] We see already a slight error creeping into the Lahore manuscript – but this does not truly constitute what might be called confusion. The copyist writes the following colophon to the autobiography of Muḥammad: 'This is the end of what has been discovered to be in the hand of the author. May peace be upon us. I have transcribed this in the City of Peace (Baghdad) in the school of al-Niẓāmiyya, on the last days of ṣafar of 556 H'; this is followed by the usual invocations. He then adds: 'he had a treatise on light and yet another on the rainbow' *'wa-lahu maqāla fī al-ḍaw' wa-ayḍan maqāla lahu fī qaws quzaḥ'* (fol. 182, line 11; Heinen, 'Ibn al-Haiṭams Autobiographie', p. 272). However, these two titles are not mentioned in the autobiography of Muḥammad and refer to two well-known titles by al-Ḥasan. This must therefore be the copyist's own addition to the Lahore manuscript and not from the author, as this phrase is missing from Ibn Abī Uṣaybiʿa's copy. It is these similar-sounding names which are the source of confusion, but as far as we are aware, Ibn Abī Uṣaybiʿa is the only one to hastily identify the two authors as one and the same person.

[48] See *Apollonius*: *Les Coniques (I–VII)*, commentaire historique et mathématique, édition et traduction du texte arabe par R. Rashed, Berlin/New York, 2008–2010.

> Al-Ḥasan ibn al-Ḥasan ibn al-Haytham transcribed this volume, diacritised and corrected from beginning to end and concluded his commentary in the month of Ṣafar, in the year 415. He wrote these lines on Saturday, the sixth day of the aforementioned month [Saturday 20 April 1024].[49]

There still exists in St. Petersburg today, a preserved manuscript, containing a collection of Ibn al-Haytham treatises and an Ibn Sahl text, copied from an original of Ibn al-Haytham's, and signed by Ibn al-Haytham himself in the same way, *al-Ḥasan ibn al-Ḥasan ibn al-Haytham.*[50] (The Ibn Sahl text had itself been copied by Ibn al-Haytham, which explains its presence in the original manuscript.) Kamāl al-Dīn al-Fārisī informs us that he undertook the editing of al-Ḥasan's treatise on the rainbow and the halo from a copy of an original which was in Ibn al-Haytham's handwriting, with the colophon 'This book has been transcribed, the whole diacritised and corrected from start to finish by reading by al-Ḥasan ibn al-Ḥasan ibn al-Haytham. He wrote these words in the month of Rajab, in the year 419 Hegira (August 1028).'[51]

II. When Ibn al-Haytham's son-in-law, Aḥmad ibn Muḥammad ibn Jaʿfar al-ʿAskarī al-Baṣrī, copied some of the books on the *Optics,* he always transcribed the name of his father-in-law as: *al-Ḥasan ibn al-Ḥasan ibn al-Haytham,* and never *Muḥammad.*[52]

III. Mathematicians and astronomers such as al-Khayyām, al-Samaw'al, and al-Fārisī, etc., who had either read or produced a commentary on Ibn al-Haytham, used the name *al-Ḥasan ibn al-Ḥasan ibn al-Haytham* or *Abū ʿAlī ibn al-Haytham,* but never *Muḥammad.*

IV. According to Muḥammad's autobiography as well as the two lists of his works up to the year 419/1027–28, all works of al-Ḥasan referred to by

[49] Cf. the manuscript of Apollonius' *Conics* transcribed by Ibn al-Haytham – ms. 2762, the Aya Sofya collection in the Süleymaniye Library. M. Nāzim Terzioğlu published a facsimile of this manuscript in the Collection of Publications of the Mathematical Research Institute no. 44, Istanbul, 1981. See also M. Schramm quoting this colophon in his *Ibn al-Haythams Weg zur Physik*, p. IX.

[50] More later on this manuscript – St. Petersburg (Leningrad) B1030.

[51] Cf. Kamāl al-Dīn al-Fārisī, *Kitāb Tanqīḥ al-manāẓir li-dhawī al-abṣār wa-al-baṣā'ir*, Osmānia Oriental Publications Bureau, Hyderabad, 1347–48/1928–30, vol. II, p. 279. This is the version of his name we find with Ibn al-Murakhkhim, al-Khayyām, al-Samaw'al, al-ʿUrḍī and many others. We have seen none at all writing his name as Muḥammad.

[52] Cf. M. Naẓīf, *Ibn al-Haytham, buḥūthuhu wa-kushūfuhu al-baṣariyya*, p. 13; cf. R. Rashed, *Geometry and Dioptrics in Classical Islam*, London, 2005, pp. 34–6. Al-ʿAskarī copied the whole of *Optics* sometime around 1083–1084. This copy was executed in Baṣra, with the seventh and final book being finished on Friday 26 January 1084.

Ibn Abī Uṣaybiʿa up to 1038, as well as in the Lahore and Kuibychev manuscripts (that is several thousand folios of mathematics, optics and astronomy comprising the most advanced research of the age and for a long time to come), would, without exception, have been composed over something like 10 years and a half (between the 29th day of Jumādā the Second 419 and the 29th day of Dhū al-ḥijja 429), which is impossible. There ought also to be a fair number of the writings of al-Ḥasan listed on one or the other of Muḥammad lists: but there are none. In fact, the number of titles common to both, which we shall go on to discuss, is precisely two – out of a grand total of ninety-two!

V. Another important fact: the two lists of the works of Muḥammad up to 25 July 1028 makes no mention of any works on the rainbow and the halo. However, we know that al-Ḥasan had finished a treatise with this title in the month of Rajab 419, which is the beginning of August 1028. If the two authors were in fact one and the same person, it might be reasonable to expect that the book would appear on at least the second of Muḥammad's lists, which dates from 25 July 1028. This is not an *ex silentio* argument: Muḥammad's mind must have been full of this treatise, which he finished at the same time as compiling this list.

VI. Yet to the best of our knowledge, there are no works attributed to al-Ḥasan – either books or dissertations, appearing on either of the Muḥammad lists, and vice versa. In the case of mathematical works on optics and astronomy (except for a few rare cases, which will be discussed later), all extant writings of al-Ḥasan can be found in catalogues of works which earlier biographers had drawn up in the name of al-Ḥasan. However, there is evidence of a different kind of error, which must have been created by various copyists – that of altering *al-Ḥasan ibn al-Ḥasan* to *al-Ḥasan ibn al-Ḥusayn* or *al-Ḥusayn ibn al-Ḥasan* by adding the letter *ī* to his first name or that of his father.[53]

VII. The cross-references in the extant works of al-Ḥasan all concern writings which appear on the lists of al-Ḥasan's works drawn up by al-Qifṭī, by Ibn Abī Uṣaybiʿa and in the Lahore manuscript, but never works grouped under the name of Muḥammad. It is the same for references which appear in the works of later mathematicians; they always refer back to the works of al-Ḥasan, on the aforementioned lists. Only one book out of the total of ninety-two causes some difficulties, and that is: *On the Configuration of the Universe.* This issue will be returned to at a later stage.

VIII. Examination of the catalogues of works by Muḥammad and by al-Ḥasan make the distinction between the two writers very clear, both in

[53] See Supplementary Notes.

terms of form and content. There are ninety titles by Muḥammad, recorded on the two lists of works; there are ninety-two titles by al-Ḥasan, according to the Ibn Abī Uṣaybiʿa list, which records works up to October 1038. Comparing Muḥammad's titles with al-Ḥasan's, there are only two which are duplicated: *Fī hay'at al-'ālam* (*On the Configuration of the Universe*) and *Fī ḥisāb al-mu'āmalāt* (*On the Arithmetic of Transactions*).[54] The way these two texts have been handed down to us creates serious problems over their authenticity. The declared aim of its author in the first text,[55] is to present the planetary orbits, based on Ptolemy's astronomy, in terms of simple and continuous movement of solid spheres. However, the author does not in any way consider the technical problems raised by such a presentation and does nothing to resolve the astronomical or mathematical difficulties which arise. But in the famous *Doubts on Ptolemy*, we have al-Ḥasan on an incomparably higher theoretical and technical level than in *The Configuration of the Universe,* criticizing the configuration of the universe as seen by Ptolemy. Within these pages, al-Ḥasan shows all the required technical skills to deal with the problem of the relation between geometric and astronomical models and the physical description of the universe. This begs the question: is this book, attributed to al-Ḥasan in Arabic as well as in Latin and Hebrew translation, really his?[56] Could he have written it in his youth? But if this were the case, he would have said so: he usually did: in his work in mathematics, his research on lunes;[57] in astronomy, his treatise on *The Movements of Each of the Seven Planets*;[58] and in optics, in his famous *Kitāb al-Manāẓir* (*Optics*).[59]

[54] See Supplementary Notes [1] and [2].

[55] This text was edited and translated into English; Y. Tzvi Langermann, *Ibn al-Haytham's On the Configuration of the World*, New York/London, 1990.

[56] Had people taken Muḥammad's work for that of al-Ḥasan, because of the nature of the titles and of the mathematical and astronomical prestige of the latter? If this substitution did take place, it is relatively early – that is well before the thirteenth century, as it was already apparent from al-Khiraqī's book *Kitāb muntahā al-idrāk fī taqāsīm al-aflāk*, ms. Paris, BN 2499 (and he died in 527/1132). This latter work describes the idea of *On the Configuration of the Universe* without actually naming the book. He attributes the idea to Abū ʿAlī ibn al-Haytham and goes on to review it. He writes (fol. 2^v): 'Abū ʿAlī ibn al-Haytham exaggerated in this description and did not prove anything of the theories he advanced but restricted himself to describing the modality of the positions of the spheres and their rotation with the planets according to the order and arrangement given in their (astronomers) books'. See also Supplementary Note [2].

[57] *Vide infra*, Chapter I.

[58] In this treatise Ibn al-Haytham revisits his previous works; he tackles afresh the problem of distances of the sun and the planets. He writes in his introduction: 'Let it be known that whosoever examines this book and other works before this and finds

Are there – apart from in these two texts – other common titles? It is tempting to consider adding a third work on analysis and synthesis to the list, but this would not really stand up to scrutiny. Al-Ḥasan did in fact write a treatise on *Analysis and Synthesis*[60] (*Fī al-taḥlīl wa-al-tarkīb*), whereas under the name of Muḥammad there is: *Kitāb fī al-taḥlīl wa-al-tarkīb al-handasiyyayni 'alā al-tamthīl li-al-muta'allimīn wa-huwa majmū' masā'il handasiyya wa-'adadiyya ḥallaltuhā wa-rakabtuhā* (*Book on Geometrical Analysis and Synthesis Including – for Students – a Collection of Geometrical and Arithmetical Problems which I have Composed and Solved*). These are two very different titles with two very different approaches. Al-Ḥasan's treatise is in fact, by the author's own admission, very closely linked to another treatise written by him immediately afterwards, *On the Knowns* (*Fī al-ma'lūmāt*).[61] In these two texts, al-Ḥasan examines the problems of the fundamental principles of mathematics – such as the existence of a general geometrical discipline – and develops the theory of demonstration; whereas Muḥammad's title tells us of his intention without any ambiguity, which is to teach students, using geometrical and numerical examples, how to proceed to the solution of problems by means

---

differences in what is said about distances should know that this is because this book records distances of planets with extreme precision whereas the previous works recorded distance according to the conventional method used by mathematicians' (see our edition and commentary in *Les Mathématiques infinitésimales*, vol. V, p. 267, 11–15).

[59] Ibn al-Haytham writes in his *Book on Optics* (*Kitāb al-Manāẓir*, Books I–II–III, ed. A. Sabra, Koweit, 1983, p. 63; *The Optics of Ibn al-Haytham*, London, 1989, p. 6): 'We formerly composed a treatise on the science of optics in which we often followed persuasive methods of reasoning; but when true demonstrations relating to all objects of vision occurred to us, we started afresh the composition of this book. Whoever, therefore, comes upon the said treatise must know that it should be discarded, for the notions expressed in it are included in the content of the present work.' This is very likely the treatise by Ibn al-Haytham mentioned as number 27 in Ibn Abī Uṣaybi'a's list and as number 26 in the Lahore manuscript, entitled *Treatise on Optics according to the Method of Ptolemy – Maqalā fī al-manāẓir 'alā ṭarīqat Baṭlamiyūs*.

[60] See our study 'L'analyse et la synthèse selon Ibn al-Haytham', in R. Rashed (ed.), *Mathématiques et philosophie de l'antiquité à l'âge classique*, Paris, 1991, pp. 131–62; and 'La philosophie des mathématiques d'Ibn al-Haytham. I: *L'analyse et la synthèse*', *MIDEO* 20, 1991, pp. 31–231, where al-Ḥasan ibn al-Haytham's treatise is edited and translated into French. See now *Les Mathématiques infinitésimales*, vol. IV: *Méthodes géométriques, transformations ponctuelles et philosophie des mathématiques*, London, 2002.

[61] Cf. our edition and French translation in 'La philosophie mathématique d'Ibn al-Haytham. II: *Les Connus*', *MIDEO* 21, 1993, pp. 87–275; *Les Mathématiques infinitésimales*, vol. IV.

of analysis and synthesis. The first is addressed to mathematicians interested in the fundamental principles of their discipline and is intended, as the title suggests, as an original study, whereas the other is written as a textbook.

It could therefore be said that, out of the ninety-two books and dissertations attributed by Ibn Abī Uṣaybiʿa to al-Ḥasan, only two of those titles also figure amongst the ninety-two titles attributed to Muḥammad, and both these titles bring with them problems of attribution and authenticity. It could therefore be concluded that the list of works attributed to Muḥammad and the list of works attributed to al-Ḥasan are totally separate.

IX. All books and dissertations attributed to al-Ḥasan are intended for research purposes: they all contain solutions to scientific problems raised either by him or by one of his predecessors. Even when he gives a commentary on previous works, his aim is to show the difficulties he encountered and to propose new solutions: one only has to read the *Commentary on Euclid's Premises* or *Resolution of the Doubts in Euclid's Book* and his *Dubitationes in Ptolemaeum* to see this. Close examination of the titles shows that they correspond exactly to the content of the books; in fact, it is in these books that al-Ḥasan reveals his views in some depth. It is also a fact that al-Ḥasan never composed summaries that aimed at facilitating the access of students to the books of contemporary writers or those of an earlier age, with the exception, perhaps, of *Discourse on Light* which condenses the major theories of his *Book on Optics.*

Another distinguishing feature of al-Ḥasan's works, and equally important, is that all the titles are about mathematics, astronomy, optics and construction of mathematical instruments. It is a completely different story for Muḥammad: his works are in the main *summaries* of and *commentaries* on the writings of earlier writers: Euclid's *Elements* and *Optics*; Apollonius' *Conics* (some of the books at least); Ptolemy's *Almagest* and *Optics*; Aristotle's *Physics*, *Meteorology*, and *De Animalibus*, etc. On the other hand, Muḥammad's writings on mathematics, astronomy and optics represent, at the most, one third of the whole of his work, with the other two thirds devoted to philosophical and medical works.

In order to better appreciate his style, let us look more closely at an example of one of the books: *The Summary by Muḥammad Ibn al-Ḥasan ibn al-Haytham of the Book of Menelaus on the Recognition of Quantities of Different Substances Mixed Together*. He writes:

> I studied the book of Menelaus on the method of distinguishing the weights of various different substances such as gold, silver and copper

> contained within a composite in order to find the quantity of each of the substances in the composite without altering its form; I subsequently found the experiments and demonstrations to be unclear and problematical for any one who would wish to use them for this purpose; I therefore summarised and rationalised this treatise and proved it in such a way so that it would be completely clear to anyone with a conceptual understanding of geometry.[62]

This kind of practice is not restricted to this treatise; it can also be seen, for example, in his commentary on the *Almagest*.

X. As mentioned above, there are at least two books known to have been written under the name of Muḥammad ibn al-Ḥasan ibn al-Haytham: the *Commentary on the Treatise of Menelaus* as well as the *Commentary on the Almagest*. The latter is especially important as it confirms certain facts related by Muḥammad in his own autobiography. This point will be discussed further later.

This commentary exists in a manuscript in the Ahmet III collection in the Topkapi Saray Museum in Istanbul, no. 3329 (2) in 124 folios, copied in 655/1257. But this unique manuscript is not complete. On the first line we see the name Muḥammad ibn al-Ḥasan ibn al-Haytham in its entirety, which then appears in the shortened version of Muḥammad ibn al-Ḥasan in the body of the text.[63] This does not in itself pose a problem for the attribution of the treatise. However this commentary does appear as a title on the list of works by Muḥammad, copied by Ibn Abī Uṣaybiʿa, and in the Lahore manuscript.

In the first of his two autobiographical lists, Muḥammad does in fact note as his third book: '*The Commentary on the Almagest*, and the summary thereof in an attempt to elucidate it, with a preliminary explanation using calculations. If God grants me life and time enough to do it, I will return again to the commentary in a more comprehensive manner and in accordance with numerical and calculatory methods.'[64] This corresponds exactly to what is contained in the preamble to his commentary:

> I found that the main intention of the majority of those who have given their commentary on the *Almagest* was to describe the chapters on

[62] Lahore manuscript, fols 44–51, under the name of Muḥammad ibn al-Ḥusayn ibn al-Haytham. A second copy can be seen under the name of Muḥammad ibn al-Ḥasan ibn al-Haytham in the ms. 81 (Medicine and Alchemy) by Nabī Khān.

[63] Fol. 121^{v}.

[64] Ibn Abī Uṣaybiʿa, *'Uyūn al-anbā'*, vol. II, p. 93. A. Heinen, 'Ibn al-Haiṭams Autobiographie', p. 262.

calculation and to expand them, revealing aspects other than those revealed by Ptolemy, without clarifying those chapters containing ideas too obscure for beginners.

Muḥammad goes on to criticize al-Nayrīzī as follows:

in this way al-Nayrīzī filled his book with endless variations of the same chapters on calculation, motivated by the desire to inflate and glorify what he wrote.

Muḥammad then sums up his own efforts in these terms:

I had the idea of setting out a proposition in the commentary of this book, the *Almagest,* where my principal objective would be to elucidate subtle ideas for the benefit of students. I would add to this a commentary on the calculation of the *zijs*; Ptolemy had not given prominence to these calculations, abridging them and even omitting them from this work, relying on agile minds to find solutions for themselves and to arrive at deductions using the principles he referred to in his book.[65]

This perfect correspondence between the autobiography and the *Commentary on the Almagest* is not the only one; there is a second coincidence, equally remarkable, again concerning the *Commentary.* Muḥammad writes, in the course of his commentary on shadows: 'Ibrāhīm ibn Sinān, mentioned this in his book, and I have myself commented on the question of shadows, their properties and all related astronomical questions, in an independent book.'[66] Let us now return to the autobiographical list of Muḥammad: he has on his list at number 21: 'a book on the instrument of shadow, which is an abridged summary of the book of Ibrāhīm ibn Sinān on the same subject (*Kitāb fī ālat al-ẓill ikhtaṣartuhu wa-lakhaṣtuhu min kitāb Ibrāhīm ibn Sinān fī dhālika*)'.[67] Not only do we see a perfect agreement between the *Commentary* and the autobiography, but he categorically states that his book on shadows is none other than an abridged version of Ibn Sinān's work.[68]

[65] Ms. Ahmet III, 3329/2, fol. 1^v.

[66] *Ibid.*, fol. 91^r.

[67] Ibn Abī Uṣaybiʿa, *'Uyūn al-anbā'*, vol. II, p. 94; A. Heinen, 'Ibn al-Haiṯams Autobiographie', p. 264.

[68] We see from an autobiographical list of Ibrāhīm ibn Sinān that he wrote a book on *The Instruments of Shadows* (*Ālāt al-aẓlāl*). See R. Rashed and H. Bellosta, *Ibrāhīm ibn Sinān. Logique et géométrie au x*^{e} *siècle*, Leiden, 2000, p. 8. It is possible that this commentary of Ibn Sinān's book has been confused with a book by al-Ḥasan ibn al-Haytham, *Maqāla fī kayfiyyat al-aẓlāl (On the Formation of Shadows)* and yet the commentary of the *Almagest* has been incorporated in the list of works of al-Ḥasan. Cf.

We now intend to leave this investigation of the correspondence between the autobiography and the *Commentary of the Almagest* in order to examine its style and composition. We can see that, just as in the *Commentary on the Book of Menelaus*, it is a summary and an explanation for teaching purposes. If further proof were needed, one has only to read Muḥammad's review of all Nayrīzī's work as well as the wording of his own project. At times, it is unmistakably a student audience that he is addressing: 'Know ye, o beginner (*I'lam ayyuhā al-mubtadi'*).' This pedagogical style seems to pervade the book. Muḥammad himself, throughout his commentary, develops his philosophical arguments at length – in true Hellenic/Islamic tradition – and it is not unusual to find him introducing a philosophical argument at the end of a piece of mathematical reasoning. It must be acknowledged that Muḥammad peppers his books with a fair number of scholars: Euclid, Archimedes, Apollonius, Autolycos of Pitania, Hypsicles, al-Nayrīzī, the Banū Mūsā, Thābit ibn Qurra, his grandson Ibn Sinān, and even Galen.

So it would seem that the only way to identify Muḥammad and al-Ḥasan would be by accepting some errors and contradictions along the way. There is no written evidence that he produced a commentary on the *Almagest* – nothing in a catalogue of his works – not even a reference within other works of his, nor is there any evidence that he wrote a commentary and summary of Ibn Sinān's book on shadows. As a generalization, one could say that there is no record of his ever having produced either an abridged or a summarized work of commentary. When he composed commentaries – such as the one on the *Elements* – it was to demonstrate the internal difficulties of the book, its inherent structure and sequence of proofs. Moreover, the stylistic markers which we have picked up in Muḥammad's work are completely foreign to al-Ḥasan's work. Al-Ḥasan did not address himself to beginners, or resort to a philosophical argument to conclude a piece of mathematical reasoning, and, apart from the introduction, where he sets out the problem, he is fairly economical with references and names.

There is more to come: this *Commentary of the Almagest* contains developments of arguments which run counter to those mentioned in the writings of al-Ḥasan, even in his youth. For example, the *Commentary of the Almagest* attempts to explain the phenomenon of the magnification of objects immersed in water – and in turn the phenomenon of 'lunar illusion' (the magnification of planets on the horizon) – in terms of reflection. This

---

A. Sabra, 'Ibn al-Haytham', *Dictionary of Scientific Biography*, vol. VI, pp. 206–8. See Supplementary Notes.

explanation would have been based to some degree on a text by al-Kindī, which indicated that the author had no knowledge of refraction.[69] As for al-Ḥasan, he belonged to a different optical tradition, which was more in tune with Ibn Sahl. He knew from a very early stage about the rules of refraction,[70] and he applied this knowledge in his early works on *The Visibility of Stars*,[71] where he deals with this same question of lunar illusion. For further evidence of style and knowledge, we can also look at Muḥammad's way of studying the isepiphanic problem: *The Sphere is the Largest of the Solid Figures with Equal Surfaces.* There are many different elements in this work which would indicate that this is not a book to attribute to al-Ḥasan, even if there had been no confusion between himself and Muḥammad.

XI. In conclusion then, the titles as well as the content of the extant works of al-Ḥasan prove that their author not only contributed to optics and astronomy (a critique of the Ptolemaic model), but also to mathematics (Archimedean mathematics, theory of conics). He was also responsible for the application of the theory of conics to geometrical construction, the theory of numbers, and the creation of geometrical instruments. All of this work provided essential building blocks for the science of mathematics. But we can find no evidence of studies in medicine or philosophy (in the Hellenic tradition), except a short tract on ethics.

Whereas with Muḥammad we encounter a philosopher, a medical theoretician, who is well versed in the current mathematical learning of his day, and particularly astronomy. This would also have been the case for philosophers in the Hellenistic/Islamic tradition such as al-Kindī, al-Fārābī and Avicenna. According to letters to contemporaries and notes on various locations where works were produced, he seems to have lived in Baghdad and in southern Iraq.

All these facts are perfectly checkable; and they illustrate how the confusion in identity arose between the mathematician and the philosopher. In fact, the misunderstanding must be laid at the door of Ibn Abī Uṣaybiʿa, since the source he used for his work does not confuse the two. (This can be clearly seen in the Lahore manuscript.) The similarity of both names, and the fact that the philosopher wrote commentaries on mathematical,

[69] R. Rashed, 'Fūthiṭos (?) et al-Kindī sur "l'illusion lunaire"', in M.-O. Goulet, G. Madec, D. O'Brien (eds), *ΣΟΦΙΗΣ ΜΑΙΗΤΟΡΕΣ. Hommage à Jean Pépin*, Paris, 1992.

[70] R. Rashed, *Geometry and Dioptrics*.

[71] Mss Lahore, fols 36–42; Tehran, University Library 493, fols 29^r–36^r.

astronomical and optical books, could well have been the source of the confusion; and once there was an amalgam of the two persons, this created an amalgam of the written works.

Obviously, the discussion of the problem which has been opened here requires more historical and biobibliographical research, and one hopes this is a task which will be undertaken in the future. It is also to be hoped that a clearer picture of al-Ḥasan will emerge, together with a better understanding of his works. Perhaps at the same time it will be possible to reveal more about Muḥammad's works on philosophy and logic, given his relationship with Ibn al-Samḥ and Ibn al-Ṭayyib of the Baghdad School, whose works are likely to contain items of great interest.[72]

3. THE WORKS OF AL-ḤASAN IBN AL-HAYTHAM ON INFINITESIMAL MATHEMATICS

Making the distinction between al-Ḥasan and Muḥammad gives us a very clear picture of the mathematician and the philosopher, and when we shed new light on their works, we raise even more questions. We can no longer avoid investigating the authenticity of the works of the mathematician. In fact, we have already given examples of books bearing his name, where the authenticity was, to say the least, doubtful. Surely these works are worth investigating in greater depth? This applies primarily to those writings which fall into the shadowy area where treatises appear on both lists under the same title. Much more surprising are those books specifically attributed to Muḥammad, which, without any hesitation or further investigation, historians have equally happily ascribed to al-Ḥasan. To the best of my knowledge, no one has examined the content and style of the *Commentary of the Almagest* by Muḥammad, nor his *Commentary on Menelaus,* nor has anyone looked into the authorship of *Asymptotes*[73] (a

[72] For further arguments, see Supplementary Notes and *Les Mathématiques infinitésimales*, vol. III, pp. 937–41 and vol. IV, pp. 957–9.

[73] We have previously commented that Muḥammad mentioned in his autobiographical list a treatise on asymptotes. As for al-Ḥasan, there is nothing anywhere in a catalogue of his writings, nor in his own statements, to suggest that he might have written a dissertation on this topic. However there is a *Treatise on the Existence of Two Lines which Draw Closer but do not Touch – Risāla fī wujūd khaṭṭayyn yaqrubān wa-lā yaltaqiyān*, ms. Cairo, Dār al-Kutub 4528, fols 15^{v}–20^{r}. This text is anonymous but the copyist writes in the colophon: 'It is possible to say from the style of writing that this is Ibn al-Haytham's work (*wa-yufham min 'ibārātihā annahā ta'līf Ibn al-Haytham*)' without explaining the reasons which led to this conclusion. We have ourselves edited, translated

work which appears on Muḥammad's list in his own handwriting), before attributing it to al-Ḥasan.

And there are those today who would attribute even more works to him that he did not write.[74]

---

and analysed this text ('Le pseudo-al-Ḥasan ibn al-Haytham: sur l'asymptote', in R. Fontaine, R. Glasner, R. Leicht and G. Veltri (eds), *Studies in the History of Culture and Science. A Tribute to Gad Freudenthal*, Leiden/Boston, 2011, pp. 7–41) and we can confirm without a shadow of a doubt that this is not the work of al-Ḥasan ibn al-Haytham. Could it be by Muḥammad ibn al-Haytham? It is actually the kind of commentary which we have come to expect from him. But this is not sufficient reason to attribute the text to him and the question remains open whilst we await the discovery of new evidence.

[74] Apart from the *Commentary on the Almagest*, A. Sabra ('Ibn al-Haytham', p. 208) attributes to al-Ḥasan ibn al-Haytham an anonymous text – Library Medicæ Laurenziana, or. 152, fols 97^v–100^r – entitled: *Kalām fī tawṭi'at al-muqaddamāt li-'amal al-quṭū' 'alā saṭḥ mā bi-ṭarīq ṣinā'ī (Introduction to the Lemmas for the Construction of Conic Sections by the Mechanical Method)*. The arguments invoked in favour of this attribution are the following: firstly, in his treatise *On Parabolic Burning Mirrors*, Ibn al-Haytham makes reference to an instrument used in the construction of conic sections; and secondly, this fragment follows on directly from a manuscript of this treatise in the same manuscript from Florence.

Al-Ḥasan ibn al-Haytham, it is true, mentions in his treatise *On Parabolic Burning Mirrors* an instrument to construct conic sections. We have discussed this question and shown that the idea of this particular instrument and this particular construction is found in Ibn Sahl's tradition (*Geometry and Dioptrics*, pp. 297ff.). But does this fragment from Florence represent part of the treatise, or is there even a small chance that it might be by Ibn al-Haytham? Examination of the text shows that this is not the case; it is full of elementary mathematical errors and is the work of a much inferior mathematician. Moreover there is no real argument to support such a conjecture: the following two examples will serve to prove that the Florence manuscript could not be by al-Ḥasan.

1) The author wants to prove:

Let AB be a segment, let C be a point on AB and let Ax, By, and Cz be perpendiculars to AB. If the two points E and D on Cz are such that

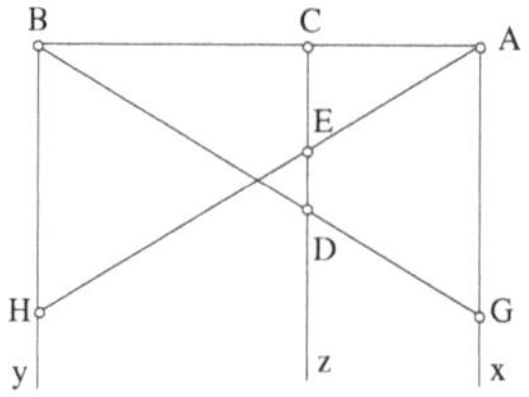

$$(1) \qquad CE \cdot CB = CA \cdot CD,$$

then the straight lines AE and BD intersect By and Ax respectively at H and G, and $AG = HB$.

The author derives from (1) that

$$\frac{CE}{CA} = \frac{CD}{CB}.$$

However

$$\frac{CE}{CA} = \frac{HB}{AB} \text{ and } \frac{CD}{CB} = \frac{AG}{AB};$$

It is true that the situation is not quite as dramatic as it seems. But we should really re-examine al-Ḥasan's scientific works, and in particular, those which fall into this confused area. The kind of rigorous critical review which was once upon a time considered the norm requires the application of various methods. The most immediate task, which is also the simplest, is to carefully compare the available catalogue of works of al-Ḥasan, and to research all cross-references made by him to his works as well as any mention of his treatises by later writers. In supplementary notes to this volume, we have set out a table which gives all the information we have gathered so far. Although not complete, this table will, in the fullness of time and with the addition of new research, become something more important. As it stands, it is more of a preliminary work, which does not deal at all with the study of each group of writings according to their content and language. And it will only be at the very end of our various lines of enquiry that we can hope to resolve the question of authenticity, which affects the whole body of work.

---

therefore $HB = AG$.

So it is clear that, if C is chosen randomly, then it is not possible, as the author wishes, to restrict the condition that $CD < CE$.

In fact, according to the hypothesis,

$CD < CE$ if $CA > CB$,
$CD = CE$ if $CA = CB$,
$CD > CE$ if $CA < CB$.

2) The first proposition is devoted to the construction of a parabola knowing the vertex B, the axis AB and a point C on the parabola.

The author gives a construction by points, beginning with the equation $y^2 = ax$.

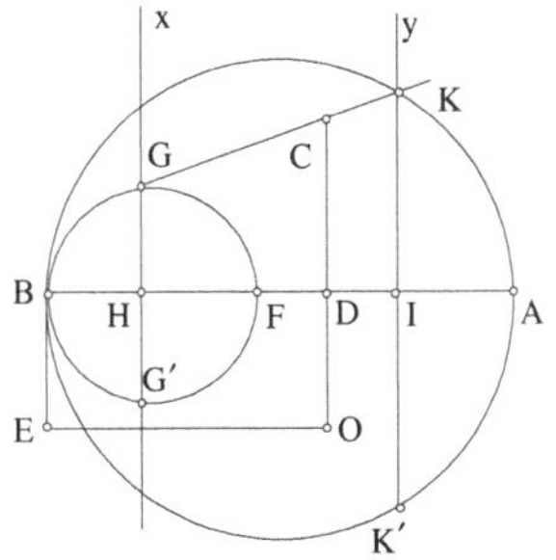

With the projection of C on the axis being point D, construct on BD a rectangle DE so that $BD \cdot BE = CD^2$; the length BE is the *latus rectum* of the parabola $BE = a$.

However, the author does not explain the construction of rectangle $BDOE$. Let Hx be a perpendicular to AB, construct F so that $HF = BE = a$; the circle with diameter BF intersects Hx at point G, so that $HG^2 = HB \cdot HF$; point G is therefore on the parabola and so is its symmetrical point G'.

The construction is repeated, taken from another point I on the axis and with points K and K'.

The proof of this lemma as well as the proposition leaves a lot to be desired and could not possibly have come from a mathematician in the same class as Ibn al-Haytham. The text is peppered with other faults of this kind, and some more serious. The fact that this anonymous fragment comes after a text by Ibn al-Haytham does not seem sufficient reason to attribute it to him.

Our aim here is more modest: and that is, to consider this problem of authenticity – but only insofar as it relates to the group of al-Ḥasan's writings on infinitesimal mathematics. In fact our task here is easier than for other groups of his writings: there is no title of al-Ḥasan on infinitesimal mathematics which appears in the grey area. There is only one title on the list of Muḥammad which deals with this subject and that is *On the Asymptotes*; this means that the writings which we know of are the most advanced research works of the day – the most difficult – and could only be the work of an eminent mathematician, that is, al-Ḥasan and not Muḥammad.

The following lists of al-Ḥasan's works indicate the kind of treatises he wrote: List I is drawn up by al-Qifṭī; List II by Ibn Abī Uṣaybiʿa, and List III (incomplete) from the Lahore manuscript.

	INFINITESIMAL MATHEMATICS[75]	I	II	III	
I	*On the Lunes*	6	20	18	*
II	*On the Quadrature of the Circle*	15	30	23	*
III	*Exhaustive Treatise on the Figures of Lunes*	—	21	21	*
IV	*On the Measurement of the Paraboloid*	5	17	20	*
V	*On the Measurement of the Sphere*	33	16	14	*
VI	*On the Division of Two Different Magnitudes*	46	40	41	*
VII	*On the Sphere which is the Largest of all the Solid Figures...*	28	26	25	*
VIII	*On the Greatest Line lying in a Segment of Circle*	—	81	Part	
IX	*On the Indivisible Part*	32	65	62	
X	*On the Sum (or all) of the Parts*	45	32	30	
XI	*On Centres of Gravity*	—	14	12	
XII	*On the Qarasṭūn (infinitesimal mechanics)*	—	67	Part	
	APPROXIMATION OF ROOTS				
1	*On the Cause of the Square Root*	25	70	Part	*
2	*On the Extraction of the Side of the Cube*	24	47	43	*

None of these titles appears on Muḥammad's own lists of his works.

The first twelve titles (to which, if ever it was written, should be added the treatise referred to in II by al-Ḥasan) seem on the face of it to divide nicely into the following four groups, to which we will return in greater detail later:

1) quadrature of lunes and of a circle;
2) measurement of curved solids;
3) isoperimetric and isepiphanic problems

With the exception of VIII, works in the above-mentioned groups have all come down to us. And although IX remains lost, we can tell from the

[75] * means: ms. available; – means: it is missing.

title that it would probably have dealt with the proof of infinite division. This was a hotly debated issue at the time, as we can see from the work of al-Sijzī, one of Ibn al-Haytham's predecessors.[76] The tenth title remains obscure: perhaps this was where Ibn al-Haytham discussed the sum of parts in infinite numbers.

4) This last group is made up of treatises XI and XII. Sadly, however, these two obviously important works remain lost and the only surviving remnant we have is a meagre abstract in the hand of al-Khāzinī. It is largely a list of definitions which he describes as being 'after Abū Sahl al-Qūhī and Ibn al-Haytham al-Miṣrī'.[77]

We will add, as an appendix to the treatises which deal with infinitesimal mathematics, two further texts: one on the extraction of the square root and the other on the extraction of the cube root. By including these treatises, however, we do not mean to imply that Ibn al-Haytham specifically proposed a clear relationship between the problem of approximation encountered in these cases and the problems of infinitesimal geometry. We will explain our reasons for this in due course.

We will now translate the nine treatises known to us.[78] The first of these texts was considered lost until now and the treatise dealing with the square root was unknown. With the exception of the treatise on the *Quadrature of the Circle*, this is the first time any of these texts have been edited. In order to clarify them, we have observed the strictest of rules of research which we have explained many times before. Let us briefly look at the manuscripts which we have used for the edition of these texts.

I. *Treatise on the Lunes – Qawl fī al-hilāliyyāt*

Al-Qifṭī cites a treatise *On the Figures of Lunes* (*Fī al-ashkāl al-hilāliyya*) written by Ibn al-Haytham. As this is the only title he quotes, it would be reasonable to wonder whether he meant this treatise or treatise III. But in the Lahore manuscript, where we see reference to both treatises, the former has the same title as in al-Qifṭī's work, which would lead one to

[76] R. Rashed, 'Al-Sijzī et Maïmonide: Commentaire mathématique et philosophique de la proposition II-14 des *Coniques* d'Apollonius', *Archives internationales d'histoire des sciences* 37.119, 1987, pp. 263–96; *id.*, *Œuvre mathématique d'al-Sijzī*. vol. I: *Géométrie des coniques et théorie des nombres au Xe siècle*, Les Cahiers du Mideo 3, Louvain-Paris, 2004.

[77] Al-Khāzinī, *Mīzān al-ḥikma*.

[78] The Arabic edition of all these treatises is published in *Les Mathématiques infinitésimales*, vol. II.

believe that this is the treatise to which al-Qifṭī refers. Ibn Abī Uṣaybi'a also cites the two treatises, attributing both to al-Ḥasan, but gives the title as the *Abridged Treatise on the Figures of Lunes* (*Maqāla mukhtaṣara fī al-ashkāl al-hilāliyya).* Perhaps the term 'abridged' (*mukhtaṣara*) – used by Ibn al-Haytham in the second treatise to describe the first one – was put in to distinguish the one from the other. (Note here that the term *mukhtaṣara* means 'succinct' and not a summary of a text already written.)

So we can see that this text had already been cited twice by Ibn al-Haytham: in the *Exhaustive Treatise on the Figures of Lunes* and in the treatise on the *Quadrature of the Circle.* Both these references can be verified, and they provide all the more evidence of an authentic work. We have had the great good fortune to have seen a single manuscript containing the entire collection of the works of Ibn al-Haytham. This has come from the 'Abd al-Ḥayy Collection, housed in the university library of Aligarh in India, no. 678/55. It was copied in 721/1321–22 in al-Sulṭāniyya, in *nasta'līq* script, in 45 folios. However, the examination of the manuscript reveals that it has been damaged, probably recently: some sheets are missing and the forty-five remaining are in disorder, and there are traces of humidity. Each folio measures 218 × 76mm, with 33 lines on each page and approximately 9 words per line. This collection includes the following works of Ibn al-Haytham: *The Measurement of the Sphere*; *The Quadrature of the Circle*; *The Cause of the Square Root* (the texts which are translated here). There are also: *The Burning Mirrors*; *The Resolution of Doubts on the First Book of the Almagest*; *The Construction of Great Circles using a Small Instrument* or *On the Compasses of the Great Circles*, and a fragment of *The Lemma on the Regular Heptagon*. It also contains the commentary by al-Ahwāzī on the 10th book of the *Elements.* The text on the lunes covers fols 14^v–16^v.

II. *The Quadrature of the Circle* (*Qawl fī tarbī' al-dā'ira*)

This title is on the three lists of works by Ibn al-Haytham (al-Qifṭī, the Lahore manuscript, and Ibn Abī Uṣaybi'a) and is cited in his treatise *On the Resolution of Doubts on Euclid's Book.*[79] However this short tract is often found as part of the *Intermediate Books* (*al-Mutawassiṭāt*), in the majority of collections of manuscripts. We have therefore edited this text using only manuscripts which we have been able to examine at first hand and not the whole range of those which we know to exist. The following is a list of those manuscripts:

[79] Ms. Istanbul, University 800, fol. 167^r.

A - Istanbul, Aya Sofya 4832 II/21, fols 39^{v}–41^{r}
B - Patna, Khudabakhsh 3692, without numbering, 3 folios
C - Istanbul, Carullah 1502/15, fols 124^{v}–126^{r}
D - Tehran, Danishgāh 1063, fols 7^{r}–9^{v}
E - Aligarh, 'Abd al-Ḥayy 678, fols 10^{r}–11^{v}, 30^{v}–30^{r}
I - Tehran, Majlis Shūrā 205/3, fols 93–101
K - Tehran, Malik 3179, fols 107^{v}–110^{r}
M - Meshhed 5395/1, fols 1^{v}–3^{r}
R - Istanbul, Beshir Aga 440, fol. 151^{r}
T - Tehran, Sepahsālār 559, fols 84^{v}–85^{r}
X - Tehran, Majlis Shūrā 2998, 1 folio without numbering
V - Roma, Vatican 320, fols 1^{v}–6^{v}

It should be noted that

1) manuscript R does not contain the text of Ibn al-Haytham, but merely a commentary added to it;

2) manuscript from Cairo, Dār al-Kutub, Taymūr – Riyāḍa 140 (fols 136–7) is not Ibn al-Haytham's text, but a summary and a later commentary;

3) two manuscripts of the text of Ibn al-Haytham found in Berlin before the Second World War – fol. 258 and quart. 559 – which were consulted by H. Suter, have since been lost;[80]

4) the objection which is to be found at the end of Ibn al-Haytham's text only appears in manuscript E.

It would be too long and tedious to give results of the examination of all these manuscripts and their comparison here. In any case, such an examination would not provide an authentic *stemma,* but merely a classification by group which would indicate the history of the manuscript tradition and which might be represented thus:

{ (B, T, K), ([D, (I, M), X, C], A) } and {E, V}.

There are two main families, the first made up of three sub-families, and the second sub-family in its turn made up of three sub-families.

[80] We are indebted to Dr. H.O. Feistel, Staatsbibliotek, Orientabteilung, for this information.

H. Suter prepared a draft copy of the text of this treatise, working from two manuscripts in Berlin (since lost) and the Vatican manuscript.[81] This work, not least as a translation of the treatise into German, has provided a valuable research tool for historians.

III. *Exhaustive Treatise on the Figures of Lunes – Maqāla mustaqṣāt fī al-ashkāl al-hilāliyya*

This is exactly how the title of this work appeared in two lists of the writings of al-Ḥasan – both in the Lahore manuscript list and in the Ibn Abī Uṣaybiʿa list. It was even used by al-Ḥasan himself in his book *On the Resolution of the Doubts on Euclid's Elements* where he writes:

> We have written a treatise on the figures of lunes in which we demonstrated that amongst lunes there are those which are equal to a right-angled triangle. The Ancients stated a part of this; however, this statement was particular; that is to say 'limited' to a single lune – constructed on the side of a square inscribed in a circle. But what we have shown here is of universal <application>; we have considered and revealed cases of all different kinds of lunes. We can say that the lune is surrounded by two arcs and yet is equal to a right-angled triangle, which is to say that the area of the lune is equal to the area of a triangle. Thus it can be shown that although the two arcs surrounding the lune are not in direct proportion to the sides of the triangle, this does not rule out the equality of their areas. We have also shown that the sum of a lune plus a complete circle equals a triangle. We also have an independent treatise (*mufrada*) in which we have shown that it is possible that a circle be equal to a square.[82]

The above description by Ibn al-Haytham and his mention of the different kinds of lunes apply to this treatise and not to treatise I. This treatise actually contains five propositions, including those of Hippocrates of Chios. Ibn al-Haytham also mentions here his own treatise on *The Quadrature of the Circle.*

This treatise has come down to us in its entirety in four manuscripts, to which we may add a very important fragment.

This text was in fact part of a very fine collection of the writings of al-Ḥasan ibn al-Haytham, held in St. Petersburg, and copied on an autograph: manuscript B1030, Oriental Institute 89, fols 50ʳ–72ᵛ, and 133ᵛ–144ʳ. This manuscript was checked against Ibn al-Haytham's autograph in 750/1349, as we can see from information on the first page, and the whole collection is

[81] H. Suter, 'Die Kreisquadratur des Ibn el-Haiṯam', *Zeitschrift für Mathematik und Physik* 44, 1899, pp. 33–47.

[82] Ms. University of Istanbul 800, fol. 167ʳ.

in the same hand, a rather mediocre *nasta'līq*. Within the text there are four omissions of exactly one phrase each and eleven omissions of one or two words each. This manuscript is denoted as L in the following *stemma*.

The second manuscript is Octavo 2970 from the Staatsbibliothek of Berlin, copied at Samarkand in *nasta'līq*. Our text, copied in 817/1414, covers fols 24r–43v (denoted as B). The geometrical figures are no longer visible – at least on the microfilm available to us. There are four omissions of one phrase each and nine omissions of one or two words.

The third manuscript is from the 'Āṭif Library in Istanbul, no. 1714, fols 158r–177v (denoted as T). We have shown that this manuscript was copied from the second, and only from this one.[83]

The fourth manuscript is from the India Office in London, no. 1270/12, Loth 734, fols 70r–78v (denoted as A). We do not have the date of its transcription, which could have been tenth century Hegira. This text has two omissions of one phrase each and nine of one or two words.

The fifth manuscript is the fragment we found. It is the famous Fātiḥ 3439 manuscript from the Süleymaniye Library in Istanbul. This manuscript had been copied in 806/1403–44 and our fragment is found in fols 115r–117r (denoted as F). It is difficult to read due to the faded ink, but we can see four omissions of one phrase and eleven omissions of one or two words.

By examining omissions found in each individual manuscript as well as omissions common to all manuscripts, along with various additions, spelling and other mistakes, we can suggest the following *stemma*:

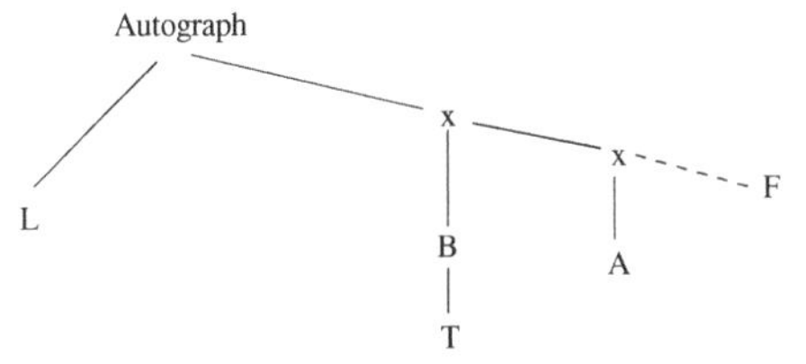

IV. *Treatise on the Measurement of the Paraboloid – Maqāla fī misāḥat al-mujassam al-mukāfī'*

Mentioned in three lists of works by Ibn al-Haytham, it is also cited by him in his *Treatise on the Measurement of the Sphere*. Our original edition was published in 1982, then slightly amended in the light of information obtained from the 1270 India Office manuscript (already referred to), fols

[83] See particularly *Geometry and Dioptrics*, p. 37.

56^{v}–69^{v}.[84] H. Suter has done a 'free' translation of this text into German.[85] By 'free' I mean that the translator often gives the sense of the text without following it to the letter, and even misses out certain paragraphs, especially those where translation would be problematic. On the whole, H. Suter explains the content of the text in a precise manner with the exception of a few paragraphs and the last part.

V. *Treatise on the Measurement of the Sphere – Qawl fī misāḥat al-kura*

This appears on Ibn al-Haytham's list of works, which he cites in his *Treatise on the Principles of Measurement.* He writes 'As for the sphere, the method of measurement is ... which we have also shown in another separate treatise (*fī qawl mufrad*).'[86]

This text was available to us in five manuscripts: the first, referred to earlier, is from the Staatsbibliothek in Berlin Octavo 2970/13, and appears in fols 145^{r}–152^{r} (this manuscript is denoted as B in the *stemma*). It was copied in 839/1435–6, as indicated in the colophon. The geometrical figures are illegible and there are three omissions of a phrase and two omissions of one word.

The second manuscript is the 1714/20, fols 211^{r}–218^{r}, from the ʿĀṭif library in Istanbul, denoted as T. This is in fact, as we have said earlier, a straight copy of the one which came before, and from no other source.

The third manuscript is the Aligarh (India) manuscript, already described, fols 1^{r}-5^{r} and 13^{v}–14^{v} (denoted as O). This manuscript has fourteen omissions of a single phrase and twenty-six omissions of one or two words. The copyist not only copied from the original, but, as is clear from the critical apparatus, also used another copy.[87]

The fourth manuscript, mentioned earlier, is B1030 from St. Petersburg – Oriental Institute 89 (denoted as L). Several pages are missing, and the

[84] R. Rashed, 'Ibn al-Haytham et la mesure du paraboloïde', *Journal for the History of Arabic Sciences* 5, 1982, pp. 191–262; *Les Mathématiques infinitésimales*, vol. II.

[85] H. Suter, 'Die Abhandlung über die Ausmessung des Paraboloides von el-Ḥasan b. el-Ḥasan b. el-Ḥaiṭam', *Bibliotheca Mathematica*, 3rd series, 12, 1911–12, pp. 289–332.

[86] *Les Mathématiques infinitésimales du IXe au XIe siècle*, vol. III: *Ibn al-Haytham. Théorie des coniques, constructions géométriques et géométrie pratique*, London, 2000, p. 610; Arabic p. 611, 8.

[87] *Les Mathématiques infinitésimales*, vol. II, p. 311, 1.

beginning of the text is incomplete: there only remains a fragment of the very end, fols 73^r–77^r.

The fifth manuscript is the 1446 (old number 176) from the National Library in Algiers, denoted here as C. The copyist had done his transcription from an original whose pages were in some disorder, and, as he was obviously unfamiliar with the topic, he has mixed up various parts of the text. It should therefore be read in the following order:

113^r–116^v (line 14) → 117^v (line 6)–118^r (middle of the last line) → 116^v (lines 15–22)–117^v (line 6) → 118^r (middle of the last line)–119^v.

This manuscript has 13 omissions of one phrase and 21 omissions of one or two words. Examination of the manuscripts as well as the copying errors would seem to suggest the following *stemma:*

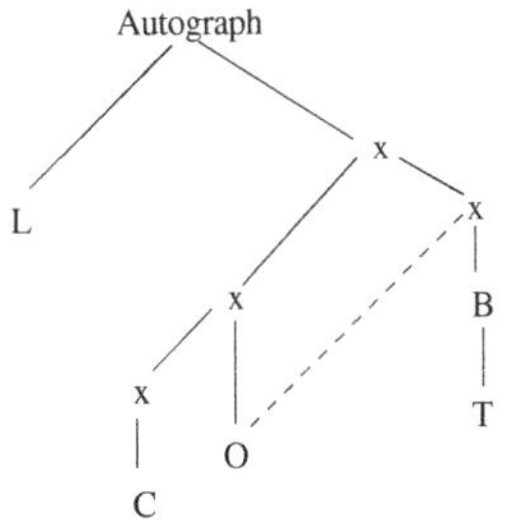

VI. *Treatise on the Division of two Different Magnitudes Mentioned in the First Proposition of the Tenth Book of Euclid – Fī qismat al-miqdārayn al-mukhtalifayn al-madhkūrayn fī al-shakl al-awwal min al-maqāla al-ʿāshira min kitāb Uqlīdis*

This short treatise appears on the three lists of Ibn al-Haytham's works, but as a shortened title on both al-Qifṭī and the Lahore manuscript lists. In this treatise Ibn al-Haytham himself makes implicit reference to two preceding treatises when he writes: 'For some geometrical notions that we have determined, we were faced with the need to take one half away from the greater of the two different magnitudes, and then one half from the remainder [...].'[88] Ibn al-Haytham also refers specifically to the treatise in his book *Resolution of the Doubts in Euclid's Elements.* He writes: 'we have devoted to this notion a treatise to demonstrate its universality, even though it is an extremely short summary and we composed it before the *Resolution of the Doubts*'.[89]

[88] *Vide infra*, p. 235.
[89] Ms. University of Istanbul 800, fol. 143^v.

Finally, Ibn al-Sarī, one of Ibn al-Haytham's successors, wrote a treatise criticizing it, quoting the exact title of the work and its introduction.[90]

This text was made available to us by the manuscript already referred to many times – Manuscript B 1030 from St. Petersburg, fols 78ᵛ–81ʳ.

VII. *Treatise on the Sphere which is the Largest of all the Solid Figures having Equal Perimeters, and on the Circle which is the Largest of all the Plane Figures having Equal Perimeters*[91] – *Qawl fī anna al-kura awsaʿ al-ashkāl al-mujassama allatī iḥāṭātuhā mutasāwiya wa-anna al-dā'ira awsāʿ al-ashkāl al-musaṭṭaḥa allatī iḥāṭātuhā mutasāwiya*

Whereas al-Qifṭī mentions this treatise under the title 'The sphere is the largest of the solid figures', in the Lahore manuscript list it appears as 'Spheres are the largest of the solids – *al-ukar awsaʿ* (*wa-sharḥ* in the manuscript) *al-mujassamā*'. Judging by these shortened titles it seems that the title found on Ibn Abī Uṣaybiʿa's list is an original. Ibn al-Haytham himself cites this treatise in his book *On the Resolution of Doubts on the Almagest* as well as in his *Treatise on Place.* He writes in the former: 'We have written an exhaustive treatise on this notion, showing by means of certain proofs that the sphere is the largest of the solid figures with similar and equal perimeters and that the circle is the largest of the plane figures with similar and equal perimeters.'[92] In his *Treatise on Place*, he also writes: 'We have shown this idea in our book: *The Sphere is the Largest of the Solid Figures having Equal Perimeters.*'[93]

This text came down to us in three manuscripts:

1) Berlin Oct 2970/9, fols 84ʳ–105ʳ;

2) the copy which was made of it, that is 'Atif 1714/18, fols 178ʳ–199ᵛ;

3) manuscript from the scientific collection in Tehran – Majlis Shura, Tugābunī 110, fols 462–502.

In the Berlin manuscript there are one omission of a phrase and seven of one word. And in the Tehran manuscript, which is a collection of 581 pages, there are nine omissions of one phrase each and six omissions of one word.

[90] See Supplementary Note [4].

[91] It is clear that in the case of volumes, the word should read 'surface', not 'perimeter'. However, as Ibn al-Haytham uses the same word for both solid and plane figures, which should be understood as 'that which surrounds', we have decided to keep the same word 'perimeter' in both cases to maintain the unity of vocabulary.

[92] Ms. Aligarh, fol. 23ᵛ.

[93] *Les mathématiques infinitésimales du IXᵉ au XIᵉ siècle*, vol. IV: *Méthodes géométriques, transformations ponctuelles et philosophie des mathématiques*, London, 2002, p. 672; Arabic p. 673, 15–16.

We have edited this treatise using both the Berlin and Tehran manuscripts, each belonging to its own very separate manuscript tradition.

Let us now come to two treatises on the extraction of roots and on approximation, which are given in the Appendix.

1. *Treatise on the Cause of the Square Root, its Doubling and its Displacement – Maqāla fī 'illat al-jadhr wa-iḍ'āfihi wa-naqlihi*

The title reported by al-Qifṭī is different from that given by Ibn Abī Uṣaybi'a: *Treatise on the Cause of Hindu Reckoning*. This difference diminishes somewhat in importance when we read the end of the treatise, and Ibn al-Haytham's proposition: 'This is what we wished to explain in relation to the causes of the *displacement* of square roots and their doubling in the Indian calculus.'[94] However, it is possible that the title reported by the biobibliographers might be more of a summary of what we have just read or of something similar. It might equally refer to a fundamentally different treatise – larger and more detailed – and whether work on the cause of the square root represented only a minor part of it. This conjecture should not be rejected *a priori*: as it stands, the treatise on roots is incomplete – without any of the preambles which Ibn al-Haytham usually gave, and without any of the introductions where he usually sets out the problem and underlines the originality of his methods.

This text belongs to the Aligarh collection mentioned earlier, fols 17^r–19^r.

2. *Treatise on the Extraction of the Side of a Cube – Fī istikhrāj ḍil' al-muka''ab*

This text figures on all three lists with several variations. Al-Qifṭī's list uses the plural 'sides (*aḍlā'*)', not the singular; and in the Lahore manuscript list there is no mention of the word 'extraction (*istikhrāj*)'. We viewed the text from a single manuscript, which formed part of the earlier-mentioned Kuibychev Library collection, fols 401^v–402^r. It breaks off abruptly at fol. 402^r. There is also a Russian translation of this text.[95]

[94] *Vide infra*, p. 356.

[95] A. Akhmedov, 'Kniga ob izvletcheni rebra kouba', *Matematika i astronomia v troudakh outchionikh srednebekovovo vostoka izdatel'stvo 'fan'*, Tashkent, 1977, pp. 113–17.

CHAPTER I

THE QUADRATURE OF LUNES AND CIRCLES

1.1. INTRODUCTION

The first group of works by Ibn al-Haytham on infinitesimal mathematics was on the quadrature of lunes and circles. The problem he wanted to solve was how to calculate in a rigorous way the area enclosed between the arcs of circles and to find out in every case – whether with lunes or circles – the exact quadrature of these curvilinear areas. The problem of the infinitesimal is ever-present in the proportion of circles under consideration or in the proportion of the squares of their diameters. No other mathematician, before or since Ibn al-Haytham, writing in Greek, Arabic or Latin, has contributed as much to this area of study, or to advance this kind of research, right up until the last decades of the seventeenth century. We are aware that this may come as a surprise, especially as the works of Ibn al-Haytham are still largely unknown.

As we have already seen, earlier biobibliographers attributed three titles which dealt with this area of study to Ibn al-Haytham, two on lunes and one on the quadrature of the circle; they are:

I. *Treatise on the Lunes.*
II. *Treatise on the Quadrature of the Circle.*
III. *Exhaustive Treatise on the Figures of Lunes.*

These are the only writings attributed to Ibn al-Haytham by early biobibliographers, and the only ones to which he himself refers in various different works. It is particularly propitious that we have been able to view all the above titles, as this has facilitated our appreciation of his contribution in this area of work as well as the development of his thought. It is interesting to note that the last of the three treatises is by far the most substantial; they are listed above in chronological order of writing. In treatise II, Ibn al-Haytham specifically refers to treatise I, from which he borrows two propositions. In treatise III, he also refers to treatise I, as a first attempt which had been superseded. In fact, treatise II must have been written before treatise III: if this were not the case, Ibn al-Haytham would have cited treatise III, which also contained relevant propositions and which was, according to the author, more complete than treatise I and destined to replace it. But that is not the case. This formal argument, albeit, connects to another reason which goes right to the heart of the content of the treatises.

Everything started with the short treatise I, conceived and compiled from the perspective of the quadrature of a circle. Ibn al-Haytham declared that he was drawn to research this area in preparation of treatise I when he came across a result 'mentioned by the Ancients' in a work on 'the figure of the lune which is equal to a triangle'; in other words, he refers here to the result ascribed to Hippocrates of Chios. Ibn al-Haytham then went on to rework two of the four propositions which, together with a technical lemma, make up treatise I, in the *Treatise on the Quadrature of the Circle.*

All the evidence seems to point to Ibn al-Haytham knowing that the quadrature of the circle was closely linked to that of certain lunes and wanting to explore this in greater depth in a preliminary treatise, where he would study the area of these lunes and even the area of a circle and lunes (cf. Proposition 5 of the first treatise). Looked at from this perspective, one could say that treatise I and treatise II form part of a tradition which stretches back as far as Hippocrates of Chios.

In fact, in the *Treatise on the Quadrature of a Circle* [II], Ibn al-Haytham does not add any more new and important mathematical results to those he had revealed in the first treatise. But if we stop at that statement, we would be missing the point of this text: we can actually see him developing a train of mathematico-philosophical thought here. We have attempted to unravel this in another piece of work.[1] There are notions here that connect with the question concerning existence in mathematics: how it relates to 'constructability', and how it was formulated based on the notion of 'the known'. The train of thought begun in this text reaches its culmination in two substantial treatises which came after this one – Ibn al-Haytham's *Analysis and Synthesis* (*al-Taḥlīl wa-al-tarkīb*) and *Known Things* (*al-Maʿlūmāt*).[2] In treatise II we can recognize expressions identical to those used in these two later treatises. This is an indication of how keen

[1] See our study 'L'analyse et la synthèse selon Ibn al-Haytham', in R. Rashed (ed.), *Mathématiques et philosophie de l'antiquité à l'âge classique*, Paris, 1991, pp. 131–62; *id.*, *Les mathématiques infinitésimales du IXe au XIe siècle*, vol. IV: *Méthodes géométriques, transformations ponctuelles et philosophie des mathématiques*, London, 2002.

[2] R. Rashed, 'La philosophie des mathématiques d'Ibn al-Haytham. I. *L'analyse et la synthèse*', *MIDEO* 20, 1991, pp. 31–231 and 'La philosophie des mathématiques d'Ibn al-Haytham. II. *Les connus*', *MIDEO* 21, 1993, pp. 87–275; *Les mathématiques infinitésimales*, vol. IV.

Ibn al-Haytham was to pursue this line of enquiry and also provides additional proof of the authenticity of these writings.[3]

These two treatises form a mathematical and historical homogeneous sub-group; and their mathematico-philosophical examination should be viewed as essential and not merely as a diversion, which may or may not be taken into account on the whim of a historian. This not only has implications for Ibn al-Haytham's other works, but as we have already shown,[4] it also reflects an unprecedented interest in the question of existence in mathematics.

In the third treatise, on figures of lunes, the orientation of Ibn al-Haytham's work changes completely; it undergoes a profound transformation, both in terms of scope and understanding. In fact, in this larger treatise, the perspective of quadrature of the circle is abandoned: the study of lunes is no longer intended to contribute, directly or indirectly, to the solution of this problem, but is presented from now on as a chapter on the quadrature of a particular class of curvilinear areas. Ibn al-Haytham expands on results obtained in the first treatise, and by increasing the number of examples, he arrives at a large number of results, which, to this day, historians insist on attributing to much later mathematicians. Briefly, it seems that this research led Ibn al-Haytham to discover a great deal about the use of $\frac{\sin^2 x}{x}$.

Of the three of Ibn al-Haytham's treatises, historians seem only to be aware of the one which deals with the quadrature of the circle.[5] As we said earlier,[6] the first of the three treatises was considered lost; as for the third, it has never been the subject of research. Such incomplete knowledge must surely colour a historian's judgement and gives a false idea of the history of this area in mathematics. Even very recently, Ibn al-Haytham's contribution

[3] On this problem, see the Introduction.

[4] We raised this question, but without sufficient emphasis, in 'La construction de l'heptagone régulier par Ibn al-Haytham', *Journal for the History of Arabic Science* 3, 1979, pp. 309–87, which we revisited as a topic in its own right in 'Analysis and Synthesis according to Ibn al-Haytham'.

[5] See for example C. J. Scriba, 'Welche Kreismonde sind elementar quadrierbar? Die 2400 jährige Geschichte eines Problems bis zur endgültigen Lösung in den Jahren 1933/1947', *Mitteilungen der mathematischen Gesellschaft in Hamburg* XI.5, 1988, pp. 517–34.

[6] Cf. Introduction, p. 29.

was characterized as a 'superficial generalization'[7] of some of the results obtained by Hippocrates of Chios, which 'did not represent any real progress'.[8] We will see that this is not the case at all and that this third treatise was in fact the beginnings of a chapter in infinitesimal mathematics.

We shall therefore examine these three treatises in order of their redaction. We will briefly transcribe the content of the first two, bringing out salient points. We will then move on to the systematic transcription of the contents of the third treatise, adding a mathematical commentary; this will allow the modern reader to follow more easily, without problems in language or overlong descriptions.

1.2. MATHEMATICAL COMMENTARY

1.2.1. *Treatise on Lunes*

This treatise is in the form of a letter sent by Ibn al-Haytham to someone addressed in the preamble in fairly formal terms – 'My Lord, Master' – which would indicate that its recipient is a man of letters or science, possibly of the upper class, and not necessarily in power. However, in his third treatise we learn that it was one of his friends 'who was contented with particular propositions';[9] we know nothing more about this individual, but a lot more about the reasons which led Ibn al-Haytham to engage in research on lunes. The knowledge he has of information from Ancient mathematicians (Hippocrates of Chios's well-known result, proving the area of a lune equal to that of a triangle) is in itself precious evidence. We discuss elsewhere the dissemination of this result in Arabic.[10]

The treatise comprises a brief preamble, four propositions and a lemma. Ibn al-Haytham used the basic tenets again twice: once in his treatise on the *Quadrature of the Circle* and then in his third treatise.

In Propositions 1, 2, 3 and 5 he takes a semicircle *ABC* and studies lunes L_1 and L_2 as limited by arcs *AB* and *BC* and a semicircle. In Propositions 1, 2 and 5 he assumes that arc *AB* equals one sixth of the circumference, and in Proposition 3, he considers *B* as any point on the circumference of the

[7] C. J. Scriba, 'Welche Kreismonde sind elementar quadrierbar?', p. 517. This judgement is quite fair if we stop at the simple mathematical results obtained in *Quadrature of the Circle*, but not if we go further and see the full value of the discussion on mathematical existence.

[8] *Ibid.*, p. 523.

[9] *Vide infra.*

[10] Cf. Volume III.

semicircle. Proposition 4 is a more or less technical lemma necessary to prove Proposition 5. Such is the structure of this short treatise of Ibn al-Haytham. Let us set out the propositions themselves.

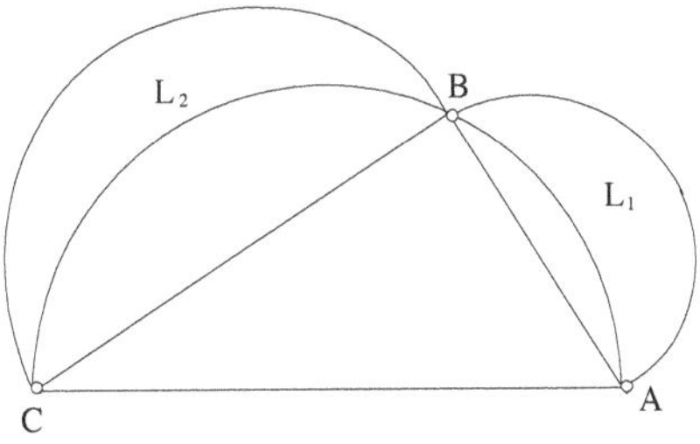

Fig. 1.1

Proposition 1:

$$L_1 + \frac{1}{24}\text{ circle (ABC)} = \frac{1}{2}\text{ tr.(ABC)}.$$

In the course of his proof, Ibn al-Haytham considers a point E on arc AB such that arc AE is equal to one eighth of the circumference, and a point I, at the intersection of radius DE and chord AB. He then proves that sect.$(ADE) = \frac{1}{2}$ circle (AGB) and deduces

$$\begin{aligned}\text{tr.}(ADI) &= \text{lune } (AGBE) + \text{port.}(EBI),\\ \text{tr.}(BAD) &= \text{lune } (AGBE) + \text{port.}(EBI) + \text{tr.}(BID),\\ \text{tr.}(BAD) &= \text{lune } (AGBE) + \text{sect.}(BED).\end{aligned}$$

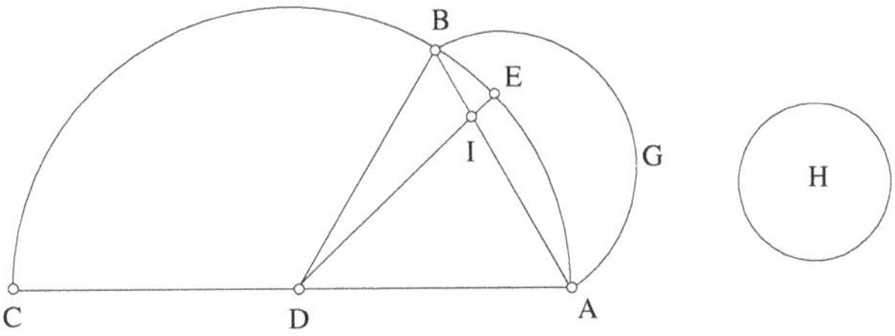

Fig. 1.2

But $\widehat{EB} = \frac{1}{24}$ of the circumference, hence sect.$(BED) = \frac{1}{24}$circle (ABC), hence the result.

However, it is not necessary to introduce these two points E and I. In fact, we have

$$\text{sect.}(ADB) = \frac{1}{6}\text{ circle } (ABC) \text{ and } \frac{1}{2}\text{ circle } (AGB) = \frac{1}{8}\text{ circle } (ABC);$$

hence

$$\text{sect.}(ADB) = \tfrac{1}{2}\text{ circle }(AGB) + \tfrac{1}{24}\text{ circle }(ABC).$$

By taking away segment (AEB) from the two members, we obtain the result.

Proposition 2:

$$L_2 = \tfrac{1}{2}\text{ tr.(ABC)} + \tfrac{1}{24}\text{ circle (ABC)}.$$

It should be noted that these first two propositions are directly linked with the case of two lunes studied in Proposition 13 of the *Exhaustive Treatise*, which, as we will see later, is an application of Propositions 8 and 9.

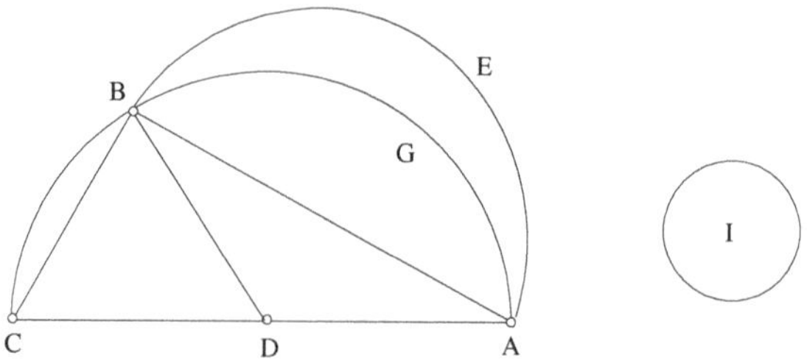

Fig. 1.3

Proposition 3:

$$L_1 + L_2 = \text{tr.(ABC)};$$

B *is any point on the circumference.*

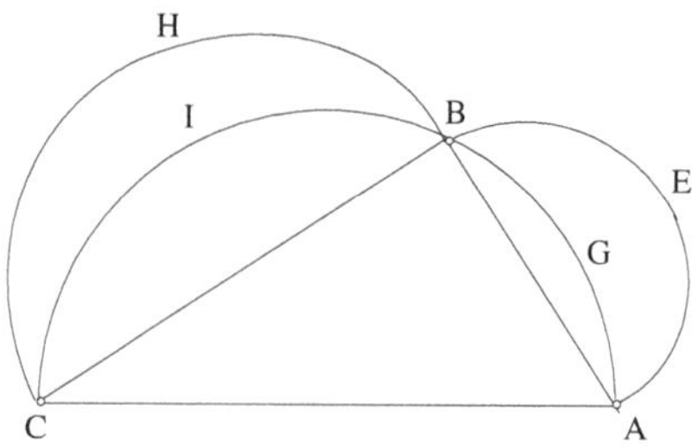

Fig. 1.4

The proof that Ibn al-Haytham constructs for this proposition is the same as that given by the Ancients (Hippocrates of Chios, according to Eudemus),[11] and is based on the ratio of the area of the circle to the square of its diameter, and also on Pythagoras' theorem. From the latter, it can be deduced:

$$\tfrac{1}{2}\text{ circle }(AEB) + \tfrac{1}{2}\text{ circle }(BHC) = \tfrac{1}{2}\text{ circle }(ABC),$$

and by taking segm.(AGB) + segm.(BIC) away from the two members, the result is obtained.

Proposition 5:

$$L_2 + \tfrac{1}{2}\text{ tr.(ABC)} = L_3 + \tfrac{1}{8}\text{ circle (ABC)},$$

with L_3 *a lune similar to* L_1, *such that* $L_3 = 2L_1$.

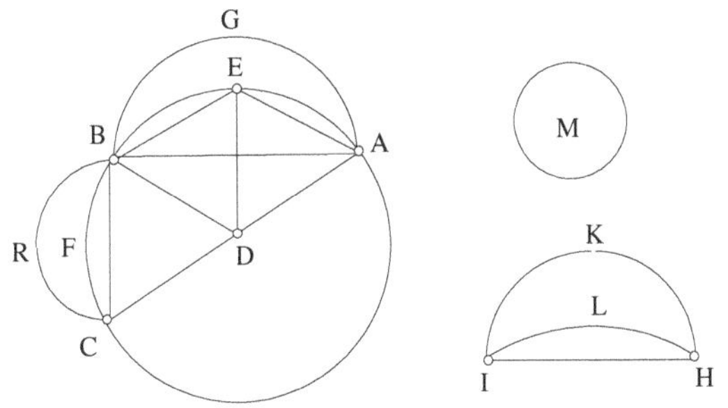

Fig. 1.5

This proof relies on Propositions 2, 3 and 4. Proposition 4 deals with the study of the ratio of two similar lunes. Ibn al-Haytham begins by proving that the ratio of similar segments is equal to the square of the ratio of their bases (a proof which is the subject of Lemma 5 in the *Exhaustive Treatise*); he then goes on to deduce from this the ratio of the two lunes.

This gives us some idea of the methods used by Ibn al-Haytham and the principal results of this short treatise.

[11] Cf. T. L. Heath, *A History of Greek Mathematics*, 2 vols, Oxford, 1921, vol. I, pp. 191–201 and O. Becker, *Grundlagen der Mathematik*, 2nd ed., Munich, 1964, pp. 29–34.

1.2.2. *Treatise on the Quadrature of the Circle*

As we have already said, this treatise depends on the previous one, and shows us Ibn al-Haytham's reasons for engaging in research on lunes. If the number of manuscripts is any indication of how well known this work was, and if the many commentaries referred to are indications of its popularity, then it goes without saying that this is Ibn al-Haytham's most widespread and most popular piece of mathematical work ever. It is true to say that in this piece of writing, Ibn al-Haytham raises a traditional yet crucial question: is it possible to square a circle exactly?

To reply to this question, he begins by recalling two results from the previous treatise: the first and the third propositions. He gives a new proof of the third proposition:

$$\frac{\text{circle }(BGC)}{\text{circle }(ABE)} = \frac{BC^2}{BA^2}, \text{ from XII.2 of the } \textit{Elements},$$

$$\frac{\text{circle }(BGC) + \text{circle }(ABE)}{\text{circle }(ABE)} = \frac{AC^2}{BA^2} = \frac{\text{circle }(ABC)}{\text{circle }(ABE)};$$

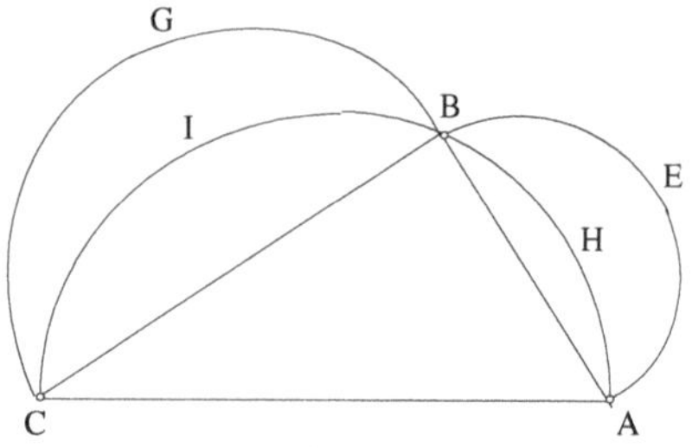

Fig. 1.6

hence

$$\tfrac{1}{2}\text{ circle }(ABC) = \tfrac{1}{2}\,(BGC) + \tfrac{1}{2}\,(ABE).$$

The sum of segment (ABH) and segment (BCI) is taken away from each side of the equality; it follows that

$$(1) \qquad L_1 + L_2 = \text{lune }(AEBH) + \text{lune }(BGCI) = \text{tr.}(ABC);$$

if B is the midpoint of the semicircle ABC, the two lunes are equal, and from (1) we derive

$$(2) \qquad L_1 = \text{tr.}(ABD).$$

Ibn al-Haytham then considers a circle with diameter HE, with H and E being the midpoints of two arcs AB which limit lune $AEBH$, and he studies the ratio of this circle to this lune, to obtain the quadrature of circle AC. He notes firstly that circle $(HE) <$ lune $(AEBH) = L_1$, he then reasons as follows: this circle is known, L_1 is also known, therefore

$$\frac{\text{circle } (HE)}{L_1} = k,$$

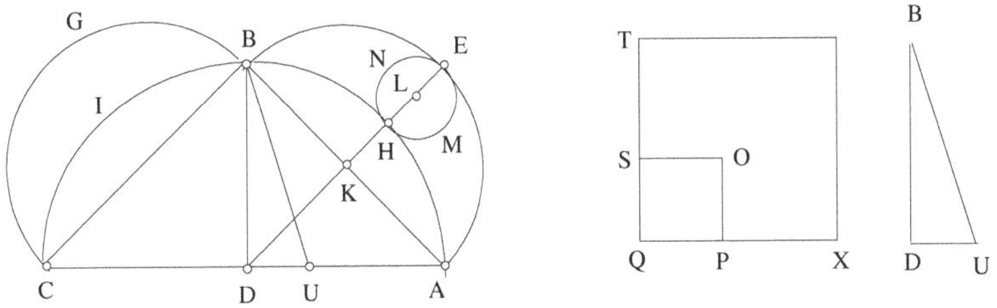

Fig. 1.7

a ratio that exists 'even if no one knows what the ratio is or even if it is impossible for anyone to know it'.[12]

Let DU be such that $\frac{DU}{DA} = k$, then $\frac{\text{tr.}(BDU)}{\text{tr.}(BDA)} = \frac{DU}{DA} = k$; therefore

$$\frac{\text{tr.}(ABD)}{L_1} = \frac{\text{tr.}(BDU)}{\text{circle } (HE)};$$

and if (2) is taken into account, then circle $(HE) = \text{tr.}(BDU)$.

But we know how to construct a square $(SPQO)$ equivalent to tr.(BDU), therefore

$$\text{tr.}(BDU) = \text{square } (SQPO) = \text{circle } (HE).$$

Then a square of side QX is constructed such that $\frac{QP}{QX} = \frac{EH}{AC}$ and XT the square constructed on QX; so we have

$$\frac{\text{square } (XT)}{\text{square } (QO)} = \frac{QX^2}{QP^2} = \frac{AC^2}{EH^2} = \frac{\text{circle } (ABC)}{\text{circle } (HE)};$$

the equality, circle $(HE) =$ square (QO), finally leads to circle $(AC) =$ square (XT).

[12] *Vide infra*, p. 101.

We can say that Ibn al-Haytham's reasoning relies on the existence of the number k, a ratio of two plane surfaces. From the existence of k, it is possible to deduce the existence of segments DU, QP and QX, even if their construction is not possible except when k is known. However, the following (not calculated by Ibn al-Haytham) gives the values as

$$HE = R\,[\sqrt{2} - 1],\ \text{circle}\ (HE) = \pi \frac{R^2}{4}\,(\sqrt{2} - 1)^2,\ k = \frac{\pi\,(\sqrt{2} - 1)^2}{2},$$

and the side of the square equivalent to the circle is $R\sqrt{\pi}$. We may notice immediately the circularity, since knowledge of k and of π are linked.

Ibn al-Haytham's text has been commented on and criticized by at least two mathematicians: Naṣīr al-Dīn al-Ṭūsī and another who might have been, according to the colophon, Ibn al-Haytham's contemporary, ʿAlī ibn Riḍwān or al-Sumaysāṭī.

The first critique by al-Ṭūsī emphasizes the length of the text and proposes another method.

Let a circle of diameter DE be inscribed in square (BC) with side AB. The circle is part of the square, therefore their ratio exists:

$$\frac{\text{square}\ (BC)}{\text{circle}\ (DE)} = k \qquad \left[\text{we have } k = \frac{4}{\pi}\right].$$

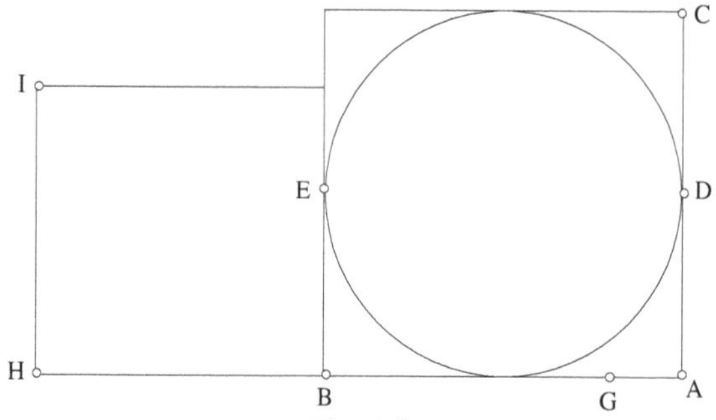

Fig. 1.8

Let BG such that $\frac{AB}{BG} = k$ and BH such that $\frac{AB}{BH} = \frac{BH}{BG}$, then $\frac{AB^2}{BH^2} = \frac{AB}{BG} = k$, hence

$$\frac{\text{square}\ (BC)}{\text{square}\ (BI)} = k.$$

And it follows that

$$\text{circle}\ (DE) = \text{square}\ (BI).$$

Al-Ṭūsī's commentary adds nothing fundamental to the text of Ibn al-Haytham. His reasoning also relies on the existence of the number k, the ratio of two plane surfaces. In fact, it is from the existence of k that the existence of segments BG and BH and therefore that of the square BI can be deduced.

The second objection, raised by ʿAlī ibn Riḍwān or al-Sumaysāṭī, is more important than the one from al-Ṭūsī, since it goes right to the heart of Ibn al-Haytham's contribution. It has its basis in mathematical philosophy and can be summarized as follows: a proof of existence does not solve the problem of construction, on whose effectiveness we rely for our knowledge of the property in question.

1.2.3. *Exhaustive Treatise on the Figures of Lunes*

This treatise, written a long time after the first, as the author himself indicates, is very different. Ibn al-Haytham describes it as 'exhaustive', whereas the first one is described as 'abridged'. This treatise is composed using apodictic methods whereas the other proceeds 'according to particular methods'.[13] The latter is destined to replace the former with new research on lunes. Our task can be therefore defined as follows: to transcribe this book, following its various twists and turns, and to note its unifying features as well as any of its shortcomings.

It must be said straightaway that in the first treatise, lunes L_1 and L_2 are associated with semicircles (ABC, AEB and BGC). In the second treatise, Ibn al-Haytham again takes up this study and generalizes the results of Propositions 1, 2 and 3, by expanding them to any given arcs AB and BC such that

$$AB + BC \leq \tfrac{1}{2} \text{ circumference.}$$

In all cases, the arcs that define lunes L_1 and L_2 are similar to the arc of a semicircle (ABC), whereby the triangle (ABC) occurs in such a way that angle B is equal to or bigger than a right angle.

The calculation of the areas of lunes involves the sums or differences between the areas of sectors or triangles, and their comparison in turn relies on the ratio of angles and the ratio of segments.

Ibn al-Haytham begins by establishing four lemmas based on various types of triangle ABC that are required for his proposed study: with a right angle at point B in Lemma 1, or with an obtuse angle at point B in Lemmas

[13] *Vide infra*, p. 107.

2, 3 and 4. Throughout the text, and wherever there is a case of two triangles similar to the initial triangle, he establishes an unequal relation between ratios of angles and ratios of segments. The results of these lemmas, which the author then goes on to use in Propositions 9 to 12, prove the role of the function f as defined by

$$f(x) = \frac{\sin^2 x}{x}$$

in the study of lunes.

We now come to a more detailed exposition of the method we have just outlined, beginning with this group of four lemmas.

Lemma 1. — *If* $\widehat{ABC} = \frac{\pi}{2}$, BA < BC *and* BD ⊥ AC, *then* $\frac{DA}{AC} < \frac{\widehat{ACB}}{\pi/2}$ *and* $\frac{DC}{AC} > \frac{\widehat{BAC}}{\pi/2}$.

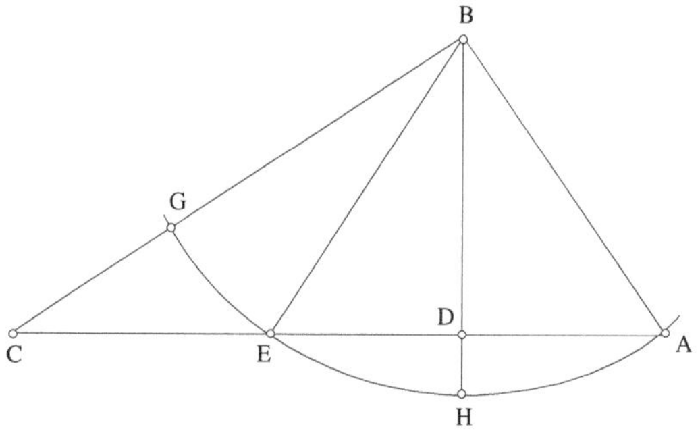

Fig. 1.9

Since $BA < BC$, then $DA < DC$ and circle (B, BA) cuts $[DC]$ at E and cuts $[BC]$ at G. The half-line BD cuts the circle at H, beyond D. We have

$$\frac{\text{tr.}(BCE)}{\text{tr.}(BDE)} > \frac{\text{sect.}(BEG)}{\text{sect.}(BEH)},$$

hence

$$\frac{\text{tr.}(BCD)}{\text{tr.}(BDE)} > \frac{\text{sect.}(BHG)}{\text{sect.}(BEH)} = \frac{\widehat{CBD}}{\widehat{EBD}}.$$

Therefore

(1) $$\frac{CD}{ED} = \frac{CD}{DA} > \frac{\widehat{CBD}}{\widehat{DBA}} \quad (\text{as } ED = DA \text{ and } \widehat{EBD} = \widehat{DBA}),$$

hence

$$\frac{CD + DA}{DA} > \frac{\widehat{CBA}}{\widehat{DBA}}.$$

But $\widehat{DBA} = \widehat{ACB}$ and $\widehat{CBA} = \frac{\pi}{2}$, so

$$\frac{DA}{AC} < \frac{\widehat{ACB}}{\pi/2}.$$

In the same way, from (1) we can deduce

$$\frac{CD}{DA + CD} > \frac{\widehat{CBD}}{\widehat{DBA} + \widehat{CBD}};$$

but $\widehat{CBD} = \widehat{BAC}$, hence

$$\frac{CD}{AC} > \frac{\widehat{BAC}}{\pi/2}.$$

Comments:

1) $$\frac{DA}{AC} = \frac{DA}{AB} \cdot \frac{AB}{AC} = \sin^2 C \text{ and } \frac{DC}{AC} = \sin^2 A.$$

Therefore the above proof gives the following result:

If $0 < C < \frac{\pi}{4} < A < \frac{\pi}{2}$, then $\frac{\sin^2 C}{C} < \frac{2}{\pi} < \frac{\sin^2 A}{A}$.

If $A = C = \frac{\pi}{4}$, then $\frac{\sin^2 A}{A} = \frac{\sin^2 C}{C} = \frac{2}{\pi}$.

This is how Lemma 1 is written, taking the radian as a unit.

2) The method used in the proof of this lemma leads to the establishment of the proposition:

$$\alpha < \beta < \frac{\pi}{2} \Rightarrow \frac{\tan \beta}{\tan \alpha} > \frac{\beta}{\alpha},$$

as the hypothesis $\widehat{ABC} = \alpha + \beta = \frac{\pi}{2}$ is not involved in the establishment of (1), only the hypothesis $\alpha < \beta < \frac{\pi}{2}$ is involved.

A very similar method is used to establish the proposition:

$$\alpha < \beta < \frac{\pi}{2} \Rightarrow \frac{\sin \beta}{\sin \alpha} < \frac{\beta}{\alpha}.$$

Let $\widehat{xOy} = \alpha$, $\widehat{xOz} = \beta$ with $\beta > \alpha$. A circle with centre O cuts Oy at A and Oz at B and Ox at D; straight line BA cuts Ox at C. Then

$$\frac{\text{tr.}(AOB)}{\text{tr.}(AOC)} < \frac{\text{sect.}(AOB)}{\text{sect.}(AOD)},$$

from which we deduce

$$\frac{\text{tr.}(BOC)}{\text{tr.}(AOC)} < \frac{\text{sect.}(BOD)}{\text{sect.}(AOD)}.$$

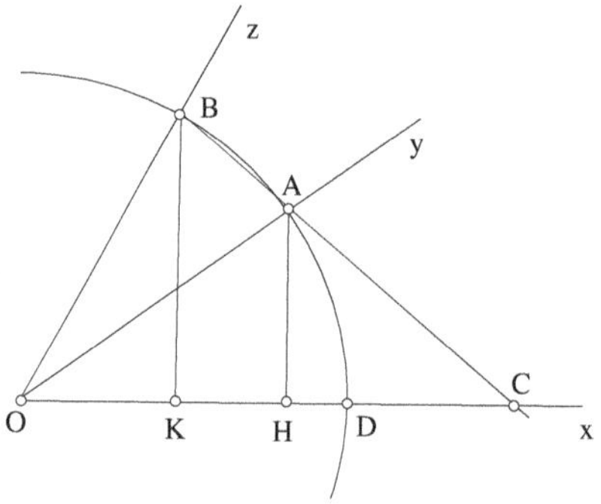

Fig. 1.10

The two triangles have the same base OC, so

$$\frac{\text{tr.}(BOC)}{\text{tr.}(AOC)} = \frac{BK}{AH} = \frac{\sin \beta}{\sin \alpha};$$

hence

$$\frac{\sin \beta}{\sin \alpha} < \frac{\beta}{\alpha} \quad \text{or} \quad \frac{\sin \beta}{\beta} < \frac{\sin \alpha}{\alpha}.$$

Therefore over interval $]0, \frac{\pi}{2}[$, function $\frac{\sin x}{x}$ is decreasing.

Lemma 2. — *If* $\widehat{ABC} > \frac{\pi}{2}$, $AB < BC$ *and* $\widehat{BDA} = \widehat{ABC}$, *then*

$$\frac{DA}{AC} < \frac{\widehat{ACB}}{\pi - \widehat{ABC}}.$$

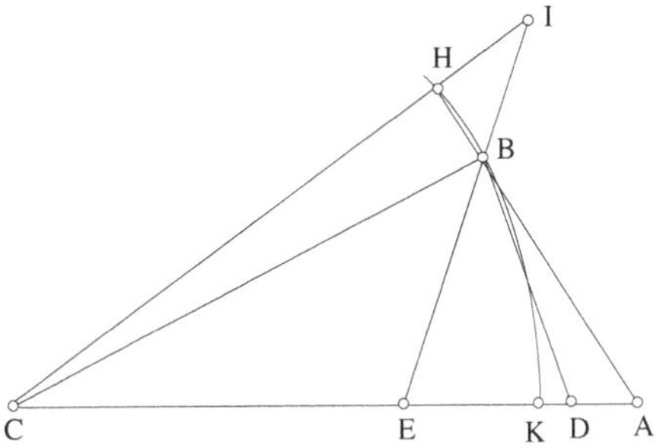

Fig. 1.11

Let E be such that $\widehat{BEC} = \widehat{BDA} = \widehat{ABC}$, so $BE = BD$ and $\frac{DA}{DB} = \frac{AB}{BC} = \frac{EB}{EC}$ (as ADB, ABC and BEC are similar triangles), hence $DA \cdot EC = BD \cdot BE = BD^2$. But $DA < DB$, $DB = BE$ and $EB < EC$, hence $DA < EC$. On the other hand, $DA \cdot DC < \left(\frac{AC}{2}\right)^2$,[14] and it follows that $DA \cdot EC < \left(\frac{AC}{2}\right)^2$, hence $BD^2 < \left(\frac{AC}{2}\right)^2$ or $BD < \frac{AC}{2}$.

We know that $EB < EC$. Let I be a point beyond B such that $EI = EC$, so $CI > CB > CE$. Therefore circle (C, CB) cuts CI at H between C and I and CE at K beyond E. Following the argument used in (1) on circular sectors HCB and BCK, and triangles ICB and BCE, then

$$\frac{\text{area }(HCB)}{\text{area }(BCK)} = \frac{\widehat{ICB}}{\widehat{BCE}} < \frac{\text{tr.}(ICB)}{\text{tr.}(BCE)};$$

and by composition

$$\frac{\widehat{ICB} + \widehat{BCE}}{\widehat{BCE}} < \frac{\text{tr.}(ICB) + \text{tr.}(BCE)}{\text{tr.}(BCE)};$$

hence

$$\frac{\widehat{ICE}}{\widehat{BCE}} < \frac{\text{tr.}(ICE)}{\text{tr.}(BCE)} = \frac{EI}{EB} \text{ or } \frac{EB}{EI} < \frac{\widehat{BCE}}{\widehat{ICE}}.$$

But

$$IE = EC \text{ and } \widehat{ICE} = \frac{1}{2}\widehat{BEA} = \frac{1}{2}(\pi - \widehat{ABC}) \text{ and } \frac{EB}{EC} = \frac{DA}{DB};$$

[14] If O is the midpoint of $[AC]$, then $DA \cdot DC = OA^2 - OD^2$; therefore $DA \cdot DC < OA^2$.

therefore

$$\frac{DA}{DB} < \frac{\widehat{ACB}}{1/2\,(\pi - \widehat{ABC})}.$$

But we have proved that $DB < \frac{AC}{2}$; therefore $\frac{DA}{AC} < \frac{\widehat{ACB}}{\pi - \widehat{ABC}}$.

Comments: $\frac{DA}{AC} = \frac{DA}{AB} \cdot \frac{AB}{AC} = \frac{AB^2}{AC^2} = \frac{\sin^2 C}{\sin^2 B}$. Lemma 2 is therefore written as

$$B > \frac{\pi}{2} \text{ and } C < \frac{\pi}{4} \Rightarrow \frac{\sin^2 C}{\sin^2 B} < \frac{C}{\pi - B}.$$

If we set $B_1 = \pi - B$, then $\sin^2 B = \sin^2 B_1$ and $B_1 > C$ (as $B_1 - C = A$), hence

$$C < B_1 < \frac{\pi}{2} \text{ and } C < \frac{\pi}{4} \Rightarrow \frac{\sin^2 C}{C} < \frac{\sin^2 B_1}{B_1}.$$

Lemma 3. — *If angle* ABC *is obtuse,* AB < BC, $\widehat{\mathrm{BAC}} \leq \frac{\pi}{4}$, *then* $\frac{\mathrm{EC}}{\mathrm{CA}} < \frac{\widehat{\mathrm{BAC}}}{\pi - \widehat{\mathrm{ABC}}}$ (E *is a point on* AC *such that* $\widehat{\mathrm{BEC}} = \widehat{\mathrm{ABC}}$).

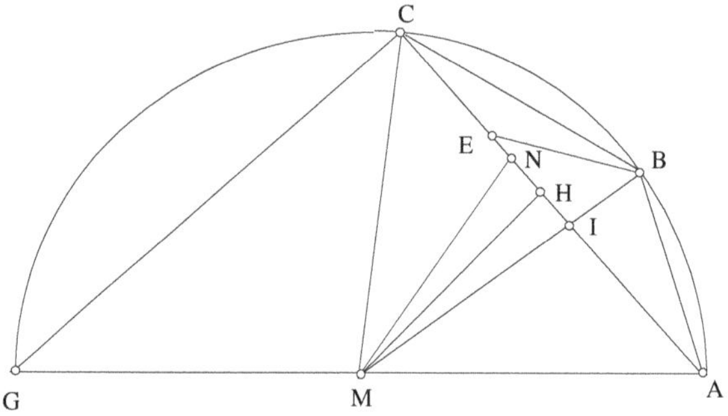

Fig. 1.12

Let M be the centre of the circumscribed circle and let G be the diametrically opposed point to A, E as defined in (2), therefore BEC is similar to ABC. Straight line BM cuts AC at I. Let MH be perpendicular to AC, then H is the midpoint of AC. Circle (M, MI) cuts AC at N. As $\widehat{ABC}$ is obtuse, then $\overset{\frown}{ABC} < \frac{1}{2}$ circle.

a) If $\widehat{ABC} \leq \frac{1}{4}$ circle, then $\widehat{AMC} \leq \frac{\pi}{2}$, $\widehat{MAC} = \widehat{MCA} \geq \frac{\pi}{4}$, $\widehat{AGC} \leq \frac{\pi}{4}$, therefore $\widehat{AGC} \leq \widehat{MAC}$. But $\widehat{MIC} > \widehat{MAC} \geq \frac{\pi}{4}$, therefore $\widehat{MIC} > \widehat{AGC}$; it follows that $\widehat{BIC} < \widehat{ABC}$. But by construction $\widehat{BEC} = \widehat{ABC}$, therefore $\widehat{BEC} > \widehat{BIC}$ and it follows that E is between I and C.

Ibn al-Haytham writes 'it is clear that $\frac{CH}{HN} > \frac{\widehat{CMH}}{\widehat{HMN}}$'.

This result can be established as in Lemma 1, based on triangles MHN and CMN and sectors described by circle (M, MI). We can deduce that

$$\frac{CH}{HI} > \frac{\widehat{CMH}}{\widehat{HMI}};$$

hence

$$\frac{IC}{CH} < \frac{\widehat{IMC}}{\widehat{HMC}} \text{ and } \frac{IC}{CA} < \frac{\widehat{BMC}}{\widehat{CMA}}.$$

But

$$\widehat{BMC} = 2\widehat{BAC} \text{ and } \widehat{CMA} = 2\,(\pi - \widehat{ABC});$$

hence

$$\frac{IC}{CA} < \frac{\widehat{BAC}}{\pi - \widehat{ABC}}.$$

It follows that

$$\frac{EC}{CA} < \frac{\widehat{BAC}}{\pi - \widehat{ABC}}.$$

b) If $\widehat{ABC} > \frac{1}{4}$ circle, then $\widehat{AMC} > \frac{\pi}{2}$. But $\widehat{BAC} \leq \frac{\pi}{4}$, therefore $\widehat{BC} \leq \frac{1}{4}$ circle.

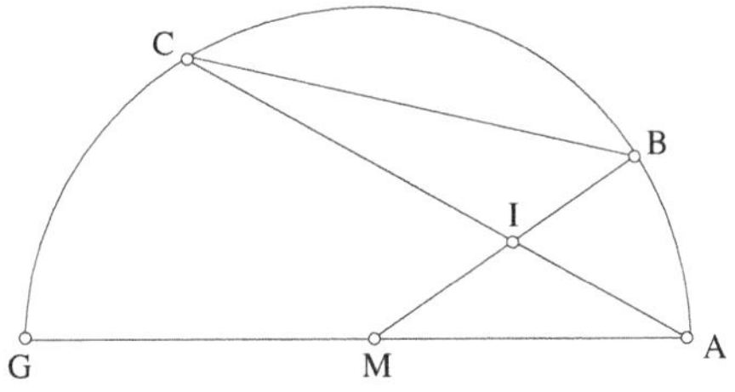

Fig. 1.13

• If $\widehat{BC} = \frac{1}{4}$ circle, then $\widehat{BMC} = \frac{\pi}{2}$ and $\widehat{MBC} = \widehat{BAC} = \frac{\pi}{4}$; therefore $\widehat{BIC} = \widehat{ABC}$, I and E thus coincide. The conclusion is as in the first part.

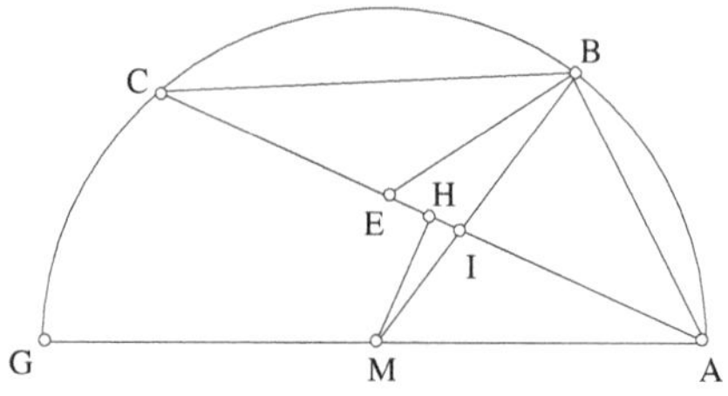

Fig. 1.14

• If $\widehat{BC} < \frac{1}{4}$ circle, then $\widehat{BMC} < \frac{\pi}{2}$ and $\widehat{MBC} > \frac{\pi}{4}$.

So $\widehat{MBC} > \widehat{BAC}$. However $\widehat{CBE} = \widehat{BAC}$; therefore $\widehat{CBE} < \widehat{CBM}$ and it follows that E is between I and C, and that H is in all cases between I and C. The result is obtained as previously.

Therefore if $\widehat{ABC}$ is an obtuse angle, $AB < BC$ and $\widehat{BAC} \leq \frac{\pi}{4}$, then

$$\frac{EC}{CA} < \frac{\widehat{BAC}}{\pi - \widehat{ABC}}.$$

Comments: The result proved in Lemma 2 for $\widehat{C}$, the smallest of the acute angles, is still valid for the acute angle $\widehat{A}$, if $\widehat{A} \leq \frac{\pi}{4}$. This result is written as

$$\frac{\sin^2 A}{A} < \frac{\sin^2 B}{\pi - B} \quad \text{or} \quad \frac{\sin^2 A}{A} < \frac{\sin^2 B_1}{B_1} \ (\text{as } B_1 > A).$$

Lemma 4. — *With the hypotheses that* $\widehat{ABC}$ *is an obtuse angle, that* AB < BC, $\widehat{BAC} > \frac{\pi}{4}$, *and with* E *such that* $\widehat{BEC} = \widehat{ABC}$, *under what conditions do we arrive at*

$$\frac{EC}{CA} > \frac{\widehat{BAC}}{\pi - \widehat{ABC}}?$$

Let there be a circle with diameter CD and centre K and let I be a point of segment KD. From point I, the perpendicular to KD cuts the circle at B and gives $\widehat{BC} > \frac{1}{4}$ of circumference, hence $\widehat{BAC} > \frac{\pi}{4}$.

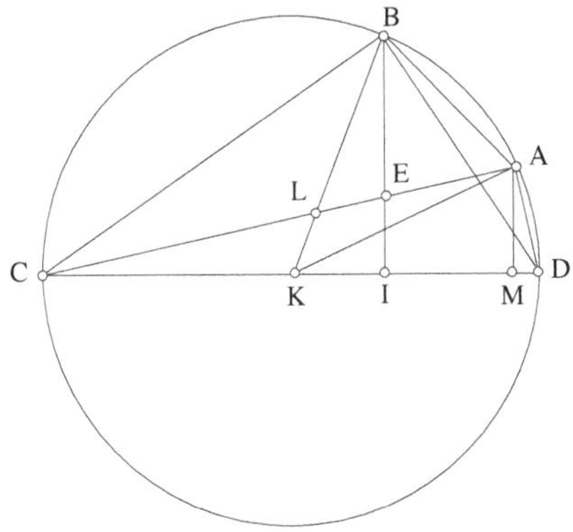

Fig. 1.15

Reciprocally, if $\widehat{BAC} > \frac{\pi}{4}$, then the projection I of B on diameter CD is between K and D.

In the right-angled triangle BDC, from Lemma 1, we have

$$\frac{IC}{CD} > \frac{\widehat{BDC}}{\pi/2},$$

hence

$$\frac{IC}{CD} > \frac{\widehat{BKC}}{\pi} = \frac{\widehat{BC}}{\widehat{CBD}}.$$

From this, we can deduce $\frac{ID}{IC} < \frac{\widehat{BD}}{\widehat{BC}}$. There exists a part $\widehat{AB}$ of $\widehat{DB}$ such that

$$\frac{ID}{IC} = \frac{\widehat{BA}}{\widehat{BC}}.$$

Let M be the projection of A on DC, between I and D, and let E be the intersection of CA and BI,[15] then $\widehat{DAC} = \widehat{CIE} = \frac{\pi}{2}$, therefore $\widehat{ADC} = \widehat{IEC} = \pi - \widehat{ABC}$, and it follows that $\widehat{BEC} = \widehat{ABC}$.

[15] Note that if L is the intersection of BK and CA, then L will be between C and E; this remark also comes into Proposition 12.

On the other hand $\frac{CI}{IM}=\frac{CE}{EA}>\frac{CI}{ID}=\frac{\widehat{BC}}{\widehat{BA}}$; we can deduce that

$$\frac{EC}{CA}>\frac{\widehat{BC}}{\widehat{CBA}}.$$

But

$$\frac{\widehat{BC}}{\widehat{CBA}}=\frac{\widehat{BKC}}{\widehat{CKA}}=\frac{\widehat{BAC}}{\widehat{ADC}}=\frac{\widehat{BAC}}{\pi-\widehat{ABC}},$$

so

$$\frac{EC}{CA}>\frac{\widehat{BAC}}{\pi-\widehat{ABC}}.$$

This argument is based on point A defined by $\frac{\widehat{BA}}{\widehat{BC}}=\frac{ID}{IC}$. If another point A' between A and D is considered, then $\frac{CI}{ID}>\frac{\widehat{BC}}{\widehat{BA'}}$.

Point E' is associated with point A' on BI and then $\frac{E'C}{E'A'}>\frac{CI}{ID}$, therefore $\frac{E'C}{E'A'}>\frac{\widehat{BC}}{\widehat{BA'}}$, and it can be concluded in the same way that $\frac{E'C}{E'A'}>\frac{\widehat{BA'C}}{\pi-\widehat{A'BC}}$.

Summary: By adding to hypotheses $\widehat{ABC}$ obtuse, $\widehat{AB}<\widehat{BC}$, $\widehat{BAC}>\frac{\pi}{4}$, E on BC such that $\widehat{BEC}=\widehat{ABC}$, the following condition: I the orthogonal projection of B on diameter CD of the circle circumscribed about ABC satisfies $\frac{ID}{IC}\leq\frac{\widehat{BA}}{\widehat{BC}}$,[16] then

$$\frac{EC}{CA}>\frac{\widehat{BAC}}{\pi-\widehat{ABC}}.$$

Comments:

1) The result proved in Lemma 4 can therefore be set out in the following form:

Given that $A>\frac{\pi}{4}$, it is possible to find an angle B_0 (which therefore depends on A) such that $B_1\geq B_0\Rightarrow\frac{\sin^2 A}{A}>\frac{\sin^2 B_1}{B_1}$.

[16] The condition laid down is therefore sufficient so that $\frac{EC}{CA}>\frac{\widehat{BAC}}{\pi-\widehat{ABC}}$. It is not necessary (see Proposition 12).

2) To understand Ibn al-Haytham's method, it is necessary to study the relation

$$(1) \qquad \frac{CE}{CA} < \frac{\widehat{BAC}}{\pi - \widehat{ABC}}.$$

Suppose $\widehat{CDA} = \alpha$, $\widehat{CDB} = \beta = \widehat{BAC}$, $\beta < \alpha < \frac{\pi}{2}$; then $\widehat{CA} = 2\alpha$, $\widehat{CB} = 2\beta$, $\widehat{AB} = 2(\alpha - \beta)$, $\widehat{BC} > \widehat{AB} \Leftrightarrow 2\beta > \alpha$.

In Lemma 3, it is assumed that $\widehat{BC} \leq \frac{1}{4}$ circle, i.e. $\beta \leq \frac{\pi}{4}$, α must therefore satisfy $\beta < \alpha < 2\beta$.

In Lemma 4, it is assumed that $\widehat{BC} > \frac{1}{4}$ circle, $\beta > \frac{\pi}{4}$; this should therefore give $\beta < \alpha < \frac{\pi}{2}$. We saw that (1) $\Leftrightarrow \frac{CI}{CM} < \frac{\widehat{BC}}{\widehat{AC}}$ (2). But

$$CI = CB \sin \beta = CD \sin^2 \beta$$
$$CM = CA \sin \alpha = CD \sin^2 \alpha.$$

Condition (2) becomes

$$\frac{\sin^2 \beta}{\sin^2 \alpha} < \frac{\beta}{\alpha}$$

or even

$$\frac{\sin^2 \alpha}{\alpha} > \frac{\sin^2 \beta}{\beta}.$$

Let

$$f(\alpha) = \frac{\sin^2 \alpha}{\alpha}, \quad 0 < \alpha < \frac{\pi}{2}.$$

$$f'(\alpha) = \frac{2\alpha \sin \alpha \cos \alpha - \sin^2 \alpha}{\alpha^2} = \frac{\sin \alpha \cos \alpha}{\alpha} \left[2 - \frac{\tan \alpha}{\alpha}\right].$$

Over interval $\left]0, \frac{\pi}{2}\right[$, $\frac{\tan \alpha}{\alpha}$ increases from 1 to $+\infty$, therefore a unique value α_0 exists such that $\frac{\tan \alpha_0}{\alpha_0} = 2$.

• For $\alpha = 60° = \frac{\pi}{3}$, we find $\frac{\tan\alpha}{\alpha} \cong 1.66$,

$\alpha = 70° = \frac{7\pi}{18}$, we find $\frac{\tan\alpha}{\alpha} \cong 2.18$;

then $\alpha_0 \cong 70° = \frac{7\pi}{18} \cong 1.22$ rd, or to be more precise $\alpha_1 = 1.16556119$ rd, say $66°\ 49'54''$.

Therefore function f presents a maximum M.

• For $\alpha = \alpha_0$, we find $M \cong 0.72$, and for $\alpha = \alpha_1$, we find $M = 0.724611354$.

$$\lim_{\alpha \to 0} f(\alpha) = 0$$

$$f\left(\frac{\pi}{4}\right) = f\left(\frac{\pi}{2}\right) = \frac{2}{\pi} \cong 0.64 \text{ and } f\left(\frac{\pi}{6}\right) = \frac{3}{2\pi} \cong 0.48.$$

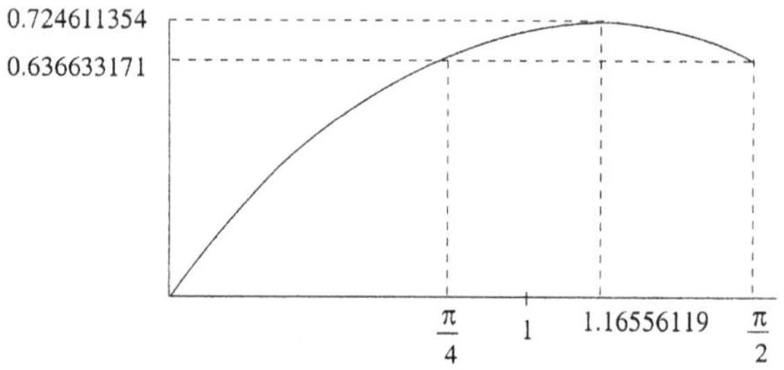

Fig. 1.16

If $\beta \leq \frac{\pi}{4}$, as in Lemma 3, then $f(\beta) < \frac{2}{\pi}$ and $2\beta < \frac{\pi}{2}$.

$$\forall\ \alpha \in\ ]\beta, 2\beta[,\ f(\alpha) > f(\beta);$$

which in this case gives $\frac{CE}{CA} < \frac{\widehat{BAC}}{\pi - \widehat{ABC}}$.

If $\beta > \frac{\pi}{4}$, as in Lemma 4, then $f(\beta) > \frac{2}{\pi}$.

• For $\frac{\pi}{4} < \beta < \alpha_0$, there exists a unique value α_1, $\alpha_1 \in \left]\alpha_0, \frac{\pi}{2}\right[$, such that $f(\alpha_1) = f(\beta)$.

a) if $\beta < \alpha < \alpha_1$, then $f(\alpha) > f(\beta)$; hence $\frac{CE}{CA} < \frac{\widehat{BAC}}{\pi - \widehat{ABC}}$.

b) If $\alpha = \alpha_1$, then $f(\alpha) = f(\beta)$; hence $\frac{CE}{CA} = \frac{\widehat{BAC}}{\pi - \widehat{ABC}}$.

c) If $\alpha_1 < \alpha < \frac{\pi}{2}$, then $f(\alpha) < f(\beta)$; hence $\frac{CE}{CA} > \frac{\widehat{BAC}}{\pi - \widehat{ABC}}$.

• For $\alpha_0 \leq \beta < \frac{\pi}{2}$, then $\forall\ \alpha \in \left]\beta, \frac{\pi}{2}\right[$, then $f(\alpha) < f(\beta)$; hence

$$\frac{CE}{CA} > \frac{\widehat{BAC}}{\pi - \widehat{ABC}}.$$

We note that for each value of $\beta > \frac{\pi}{4}$ (for $\widehat{BC} > \frac{1}{4}$ circle), there is an associated value α_1 of α for which $\frac{CE}{CA} = \frac{\widehat{BAC}}{\pi - \widehat{ABC}}$ which leads to a lune equivalent to a triangle (see *Comments* on Proposition 12). The case of Proposition 13, corresponding to $\beta = \frac{\pi}{4}$ and $\alpha_1 = \frac{\pi}{2}$, is the limit case; this leads to two equal lunes, each one equivalent to a triangle.

3) The determination of point A such that $\frac{\widehat{AB}}{\widehat{BC}} = \frac{ID}{IC}$ is possible because they are arcs of the same circle.

If $\frac{DI}{IC} = \frac{1}{2}$ or $\frac{DI}{IC} = \frac{1}{2^n}$, then we successively divide an arc in half up to the point where one part of DC is homologous with DI.

Ibn al-Haytham does not give any indication of the construction of A if $\frac{ID}{IC} = k$, where k is any ratio.

If for example $\frac{ID}{IC} = \frac{1}{3}$, the construction of point A on arc $\widehat{BD}$ is effected by trisection of angle $\widehat{BKC}$.

Three technical lemmas 5, 6 and 7 follow this group of four lemmas.

Lemma 5. — *A circle equivalent to a sector of a circle of a given diameter* AD; *let* ACB *be the sector of the circle. Then*

$$\frac{\text{sect.}(ABC)}{\text{circle }(ABD)} = \frac{\widehat{AB}}{\text{the complete circumference}}.$$

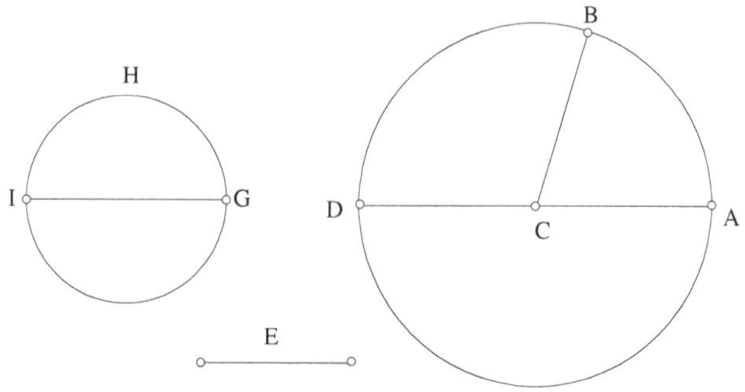

Fig. 1.17

Arc $\widehat{AB}$ and the complete circumference are of the same type of magnitude, and their ratio exists whether it is a known ratio or not; whatever this ratio is, a straight line E exists such that

$$\frac{E}{AD} = \frac{\widehat{AB}}{\text{the complete circumference}}.$$

A straight line GI is then defined such that $\frac{GI}{E} = \frac{AD}{GI}$, which then gives $\frac{E}{AD} = \frac{GI^2}{AD^2}$. Let GHI be a circle with diameter GI, then

$$\frac{\text{circle }(GHI)}{\text{circle }(ABD)} = \frac{GI^2}{AD^2} = \frac{\text{sect.}(ABC)}{\text{circle }(ABD)},$$

then sector (ABC) and circle (GHI) have the same area.

Comments:

1) Ibn al-Haytham gives no explanation for the construction of E and GI.

If the ratio $\frac{E}{AD}$ is a known rational number, the construction of E from AD is immediate as is the construction of GI mean proportional between E and AD.

If the ratio $\frac{E}{AD}$ is not a known number, the fact which interests Ibn al-Haytham is that E exists and, consequently, GI and the circle of diameter GI also exist.

2) The argument here should be referred back to that used in his treatise *On the Quadrature of the Circle*.

Lemma 6. — *The ratio of surfaces of two similar segments in two different circles is equal to the ratio of surfaces of the two circles and the ratio of the squares to the bases.*

Let *ABC* and *EGH* be two similar segments in two different circles with centres *D* and *I*.

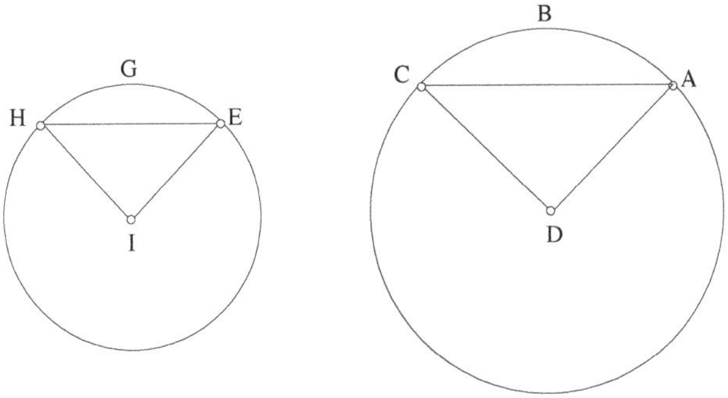

Fig. 1.18

Then $\widehat{ADC} = \widehat{EIH}$ and triangles *ADC* and *EIH* are similar and

$$\frac{\text{tr.}(ADC)}{\text{tr.}(EIH)} = \frac{AD^2}{IE^2} = \frac{AC^2}{EH^2} = \frac{\text{circle } (ABC)}{\text{circle } (EGH)} = \frac{\text{sect.}(ADC)}{\text{sect.}(EIH)} .$$

From this, we can deduce

$$\frac{\text{segm.}(ABC)}{\text{segm.}(EGH)} = \frac{\text{circle } (ABC)}{\text{circle } (EGH)} = \frac{AC^2}{EH^2} .$$

Comment: Ibn al Haytham makes use of Proposition XII.2 of the *Elements* here, which he has cited in his treatise *On the Quadrature of the Circle.*

Lemma 7. — *If segments* ABC *and* AEB *are similar and if the small arc* $\overset{\frown}{\text{AB}}$ *of the first circle and arc* $\overset{\frown}{\text{AEB}}$ *of the second are on the same side of the straight line* AB, *then arc* $\overset{\frown}{\text{AEB}}$ *is outside the first circle.*

Segments *ABC* and *AEB* are similar, if *AK* is the tangent at *A* to circle *ABC* and *AK′* the tangent at *A* to circle *AEB*, then $\widehat{KAC} = \widehat{K'AB}$; but $\widehat{KAB} < \widehat{KAC}$, hence $\widehat{KAB} < \widehat{K'AB}$, therefore *AK* is in angle $\widehat{K'AB}$, and *AK* cuts arc $\overset{\frown}{AEB}$. Point *K* of arc *AEB* is outside circle *ABC*, and any point of segment [*AK*] is in the lune between arcs *AEB* and *AIB*.

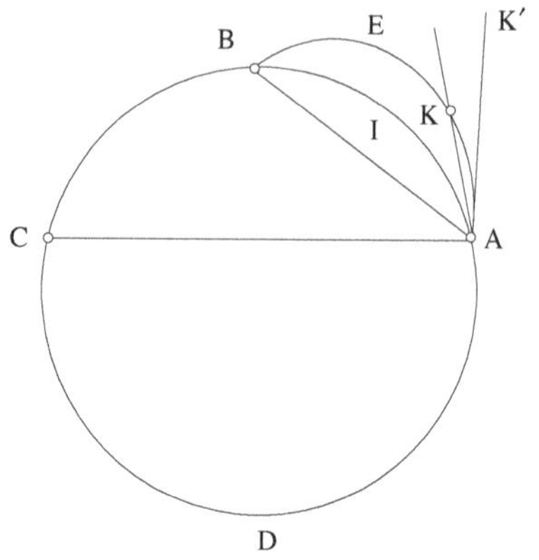

Fig. 1.19

Points *A* and *B* are common to the two circles, the small circle therefore has an arc *AB* outside the larger one, therefore this arc is arc *AKB*.

Comment: If arc *AEB* and arc *AIB* were on either side of straight line *AB*, arc *AEB* would then be inside the larger circle.

Proposition 8. — *If* B *is any point of a semicircle* ABC, *and* ADB *and* BEC *the two semicircles constructed on* AB *and* BC, *then*

$$\text{lune (ADBGA)} + \text{lune (BECHB)} = \text{tr.(ABC)}.$$

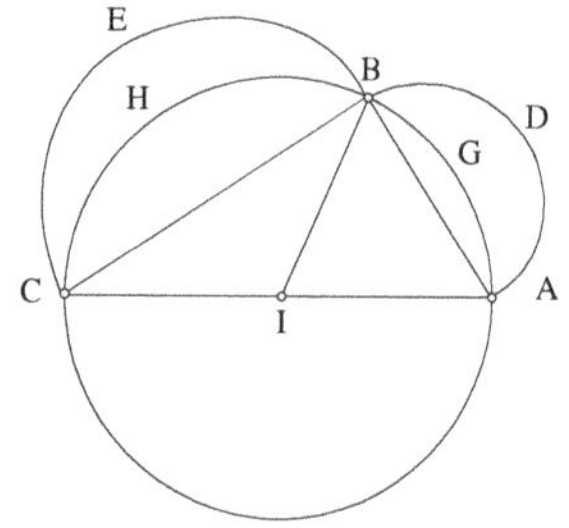

Fig. 1.20

Same proof as that given in the treatise *On the Quadrature of the Circle.*

Proposition 9. — *If* $\widehat{\text{BA}} = \widehat{\text{BC}} = \frac{1}{4}$ *circle, then*

$$\text{lune (ADBGA)} = \text{lune (BECHB)} = \text{tr.(ABI)} = \text{tr.(BIC)}.$$

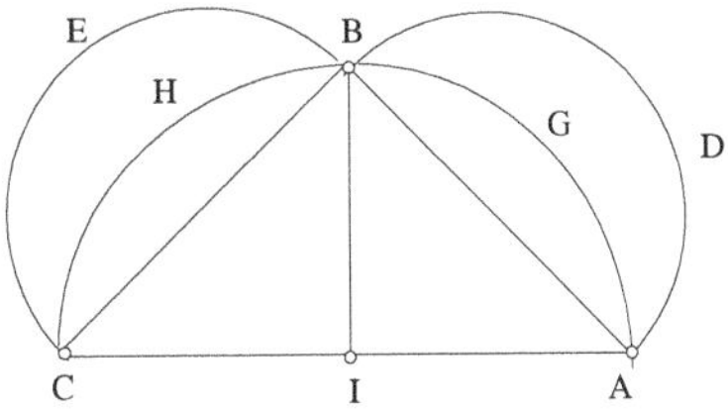

Fig. 1.21

If $\widehat{AB} < \widehat{BC}$, then circle (N) exists such that

$$\text{lune }(ADBGA) + (N) = \text{tr.}(ABI)$$
$$\text{lune }(BECHB) - (N) = \text{tr.}(BCI).$$

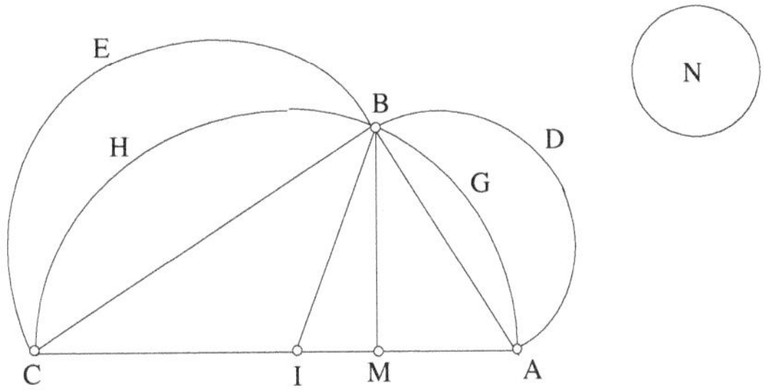

Fig. 1.22

Let $BM \perp AC$; $AB^2 = AM \cdot AC$, therefore

$$\frac{MA}{AC} = \frac{AB^2}{AC^2} = \frac{1/2 \text{ circle } (ADB)}{1/2 \text{ circle } (ABC)}.$$

On the other hand, according to Lemma 1:

$$\frac{MA}{AC} < \frac{\widehat{ACB}}{\pi/2},$$

hence

$$\frac{MA}{AC} < \frac{\widehat{AIB}}{\pi} = \frac{\text{sect.}(AIBG)}{1/2 \text{ circle } (ABC)};$$

therefore

$$\text{sect.}(AIBG) > \frac{1}{2} \text{ circle } (ADB).$$

According to Lemma 5, any circular sector is equal to a circle, therefore a circle (C_1) exists and a circle (C_2) exists, and they are equivalent respectively to sector $(AIBG)$ and semicircle (ADB), which gives $(C_1) > (C_2)$.

Ibn al-Haytham deduces from this that a circle (N) exists such that $(C_1) = (C_2) + (N)$, therefore sector $(AIBG)$ = semicircle $(ADB) + (N)$; by taking away segment (AGB) from the two members, this becomes

$$\text{tr.}(AIB) = \text{lune } (ADBGA) + (N).$$

Taking into account Proposition 8 and the equality

$$\text{tr.}(AIB) = \text{tr.}(BIC) = \frac{1}{2}\,\text{tr.}(ABC),$$

we have

$$\text{tr.}(BCI) = \text{lune } (BECHB) - (N).$$

Proposition 10. — *Let* B *be a point on an arc* $\widehat{\text{ABC}} < \frac{1}{2}$ *circle, and construct on* AB *and* BC *segments similar to segment* ABC *and let points* N *and* O *on* AC *be such that* $\widehat{\text{BNA}} = \widehat{\text{BOC}} = \widehat{\text{ABC}}$. *Circle* K *exists such that*

$$\text{lune (ADBHA)} + \text{lune (BICMB)} + \text{(K)} = \text{tr.(ABC)} + \text{tr.(ENO)}.$$

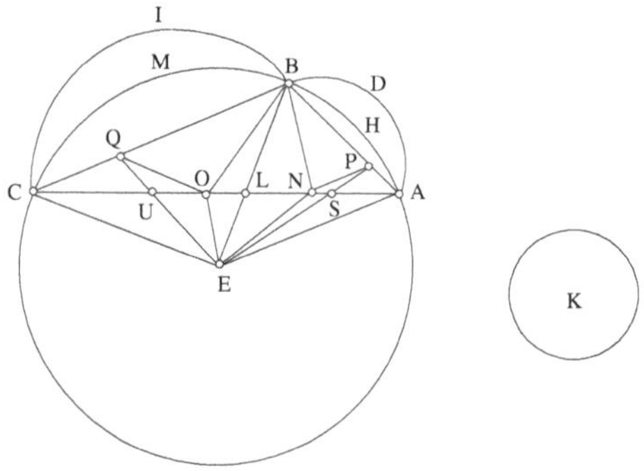

Fig. 1.23

Triangles ABC, ANB and BOC are similar. Then

$$\frac{CA}{AN} = \frac{CA^2}{AB^2} = \frac{\text{segm.}(ABC)}{\text{segm.}(ADB)}$$

and

$$\frac{AC}{CO} = \frac{AC^2}{CB^2} = \frac{\text{segm.}(ABC)}{\text{segm.}(BIC)};$$

hence

$$\frac{AC}{AN + CO} = \frac{\text{segm.}(ABC)}{\text{segm.}(ADB) + \text{segm.}(BIC)}.$$

But

$$\frac{AC}{AN + CO} = \frac{\text{tr.}(AEC)}{\text{tr.}(EAN) + \text{tr.}(ECO)}.$$

From this, we can deduce

$$\frac{AC}{AN + CO} = \frac{\text{sect.}(AECB)}{\text{segm.}(ADB) + \text{segm.}(BIC) + \text{tr.}(AEN) + \text{tr.}(CEO)}.$$

We have $AC > AN + CO$, therefore

$$\text{sect.}(AECB) > \text{segm.}(ADB) + \text{segm.}(BIC) + \text{tr.}(AEN) + \text{tr.}(CEO).$$

X exists as a part of sector $(AECB)$ such that

$$\frac{AC}{AN + CO} = \frac{\text{segm.}(AECB)}{X},$$

with $X = \text{segm.}(ADB) + \text{segm.}(BIC) + \text{tr.}(AEN) + \text{tr.}(CEO)$.

The difference: sect.$(AECB) - X$ is a sector of circle (E, EA) and a circle (K) exists which is equal to this sector. Therefore

$$\text{sect.}(AECB) = \text{segm.}(ADB) + \text{segm.}(BIC) + \text{tr.}(AEN) + \text{tr.}(CEO) + (K).$$

The sum: segm.(AHB) + segm.(BMC) + tr.(AEN) + tr.(CEO) is common to the two members; there remains:

$$\text{lune } (ADBHA) + \text{lune } (BICMB) + (K) = \text{tr.}(ABC) + \text{tr.}(ENO).$$

Let $PN \parallel EA$, P on AB, let PE cut AC at S, and let $OQ \parallel EC$, Q on BC; QE cuts AC at U, this gives tr.(ASP) = tr.(ESN) and tr.(CUQ) = tr.(EUO), therefore $(EPBQ)$ = tr.(ABC) + tr.(ENO), and it follows that

$$\text{lune } (ADBHA) + \text{lune } (BICMB) + (K) = (EPBQ).$$

Proposition 11. — This proposition takes up the hypotheses of Proposition 10, as well as the notations and the figure. Note that the hypothesis $\widehat{ABC} < \frac{1}{2}$ circle implies $\widehat{ABC}$ is obtuse, and it follows $\widehat{BAC} + \widehat{BCA} < \frac{\pi}{2}$. It is assumed that $\widehat{BAC} \geq \widehat{BCA}$, therefore $\widehat{BCA} \leq \frac{\pi}{4}$; but equally $\widehat{BAC} \leq \frac{\pi}{4}$ (11c) or $\widehat{BAC} > \frac{\pi}{4}$ (12).

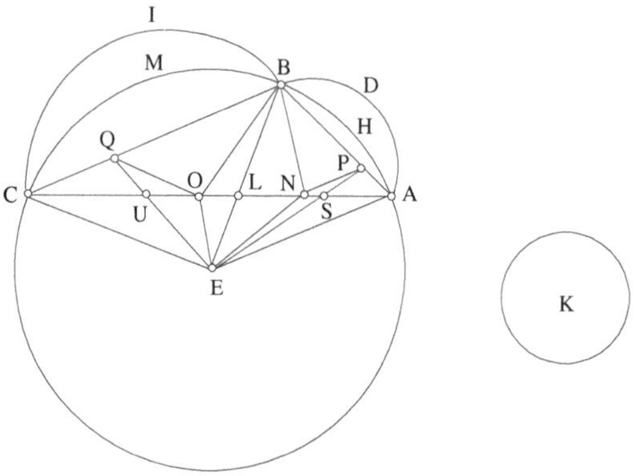

Fig. 1.24

a) If $\widehat{ABC} < \frac{1}{2}$ circle and $\widehat{AB} = \widehat{BC}$, then lune $(ADBHA)$ = lune $(BICMB)$ and tr.(PEB) = tr.(QEB); hence

$$\text{lune }(ADBHA) + \frac{1}{2}(K) = \text{tr.}(PEB),$$
$$\text{lune }(BICMB) + \frac{1}{2}(K) = \text{tr.}(QEB),$$

(immediate result as per Proposition 10).

b) If $\widehat{ABC} < \frac{1}{2}$ circle and $\widehat{AB} < \widehat{BC}$, then a complete circle (Z) exists, $(Z) \neq \frac{1}{2}(K)$, such that

$$\text{lune }(ADBHA) + (Z) = \text{tr.}(PEB).$$

c) If $\widehat{BAC} \leq \frac{\pi}{4}$, therefore a complete circle (Z') exists such that

$$\text{lune }(BICMB) + (Z') = \text{tr.}(QEB),\ (Z') = K - (Z).$$

Proof: We have seen before (see Comment 2 after Lemma 4) that

$$\frac{NA}{AC} < \frac{\widehat{BCA}}{\pi - \widehat{ABC}}.$$

But $\widehat{BEA} = 2\widehat{BCA}$ and $\widehat{AEC} = 2(\pi - \widehat{ABC})$, hence

$$\frac{NA}{AC} < \frac{\widehat{BEA}}{\widehat{AEC}},$$

or even

$$\frac{NA}{AC} < \frac{\text{sect.}(BEA)}{\text{sect.}(CEA)}.$$

But

$$\frac{NA}{AC} = \frac{\text{segm.}(ADB)}{\text{segm.}(ABC)},$$

hence

$$\frac{\text{segm.}(ADB)}{\text{segm.}(ABC)} < \frac{\text{sect.}(BEA)}{\text{sect.}(CEA)}.$$

Therefore a sector Y exists, which is part of sector BEA, such that

$$\frac{\text{segm.}(ADB)}{\text{segm.}(ABC)} = \frac{Y}{\text{sect.}(CEA)} = \frac{AN}{AC} = \frac{\text{tr.}(AEN)}{\text{tr.}(AEC)};$$

hence

$$\frac{\text{segm.}(ADB)}{\text{segm.}(ABC)} = \frac{Y - \text{tr.}(AEN)}{\text{sect.}(CEA) - \text{tr.}(AEC)} = \frac{Y - \text{tr.}(AEN)}{\text{segm.}(ABC)},$$

therefore $Y - \text{tr.}(AEN) = \text{segm.}(ADB)$; but $\text{tr.}(AEN) = \text{tr.}(APE)$, hence

$$\text{segm.}(ADB) + \text{tr.}(APE) = Y.$$

A circle Z exists such that $Y + Z = \text{sect.}(AEB)$, therefore

$$\text{segm.}(ADB) + \text{tr.}(APE) + Z = \text{sect.}(AEB).$$

The sum $\text{segm.}(AHB) + \text{tr.}(APE)$ is common to the two members, and there remains lune $(ADBHB) + Z = \text{tr.}(PEB)$.

Moreover, if $\widehat{BAC} \leq \frac{\pi}{4}$, then $\frac{OC}{CA} < \frac{\widehat{BAC}}{\pi - \widehat{ABC}}$, and we show, as for the small lune, that a circle Z' exists such that lune $(BICMB) + Z' = \text{tr.}(QEB)$.

According to Proposition 10, the two circles Z and Z' associated with the lunes are such that $Z + Z' = K$, with K being the circle associated with the sum of the two lunes.

Proposition 12. — *Same problem assuming* $\widehat{BAC} > \frac{\pi}{4}$.

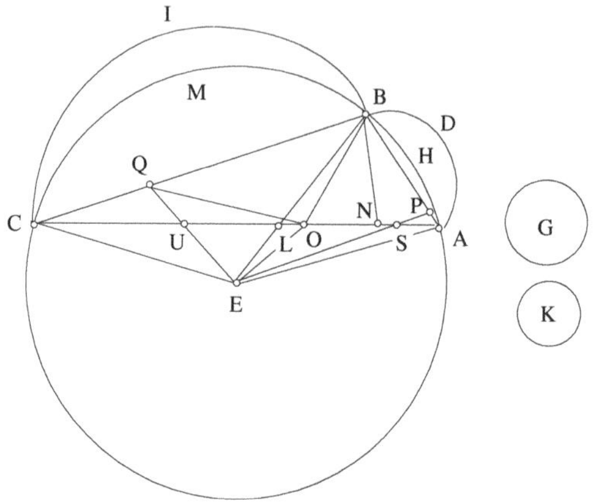

Fig. 1.25

As in Proposition 11, $\frac{NA}{AC} < \frac{\widehat{BCA}}{\pi - \widehat{ABC}}$, therefore the conclusion (b) remains true.

But O is between L and A (see later) and it is possible that $\frac{OC}{CA} < \frac{\widehat{BAC}}{\pi - \widehat{ABC}}$, as in Proposition 11 and, in this case, the conclusion (c) remains true; or $\frac{OC}{CA} > \frac{\widehat{BAC}}{\pi - \widehat{ABC}}$, which is the author's hypothesis.

With the hypotheses defined in this way and (K) as the complete defined circle in Proposition 10, a complete circle (G) exists such that

$$\text{lune } (BICMB) = \text{tr.}(BEQ) + (G)$$
$$\text{lune } (ADBHA) + (G) + (K) = \text{tr.}(BEP).$$

Proof: We know that

$$\frac{OC}{CA} = \frac{BC^2}{AC^2} = \frac{\text{segm.}(BIC)}{\text{segm.}(CBA)} = \frac{\text{tr.}(OEC)}{\text{tr.}(CEA)} = \frac{\text{segm.}(BIC) + \text{tr.}(OEC)}{\text{sect.}(ECBA)}.$$

But

$$\frac{\widehat{BAC}}{\pi-\widehat{ABC}}=\frac{\widehat{BEC}}{\widehat{CEA}}=\frac{\text{sect.}(BECM)}{\text{sect.}(ECBA)} \text{ and } \text{tr.}(OEC)=\text{tr.}(QEC);$$

hence

$$\frac{\text{segm.}(BIC)+\text{tr.}(QEC)}{\text{sect.}(ECBA)}>\frac{\text{sect.}(BECM)}{\text{sect.}(ECBA)}.$$

A complete circle (G) exists such that

$$\text{segm.}(BIC)+\text{tr.}(QEC)=\text{sect.}(BECM)+(G).$$

Taking away the two members segm.(BMC) + tr.(QEC), this leaves

$$\text{lune } (BICMB)=\text{tr.}(BEQ)+(G).$$

In Proposition 10 we saw that a complete circle (K) existed such that

$$\text{lune } (ADBHA)+\text{lune } (BICMB)+(K)=\text{tr.}(BEP)+\text{tr.}(BEQ);$$

hence

$$\text{lune } (ADBHA)+(G)+(K)=\text{tr.}(BEP).$$

Comments:

1) Suppose $(G)+(K)=(Z)$; $(Z)>(K)$, then

$$\text{lune } (ADBHA)+(Z)=\text{tr.}(BEP),$$

$$\text{lune } (BICMB)=\text{tr.}(BEQ)+(Z)-(K).$$

2) If $\frac{OC}{CA}=\frac{\widehat{BAC}}{\pi-\widehat{ABC}}$, then $(G)=(O)$, $(Z)=(K)$ and then

$$\text{lune } (BICMB)=\text{tr.}(BEQ).$$

Note that Ibn al-Haytham did not indicate this result which leads to a lune equivalent to a triangle and, consequently, equivalent to a square.

Comment on Lemmas 3 and 4 and Propositions 11 and 12:

Lemmas 3 and 4 and Propositions 11 and 12 use a triangle ABC satisfying $\widehat{ABC}$ as obtuse and $BC>BA$ and where point E of BC is such that $\widehat{BEC}=\widehat{ABC}$.

In 3 and in 11, it is assumed that $\widehat{BAC}\leq\frac{\pi}{4}$ [Figs 1.26 and 1.27].

In 4 and in 12, it is assumed that $\widehat{BAC}>\frac{\pi}{4}$ [Fig. 1.28].

$\overset{\frown}{BC} < \frac{1}{4}$ circle, $CE < CL$ [Fig. 1.26].

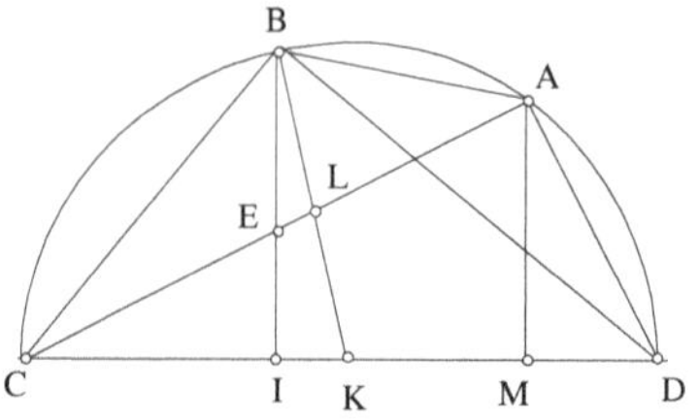

Fig. 1.26

$\overset{\frown}{BC} = \frac{1}{4}$ circle, $CE = CL$ [Fig. 1.27].

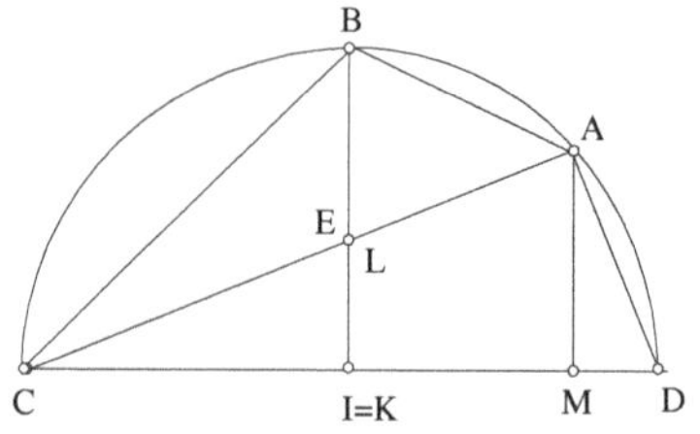

Fig. 1.27

$\overset{\frown}{BC} > \frac{1}{4}$ circle, $CE > CL$ [Fig. 1.28].

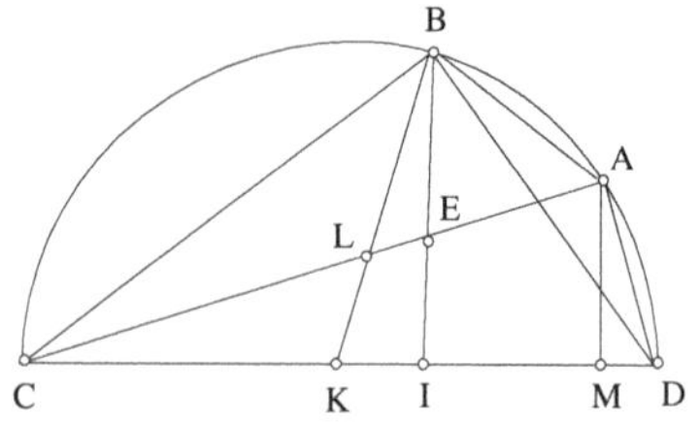

Fig. 1.28

Let K be the centre of the circle circumscribed at ABC, let CD be the diameter from point C and let I be the intersection of BE and CD. We have $\widehat{ABC} + \widehat{ADC} = \pi$ and by hypothesis $\widehat{BEC} = \widehat{ABC}$, therefore $\widehat{AEI} + \widehat{ADC} = \pi$, and it follows $\widehat{EID} = \frac{\pi}{2}$, since $\widehat{DAE} = \frac{\pi}{2}$; therefore in the three cases above, $BI \perp DC$.

Let L be the intersection of BK and CA, and from this:

If $\widehat{BAC} < \frac{\pi}{4}$, $\widehat{BC} < \frac{1}{4}$ circle, I is between C and K, therefore E is between C and L, $CE < CL$ [Fig. 1.26].

If $\widehat{BAC} = \frac{\pi}{4}$, $\widehat{BC} = \frac{1}{4}$circle, $I = K$, therefore $E = L$, $CE = CL$ [Fig. 1.27].

If $\widehat{BAC} > \frac{\pi}{4}$, $\widehat{BC} > \frac{1}{4}$ circle, I is between K and D, therefore E is between L and A and $CE > CL$ [Fig. 1.28].

In the three cases: $\frac{LC}{CA} < \frac{B\widehat{A}C}{\pi - A\widehat{B}C}$.

In Lemma 3, by hypothesis, $\widehat{BAC} \le \frac{\pi}{4}$, therefore $CE \le CL$ and it follows $\frac{CE}{CA} < \frac{B\widehat{A}C}{\pi - A\widehat{B}C}$.

But, in Lemma 4, $\widehat{BAC} > \frac{\pi}{4}$, therefore $CE > CL$ and it is not possible to conclude anything about arcs $\widehat{BC}$ and $\widehat{BA}$ without a complementary hypothesis.

The condition for

$$(1) \qquad \frac{CE}{CA} < \frac{B\widehat{A}C}{\pi - A\widehat{B}C}:$$

we have $\pi - \widehat{ABC} = \widehat{ADC}$ and $\widehat{BAC} = \widehat{BDC}$, and, on the other hand, if $AM \perp CD$, then $\frac{CE}{CA} = \frac{CI}{CM}$, and it follows

$$(1) \qquad \Leftrightarrow \frac{CI}{CM} < \frac{B\widehat{D}C}{A\widehat{D}C} \Leftrightarrow \frac{CI}{CM} < \frac{\widehat{BC}}{\widehat{AC}} \quad (2).$$

But $CM = CI + IM$ and $\widehat{AC} = \widehat{AB} + \widehat{BC}$, from which we can deduce

$$(3) \qquad \frac{EC}{CA} < \frac{B\widehat{A}C}{\pi - A\widehat{B}C} \Leftrightarrow \frac{IM}{IC} > \frac{\widehat{AB}}{\widehat{BC}}$$

and

$$(4) \qquad \frac{EC}{CA} \geq \frac{B\widehat{A}C}{\pi - A\widehat{B}C} \Leftrightarrow \frac{IM}{IC} \leq \frac{\widehat{AB}}{\widehat{BC}}.$$

Note that (4) is therefore a necessary condition and sufficient so that the inequality $\frac{EC}{CA} \geq \frac{B\widehat{A}C}{\pi - A\widehat{B}C}$ is satisfied.

Ibn al-Haytham showed that $\frac{EC}{CA} > \frac{B\widehat{A}C}{\pi - A\widehat{B}C}$ (strict inequality) is satisfied if

$$(5) \qquad \frac{ID}{IC} \leq \frac{\widehat{AB}}{\widehat{BC}}.$$

Then $ID > IM$, hence $\frac{ID}{IC} > \frac{IM}{IC}$, and it follows that condition (5) given by Ibn al-Haytham is sufficient to have $\frac{EC}{CA} > \frac{B\widehat{A}C}{\pi - A\widehat{B}C}$, but is not necessary.

Ibn al-Haytham then goes on to examine particular cases for which it is possible to determine the circles which are used in the expression of the surface of lunes as their ratio to a given circle.

Proposition 13. — *If* $\widehat{ABC} = \frac{\pi}{2}$ *and* $\widehat{AB} = \frac{1}{2}\widehat{BC}$, *then if* circle (K) = $\frac{1}{24}$ circle (ABC) *and* circle (M) = $\frac{1}{12}$ circle (ABC), *then*

lune (AEBNA) + (K) = tr.(ABD),
lune (BHCIB) – (K) = tr.(BCD),
lune (AEBNA) + (M) = lune (BHCIB).

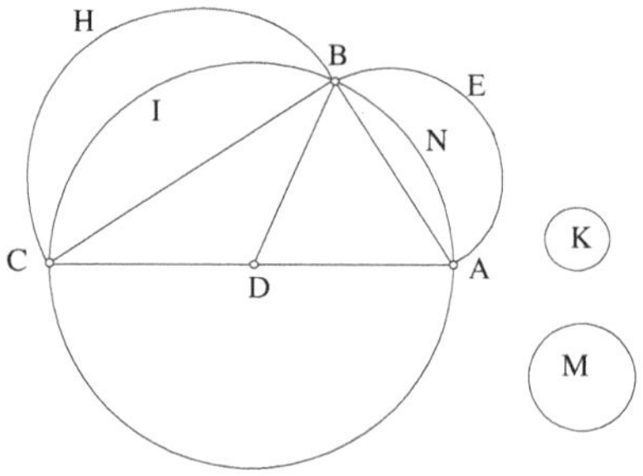

Fig. 1.29

$AB = AD = DC$. Sector (ADB) is a third of semicircle (ABC) and semicircle (AEB) is a quarter of semicircle (ABC); hence

$$\text{sect.}(ADB) - \tfrac{1}{2}\,\text{circle } (AEB) = \tfrac{1}{24}\,\text{circle } (ABC) = (K).$$

If segment BNA is taken away from the two terms of the difference, then tr.(ABD) – lune $(AEBNA) = (K)$, hence lune $(AEBNA) + (K) =$ tr.(ABD).

Next, according to Proposition 8: tr.$(BCD) + (K) =$ lune $(BHCIB)$. And since tr.$(BCD) =$ tr.(ABD), then lune $(AEBNA) + (M) =$ lune $(BHCIB)$.

Proposition 14. — Particular case of Proposition 11a with $\widehat{ABC} = \frac{1}{3}$ of the circumference.

Let (S) *and* (U) *be two complete circles such that* (S) $= \frac{1}{9}$ circle (ABC) *and* (U) $= \frac{1}{2}$ (S), *then*

lune (AHBIA) + lune (BKCMB) + (S) = (GPBQ),
lune (AHBIA) + (U) = tr.(PBG),
lune (BKCMB) + (U) = tr.(QBG).

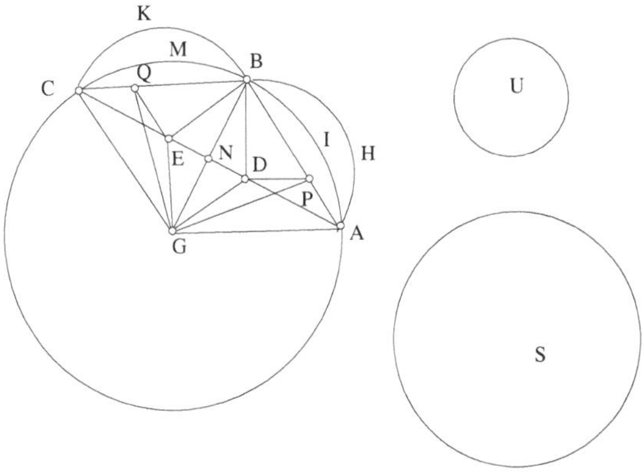

Fig. 1.30

AC is the side of the equilateral triangle inscribed in circle ABC and AB the side of the hexagon, hence

$$AB^2 = \tfrac{1}{3}AC^2 = BC^2.$$

Similarly

$$DA^2 = \tfrac{1}{3}AB^2 = \tfrac{1}{9}AC^2;$$

hence

$$DA = \tfrac{1}{3}AC,$$

and for the same reason

$$CE = \tfrac{1}{3}AC, \quad DE = \tfrac{1}{3}AC = DB = EB.$$

Therefore

$$AB^2 + BC^2 = \frac{2}{3}AC^2;$$

hence

$$\text{segm.}(AHB) + \text{segm.}(BKC) = \frac{2}{3}\ \text{segm.}(ABC).$$

But

$$\text{tr.}(AGD) + \text{tr.}(EGC) = \frac{2}{3}\ \text{tr.}(AGC),$$

therefore

$$\text{segm.}(AHB) + \text{segm.}(BKC) + \text{tr.}(AGD) + \text{tr.}(EGC) = \tfrac{2}{3}\,\text{sect.}(AGCB).$$

But

$$(S) = \tfrac{1}{9}\,\text{circle}\ (ABC) = \tfrac{1}{3}\,\text{sect.}(AGCB),$$

$$\text{tr.}(AGD) = \text{tr.}(AGP) \ \text{ and } \ \text{tr.}(EGC) = \text{tr.}(GQC);$$

hence

$$\text{segm.}(AHB) + \text{segm.}(BKC) + \text{tr.}(AGP) + \text{tr.}(GQC) + (S) = \text{sect.}(AGCB).$$

By taking away from the two members the sum:

$$\text{segm.}(AIB) + \text{segm.}(BMC) + \text{tr.}(AGP) + \text{tr.}(GQC),$$

this leaves

$$\text{lune}\ (AHBIA) + \text{lune}\ (BKCMB) + (S) = \text{quad.}(BPGQ).$$[17]

[17] Quad.$(BPGQ)$ = quad.$(ABCG)$ – [tr. (APG) + tr. (GQC)]
$= 2\ \text{tr.}(ABC) - \tfrac{2}{3}\,\text{tr.}(ABC) = \tfrac{4}{3}\ \text{tr.}(ABC).$

But the two lunes are equal and moreover

$$\text{tr.}(PBG) = \text{tr.}(BGQ) = \frac{1}{2}\text{quad.}(BPGQ),\ (U) = \frac{1}{2}(S);$$

therefore

$$\text{lune }(AHBIA) + (U) = \text{tr.}(PBG)$$
$$\text{lune }(BKCMB) + (U) = \text{tr.}(BGQ).$$

Proposition 15. — This proposition applies to two cases; the first (15a) is a particular case of Proposition 10.

Let $\widehat{\mathrm{AC}} = \frac{1}{3}$ *circumference,* $\widehat{\mathrm{AB}} = \frac{1}{4}$ *circumference, and let* (AEB) *and* (BHC) *be similar segments to* (ABC)*; let* BD *cut* AC *at* I *and let* L *on* AC *be such that* $\widehat{\mathrm{BLC}} = \widehat{\mathrm{ABC}}$. *And let there be two circles of diameters* NP *and* QS *such that*

$$\text{circle (NP)} = \frac{1}{3}\text{circle (ABC)} \ \textit{and}\ \frac{\mathrm{NP}^2}{\mathrm{QS}^2} = \frac{\text{circle (NP)}}{\text{circle (QS)}} = \frac{\mathrm{AC}}{\mathrm{IL}},$$

therefore

$$\text{lune (AEBMA)} + \text{lune (BHCKB)} + \text{circle (QS)} = \text{tr.(ABC)} + \text{tr.(DIL)}.$$

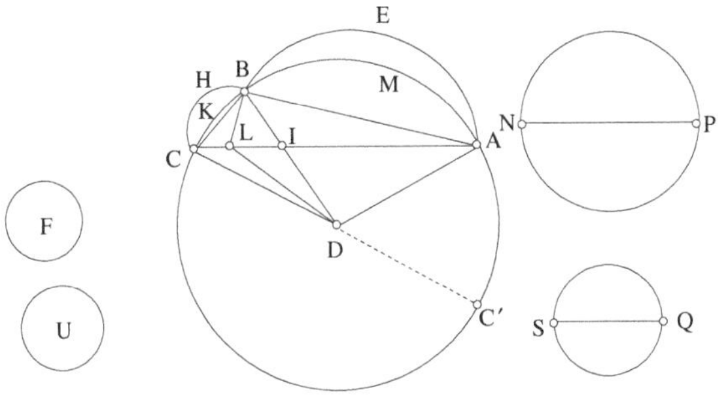

Fig 1.31

Proof: According to the data, $\widehat{AIB} = \widehat{ABC} = \frac{2\pi}{3}$,[18] triangle *AIB* and *ABC* on the one hand, and *BLC* and *ABC* on the other hand are similar, then

[18] We have $\widehat{ABC} = \frac{2\pi}{3} = \widehat{ADC}$. As $\widehat{ADB}$ is a right angle, angle $\widehat{IDC}$ is $\frac{2\pi}{3} - \frac{\pi}{2} = \frac{\pi}{6}$; moreover, $\widehat{DCI}$ subtends arc $\widehat{AC'} = \pi - \frac{2\pi}{3} = \frac{\pi}{3}$, then $\widehat{DCI} = \frac{\pi}{6}$ and $\widehat{DIC} = \pi - 2 \cdot \frac{\pi}{6} = \frac{2\pi}{3}$.

$$AB^2 = CA \cdot AI \quad \text{and} \quad CB^2 = AC \cdot CL.$$

From this, we can deduce

$$\frac{IA + CL}{AC} = \frac{AB^2 + BC^2}{AC^2} = \frac{\text{segm.}(AEB) + \text{segm.}(BHC)}{\text{segm.}(ABC)}$$

$$= \frac{\text{tr.}(AID) + \text{tr.}(LDC)}{\text{tr.}(ADC)}$$

$$= \frac{\text{segm.}(AEB) + \text{segm.}(BHC) + \text{tr.}(AID) + \text{tr.}(LDC)}{\text{sect.}(ADCB)}.$$

Therefore

$$\frac{IA + CL}{AC} = \frac{\text{segm.}(AEB) + \text{segm.}(BHC) + \text{tr.}(AID) + \text{tr.}(LDC)}{\text{circle }(NP)}.$$

But

$$\frac{\text{circle }(QS)}{\text{circle }(NP)} = \frac{IL}{AC} \text{ and } IA + CL + IL = AC;$$

hence

$$\text{segm.}(AEB) + \text{segm.}(BHC) + \text{tr.}(ADI) + \text{tr.}(LDC) + \text{circle }(QS)$$
$$= \text{circle }(NP) = \text{sect.}(ADCB).$$

The parts common to the two members are: segm.(AMB), segm.(BKC), tr.(ADI), tr.(LDC), and it follows

(1) lune $(AEBMA)$ + lune $(BHCKB)$ + circle (QS) = tr.(ABC) + tr.(DIL).

If d is the diameter of circle ABC, then

$$AC^2 = \frac{3}{4}d^2,\ AB^2 = \frac{1}{2}d^2;$$

hence

$$AB^2 = \frac{2}{3}AC^2,$$

hence

$$IA = \frac{2}{3}AC \text{ and } IC = \frac{1}{3}AC, \text{ since } \frac{IA}{AC} = \frac{AB^2}{AC^2}.$$

But

$$\widehat{AIB} = \widehat{BLC} = \widehat{ABC} = \frac{2\pi}{3};$$

hence

$$\widehat{BIL} = \widehat{BLI} = \frac{\pi}{3},$$

triangle *BLI* is therefore equilateral and $BL = IL$. But

$$\frac{BL}{LC} = \frac{AB}{BC} \quad \text{and } AB > BC, \text{ since } \widehat{BC} = \frac{1}{3}\widehat{AB};$$

hence $BL > LC$, from which we can deduce $LI > LC$, $2\,LI > IC$, and it follows

$$LI > \frac{1}{6}AC,\text{[19]} \quad \text{and circle } (QS) > \frac{1}{6}\text{circle } (NP);$$

therefore

$$\text{circle } (QS) > \frac{1}{18}\text{circle } (ABC).$$

15b. — *If circles* (F) *and* (U) *are defined by* $(F) = \frac{1}{36}$ circle (ABC), (U) = (QS) – (F) [(QS) > $\frac{1}{18}$ circle (ABC) *has already been seen*], *then*

$$\text{lune (AEBMA)} + (F) = \text{tr.(ABI)}$$
$$\text{lune (BHCKB)} + (U) = \text{tr.(BIC)} + \text{tr.(IDL)}.$$

We know that

$$\text{sect.}(ADBM) = \frac{1}{4}\text{circle } (ABC), \ \text{sect.}(ADCB) = \frac{1}{3}\text{circle } (ABC),$$

therefore

$$\text{sect.}(ADBM) = \frac{3}{4}\text{sect.}(ADCB).$$

On the other hand

[19] The calculus of BC in the isosceles triangle BDC, $\left(B\hat{D}C = \frac{\pi}{6}\right)$, gives $BC^2 = \frac{d^2}{4}(2 - \sqrt{3})$. Moreover, $\frac{CL}{AC} = \frac{CB^2}{AC^2}$, hence $\frac{CL}{AC} = \frac{2-\sqrt{3}}{3}$. But $\frac{IC}{AC} = \frac{1}{3}$, so $\frac{LI}{AC} = \frac{IC - CL}{AC} = \frac{\sqrt{3}-1}{3}$ and $\frac{\text{circle } (QS)}{\text{circle } (NP)} = \frac{\sqrt{3}-1}{3}$, hence circle $(QS) = \frac{\sqrt{3}-1}{9}$ circle (ABC).

$$AB^2 = \frac{2}{3} AC^2,$$

hence

$$\text{segm.}(AEB) = \frac{2}{3}\ \text{segm.}(ABC),$$

and

$$AI = \frac{2}{3} AC;$$

hence

$$\text{tr.}(ADI) = \frac{2}{3}\ \text{tr.}(ADC).$$

It follows that

$$\text{segm.}(AEB) + \text{tr.}(ADI) = \frac{2}{3}\ \text{sect.}(ADCB)$$

and

$$\begin{aligned}&\text{sect.}(ADBM) - [\text{segm.}(AEB) + \text{tr.}(ADI)]\\ &= \frac{1}{12}\,\text{sect.}(ADCB) = \frac{1}{36}\,\text{circle}\ (ACB) = (F);\end{aligned}$$

hence

$$\text{sect.}(ADBM) = \text{segm.}(AEB) + \text{tr.}(ADI) + (F).$$

By taking away from the two members the sum of segm.(ABM) + tr.(ADI), this leaves

$$\text{lune}\ (AEBMA) + (F) = \text{tr.}(ABI)$$

or

$$\text{(2)} \qquad \text{lune}\ (AEBMA) + \frac{1}{36}\,\text{circle}\ (ABC) = \text{tr.}(ABI).$$

From (1) and (2) we can deduce

$$\text{lune}\ (BHCKB) + (QS) - (F) = \text{tr.}(ABC) + \text{tr.}(DIL) - \text{tr.}(ABI),$$

therefore

$$\text{lune}\ (BHCKB) + (U) = \text{tr.}(BIC) + \text{tr.}(DIL).$$

Proposition 16. — *Let* $\widehat{\mathrm{AB}} = \frac{1}{6}$ *circle,* $\widehat{\mathrm{BC}} = \frac{1}{3}$ *circle,* E *midpoint of* AD *and* G *such that* AG = $\frac{3}{8}$ GC, [AG = $\frac{3}{11}$ AC, *hence* G *between* E *and* C].

If $\widehat{\mathrm{AH}} = \frac{1}{4}\widehat{\mathrm{AB}}$, *then* $\widehat{\mathrm{HB}} = \frac{3}{8}\widehat{\mathrm{BC}}$.

The straight line CH cuts BE at I; so $\widehat{BIC} = \widehat{HBC}$.[20] Let $GK \parallel AH$, K between I and C, hence

$$\frac{KH}{KC} = \frac{GA}{GC} = \frac{\widehat{HB}}{\widehat{BC}} = \frac{3}{8},$$

from which we can deduce

$$\frac{CK}{HC} = \frac{\widehat{BC}}{\widehat{HBC}} \quad \text{and} \quad \frac{IC}{HC} > \frac{\widehat{BC}}{\widehat{HBC}} \quad (\text{since } IC > KC);$$

but

$$\frac{\widehat{BC}}{\widehat{HBC}} = \frac{\widehat{BDC}}{\widehat{CDH}} = \frac{\widehat{BHC}}{\pi - \widehat{HBC}} = \frac{\text{sect.}(BDCM)}{\text{sect.}(CDHB)},$$

therefore

$$\frac{IC}{HC} > \frac{\text{sect.}(BDCM)}{\text{sect.}(CDHB)}.$$

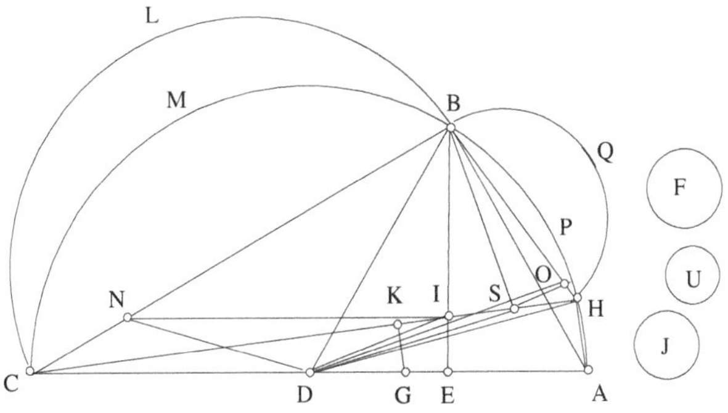

Fig. 1.32

[20] $\widehat{BIC} = \widehat{HIE}$, and in the inscribed quadrilaterals $AHBC$ and $AHIE$, the angles $\widehat{HBC}$ and $\widehat{HIE}$ have the same complement $\widehat{HAC}$, then $\widehat{HBC} = \widehat{HIE} = \widehat{BIC}$.

Let S on IH be such that $\widehat{BSH} = \widehat{HBC}$ and let $SO \parallel DH$ and $IN \parallel DC$. Sector $(CDHB)$ is a known part of circle (ABC), [in fact $\widehat{AH} = \frac{1}{4}\widehat{AB} = \frac{1}{24}$ of the circumference, therefore sect.$(CDHB) = \frac{11}{24}$ of the circle (ABC)].

Let (F) be a circle equivalent to this sector, therefore to $\frac{11}{24}$ of (ABC); let (U) and (J) be two circles such that

$$\frac{(U)}{(F)} = \frac{SI}{HC} \quad \text{and} \quad \frac{(J)}{(F)} = \frac{IK}{HC}.$$

Let (HQB) and (BLC) be similar segments to segment (ABC) constructed on HB and BC, then according to Proposition 10

(1) lune $(HQBPH)$ + lune $(BLCMB) + (U) =$ tr.(DOB) + tr.(DNB).

On the other hand

$$\frac{IK}{CH} = \frac{IC}{CH} - \frac{KC}{CH} = \frac{IC}{CH} - \frac{\widehat{BC}}{\widehat{HBC}} = \frac{IC}{CH} - \frac{\text{sect.}(BDCM)}{\text{sect.}(CDHB)}$$

and

$$\frac{IK}{CH} = \frac{(J)}{(F)} = \frac{(J)}{\text{sect.}(CDHB)},$$

therefore

$$\frac{IC}{CH} = \frac{\text{sect.}(BDCM) + (J)}{\text{sect.}(CDHB)}.$$

But because of similar triangles BIC and HBC

$$\frac{IC}{CH} = \frac{BC^2}{CH^2} = \frac{\text{segm.}(BLC)}{\text{segm.}(CBH)} = \frac{\text{tr.}(IDC)}{\text{tr.}(CDH)} = \frac{\text{tr.}(DNC)}{\text{tr.}(CDH)},$$

$$\frac{IC}{CH} = \frac{\text{segm.}(BLC) + \text{tr.}(DNC)}{\text{sect.}(CDHB)},$$

from which we can deduce

$$\text{segm.}(BLC) + \text{tr.}(DNC) = \text{sect.}(BDCM) + (J).$$

By taking away from the two members the sum of segm.(BMC) + tr.(DNC), this leaves

(2) lune $(BLCMB) =$ tr.$(BDN) + (J)$.

From (1) and (2), we deduce

$$\text{lune } (HQBPH) + (U) + (J) = \text{tr.}(DOB).$$

In summary: In a circle with centre D, where the givens are $\widehat{BC} = \frac{1}{3}$ circle and $\widehat{BH} = \frac{1}{8}$ circle, and segments similar to segment HBC are constructed on BC and BH, in this way lunes $(HQBPH)$ and $(BLCMB)$ can be determined. If circles (F), (U) and (J) are defined by $(F) = \frac{11}{24}$ of the given circle, then $\frac{(U)}{(F)} = \frac{SI}{HC}$ and $\frac{(J)}{(F)} = \frac{IK}{HC}$, with points K, I and S as points of HC which satisfy

$$\frac{KH}{KC} = \frac{3}{8}, \; \widehat{BIC} = \widehat{BSH} = \widehat{CBH},$$

and if O is the point of BH and N the point of BC such that $SO \parallel DH$, $IN \parallel DC$, then

$$\text{lune } (BLCMB) = \text{tr.}(BDN) + (J),$$
$$\text{lune } (HQBPH) + (U) + (J) = \text{tr.}(DOB).$$

Proposition 17. — *Let* $\widehat{\text{AC}} = \frac{1}{3}$ *circumference,* $\widehat{\text{AB}} = \frac{1}{4}$ *circumference,* (AEB) *segment similar to* (ABC), (ANB) *semicircle, circle* (K) = $\frac{1}{36}$ *circle* (ABC), *then* lune (ANBEA) = tr.(ADI) + (K).

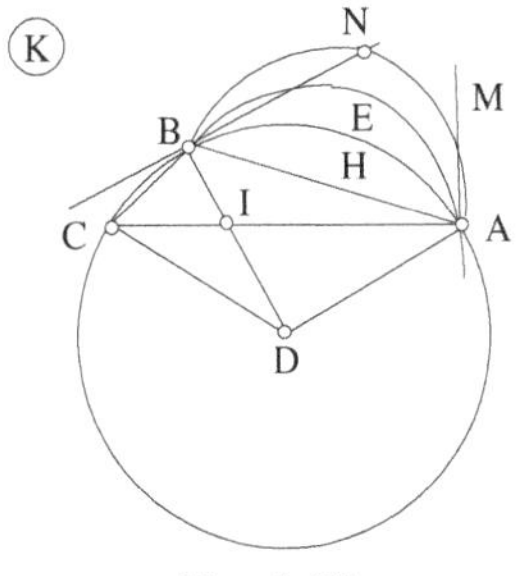

Fig. 1.33

If AM is a tangent at A to arc $\widehat{AEB}$, then $\widehat{MAB} = \frac{\pi}{3}$. However the tangent at A to arc $\widehat{ANB}$ is perpendicular to AB, therefore AM cuts arc $\widehat{ANB}$, and this arc is wholly outside the circle AEB.

According to Proposition 15: lune $(AEBHA) + (K) =$ tr.(ABI); hence by adding tr.(ADI) to the two members: lune $(AEBHA) + (K) +$ tr.$(ADI) =$ tr.(ABD).

But according to Proposition 9: lune $(ANBHA) =$ tr.(ABD); from which can be deduced: lune $(ANBEA) =$ tr.$(ADI) + (K)$.

Let (LGP) be a right isosceles triangle which is equivalent to triangle (ADI). Arc $\overset{\frown}{LQP}$ of circle (G, GP) and semicircle (LSP) determine the lune $(LSPQL)$.

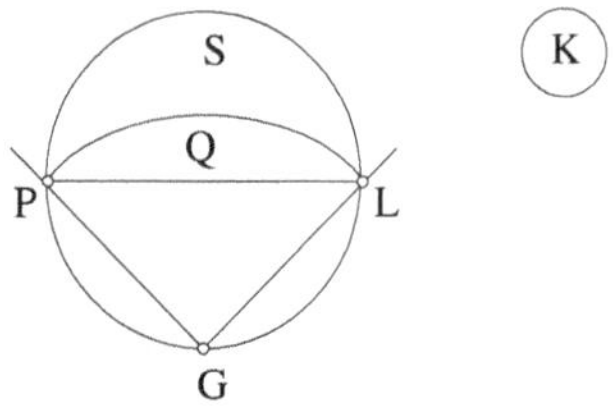

Fig. 1.34

According to Proposition 9: lune $(LSPQL) =$ tr.(LGP); however tr.$(LGP) =$ tr.(ADI), therefore lune $(ANBEA) =$ lune $(LSPQL) + (K)$.

Proposition 18. — (D) *designates the surface of circle* (D, DA). *Let* $\overset{\frown}{AB} = \overset{\frown}{BC} = \frac{1}{6}$ *of the circumference, let segments* (AEB) *and* (BIC) *be similar to* segm.(ABC), (AFC) *semicircle of diameter* AC, L *and* M *on* AC *such that* $\widehat{BLA} = \widehat{BMC} = \widehat{ABC} = \frac{2\pi}{3}$, *with* (N) *and* (U) *as given circles* (N) $= \frac{1}{9}$ (D), *and* (U) $= \frac{1}{24}$ (D), *and* (P) = (N) + (U), *therefore*

$$\text{fig.(AFCIBEA)} + \text{tr.(DLM)} = \text{(P)}.$$

Let AS be a tangent at A to arc $\overset{\frown}{AEB}$, then

$$\widehat{SAL} = \widehat{SAB} + \widehat{BAC} = \frac{\pi}{3} + \frac{\pi}{6} = \frac{\pi}{2}.$$

Therefore arcs $\overset{\frown}{AEB}$ and $\overset{\frown}{AFC}$ are tangents at A. We know that B is inside circle AFC, therefore arc $\overset{\frown}{AEB}$ is inside circle AFC,[21] and similarly

[21] According to *Elements* III.13.

for arc $\widehat{BIC}$. The straight line AD cuts the circle (D, DA) again, at G, then $\widehat{CG} = \widehat{AB}$. We have tr.$(ADC)$ = tr.(CDG) < sect.$(CDG) = \frac{1}{6}(D)$ and tr.(DLM) $= \frac{1}{3}$ tr.(ADC). Therefore tr.$(DLM) < \frac{1}{2}(N)$. But according to Proposition 14:

(1) lune $(AEBHA)$ + lune $(BICKB)$ + (N) = tr.(ABC) + tr.(DLM).

Let $(U) = \frac{1}{24}(D)$; we know according to Proposition 13 that

(2) lune $(AFCBA)$ = tr.(ADC) + (U) = tr.(ABC) + (U).

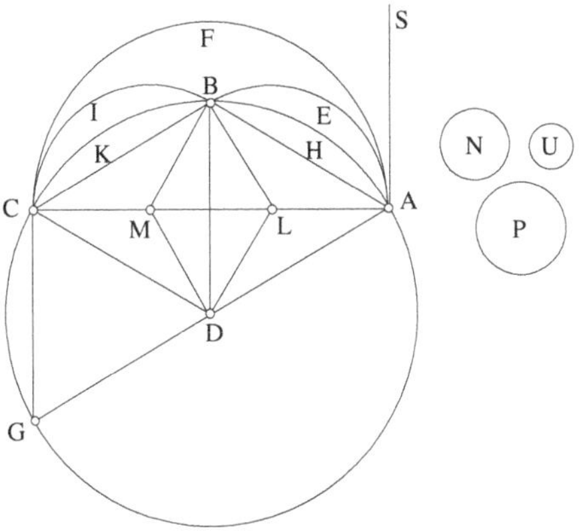

Fig. 1.35

From (1) and (2), we deduce

lune $(AEBHA)$ + lune $(BICKB)$ + (N) – tr.(DLM) + (U) = lune $(AFCBA)$,

hence

$$(N) - \text{tr.}(DLM) + (U) = \text{fig.}(AFCIBEA).$$

We set $(N) + (U) = (P)$, then (P) = fig.$(AFCIBEA)$ + tr.(DLM).

Note that $(P) = \frac{1}{9}(D) + \frac{1}{24}(D) = \frac{11}{72}$(D) and tr.$(DLM) = \frac{1}{3}$ tr.(ABC).

Proposition 19. — *A portion of a circle comprised between two parallels and equal to a quadrant of a circle is to be constructed.*

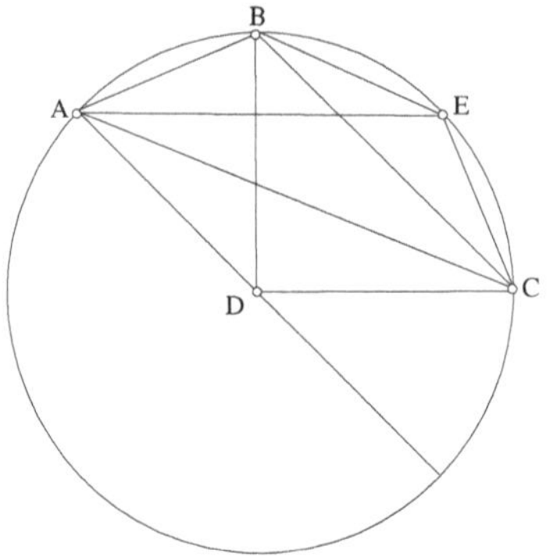

Fig. 1.36

On a circle with centre D, *take points* A, B, C *such that* $\widehat{BDC} = \frac{\pi}{2}$ *and* DA || BC. *If* E *is the midpoint of arc BC, then straight lines* BE *and* CA *satisfy the problem.*

If $\widehat{BDC} = \frac{\pi}{2}$, then $\widehat{BCD} = \widehat{CBD} = \widehat{BDA} = \frac{\pi}{4}$. Therefore $\widehat{AB} = \widehat{EC} = \widehat{EB}$, hence

1) segm.(AB) = segm.(EC) = segm.(BE)
2) $\widehat{EBC} = \widehat{BCA}$, therefore straight lines BE and AC are parallel.

On the other hand, since $AD \parallel BC$, then tr.(BAC) = tr.(BDC); hence

$$\begin{aligned}
&\text{tr.}(BAC) + \text{tr.}(BEC) = \text{tr.}(BDC) + \text{tr.}(BEC),\\
&\text{quadr.}(ABEC) = \text{quadr.}(DBEC),\\
&\text{quadr.}(ABEC) + \text{segm.}(AB) + \text{segm.}(EC)\\
&\qquad = \text{quadr.}(DBEC) + \text{segm.}(BE) + \text{segm.}(EC),\\
&\text{portion } (EBAC) = \text{sector } (BDCE) = \tfrac{1}{4}\text{ circle } (ABC).
\end{aligned}$$

Proposition 20. — *On the first circle* (ABH), *are points* A, B, C, D, *such that* $\widehat{AB} = \widehat{BC} = \widehat{CD} = \frac{1}{8}$ *of the circumference. Arc* AEGD *is constructed symmetrically to* ABCD *with* $\widehat{AE} = \widehat{EG} = \widehat{GD}$.

Therefore according to Proposition 19: $BC \parallel AD \parallel EG$, and each of the portions of circles $(ABCD)$ and $(AEGD)$ between parallel lines is equal to a quadrant of a circle.

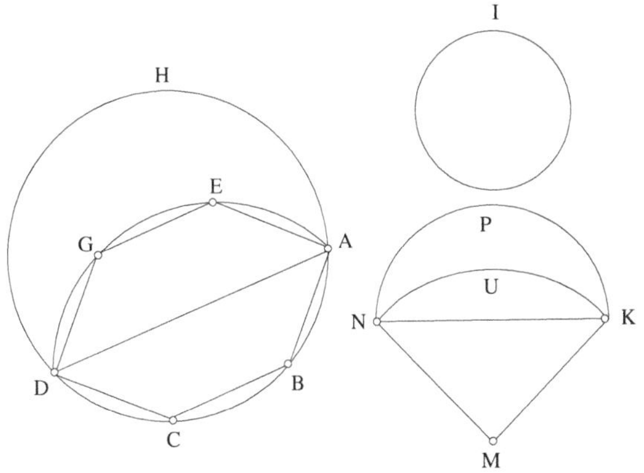

Fig 1.37

Therefore

$$\text{portion } (ABCD) + \text{portion } (AEGD) = \tfrac{1}{2} \text{ circle } (ABH);$$

it follows that

$$\text{lune } (AHDGE) + \text{segm.}(GE) + \text{segm.}(CB) = \tfrac{1}{2} \text{ circle } (ABH).$$

But

$$\text{segm.}(GE) = \text{segm.}(BC) = \text{segm.}(AE) = \text{segm.}(DG),$$

therefore

$$\text{lune } (AHDGE) = \text{portion } (ABCD) + \text{quadr.}(AEGD).$$

Suppose $(I) = \frac{1}{4}$ circle (ABH) and let KMN be a right isosceles triangle, and $\widehat{M} = \frac{\pi}{2}$, such that tr.$(KMN)$ = quadr.$(AEGD)$. Then

$$\text{lune } (AHDGEA) = (I) + \text{tr.}(KMN).$$

Arc $\widehat{KUN}$ of circle (M, MK) and semicircle (KPN) determine the lune $KPNUK$, so

$$\text{lune } (KPNUK) = \text{tr.}(KMN);$$

therefore

$$\text{lune } (AHDGEA) = \text{lune } (KPNUK) + (I).$$

Proposition 21. — Property of lunes where two arcs form the sum of a complete circle [lune from Proposition 20 has this property].

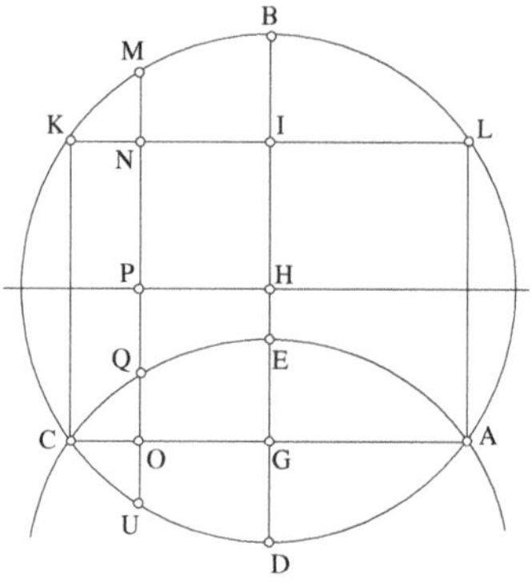

Fig. 1.38

Let (H, HA) *be a circle, with a chord* AC *of this circle separating arcs* $\overset{\frown}{ADC}$ and $\overset{\frown}{ABC}$ ($\overset{\frown}{ADC} < \overset{\frown}{ABC}$). *Let* $\overset{\frown}{AEC}$ *be the symmetrical arc of circle* $\overset{\frown}{ADC}$. *The perpendiculars to* AC *at* A *and* C, *at its midpoint* G *and at any point* O, *determine in lune* (ABCEA) *the segments* AL, CK, EB *and* MQ *which are all equal.*

Comment: The perpendicular to the chord AC at its midpoint G cuts the arcs of the lune at E and B. The two arcs $\overset{\frown}{AEC}$ and $\overset{\frown}{LBK}$ correspond in a translation of vector $\overrightarrow{EB}$. Hence the result announced as $AL = EB = QM = CK$.

Note that this property of the translation associated with two equal circles is studied by Ibn al-Haytham in his treatise *On Known Things*, Proposition 11.[22]

Proposition 22. — *If lunes constructed on similar arcs* $\overset{\frown}{ANB}$ *and* $\overset{\frown}{DOE}$ *of two circles* (H) *and* (I) *are limited by similar arcs* $\overset{\frown}{AKB}$ *and* $\overset{\frown}{DME}$, *then*

$$\frac{\text{lune (AKBNA)}}{\text{lune (DMEOD)}} = \frac{(\text{H})}{(\text{I})}.$$

[22] R. Rashed, *Les mathématiques infinitésimales*, vol. IV.

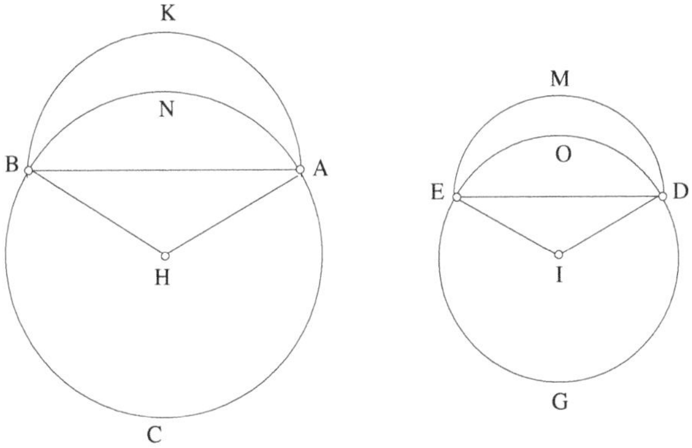

Fig. 1.39

Similar segments are associated to similar arcs, and therefore according to Proposition 6:

$$\frac{AB^2}{ED^2} = \frac{(H)}{(I)} = \frac{\text{segm.}(ANB)}{\text{segm.}(DOE)} = \frac{\text{segm.}(AKB)}{\text{segm.}(DME)},$$

and it follows that

$$\frac{AB^2}{ED^2} = \frac{(H)}{(I)} = \frac{\text{lune }(AKBNA)}{\text{lune }(DMEOD)}.$$

Proposition 23. — *Let two circles* (K) *and* (I) *such that* (I) = 3(K) *and in each of them a chord on the side of hexagon* AB *and* EG *respectively.*

On AB *we construct an equilateral triangle* ABD, D *outside the circle, and on* EG *an arc* $\widehat{\text{EPG}}$ *equal to a third of the circumference, therefore*

lune (EPGOE) = fig.(ADBMA).

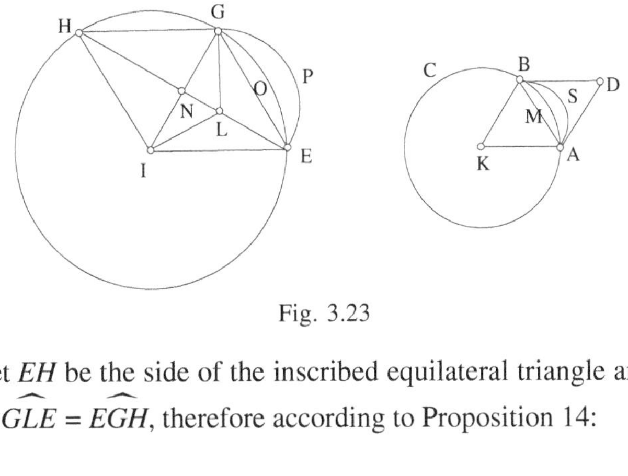

Fig. 3.23

Proof: Let *EH* be the side of the inscribed equilateral triangle and *L* on *EH* such that $\widehat{GLE} = \widehat{EGH}$, therefore according to Proposition 14:

$$\begin{aligned}\text{lune } (EPGOE) + \frac{1}{18}(I) &= \text{tr.}(EGN) + \text{tr.}(ILN)\\ &= \text{tr.}(HLI) = \frac{2}{3}\,\text{tr.}(EIH)\\ &= \frac{2}{3}(EIG).\end{aligned}$$

But we know that $(I) = 3(K)$, therefore $EG^2 = 3AB^2$, and it follows that

$$\begin{aligned}\text{tr.}(EIG) &= 3\ \text{tr.}(AKB),\\ \frac{2}{3}\text{tr.}(EIG) &= 2\ \text{tr.}(AKB) = \text{lozenge } (ADBK);\end{aligned}$$

hence

$$\text{lune } (EPGOE) + \frac{1}{18}(I) = \text{lozenge } (ADBK).$$

On the other hand,

$$\text{sect.}(AKBM) = \frac{1}{6}(K) = \frac{1}{18}(I);$$

therefore

$$\begin{aligned}&\text{lune } (EPGOE) = \text{lozenge } (ADBK) - \text{sect.}(AKBM)\\ &\text{lune } (EPGOE) = \text{fig.}(ADBMA).\end{aligned}$$

If arc $\widehat{ASB}$ is constructed on AB equal to one third of the circumference, the lunes ($EPGOE$) and ($ASBMA$) are similar, therefore

$$3 \cdot \text{lune } (ASBMA) = \text{fig.}(ADBMA)$$

and

$$2 \cdot \text{lune } (ASBMA) = \text{fig.}(ADBSA).$$

This proposition ends the *Exhaustive Treatise* of Ibn al-Haytham, the most substantial treatise on lunes ever to be known before the eighteenth century.

1.3. *Translated texts*

Al-Ḥasan ibn al-Haytham

1.3.1. *Treatise on Lunes*

1.3.2. *Treatise on the Quadrature of the Circle*

1.3.3. *Exhaustive Treatise on the Figures of Lunes*

In the name of God, the Merciful, the Compassionate

TREATISE OF AL-ḤASAN IBN AL-ḤASAN IBN AL-HAYTHAM

On Lunes

Upon my examining – may God bless our Lord, the master, with permanence and may He bestow upon him everlasting happiness and prosperity – the shape of the lune, equal to a triangle as mentioned by the Ancients, its admirable property and its astonishing composition lead me to ponder on the properties of lunes and on the curious notions that occur in it. It is thus that I have deduced the propositions I enclose in this treatise. I disclose them in his presence so that he may know them, meditate and use them to show the virtues of geometry and its hidden notions. I implore God to grant me his good assistance so that I may be faithful to my promises. Of them, He is the Master.

<1> If we draw in a circle any diameter and a chord equal to half the diameter, if we join the centre to the extremities of the chord and construct on the chord a semicircle, then the sum of the generated lune and the circle forming one part of twenty-four parts of the first circle is equal to the generated triangle.

Example: We produce in the circle *ABC*, of centre *D*, the diameter *AC* and the chord *AB* equal to half the diameter and we join *DB*. We construct on the straight line *AB* the semicircle *AGB* and we draw a circle *H* equal to one part of twenty-four parts of the circle *ABC*.

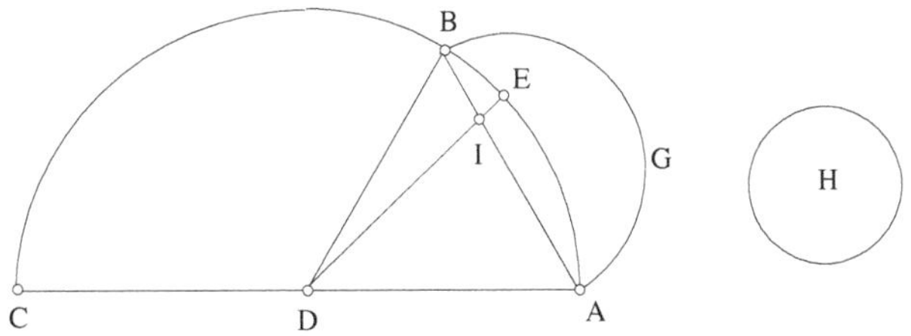

Fig. I.1.1

I say that the sum of the lune AGBE *and the circle* H *is equal to the triangle* ABD.

Proof: We separate the arc *AE* equal to an eighth of the circle and we join *DIE*. Since the straight line *AB* is half of diameter *AC*, the square of *AB* is equal to a quarter of the square of *AC* and the ratio of the square of *AB* to the square of *AC* is equal to the ratio of the circle to the circle, because the ratio of the circle to the circle is equal to the ratio of the square of the diameter to the square of the diameter. Therefore, the circle of diameter *AB* is a quarter of the circle *ABC*, and half of the circle *AGB* is an eighth of the circle *ABC*. However, the arc *AE* is an eighth of the circle *ABC*, therefore the sector *AED* is an eighth of the circle *ABC*. Yet, we have shown that half of the circle *AGB* is an eighth of the circle *ABC*, therefore the sector *AED* is equal to half of the circle *AGB*. Subtracting the common portion *AEI*, there remains the triangle *AID* equal to the lune *AGBE* plus the portion *BEI*. We take the common triangle *BID*; the triangle *ABD* will be equal to the lune *AGBE* plus the portion *BEI* plus the triangle *BID*. However, the portion *BEI* plus the triangle *BID* is the sector *BED*; therefore, the triangle *ABD* is equal to the lune *AGBE* plus the sector *BDE*. However, *AE* is an eighth of the circle and *AB* is a sixth of the circle; therefore, *EB* is one part of twenty-four parts of the circumference of the circle, and the sector *BED* is one part of twenty-four parts of the circle. But the circle *H* is one part of twenty-four parts of the circle, so the sector *BED* is equal to circle *H*; consequently, the sum of the lune *AGBE* and of the circle *H* is equal to the triangle *ABD*. That is what we wanted to prove.

<2> If we produce in a circle one of its diameters and the side of an equilateral triangle, join the centre to the extremities of the chord and construct on the chord a semicircle, then the generated lune is equal to the generated triangle plus the circle forming one part of twenty-four parts of the circle.

Example: We produce in the circle *ABC*, of centre *D*, the diameter *AC*, and a chord *AB* equal to the side of the equilateral triangle and we join *DB*. We construct on *AB* a semicircle and we draw the circle *I* equal to one part of twenty-four parts of the circle *ABC*.

I say that the lune AEBG *is equal to the triangle* ABD *plus the circle* I.

Proof: We join *BC*. As *AB* is the chord of the triangle and *ABC* is a semicircle, the arc *BC* is one sixth of the circle. The straight line *BC* is therefore half a diameter and *ABC* is a right angle because it lies in a semicircle. The square of *AC* is therefore equal to the square of *AB* plus the square of *BC*. But the square of *BC* is a quarter of the square of *AC*; there remains the square of *AB* which is three quarters of the square of *AC*. The circle, of diameter *AB*, is therefore three quarters of the circle *ABC*, and the semicircle *AEB* is three quarters of the semicircle *ABC*. However, the arc *AB* is two thirds of the arc *ABC*, therefore the sector *AGBD* is two thirds of the semicircle *ABC*. But the circle *I* is one part of twenty-four parts of the circle, so the circle *I* is half of the sixth of the semicircle *ABC*, the sector *AGBD* is two thirds of this half, and the sum of the sector *AGBD* and of the circle *I* is three quarters of half of the circle *ABC*. We have thus shown that half of the circle *AEB* is equal to the sector *AGBD* plus the circle *I*. Subtracting the common portion *AGB*, there remains the lune *AEGB* equal to the triangle *ABD* plus the circle *I*. That is what we wanted to prove.

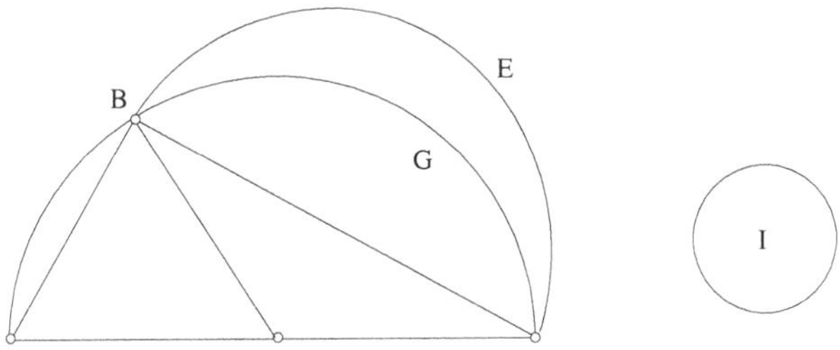

Fig. I.1.2

<3> If we produce any diameter in a circle and join any point marked on its circumference to the two extremities of the diameter by two straight lines on which we construct two semicircles, then the sum[1] of the generated two lunes is still equal to the generated triangle.

Example: We produce in the circle *ABC* any diameter *AC*, we mark any point on its circumference – let it be point *B* – and we join the two straight lines *AB* and *BC* on which we construct the two semicircles *AEB* and *BHC*.

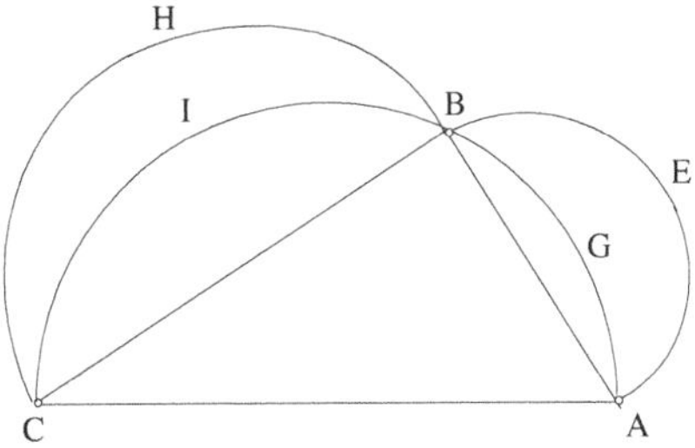

Fig. 1.3

[1] The word sum has sometimes been inserted to serve a clearer translation.

I say that the sum of the two lunes AEB *and* BHC *is equal to the triangle* ABC.

Proof: The angle *ABC* is a right angle, therefore the square of *AC* is equal to the sum of the squares of *AB* and *BC*. The circle *ABC* is therefore equal to the sum of the two circles whose diameters are *AB* and *BC*, and the semicircle *ABC* is equal to the sum of the two semicircles *AEB* and *BHC*. Subtracting the two common segments *AGB* and *BIC*, there remains the triangle *ABC* equal to the sum of the two lunes *AEBG* and *BHCI*. That is what we wanted to prove.

<4> If two lunes are <defined> from similar arcs,[2] then the ratio of one to the other is equal to the ratio of the squares of their bases, one to the other.

Example: The two lunes *ABCD* and *HIKL* <are defined> from similar arcs; their bases are *AC* and *HK*.

I say that the ratio of the lune ABCD *to the lune* HIKL *is equal to the ratio of the square of* AC to *the square of* HK.

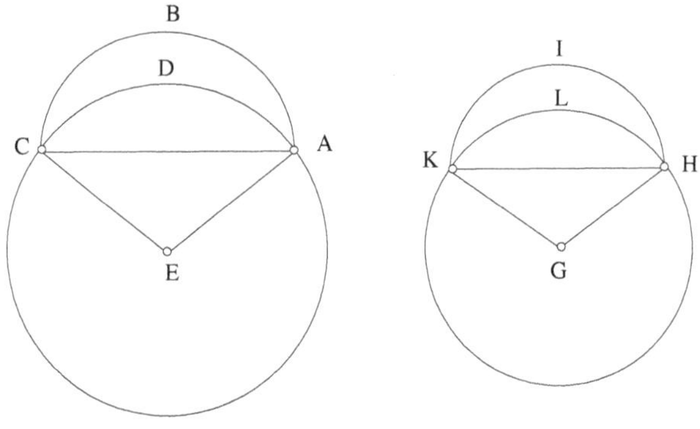

Fig. I.1.4

Proof: We draw two circles *ADC* and *HLK*; let their centres be the two points *E* and *G*. We join *EA*, *EC*, *GH* and *GK*. Since the arc *ADC* is similar to the arc *HLK*, the ratio of the circle to the circle is equal to the ratio of the sector *ACE* to the sector *HKG*. However, the ratio of the circle to the circle is equal to the ratio of the square of *AC* to the square of *HK*. Therefore, the ratio of the sector to the sector is equal to the ratio of the square of *AC* to the square of *HK*. But the ratio of the triangle *ACE* to the triangle *HKG* is equal to the ratio of the square of *AC* to the square of *HK* also, because of

[2] 'Similar' is understood here to mean: in relation of one to the other.

the similarity of the two triangles. There remains the ratio of segment *ADC* to segment *HLK* equal to the ratio of the square of *AC* to the square of *HK*. Likewise, the segment *ABC* is similar to the segment *HIK*; therefore the ratio of the segment *ABC* to the segment *HIK* is equal to the ratio of the square of *AC* to the square of *HK*. We subtract the two segments *ADC* and *HLK*, which follow the ratio of the square of *AC* to the square of *HK*; there remains the ratio of the lune *ABCD* to the lune *HIKL* equal to the ratio of the square of *AC* to the square of *HK*. That is what we wanted to prove.

<5> If we draw in a circle the side of an equilateral triangle, on which we construct a semicircle, if we then divide the arc <associated with> the triangle into two halves and if we join the two straight lines, then the sum of the lune and the triangle thus generated is equal to the sum of another lune and a circle.

Example: We draw in the circle *ABC* the straight line *AB*, equal to the side of an equilateral triangle, on which we construct a semicircle *AGB*; we divide <the arc> *AEB* into two halves at point *E* and we join *AE* and *EB*.

I say that the sum of the lune AGBE *and the triangle* AEB *is equal to the sum of another lune and a circle.*

Proof: We define the centre, let it be *D*; we draw the diameter *ADC*, we join *BD*, *DE* and *BC*, we construct on the straight line *BC* the semicircle *BRC*, we construct *HI* whose square is equal[3] to twice the square of *BC*, and we construct on *HI* a lune from two arcs similar to the two arcs *BFC* and *BRC*; let it be the lune *HIKL*. We construct the circle *M* equal to an eighth of the circle *ABC*.

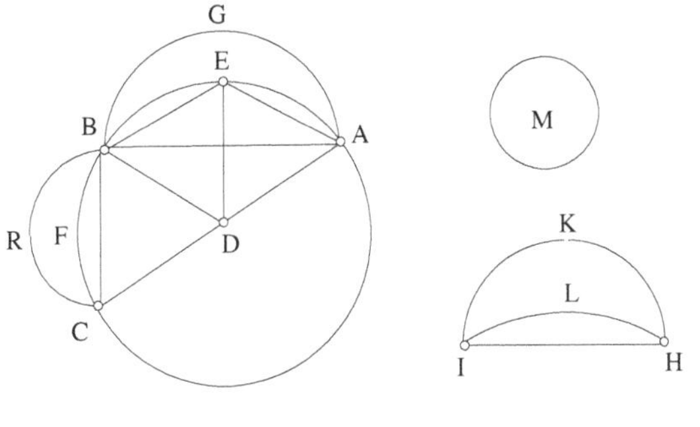

Fig. I.1.5

I say that the sum of the lune AGBE *and the triangle* AEB *is equal to the sum of the lune* HKIL *and the circle* M.

[3] Lit.: we construct the square of *HI* equal…

As each of <the straight lines> *AE*, *EB* and *BC* intercepts a sixth <of the circle>, the lunes constructed on them, which are similar to the lune *BRCF*, are equal and the three triangles *AED*, *EDB* and *BDC* are equal. But the sum of the lune *BRCF* and the circle, which is one part of twenty-four parts of the circle *ABC*, is equal to the triangle *DBC*; the sum of the three lunes constructed on the straight lines *AE*, *EB*, *BC* – which is three times the lune *BRCF* plus the circle, which is an eighth of the circle *ABC* – is therefore equal to the sum of the three triangles *AED*, *EDB* and *DBC*. But the lune *HKIL* is twice the lune *BRCF*; the sum of the two lunes *HKIL* and *BRCF* is therefore three times the lune *BRCF*. But the circle *M* is one eighth of the circle *ABC*, so the sum of the two lunes *HKIL* and *BRCF* and of the circle *M* is equal to the sum of the three triangles, which are *AED*, *EDB* and *BDC*. But the three triangles are the quadrilateral *AEBC*, and the quadrilateral *AEBC* is the sum of the two triangles *AEB* and *ABC*, so the sum of the two lunes *HIKL* and *BRCF* and of the circle *M* is equal to the sum of the two triangles *AEB* and *ABC*. But since the point *B* lies on the circumference of the circle and since we have drawn the two straight lines *AB* and *CB* from the two extremities of the diameter on which we have constructed the two lunes *AGBE* and *BRCF*, the sum of the lunes is equal to the triangle *ABC*. But the sum of the two lunes *HKIL* and *BRCF* and of the circle *M* is equal to the sum of the two triangles *ABC* and *AEB*, so the sum of the two lunes *HKIL* and *BRCF* and of the circle *M* is equal to the sum of the two lunes *BRCF* and *AGBE* and of the triangle *AEB*. We then subtract the lune *BRCF* common to both sides; there remains the sum of the lune *AGBE* and the triangle *AEB* equal to the sum of the lune *HKIL* and the circle *M*. That is what we wanted to prove.

The treatise on the lunes is completed. Thanks be to God, Lord of the worlds.

In the Name of God, the Merciful, the Compassionate

TREATISE BY AL-ḤASAN IBN AL-ḤASAN IBN AL-HAYTHAM

On the Quadrature of the Circle

Many philosophers have believed that the area of a circle cannot be equal to the area of a square limited by straight lines. This notion has been revisited many times in their dialogues and controversies, but we have discovered no early or modern work containing a polygonal figure exactly equal to the area of a circle. Archimedes made use of a certain approximation[1] in his work on the measurement of the circle, and this latter notion is among those that have reinforced the opinion of the philosophers in their conviction. Given this, we have thought deeply about this notion, and it has been revealed to us that it is possible, not difficult, and has analogues: namely, that there can exist a lune enclosed by two arcs of two circles that is also equal to a triangle, and there can also exist a lune and a circle whose sum is equal to a triangle. We have shown several different figures of this species in our book *On Lunes*. But as this is so for figures of lunes, we have become more and more convinced that it is possible for the area of a circle to be equal to the area of a quadrilateral with straight sides. We have therefore given our utmost attention to this notion, until it was made apparent to us through a proof that it is possible and that there is no ambiguity as to this possibility. We have therefore written this treatise.

We say that, if one of the diameters is drawn in a circle, and any point is then marked on one of its halves so formed, and if we join this point to the extremities of the diameter by two straight lines, and if we then construct two semicircles on these two straight lines, then the sum of the two lunes formed by the circumference of these two semicircles and the circumference of the first circle is equal to the triangle formed in the first circle. We have proved this notion in our book *On Lunes*, and we repeat the proof here.

Let there be a circle on which lie A, B and C, and let D be its centre. We draw the straight line ADC through the point D; AC will then be the diameter of the circle. We mark a point B on the circumference, we join the

[1] Lit.: a certain simplification.

two straight lines *AB* and *BC*, and we construct two semicircles *AEB* and *BGC* on the two straight lines *AB* and *BC*.

I say that the sum of the two lunes AEBHA *and* BGCIB *is equal to the triangle* ABC.

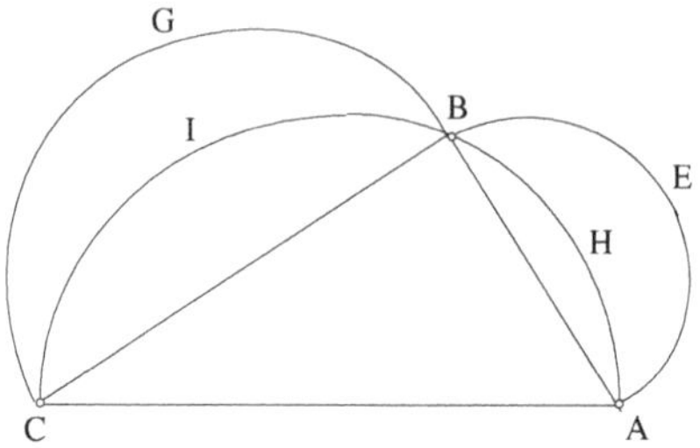

Fig. II.2.1

Proof: The ratio of any two circles, one to the other, is equal to the ratio of the square of the diameter of one to the square of the diameter of the other, as has been shown in Proposition 2 of Book 12 of the *Elements*. The ratio of the circle *BGC* to the circle *BEA* is therefore equal to the ratio of the square of *CB* to the square of *BA*. Composing, the ratio of the sum[2] of the squares of *CB* and *AB* to the square of *AB* is equal to the ratio of the sum of the circles *BGC* and *BEA* to the circle *BEA*. But the sum of the squares of *CB* and *AB* is equal to the square of *AC*. Therefore, the ratio of the square of *AC* to the square of *AB* is equal to the ratio of the sum of the circles *BGC* and *BEA* to the circle *BEA*. But the ratio of the square of *AC* to the square of *AB* is equal to the ratio of the circle *ABC* to the circle *BEA*. The ratio of the sum of the circles *BGC* and *BEA* to the circle *BEA* is therefore equal to the ratio of the circle *ABC* to the circle *BEA*. The circle *ABC* is therefore equal to the sum of the circles *BGC* and *BEA*. The semicircle *ABC* is therefore equal to the sum of the semicircles *AEB* and *BGC*. If we remove the two portions *AHB* and *BIC* which are common to the circle *ABC* and the sum of the circles *AEB* and *BGC*, there remains the triangle *ABC* which is equal to the sum of the two lunes *AEBHA* and *BGCIB*. That is what we wanted to prove.

If the two arcs *AHB* and *BIC* are equal, then the two straight lines *AB* and *BC* will also be equal, the two circles *AEB* and *BGC* will be equal, their halves will be equal, and the two lunes *AEBHA* and *BGCIB* will be equal. We join *BD*; the two triangles *ABD* and *BDC* will be equal. But we have shown that the sum of the two lunes is equal to the triangle *ABC*. If the two lunes are equal and the two triangles *ABD* and *BCD* are equal, then each of

[2] We have occasionally inserted the word 'sum' where required in the translation.

the lunes is equal to each of the triangles, and the lune *AEBHA* is equal to the triangle *ABD*.

Now that we have proved this, we go back to the circle, the lune *AEBHA* and the triangle *ABD*, and we divide the straight line *BA* into two halves at the point *K*. The point *K* will then be the centre of the circle *AEB*. We join *DK* and we extend it so that it cuts the two arcs *AHB* and *AEB* at the points *H* and *E*. The straight line *DKHE* will then be a diameter of the circle *ABC* and a diameter of the circle *AEB*, as it passes through their centres. We divide the straight line *EH* into two halves at the point *L*. With *L* at the centre, we draw a circle at a distance *LH* and let this be the circle *HMEN*. This circle will then be tangent to the circle *ABC* on the outside and tangent to the circle *AEB* on the inside, as it meets both of these circles at the extremity of a diameter that is common to these two circles and to the circle which is tangent to them. The circle *HMEN* therefore lies entirely within the lune *AEBHA*. Consequently, this circle is itself a part of this lune. But any magnitude has a ratio to any other magnitude of which it is a part, even if no one knows what the ratio is or even if it is impossible for anyone to know it, as the ratio between magnitudes does not depend upon them being known by anyone, nor upon anyone's ability to determine them and hence know them. The ratio between the magnitudes is a specific notion to those that are of the same type (*jins*). Therefore, if two magnitudes are of the same type, and if each of them is limited, finite and fixed in its magnitude and does not change in any way, either by increasing, reducing, or changing its type, then the ratio of one to the other is one single fixed ratio which does not change and which does not modify its form in any manner whatsoever.

For any magnitude, a part of that magnitude is of the same type if the part is limited, finite and does not change in its magnitude, or outline, or shape, and if the whole magnitude is also fixed in its state and does not change in its type, or magnitude, or outline, or shape. If both the magnitude and the part of the magnitude have this property, then the whole magnitude has a single fixed ratio to the part of the magnitude, which does not change or vary in any manner whatsoever.

If the circle *ABC* is known in magnitude,[3] then its circumference is known, its diameter is also known, and its centre is known. The diameter *AC* is therefore known, the arc *AB* which is one quarter of the circumference is known, the straight line *AB* is known, the straight line *BD*

[3] Ibn al-Haytham's terms in this paragraph on 'known in magnitude' should be looked at in conjunction with his treatise on *The Knowns*. Cf. 'La philosophie des mathématiques d'Ibn al-Haytham. II. *Les Connus*', *MIDEO* 22, 1993, pp. 87–275, see pp. 97ff.

is known, and the triangle *ABD* is known. By the term 'known' I wish to express that which I stated in describing the circle *ABC*, i.e. that it is fixed in its state and does not change, as mathematicians use the word 'known' to mean 'that which does not change'. And the semicircle *AEB* will be known as the straight line *AB*, which is its diameter, is known. The arc *AEB* is known as it does not change, and the arc *AHB* is also known. Therefore, the lune *AEBHA* is known, i.e. it is fixed according to the same property and it does not change, neither in its type, nor in its magnitude, nor in its outline. By its 'type', I mean that it is a plane surface. The straight line *KE*, which is the half-diameter of the circle, is known, and the straight line *KH* is known as the two points *K* and *H* are known. The remaining straight line *EH* is also known as it does not change, neither in its magnitude, nor its type, nor its outline. But the straight line *EH* is the diameter of the circle *HMEN*, and therefore the circle *HMEN* is known. Its magnitude, outline and shape do not change. But the circle *HMEN* is a part of the lune *AEBHA*, and neither the lune *AEBHA* nor the circle *HMEN* change in any way. They are also of the same type as one is a part of the other. Therefore, the lune *AEBHA* has a fixed ratio with the circle *HMEN* according to the same property, and this ratio does not change in any way. But the ratio of any one magnitude to one of its parts is the same as the ratio of any other magnitude to its similar part. The ratio of the lune *AEBHA* to the circle *HMEN* is therefore equal to the ratio of the straight line *AD* to one of its parts, whether or not we know the magnitude of this part, or even if we are unable to determine it or to succeed in finding it. Let this part be *DU*. Then the ratio of *AD* to *DU* will be equal to the ratio of the lune *AEBHA* to the circle *HMEN*. Therefore the ratio of *AD* to *DU* is a fixed ratio that never changes, as the ratio of the lune to the circle is a fixed ratio that does not change. If the ratio of *AD* to *DU* is a fixed ratio that never changes, then the straight line *DU* is a unique straight line that does not change, as the straight line *AD* is a straight line with a known magnitude whose magnitude does not change. We join *BU* so that *BUD* is a triangle. The ratio of the triangle *ABD* to the triangle *BDU* is equal to the ratio of the straight line *AD* to the straight line *DU*. But the ratio of *AD* to *DU* is equal to the ratio of the lune *AEBHA* to the circle *HMEN*, and therefore the ratio of the triangle *ABD* to the triangle *BDU* is equal to the ratio of the lune *AEBHA* to the circle *HMEN*. By permutation, the ratio of the triangle *ABD* to the lune *AEBHA* is equal to the ratio of the triangle *BDU* to the circle *HMEN*. But we have already shown that the lune *AEBHA* is equal to the triangle *ABD*. Therefore the circle *HMEN* is equal to the triangle *BDU*. But any triangle is equal to a square, as has been shown at the end of the second Book of Euclid's *Elements*.[4]

[4] Euclid, *Elements*, II.14.

Let us construct a square equal to the triangle *BDU*, and let this be the square *SQPO*. The circle *HMEN* will be equal to this square *SQPO*. But the ratio of the diameter *AC* to the diameter *EH* is a known ratio,[5] as each of these two diameters has a known magnitude. Let the ratio of *AC* to *EH* be equal to the ratio of *XQ* to *QP*. Then the ratio of the square of *AC* to the square of *HE* is equal to the ratio of the square of *XQ* to the square of *QP*. We construct a square on the straight line *XQ*, and let this square be the square *XT*. The ratio of the square of *AC* to the square of *EH* is equal to the ratio of the square *XT* to the square *QO*. But the ratio of the square of *AC* to the square of *HE* is the ratio of the circle *ABC* to the circle *HMEN*. Therefore, the ratio of the square *XT* to the square *QO* is equal to the ratio of the circle *ABC* to the circle *HMEN*. But the square *QO* is equal to the circle *HMEN*. Therefore, the square *XT* is equal to the circle *ABC*.

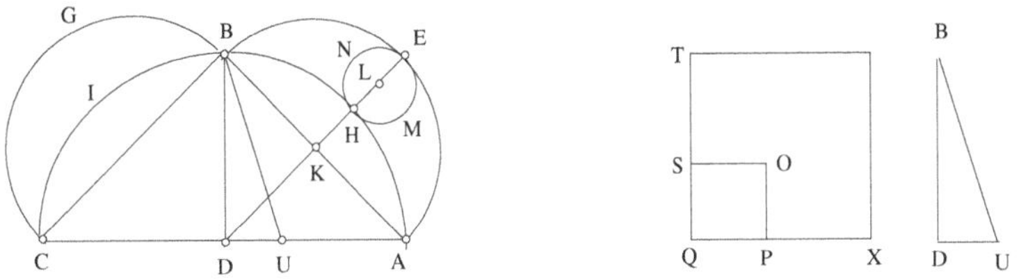

Fig. II.2.2

In this proof, we have shown that any circle is equal to a quadrilateral with straight sides.

But how can we find this square? We will discuss this in a separate treatise,[6] as our aim in this treatise was simply to prove that this notion is possible, and to show that the opinion of anyone who believes that a circle cannot be equal to a quadrilateral with straight sides is wrong. We have shown, by means of the proofs described in this treatise, that any circle is

5 $\frac{AC}{EH} = \frac{2R}{R(\sqrt{2}-1)} = 2(\sqrt{2}+1).$

6 It is easy to understand that this treatise on the construction of a square equal to a circle was never written. And one would search in vain for any mention in the writings of ancient bibliographers or in those of Ibn al-Haytham himself. The author of the Objection (pp. 105–6) notes that the treatise of Ibn al-Haytham 'has not yet appeared, and it is not mentioned in the list of his writings'. Let us note also that al-Ḥasan ibn al-Haytham's critic, despite the validity of this objection, did not grasp Ibn al-Haytham's real intention, which was to inscribe the circle in the square to perfection and to revive the method proposed by his critic. His subsequent work on lunes was really a way of avoiding work on ratios of rectilinear figures to curved figures in order to concentrate on ratios between homogeneous figures – circles and lunes.

equal to a quadrilateral with straight sides. It has therefore been shown from this that the belief of this group is false, and that it is true that any circle is equal to a quadrilateral with straight sides. The truths of these intelligible notions does not depend on their being found or determined in act by any human being. If it can be proved that such a notion is possible, then the notion becomes true, whether or not it is determined in act by a human being. That which we have written in order to identify this notion is sufficient. That is the objective that we have sought in this treatise.

The treatise on the quadrature of the circle is complete.

I make the following addition to this treatise:

While it was sufficient to establish that which was sought by establishing the possibility of it in the manner described, it would have been possible to avoid such a long proof by means of a shorter explanation, saying the following:

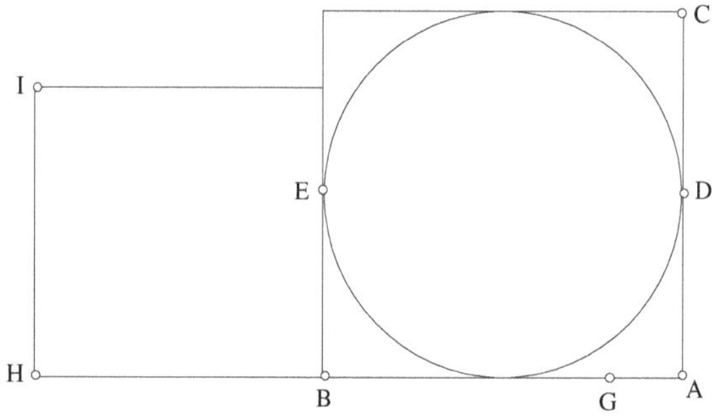

Fig. II.2.3

Let *AB* be a known straight line on which is constructed the square *BC*, which is also known, and within it the circle *DE*, which is also known as its diameter *DE*, which is equal to *AB*, is known. As the circle is a known part of a known whole, which is the square, then it has a ratio to it. Let this ratio be equal to the ratio of *BA* to *BG*. We draw *BH* between them, in continuous proportion such that the ratio of *AB* to *BH* is equal to the ratio of *BH* to *BG* and we construct the square *BI* on *BH*. The ratio of *AB* to *BG*, i.e. the ratio of the square *BC* to the circle *DE*, is therefore equal to the ratio of the square *BC* to the square *BI*. Therefore, the ratio of the square *BC* to the circle *DE* and its ratio to the square *BI* are the same. Therefore, the circle *DE* is equal to the square *BI*.

We have found that which we sought, and without any need for the perplexity offered by some other earlier and modern authors.

Objection

If the notion mentioned by the Shaykh Abū ʿAlī in this treatise were proved by means of his proof, then it would have been possible to prove it using the same technique in a much easier fashion than that which he employed: If we draw in any circle a square, then this square is a part of the circle and the part has a certain ratio to the whole, as he has described, even if this ratio is not known. Let this ratio be equal to the ratio of this square to another square. Then the ratio of the square constructed within the circle to the circle and its ratio to another square are the same, and therefore the circle is equal to this other square.

However, to my eyes he has done nothing in this treatise, as that which was sought was to construct a square equal to the circle. Whether this is possible or not in the divine knowledge does not help us in finding that which is sought. In saying that this is possible without our having the ability <to construct it> adds nothing to the belief held by the earlier authors, as their affirmation was in fact simply that, until now, this has not been found by means of a proof. This explanation is no clearer than the affirmation made regarding the chord of one degree, as if the chord of one and a half degrees is known and the chord of one half-degree plus one quarter is known, then the chord of one degree exists, but its ratio to the diameter is not known at the present time, despite the fact that they are of the same type.

If the knowledge of something is inaccessible to us, then that thing is inaccessible, and our conviction that knowledge of it is possible is of no use whatsoever. If one examines the writings on these notions, they can be divided into three groups as follows: The notion is known, i.e. it has been established by means of a proof; it is known or knowledge of it is inaccessible; or that knowledge of the notion is inaccessible, i.e. that it has not been established by a proof that it is either known or that its knowledge is inaccessible, like knowledge of the chord of one ninth part of a circle, or the knowledge of the chord of one degree. There are many similar cases to these latter two in this third group. Neither has he shown that knowledge of the squaring of the circle is necessary, and neither has he produced that which he promised. No treatise on this subject has yet appeared, and it is not mentioned in the list of his writings.

I found this objection transcribed at the end of the treatise. I think that it was written by either Ibn al-Sumaysāṭī or the doctor ʿAlī ibn Riḍwān.[7]

[7] Doctor Ibn Riḍwān is not unknown. His biobibliography is to be found in al-Qifṭī, *Ta'rīkh al-ḥukamā'*, and Ibn Abī Uṣaybiʿa, *'Uyūn al-anba'*. See also J. Schacht and Max Meyerhof, *The Medico-Philosophical Controversy between Ibn Butlan of Baghdad and Ibn Ridwan of Cairo*, Cairo, 1937, p. 12. The titles of these works show that he is concerned with philosophy, not mathematics. Al-Sumaysāṭī, as his name indicates, is a Persian from Sumaysāṭ, the same provenance as various other scholars. He is known for his text *al-Dā'ira awsaʿ al-ashkāl*, see *Founding Figures and Commentators in Arabic Mathematics*, vol. I, p. 546.

In the Name of God, the Merciful, the Compassionate

EXHAUSTIVE TREATISE BY AL-ḤASAN IBN AL-ḤASAN IBN AL-HAYTHAM

On the Figures of Lunes

Some of my friends have questioned me regarding the figure of the lune constructed on the circumference of a circle. I therefore wrote a brief treatise on the figures of lunes according to particular methods, as he that questioned me was in haste and would be content with particular propositions.

After a certain time had passed, an idea came to me in relation to this notion. I then arrived at its determination using scientific methods and, using this notion, I also arrived at the determination of other species of lunes which were not included in the first treatise. I then decided to write a treatise on these figures, in which I made an exhaustive treatment of everything that could be said about this notion. As a result, I wrote this treatise, beginning with the lemmas used in the proofs.

Lemmas

– **1** – Let there be any right-angled triangle such that the sides enclosing the right angle are of different lengths, and such that, if a perpendicular is dropped from the right angle onto the base, which is the hypotenuse, then the ratio of the lesser part of the two parts of the base to the entire base is less than the ratio of the angle – of all the angles of the triangle – which intercepts the shorter side, to the right angle, and the ratio of the greater part of the two parts of the base to the entire base is greater than the ratio of the angle which intercepts the longer side to the right angle.

Example: Let there be a triangle ABC, in which the angle ABC is a right angle, and let the side AB be less than the side BC. Draw the perpendicular BD.

I say that the ratio of DA *to* AC *is less than the ratio of the angle* ACB *to a right angle, and that the ratio of* DC *to* CA *is greater than the ratio of the angle* BAC *to a right angle.*

Proof: We make *DE* equal to *DA* and we join *BE*; *CB* will then be greater than *BE*, and *BE* will be greater than *BD*, as the angle *BDC* is a right angle. Taking the point *B* as the centre, we draw the arc of a circle at a distance *BE.* This will cut the straight line *BC* and lie outside the straight line *BD*. Let this be the arc *GEH*. The ratio of the triangle *BCE* to the triangle *BED* will therefore be greater than the ratio of the sector *BGE* to the sector *BEH*. By composition, the ratio of the triangle *BCD* to the triangle *BDE* is greater than the ratio of the sector *BGH* to the sector *BHE*. The ratio of the straight line *CD* to the straight line *DE* is therefore greater than the ratio of the angle *HBG* to the angle *HBE*. But *ED* is equal to *DA*, and the angle *DBE* is equal to the angle *DBA*. The ratio of *CD* to *DA* is therefore greater than the ratio of the angle *CBD* to the angle *DBA*. By composition, the ratio of *CA* to *AD* is greater than the ratio of the angle *CBA* to the angle *DBA*. By inversion, the ratio of *DA* to *AC* is therefore less than the ratio of the angle *ABD* to the angle *ABC*. But the angle *ABD* is equal to the angle *ACB*, and the angle *ABC* is a right angle. Therefore, the ratio of *DA* to *AC* is less than the ratio of the angle *ACB* to a right angle.

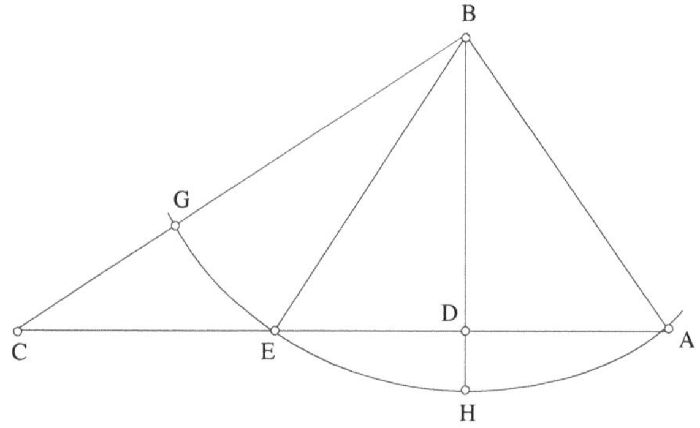

Fig. 3.1

Similarly, since the ratio of *CD* to *DA* is greater than the ratio of the angle *CBD* to the angle *ABD*, the ratio of *AD* to *DC* is, by inversion, less than the ratio of the angle *ABD* to the angle *DBC*. Composing, the ratio of *AC* to *CD* is less than the ratio of the angle *ABC* to the angle *DBC*. By inversion, the ratio of *DC* to *CA* is therefore greater than the ratio of the angle *CBD* to the angle *CBA*. But the angle *CBD* is equal to the angle *BAC*. Therefore, the ratio of *DC* to *CA* is greater than the ratio of the angle *BAC* to a right angle. That is what we wanted to prove.

– **2** – *We also say that, if the triangle* ABC *has an obtuse angle such that the angle* ABC *is obtuse and the straight line* AB *is less than the straight line* BC, *and if a straight line* BD *is drawn such that the angle* BDA *is equal to the angle* ABC, *then the ratio of* DA *to* AC *is less than the ratio of the angle* ACB *to the supplement*[1] *of the angle* ABC.

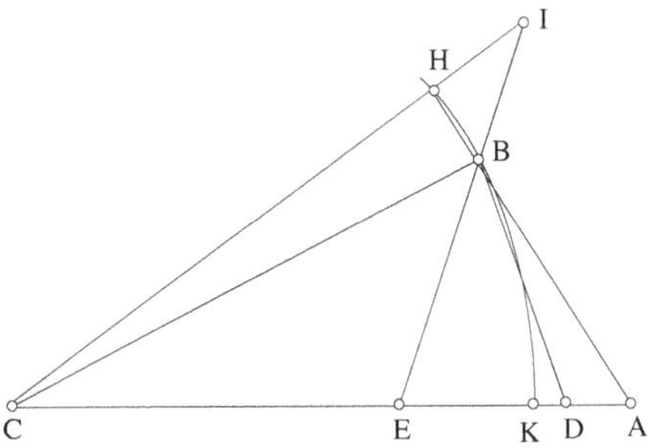

Fig. 3.2

Proof: We also draw *BE* such that the angle *BEC* is equal to the angle *BDA*. The angle *BED* will then be equal to the angle *BDE*, the two straight lines *BD* and *BE* will be equal, and the ratio of *AD* to *DB* will be equal to the ratio of *AB* to *BC*. Similarly, the ratio of *BE* to *EC* will be equal to the ratio of *AB* to *BC*. The ratio of *AD* to *DB* is therefore equal to the ratio of *BE* to *EC*, so the product of *CE* and *DA* is equal to the square of *DB*. But the straight line *AD* is less than the straight line *DB*, and it is therefore less than the straight line *EC*. It is therefore very much less than the straight line *DC*. But the product of *CD* and *DA* is very much less than the square of one half of *AC*. Therefore, the product of *CE* and *DA* is very much less than the square of one half of *AC*, and the square of *DB* is <very much> less than the square of one half of *AC*. Therefore, *DB* is less than one half of *AC*, so the ratio of *DA* to one half of *AC* is less than the ratio of *DA* to *DB*. But, as the ratio of *BE* to *EC* is equal to the ratio of *AB* to *BC*, *BE* is less than *EC*. We extend *EB*, we make *IE* equal to *EC*, and we join *CI*; *IC* will then be greater than *CB*. But *CB* is greater than *CE* as the angle *BEC* is obtuse. Taking the point *C* as the centre, we draw the arc of a circle at a distance *CB*; let it be the arc *KBH*. The ratio of the straight line *IE* to the straight line *EB* will then be greater than the ratio of the angle *ICE* to the angle *BCE*. By inversion, the ratio of *BE* to *EI* is less than the ratio of the angle *BCE* to the angle *ICE*. But the angle *ICE* is one half of the angle *BED* which is equal to the supplement of the angle *ABC*. But the straight line *IE* is equal

[1] Lit.: to the angle which is the supplement.

to the straight line *EC*. The ratio of *BE* to *EC* is therefore less than the ratio of the angle *ACB* to one half of the supplement of the angle *ABC*. But the ratio of *BE* to *EC* is equal to the ratio of *AD* to *DB*. The ratio of *AD* to *DB* is therefore less than the ratio of the angle *ACB* to one half of the supplement of the angle *ABC*. But we have shown that the straight line *DB* is less than one half of *AC*. The ratio of *DA* to one half of *AC* is therefore very much less than the ratio of the angle *ACB* to one half of the supplement of the angle *ABC*. The ratio of *DA* to the whole of *AC* is therefore less than the ratio of the angle *ACB* to the whole supplement of the angle *ABC*. That is what we wanted to prove.

– **3** – *We also say that, if the angle* BAC *is not greater than half a right angle, then the ratio of* EC *to* CA *is less than the ratio of the angle* BAC *to the supplement of the angle* ABC.

We take up the shape of the triangle in order to not increase the drawings of lines; let it be the triangle *ABC*. We draw the circumscribed circle; let it be the circle *ABCG* and let its centre be *M*. We join the straight line *AM* and we extend it as far as *G*. We join the straight lines *MIB*, *MC* and *CG*. As the angle *ABC* is obtuse, the arc *ABC* is less than one half of a circle. It is therefore either greater than one quarter of a circle or not greater than one quarter of a circle.

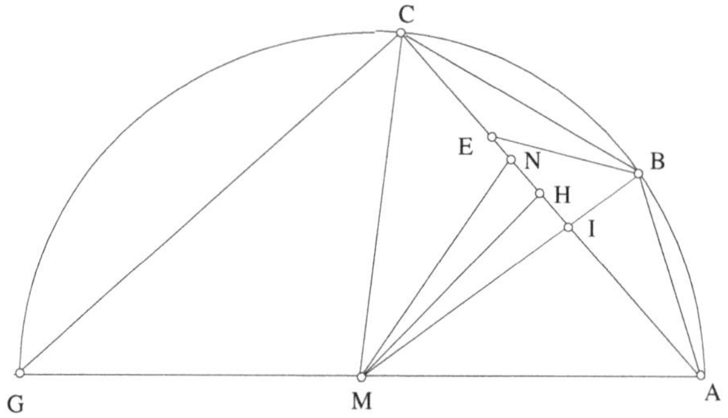

Fig. 3.3

If the arc *ABC* is not greater than one quarter of a circle, then the angle *AMC* is not greater than a right angle. The angle *AGC* is therefore not greater than one half of a right angle. If the angle *AMC* is not greater than a right angle, then each of the two angles *MAC* and *MCA* is not less than one half of a right angle; they are either equal to one half of a right angle or greater than one half of a right angle. But the angle *AGC* is either one half of a right angle or is less than one half of a right angle. Therefore, the angle

AGC is not greater than the angle *MAC*. But the angle *MIC* is greater than the angle *MAC*. Therefore, the angle *MIC* is greater than the angle *AGC*, so the angle *BIC* is less than the angle *ABC*. The angle *BEC* is therefore greater than the angle *BIC*. The point *E* therefore lies between the two points *I* and *C*. We drop a perpendicular from the point *M* onto *AC*; let it be *MH*. The point *H* therefore lies between the two points *I* and *C* as *MH* divides the arc *ABC* into two halves if it is extended. We make *HN* equal to *HI*.

If we take the point *M* as the centre and draw an arc of a circle at a distance *MN*, it can clearly be seen that the ratio of *CH* to *HN*, which is equal to *HI*, is greater than the ratio of the angle *CMH* to the angle *HMN*, which is equal to the angle *HMI*. The ratio of *CH* to *HI* is therefore greater than the ratio of the angle *CMH* to the angle *HMI*. By inversion, the ratio of *IH* to *HC* is therefore less than the ratio of the angle *IMH* to the angle *HMC*. By composition, the ratio of *IC* to *CH* is less than the ratio of the angle *IMC* to the angle *HMC*, and the ratio of *IC* to *CA* is less than the ratio of the angle *BMC* to the angle *CMA*. But the angle *BMC* is twice the angle *BAC*, and the angle *CMA* is twice the angle *AGC*, which is equal to the supplement of the angle *ABC*. The ratio of *IC* to *CA* is therefore less than the ratio of the angle *BAC* to the supplement of the angle *ABC*. The ratio of *EC* to *CA* is therefore very much less than the ratio of the angle *BAC* to the supplement of the angle *ABC*.

If the arc *ABC* is greater than one quarter of a circle, then the angle *AMC* is greater than a right angle. But we have stated in the hypothesis that the angle *BAC* is not greater than one half of a right angle, and it must therefore be either one half of a right angle or less than one half of a right angle. The arc *BC* is therefore either one quarter of a circle or less than one quarter of a circle. If the arc *BC* is one quarter of a circle, then the angle *BMC* is a right angle, the angle *MBC* is one half of a right angle, and the angle *BAC* is one half of a right angle. The angle *CBI* is therefore equal to the angle *BAC*. But the angle *ACB* is common, and therefore the angle *BIC* is equal to the angle *ABC* and the point *E* is the same as the point *I*.

We can show – as we did in the first part – that the ratio of *IC* to *CA* is less than the ratio of the angle *BAC* to the supplement of the angle *ABC*, and that the point *E* is the same as the point *I*. The ratio of *EC* to *CA* is therefore less than the ratio of the angle *BAC* to the supplement of the angle *ABC*.

If the arc *BC* is less than one quarter of a circle, then the angle *BMC* is less than a right angle. The angle *MBC* is therefore greater than one half of a right angle, and the angle *BAC* is less than one half of a right angle. Therefore, the angle *MBC* is greater than the angle *BAC*, the angle *CBE* is less than the angle *MBC*, and the point *E* lies between the two points *I* and *C*.

Using this method, we have shown that, in all cases, the ratio of *IC* to *CA* is less than the ratio of the angle *IMC* to the angle *CMA*, as the perpendicular *MH* always falls between the two points *I* and *C*, since the arc *CB* is greater than the arc *BA*. If the point *E* lies between the two points *I* and *C*, the straight line *EC* is less than the straight line *IC*, and the ratio of *EC* to *CA* will then be very much less than the ratio of the angle *BMC* to the angle *CMA*. But the angle *CMA* is twice the supplement of the angle *ABC* and the angle *BMC* is twice the angle *BAC*. The ratio of *EC* to *CA* is therefore less than the ratio of the angle *BAC* to the supplement of the angle *ABC*. If the angle *BAC* of the obtuse-angled triangle *ABC* is not greater than one half of a right angle, then the ratio of *EC* to *CA* is less than the ratio of the angle *BAC* to the supplement of the angle *ABC*. That is what we wanted to prove.

– **4** – *We also say that, if the angle* BAC *is greater than one half of a right angle, then the ratio of* EC *to* CA *may be greater than the ratio of the angle* BAC *to the supplement of the angle* ABC.[2]

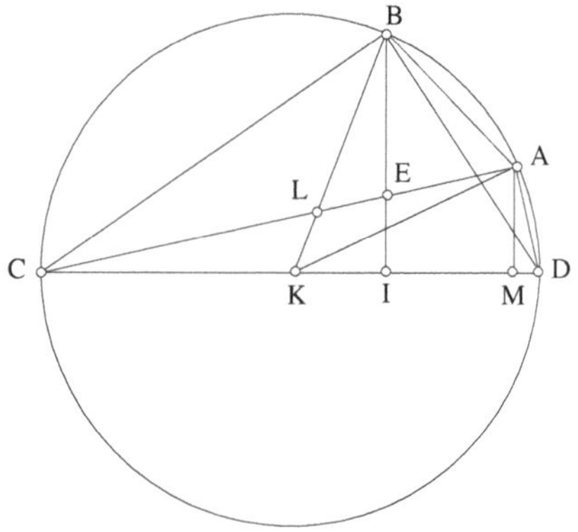

Fig. 3.4

Let us draw a circle on which lie *A*, *B*, *C* and *D*. Let *K* be its centre. We draw the diameter *CKD*, and we assume a point anywhere on the straight line *KD*; let it be *I*, from which we draw the perpendicular *IB*, and we join the straight lines *CB*, *BD* and *KB*. The triangle *DBC* will then be a right-angled triangle. The ratio of *IC* to *CD* is therefore greater than the ratio of the angle *BDC* to a right angle, as has been shown in the first proposition of this treatise. The ratio of *IC* to *CD* is therefore greater than the ratio of the

[2] Here, the author is considering a condition that is sufficient to verify the result.

angle *BKC* to two right angles; it is therefore greater than the ratio of the arc *BC* to the arc *CBD*. By inversion, the ratio of *DC* to *CI* is less than the ratio of the arc *DBC* to the arc *CB*. By separation, the ratio of *DI* to *IC* is therefore less than the ratio of the arc *DB* to the arc *BC*. The ratio of *DI* to *IC* is therefore equal to the ratio of a part of the arc *DB* to the arc *BC*. Let this part be the arc *AB*.[3] We join the straight lines *CEA*, *AB* and *AD*. The angle *DAC* is therefore a right angle, and the angle *ABC* is obtuse. As the angle *DAC* is a right angle, it is equal to the angle *EIC*. But the angle *ACD* is common to the two triangles *ADC* and *EIC*. Therefore, the angle *ADC* is equal to the angle *IEC*. But the angle *ADC* is equal to the supplement of the angle *ABC*, so the angle *IEC* is equal to the supplement of the angle *ABC*. The angle *BEC* is therefore equal to the angle *ABC*. We draw the perpendicular *AM*. The ratio of *CI* to *IM* will then be equal to the ratio of *CE* to *EA*. But the ratio of *CI* to *IM* is greater than the ratio of *CI* to *ID*, so the ratio of *CE* to *EA* is greater than the ratio of *CI* to *ID*. But the ratio of *CI* to *ID* is equal to the ratio of the arc *CB* to the arc *BA*. Therefore, the ratio of *CE* to *EA* is greater than the ratio of the arc *CB* to the arc *BA*. By inversion, the ratio of *AE* to *EC* is less than the ratio of the arc *AB* to the arc *BC*. Composing, the ratio of *AC* to *CE* is less than the ratio of the arc *ABC* to the arc *CB*. By inversion, the ratio of *EC* to *CA* is therefore greater than the ratio of the arc *BC* to the arc *CBA*. We join *AK*. The ratio of the arc *BC* to the arc *CBA* is equal to the ratio of the angle *BKC* to the angle *CKA*. The ratio of *EC* to *CA* is therefore greater than the ratio of the angle *BKC* to the angle *CKA*. But the angle *BKC* is twice the angle *BAC*, and the angle *CKA* is twice the angle *ADC* which is equal to the supplement of the angle *ABC*. The ratio of *EC* to *CA* is therefore less than the ratio of the angle *BAC* to the supplement of the angle *ABC*.

It must necessarily be the same if the ratio of the arc *AB* to the arc *BC* is greater than the ratio of *DI* to *IC*, as the ratio of *CI* to *ID* will be greater than the ratio of the arc *CB* to the arc *BA*. But the ratio of *CE* to *EA* is greater than the ratio of *CI* to *ID*, so the ratio of *CE* to *EA* is greater than the ratio of the arc *CB* to the arc *BA*. We can show – as we have shown previously – that the ratio of *EC* to *CA* is greater than the ratio of the angle *BAC* to the supplement of the angle *ABC*.

From all this we can show that, if the ratio of the arc *AB* to the arc *BC* is not less than the ratio of the straight line *DI* to the straight line *IC*, then the ratio of *EC* to *CA* is greater than the ratio of the angle *BAC* to the supplement of the angle *ABC*.

It is clear that this ratio is possible – that is, that the ratio of the arc *AB* to the arc *BC* is equal to, or greater than, the ratio of the straight line *DI* to

[3] This therefore defines a point *A* on the arc *BD*.

the straight line *IC*. This is absolutely possible as the two arcs *AB* and *BC* belong to the same circle. Therefore the ratio between them can be any ratio between two homogeneous magnitudes. The existence of this can effectively be shown by a construction, and what is more, by an easy construction: If the ratio of the straight line *DI* to the straight line *IC* is a ratio of halves – that is, if *DI* is one half of *IC*, or one half of one half, or one half of one half of one half, or so on to infinity – then the existence of a portion of the arc *DB* such that its ratio to the arc *BC* is this ratio is both possible and easy, and is also so if the arc *BC* is divided into two halves and its half into two halves until the division results in the part that is homologous to the part *DI* of *IC*. Now we make the arc *AB*, which is a part of *DB*, equal to this part obtained following the division. The ratio of the arc *AB* to the arc *BC* will then be equal to the ratio of the straight line *DI* to the straight line *IC*.

If *DI* is less than *IC*, then the arc *CB* is greater than the arc *BD*, and the angle *BAC* is then greater than one half of a right angle, the straight line *CB* is greater than the straight line *BA*, and the ratio of *EC* to *CA* is greater than the ratio of the angle *BAC* to the supplement of the angle *ABC*.

It follows from what we have shown, that it is possible to find an obtuse-angled triangle, in which the two straight lines enclosing the obtuse angle are different, such that a straight line – drawn from the obtuse angle to its chord and which encloses with the chord an angle equal to the obtuse angle adjacent to the greatest side – separates, from the chord of the obtuse angle adjacent to the greatest side, a straight line whose ratio to the entire chord is greater than the ratio of the angle which intercepts the greatest side to the supplement of the obtuse angle. That is what we wanted to prove.

– **5** – *We also say that any sector of a circle whose vertex is at the centre of the circle is equal to a complete circle.*

Example: In the circle *ABD*, let there be a sector *ABC* whose vertex lies at the point *C*, which is the centre of the circle.

I say that it is equal to a complete circle.

Proof: The ratio of the arc *AB* to the circumference of the circle is equal to the ratio of the sector *ABC* to the entire circle. But the arc *AB* and the circumference of the circle are two magnitudes of the same type, and superposition and difference are possible between them. Any ratio between two homogeneous magnitudes between which superposition and difference are possible exists between any two homogeneous magnitudes between which superposition and difference are possible. If the ratio is numeric, this is obvious. But if the ratio is not numeric, it is still between two homogeneous magnitudes and we have either found this ratio or we have not found this ratio, as a ratio is a concept which is specific to homogeneous magnitudes, and it does not depend on our knowledge of it, nor on the fact that we have found it. And among these ratios, there is none that is more suited to homogeneous magnitudes than any of the others. Therefore, the ratio of the arc *AB* to the circumference of the circle is equal to the ratio of a straight line to the diameter of the circle, which is the straight line *AD* – we have either found this line, or we have not found it. Let this straight line be the straight line *E* and let the straight line *GI* be a proportional mean between the straight line *E* and the straight line *AD*. On the straight line *GI*, we draw a circle whose diameter is *GI*; let it be *GHI*. Since the ratio of the straight line *E* to the straight line *GI* is equal to the ratio of *GI* to *AD*, the ratio of *E* to *AD* is equal to the ratio of the square of *GI* to the square of *AD*. But the ratio of the square of *GI* to the square of *AD* is equal to the ratio of the circle *GHI* to the circle *ABD*, so the ratio of the circle *GHI* to the circle *ABD* is equal to the ratio of the straight line *E* to the straight line *AD*. But the ratio of *E* to *AD* is equal to the ratio of the arc *AB* to the circumference of the circle, and the ratio of the arc *AB* to the circumference of the circle is equal to the ratio of the sector *ACB* to the whole circle. Therefore, the ratio of the circle *GHI* to the circle *ABD* is equal to the ratio of the sector *ACB* to the circle *ABD*. The circle *GHI* is therefore equal to the sector *ACB*. That is what we wanted to prove.

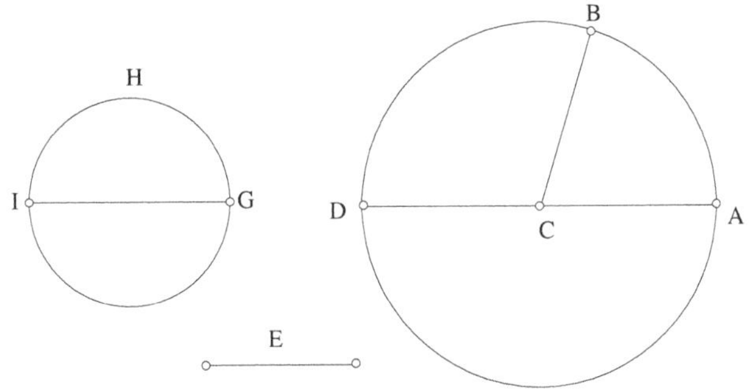

Fig. 3.5

– **6** – *We also say that, if we have two similar segments of two different circles, then the ratio of one to the other is equal to the ratio of the circle to the circle and is equal to the ratio of the square of the base of the segment to the square of the base of the segment.*

Example: Let the two segments *ABC* and *EGH* be two similar segments of two different circles.

I say that the ratio of the segment ABC *to the segment* EGH *is equal to the ratio of the circle to the circle.*

Proof: Let us complete the two circles, let the point *D* be the centre of the circle *ABC* and let the point *I* be the centre of the circle *EGH*. We join the straight lines *AD*, *CD*, *EI* and *HI*. As the two segments *ABC* and *EGH* are similar, the two angles *ADC* and *EIH* are equal. Therefore, the two triangles *ADC* and *EIH* are similar, and the ratio of the triangle *ADC* to the triangle *EIH* is equal to the ratio of the square of *AD* to the square of *EI* and equal to the ratio of the square of *AC* to the square of *EH*. But the ratio of the square of *AD* to the square of *EI* is equal to the ratio of the circle *ABC* to the circle *EGH*, and the ratio of the circle *ABC* to the circle *EGH* is equal to the ratio of the sector *ADCB* to the sector *EIHG*, since the ratio of any sector to its circle is equal to the ratio of any similar sector to its circle. The ratio of the sector *ADCB* to the sector *EIHG* is therefore equal to the ratio of the triangle *ADC* to the triangle *EIH* and is equal to the ratio of the remainder to the remainder. The ratio of the segment *ABC* to the segment *EGH* is therefore equal to the ratio of the sector *ADCB* to the sector *EIHG*. But the ratio of the sector to the sector is equal to the ratio of the circle to the circle, so the ratio of the segment *ABC* to the segment *EGH* is equal to the ratio of the circle *ABC* to the circle *EGH* and is equal to the ratio of the square of *AC* to the square of *EH*, since the ratio of these two squares is equal to the ratio of the square of the diameter to the square of the diameter. That is what we wanted to prove.

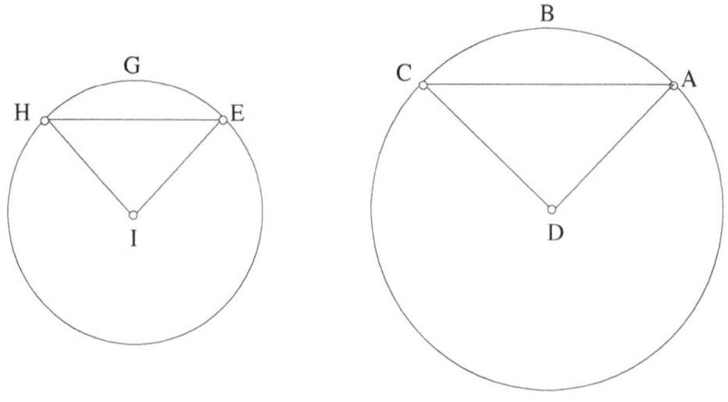

Fig. 3.6

– **7** – *We also say that, if any chord is drawn within a segment of a circle, and if a segment similar to the first segment is constructed on it, then the entire circumference*[4] *of the second segment will lie outside the first circle.*

[4] i.e. the arc.

Example: Let there be a segment *ABC* of the circle *ABCD*, within which is drawn the chord *AB* on which is constructed the segment *AEB* such that it is similar to the segment *ABC*.

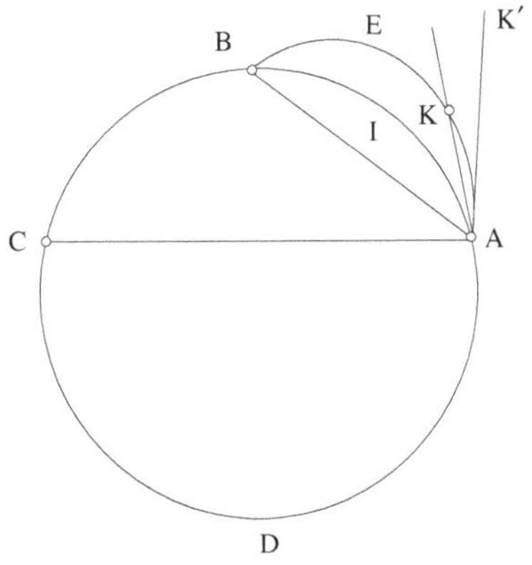

Fig. 3.7

I say that the circumference of the segment AEB *lies entirely outside the segment* ABC.

Proof: We draw the straight line *AK* as a tangent to the circle *ABCD*. The angle *KAC* is then equal to the angle that lies on the segment *ADC*, and the angle *KAB* is less than the angle which lies on the completed segment *AEB*. Therefore, the straight line *AK* cuts the circumference of the segment *AEB* and it is a tangent to the arc *ABC*; the straight line *AK* lies between the two arcs, the angle *EAI* lies outside the arc *AB*, the angle *EAB* is equal to the angle *EBA*, and the angle *IAB* is equal to the angle *IBA*. The remaining angle *EBI* is therefore equal to the angle *EAI*, the entire arc *AEB* therefore lies outside the segment *ABC* and, as a result, the arc *AEB* and the arc *AIB* enclose a figure of lune that lies entirely outside the segment *ABC*. That is what we wanted to prove.

<Propositions>

– **8** – Now that these lemmas have been proved, we say that, if a diameter is drawn in a circle, and if a point is marked anywhere on either half of the circumference, and if this point is joined to the ends of the diameter by two straight lines, and if a semicircle is constructed on each of these straight lines, then the sum of the two lunes formed by the circumferences of these two semicircles and the circumference of the first semicircle is equal to the triangle formed inside the semicircle.

Example: Let there be a circle *ABC* in which is drawn the diameter *AC*. We assume a point *B* on one half of *ABC*. We join the two straight lines *AB* and *BC* and we construct two semicircles on the two straight lines *AB* and *BC*, which are *ADB* and *BEC*.

I say that the sum of the two lunes ADBGA *and* BECHB *is equal to the triangle* ABC.

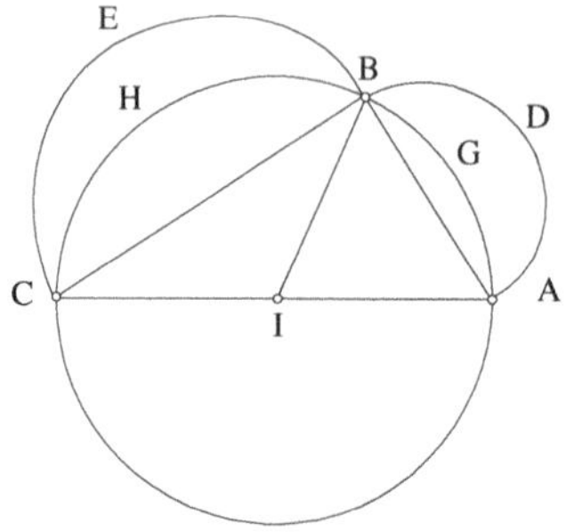

Fig. 3.8

Proof: The ratio of any circle to any other circle is equal to the ratio of the square of its diameter to the square of its diameter. Therefore the ratio of the sum of the two circles *ADB* and *BEC* to the circle *ABC* is equal to the ratio of the sum[5] of the two squares of *AB* and *BC* to the square of *AC*. But the sum of the two squares of *AB* and *BC* is equal to the square of *AC*, so the sum of the two circles *ADB* and *BEC* is equal to the circle *ABC*. The sum of the two halves of *ADB* and *BEC* is therefore equal to the semicircle *ABC*. Removing the two common segments *AGB* and *BHC* leaves the sum of the two lunes *ADBGA* and *BECHB* equal to the triangle *ABC*. That is what we wanted to prove.

– **9** – Let us take up the same figure. Let the point *I* be the centre of the circle, and join *IB*. If the two arcs *AB* and *BC* are equal, then the two segments *ADB* and *BEC* will be equal. The triangles *ABI* and *BIC* are equal, and therefore the two lunes are equal and each lune is equal to the adjacent triangle.[6]

[5] We occasionally add the word 'sum' for the purposes of translation.
[6] Lit.: which follows.

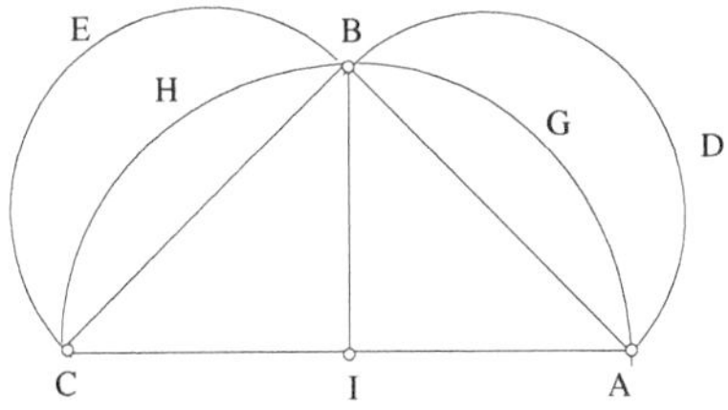

Fig. 3.9.1

If the two arcs *AB* and *BC* are different, then the two lunes are different. But as the two triangles are equal, we say that the sum of the smaller of the two lunes and a complete circle is equal to the triangle that is adjacent to the smaller lune, and that the remaining triangle plus the same circle is equal to the remaining lune.

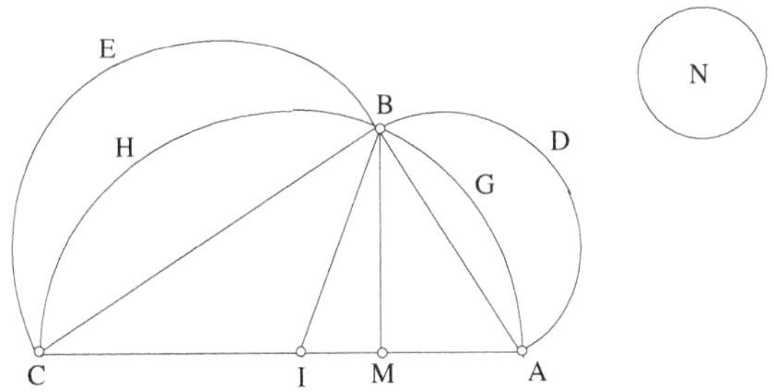

Fig. 3.9.2

Let the arc *AB* be less than the arc *BC*. From the point *B*, we draw the perpendicular *BM*. The ratio of *MA* to *AC* is then equal to the ratio of the square of *AB* to the square of *AC*, and is equal to the ratio of the semicircle *ADB* to the semicircle *ABC*. But we have shown in the first proposition in the lemmas that the ratio of *MA* to *AC* is less than the ratio of the angle *ACB* to a right angle. But the angle *AIB* is twice the angle *ACB*, so the ratio of *MA* to *AC* is less than the ratio of the angle *AIB* to two right angles. But the ratio of the angle *AIB* to two right angles is equal to the ratio of the sector *AIBG* to the semicircle *ABC*. The ratio of the semicircle *ADB* to the semicircle *ABC* is therefore less than the ratio of the sector *AIB* to the semicircle *ABC*, and therefore the sector *AIB* is greater than the semicircle *ADB*. But every sector is equal to a complete circle, every semicircle is equal to a complete circle, and the difference between the greater and the lesser of two different circles is also equal to a complete circle. Therefore, the sector *AIB* exceeds the semicircle *ADB* by a complete circle. Let this circle be the circle *N*. The sum of the semicircle *ADB* and the circle *N* is therefore equal to the sector *AIB*. Removing the common sector *BGA* leaves the lune *ADBGA* and the circle *N*, whose sum is equal to the triangle *AIB*. Yet, we have shown that the sum of the two lunes is equal to the triangle *ABC*. If the triangle *AIB* exceeds the lune *ADBGA* by the circle *N*, then the triangle *BIC* is less than the lune *BECHB* by the circle *N*, and the sum of the triangle *BIC* and the circle *N* is equal to the lune *BECHB*. That is what we wanted to prove.

<Conclusion> It clearly follows from that which we have shown at the start of this section, that any lune constructed on a quarter of a circle such that its circumference is a semicircle is equal to the right-angled triangle inscribed within the quarter circle.

– **10** – Let us draw another circle – on which lie *A*, *B* and *C*, and whose centre is *E* – in which any chord is drawn cutting off a segment that is less than a semicircle; let it be the segment *ABC*. We assume any point on the arc *ABC* – let it be the point *B* – and we join the two straight lines *AB* and *BC*. On each of the straight lines *AB* and *BC*, we construct a segment similar to the segment *ABC*; let these two segments be *ADB* and *BIC*. We draw two straight lines *BN* and *BO* such that each of the angles *BNA* and *BOC* is equal to the angle *ABC*. We join the straight lines *EA*, *EC*, *EN* and *EO*.

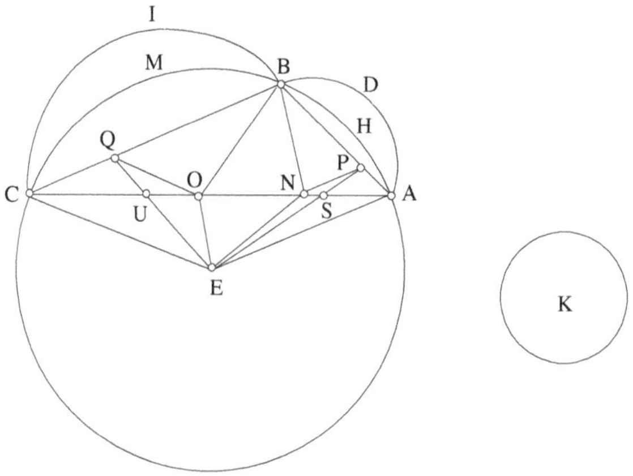

Fig. 3.10

I say that the sum of the two lunes ADBHA *and* BICMB *and a complete circle is equal to the sum of the triangle* ABC *and the triangle* *ENO*.

Proof: The ratio of *CA* to *AN* is equal to the ratio of the square of *CA* to the square of *AB*, it is equal to the ratio of the square of the diameter of the circle *ABC* to the square of the diameter of the circle *ADB*, and is equal to the ratio of the segment *ABC* to the segment *ADB*. Similarly, the ratio of *AC* to *CO* is equal to the ratio of the square of *AC* to the square of *CB*, and is equal to the ratio of the segment *ABC* to the segment *BIC*. Therefore, the ratio of *AC* to the sum of the two straight lines *AN* and *CO* is equal to the ratio of the segment *ABC* to the sum of the two segments *ADB* and *BIC*. But the ratio of *AC* to the sum of the two straight lines *AN* and *CO* is equal to the ratio of the triangle *AEC* to the sum of the two triangles *AEN* and *CEO*, so the ratio of the segment *ABC* to the sum of the two segments *ADB* and *BIC* is equal to the ratio of the triangle *AEC* to the sum of the two triangles *AEN* and *CEO*, and is equal to the ratio of the whole to the whole. The ratio of *AC* to the sum of *AN* and *CO* is therefore equal to the ratio of the sector *AECB* to the sum of the segments *ADB* and *BIC* and the two triangles *AEN* and *CEO*. But *AC* is greater than the sum of the two straight lines *AN* and *CO*, so the sector *AECB* is greater than the sum of the two segments *ADB* and *BIC* and the two triangles *AEN* and *CEO*. But the ratio of *AC* to the sum of the two straight lines *AN* and *CO* is the ratio of the arc *ABC* to one of its parts, and is equal to the ratio of the sector *AECB* to the sector whose base is this arc and which forms a part of the arc *ABC*. This sector whose base is a part of the arc *ABC* is therefore equal to the sum of the two segments *ADB* and *BIC* and the two triangles *AEN* and *CEO*. But the amount by which the sector *AEC* exceeds this sector is a sector within the circle *ABC* whose vertex is at *E*. It is therefore equal to a complete circle. Let this circle be the circle *K*. The sector *AECB* is therefore equal to the sum of the segments *ADB* and *BIC*, the triangles *AEN* and *CEO*, and the circle *K*. Removing the common parts, which are the segments *AHB* and *BMC* and the two triangles *AEN* and *CEO*, leaves the two lunes *ADBHA* and *BICMB* and the circle *K*, the sum of which is equal to that of the triangles *ABC* and *ENO*. We draw the straight line *NP* parallel to the straight line *AE*, and we join *ESP*. The triangle *ASP* will then be equal to the triangle *ESN*. We draw the straight line *OQ* parallel to the straight line *EC*, and we join *EUQ*. The triangle *CUQ* will then be equal to the triangle *EUO*. Removing the two triangles *ASP* and *CUQ* from the triangle *ABC*, and adding the two triangles *ESN* and *EUO* to the triangle *ENO*, makes the quadrilateral *EPBQ* equal to the sum of the two triangles *ABC* and *ENO*. The sum of the two lunes *ADBHA* and *BICMB* and the circle *K* is therefore equal to the quadrilateral *EPBQ*. That is what we wanted to prove.

– **11** – Let us take up the same figure and join the straight line *ELB*. If the two arcs *AB* and *BC* are equal, then the two straight lines *AB* and *BC* are equal, the two straight lines *AN* and *CO* are equal, the two straight lines *PB* and *QB* are equal, the two triangles *PEB* and *QEB* are equal, and the two lunes are equal. The sum of each of these lunes and a complete circle equal to the semicircle *K* is equal to the triangle adjacent to the lune, either the triangle *PEB* or the triangle *QEB*.

If the two arcs *AB* and *BC* are different, and if the arc *AB* is the lesser of the two arcs, then the sum of the lune *ADBHA* and a complete circle is also equal to the triangle *PBE*.

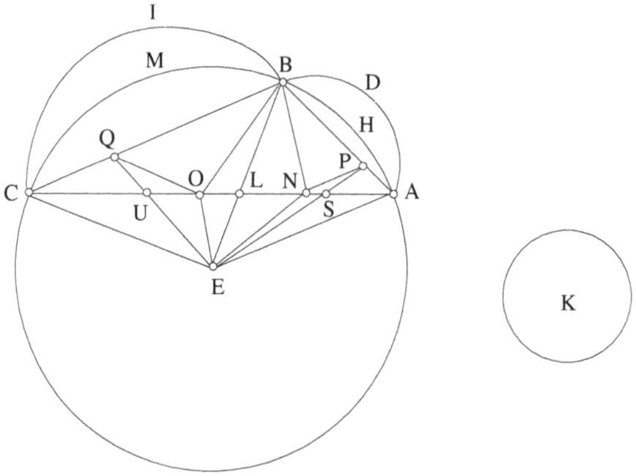

Fig. 3.11

Proof: We have shown that the ratio of *NA* to *AC* is less than the ratio of the angle *BCA* to the supplement of the angle *ABC*. But the angle *BEA* is twice the angle *BCA* and the angle *AEC* is twice the supplement of the angle *ABC*. The ratio of *NA* to *AC* is therefore less than the ratio of the angle *BEA* to the angle *AEC*; so it is less than the ratio of the sector *BEA* to the sector *AEC*. But the ratio of *NA* to *AC* is equal to the ratio of the square of *BA* to the square of *AC*, and is equal to the ratio of the segment *ADB* to the segment *ABC*. The ratio of the segment *ADB* to the segment *ABC* is therefore less than the ratio of the sector *BEA* to the sector *AEC*. The ratio of the segment *ADB* to the segment *ABC* is therefore equal to the ratio of a sector that is less than the sector *BEA* to the sector *AEC*. The ratio of the sector that is less than the sector *BEA* to the sector *AEC* is therefore equal to the ratio of *NA* to *AC*, it is equal to the ratio of the triangle *AEN* to the triangle *AEC*, and is equal to the ratio of the amount by which the small sector exceeds the triangle *AEN* to the segment *ABC*. The ratio of the segment *ADB* to the segment *ABC* is therefore equal to the ratio of the amount by which the small sector exceeds the triangle *AEN* to the segment *ABC*. The amount by which the small sector exceeds the triangle *AEN* is therefore equal to the segment *ADB*. The segment *ADB* plus the triangle *AEN* – that is, the triangle *APE* – is equal to the small sector whose ratio to the sector

AEC is equal to the ratio of *NA* to *AC*. But the amount by which the sector *AEB* exceeds the small sector is a complete circle, and therefore the sum of the small sector and the circle is equal to the sector *AEB*. The sum of the segment *ADB* and the triangle *AEP* and the complete circle is equal to the sector *AEB*. Removing the common parts, which are the segment *AHB* and the triangle *APE*, leaves the lune *ADBHA* and the complete circle, whose sum is equal to the triangle *PEB*.

If the angle *BAC* is not greater than one half of a right angle, then the ratio of *OC* to *CA* is less than the ratio of the angle *BAC* to the supplement of the angle *ABC*. We can show – as we have shown for the straight line *NA* – that the sum of the lune *BICMB* and a complete circle is equal to the triangle *QEB*. But we have shown that, in all cases, the sum of the two lunes and a complete circle is equal to the quadrilateral *PEQB*, which is the sum of the two triangles *PEB* and *BEQ*. The sum of the two circles that are associated with the two lunes is therefore equal to the circle that is associated with the sum of the two lunes.

It follows from that which we have shown that the sum of each of the two lunes and a complete circle is equal to a known triangle – provided that the angle *BAC* is not greater than one half of a right angle. That is what we wanted to prove.

– **12** – Let us take up the same figure, let the angle *BAC* be greater than one half of a right angle, and let the ratio of *OC* to *CA* be greater than the ratio of the angle *BAC* to the supplement of the angle *ABC*.[7] The point *O* will then lie outside the triangle *BEC* since we have shown that the ratio of *LC* to *CA* is less than the ratio of the angle *BAC* to the supplement of the angle *ABC*.

I say that the lune BICMB *exceeds the triangle* QEB *by a complete circle, and that the lune* ADBHA *plus the circle by which the quadrilateral* BPEQ *exceeds the sum of the two lunes, plus the circle by which the lune* BICMB *exceeds the triangle* QEB *is equal to the triangle* PEB.

Proof: The ratio of *OC* to *CA* is equal to the ratio of the square of *BC* to the square of *CA*, it is equal to the ratio of the segment *BIC* to the segment *CBA*, and is equal to the ratio of the triangle *OEC* to the triangle *CEA*, and is equal to the ratio of the sum of the segment *BIC* and the triangle *EOC* to the sector *ECBA*. But the ratio of the angle *BAC* to the supplement of the angle *ABC* is equal to the ratio of the angle *BEC* to the angle *CEA*, and is equal to the ratio of the sector *BECM* to the sector *ECBA*. The ratio of the sum of the segment *BIC* and the triangle *OEC* – that is, the triangle *QEC* – to the sector *ECBA* is greater than the ratio of the sector *BECM* to

[7] See Proposition 4.

the sector *ECBA*. It is therefore equal to the ratio of the sector that is greater than the sector *BECM* to the sector *ECBA*. This large sector exceeds the sector *BECM* by a complete circle. Let this circle be the circle *G*. The sum of the segment *BIC* and the triangle *QEC* is then equal to that of the sector *BECM* and the circle *G*. Removing the common parts, which are the segment *BMC* plus the triangle *QEC*, leaves the lune *BICMB*, which is equal to the triangle *BEQ* plus the circle *G*. The lune *BICMB* therefore exceeds the triangle *BEQ* by the circle *G*.

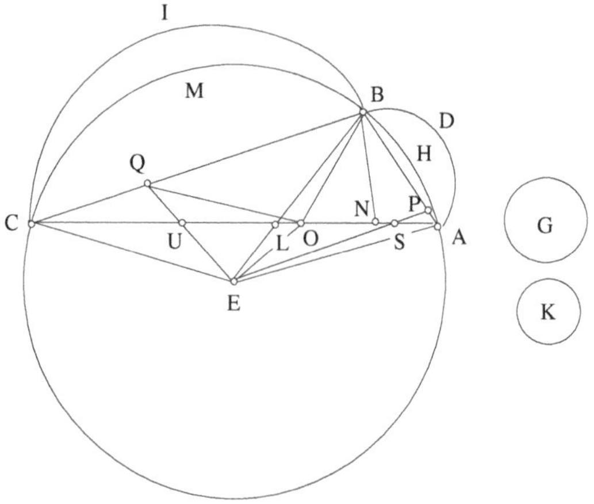

Fig. 3.12

Similarly, we have shown in the tenth proposition of this treatise that the two lunes *ADBHA* and *BICMB* plus the circle *K* are equal to the sum of the two triangles *BPE* and *BEQ*. Therefore, the two triangles *PEB* and *BEQ* exceed the two lunes by the circle *K*. If the triangle *BEQ* is less than the lune *BICMB* by the circle *G*, then the triangle *BPE* exceeds the lune *ADBHA* plus the circle *K* by the circle *G*. Therefore, the sum of the lune *ADBHA* and the two circles *K* and *G* is equal to the triangle *PEB*. That is what we wanted to prove.

– **13** – In order for this notion to be obvious, and for it to be rational, we assume that there is a numerical ratio between the two arcs. We draw a circle on which lie *A*, *B* and *C*; let *D* be its centre. We draw the diameter *ADC*, we draw the chord *AB*, and we make it equal to the half-diameter. We join *BC* and *BD*, and we construct a semicircle on each of the straight lines *AB* and *BC*. Let these be the semicircles *AEB* and *BHC*. We make the circle *K* equal to one part of twenty-four parts of the circle *ABC*. This is possible and easy. Make the circle *M* equal to one part of twelve parts of the circle *ABC*.

I say that the sum of the lune AEBNA *and the circle* K *is equal to the triangle* ADB, *that the sum of the triangle* BDC *and the circle* K *is equal to the lune* BHCIB, *and that the sum of the lune* AEBNA *and the circle* M *is equal to the lune* BHCIB.

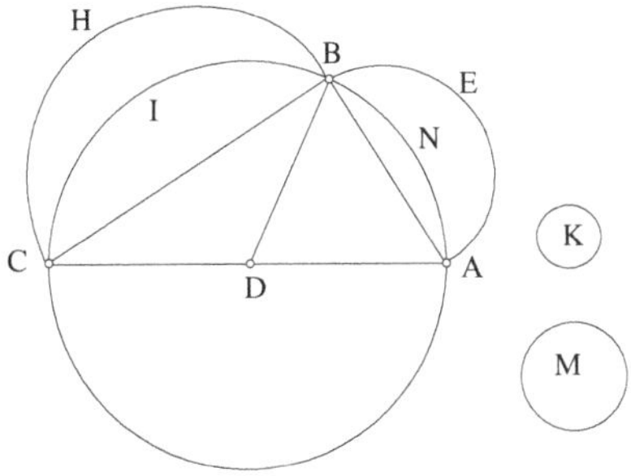

Fig. 3.13

Proof: The square of *AB* is one quarter of the square of *AC*, and therefore the semicircle *AEB* is one quarter of the semicircle *ABC*. But, since *AB* is equal to the half-diameter, the sector *ADB* is one sixth of the circle, and one half of the circle *ABC* is therefore equal to three times the sector *ADB* and four times the semicircle *AEB*. The sector *ADB* therefore exceeds the semicircle *AEB* by one half of one sixth of the semicircle *ABC*. But one half of one sixth of the half is one part of twenty-four parts of the whole. The sector *ADB* therefore exceeds the semicircle *AEB* by the circle *K*, and therefore the sum of the semicircle *AEB* and the circle *K* is equal to the sector *ADB*. Removing the common segment *ANB* leaves the lune *AEBNA* and the circle *K*, whose sum is equal to the triangle *ADB*. But, if the sum of the lune *AEBNA* and the circle *K* is equal to the triangle *ADB*, then the triangle *ADB* exceeds the lune *AEBNA* by the circle *K*. But we have shown that the triangle *ABC* is equal to the sum of the two lunes *AEBNA* and *BHCIB*, and the triangle *BDC* is therefore less than the lune *BHCIB* by the circle *K*. The sum of the triangle *BDC* and the circle *K* is therefore equal to the lune *BHCIB*. So the lune *BHCIB* exceeds the triangle *BDC* by the circle *K*. But the triangle *BDC* exceeds the lune *AEBNA* by the circle *K* as the triangle *BDC* is equal to the triangle *ADB*. The lune *BHCIB* therefore exceeds the lune *AEBNA* by twice the circle *K*. But the circle *M* is twice the circle *K*, and therefore the lune *BHCIB* exceeds the lune *AEBNA* by the circle *M*. That is what we wanted to prove.

<**14**> Let us draw another circle on which lie *A*, *B* and *C*, and in which is drawn a chord equal to the side of an equilateral triangle inscribed within the circle; let it be the straight line *AC*, and let the smaller arc be *ABC*. We divide it into two halves at the point *B*, we join *AB* and *BC*, and we draw two straight lines *BD* and *BE* such that each of the two angles *BDA* and *BEC* is equal to the angle *ABC*. On each of the straight lines *AB* and *BC*, we construct a segment similar to the segment *ABC*; let them be the two segments *AHB* and *BKC*. We make the circle *S* one ninth of the circle *ABC*, and the circle *U* one half of the circle *S*. Let the point *G* be the centre of the circle *ABC*. We join *AG*, *CG*, *BG*, *DG* and *EG*, we draw the straight line *DP* parallel to the straight line *GA* and the straight line *EQ* parallel to the straight line *GC*, and we join *PG* and *QG*.

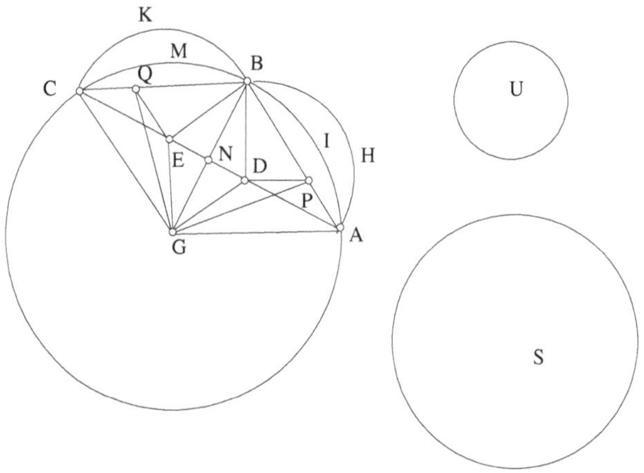

Fig. 3.14

I say that the sum of the two lunes AHBIA *and* BKCMB *and the circle* S *is equal to the quadrilateral* GPBQ, *and that the sum of each of the lunes and the circle* U *is equal to one of the triangles* PBG *or* QBG.

Proof: The arc *ABC* is one third of the circle, and therefore the arc *AB* is one sixth of the circle. The square of *AB* is then one third of the square of *AC*, and the straight line *DA* is one third of *AC*. Similarly, *EC* is one third of *AC*, therefore the sum of the two straight lines *DA* and *EC* is two thirds of *AC*, and the sum of the squares of *AB* and *BC* is two thirds of the square of *AC*. The sum of the two segments *AHB* and *BKC* is two thirds of the segment *ABC*. But the sum of the two triangles *AGD* and *EGC* is two thirds of the triangle *AGC*, so the sum of the two segments *AHB* and *BKC* and the two triangles *AGD* and *CGE* is equal to two thirds of the sector *AGCB*. But the sector *AGCB* is one third of the circle *ABC* and the circle *S* is one ninth of the circle *ABC*, so the circle *S* is one third of the sector *AGCB*. The sum of the two segments *AHB* and *BKC*, the two triangles *AGD* and *EGC*, and the circle *S* is equal to the sector *AGCB*. But the sum of the two triangles *AGD* and *EGC* is equal to the sum of the two triangles *AGP* and *CGQ*. The sum of the two segments *AHB* and *BKC*, the two triangles *AGP* and *CGQ*, and the circle *S* is equal to the sector *AGCB*. Removing the common parts leaves the two lunes *AHBIA* and *BKCMB* and the circle *S*, whose sum is equal to the quadrilateral *BPGQ*. But, since the two straight lines *AB* and *BC* are equal, the two lunes are equal, the two triangles *ABN* and *CBN* are equal, the two triangles *GDN* and *GEN* are equal, the two triangles *PBG* and *BGQ* are equal, the triangle *GPB* is equal to the sum of the two triangles *ABN* and *GDN*, and the triangle *GQB* is equal to the sum of the triangles *CBN* and *GEN*. Therefore, the sum of each of the two lunes and the circle *U*, which is one half of the circle *S*, is equal to one of the two triangles *PBG* and *BGQ*, and is equal to one of the two triangles *ABN* or *CBN* plus one of the two triangles *GDN* or *GEN*. That is what we wanted to prove.

It follows clearly from this proof that, if an arc of a circle that is less than one quarter of the circle is intercepted by a straight line, and if a segment is constructed on this straight line that is similar to the segment contained within twice this arc, then the sum of the generated lune and a known circle will be equal to a known triangle.[8]

– **15** – Let us draw another circle on which lie *A*, *B* and *C*, and within which is drawn the chord *AC* which cuts off one third, and draw *AB* which cuts off one quarter. Let *D* be the centre of the circle. We join the straight lines *AD*, *CD*, *BID* and *BC* and, on each of the two straight lines *AB* and *BC* we construct two segments similar to the segment *ABC*; let them be *AEB* and *BHC*. We draw *BL* such that the angle *BLC* is equal to the angle *ABC* and join *DL*. We make the circle *NP* equal to the circle *ABC*, and we draw its diameter; let it be *NP*. We make the ratio of the square of *NP* to the square of the straight line *QS* equal to the ratio of *AC* to *IL*. We draw a circle with *QS* as its diameter; let it be the circle *QS*. The ratio of the circle *QS* to the circle *NP* is therefore equal to the ratio of *IL* to *AC*.

I say that the sum of the two lunes AEB *and* BHC *and the circle* QS *is equal to the sum of the two triangles* ABC *and* DIL.

[8] This result was demonstrated in Proposition 11a. In Proposition 15, it is shown that, in the case where $AB = \frac{1}{6}$ of a circle, the 'known circle' is equal to $\frac{1}{18}$ of the circle *ABC*.

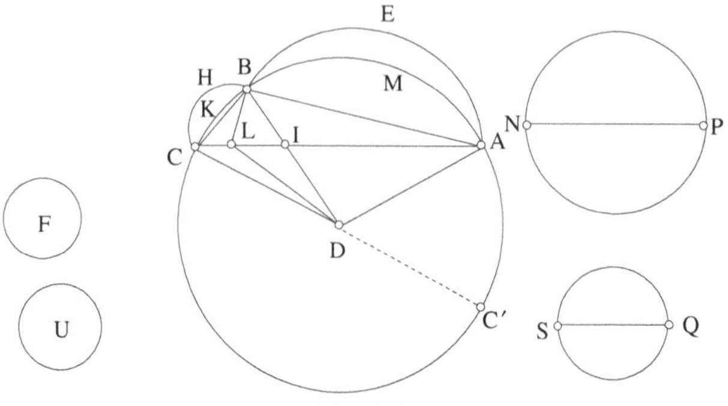

Fig. 3.15

Proof: The arc *ABC* is one third of the circle, and therefore the angle *ABC* is a right angle[9] plus one third, and the angle *DAC* is one third of a right angle. The arc *AB* is one quarter of a circle, and therefore the angle *ADB* is a right angle, and the angle *AIB* is equal to the sum of the two angles *ADI* and *DAI*. Therefore, the angle *AIB* is a right angle plus one third. But the angle *ABC* is a right angle plus one third, as it is inscribed within one third of a circle. The angle *AIB* is therefore equal to the angle *ABC* and, as a result, the product of *CA* and *AI* is equal to the square of *AB*. Similarly, the product of *AC* and *CL* is equal to the square of *CB*, as the angle *BLC* is equal to the angle *ABC*. The ratio of the sum of the two straight lines *IA* and *LC* to the straight line *AC* is therefore equal to the ratio of the sum of the squares of *AB* and *BC* to the square of *AC*. It is equal to the ratio of the sum of the two segments *AEB* and *BHC* to the segment *ABC*, and to the ratio of the sum of the two triangles *ADI* and *LDC* to the triangle *ADC*. It is therefore equal to the ratio of the sum of the two segments *AEB* and *BHC* and the two triangles *ADI* and *LDC* to the sector *ADCB*. The ratio of the sum of the two straight lines *IA* and *LC* to the straight line *AC* is equal to the ratio of the sum of the two segments *AEB* and *BHC* and the two triangles *ADI* and *LDC* to the circle *NP*. But the ratio of the circle *QS* to the circle *NP* is equal to the ratio of *IL* to *AC*, so the ratio of the sum of the straight lines *IA*, *IL* and *LC* to the straight line *AC* is equal to the ratio of the sum of the two segments *AEB* and *BHC* and the two triangles *ADI* and *LDC* and the circle *QS* to the circle *NP*, which is equal to the sector *ADC*. The sum of the two segments *AEB* and *BHC*, the two triangles *ADI* and *LDC*, and the circle *QS* is equal to the sector *ADCB*. Removing the common parts, which are the two segments *AMB* and *BKC*,

[9] We add the word 'angle' for the purposes of translation, in this proposition and in the following ones.

and the two triangles *ADI* and *LDC*, leaves the two lunes *AEBMA* and *BHCKB* and the circle *QS*, whose sum is equal to that of the two triangles *ABC* and *DIL*.

Since the straight line *AC* is the side of an equilateral triangle, its square is three quarters of the square of the diameter of the circle, and since the straight line *AB* is the side of the square, its square is one half of the square of the diameter of the circle. The square of *AB* is therefore two thirds of the square of *AC*. The straight line *IA* is then two thirds of the straight line *AC*, and the straight line *IC* is one third of *AC*. But each of the angles *AIB* and *BLC* is equal to a right angle plus one third, so each of the angles *BIL* and *BLI* is two thirds of a right angle. The triangle *IBL* is thus equilateral, and the ratio of *BL* to *LC* is equal to the ratio of *AB* to *BC*. But *AB* is greater than *BC*, as the arc *AB* is a quarter of a circle and the arc *ABC* is one third of it, so the arc *AB* is three times the arc *BC*. The straight line *BL* is greater than the straight line *LC*, and the straight line *IL* is therefore greater than the straight line *LC*. But the straight line *IC* is one third of *AC*, so the straight line *IL* is greater than one sixth of *AC*, and the circle *QS* is greater than one sixth of the sector *ADCB*, and it is therefore greater than one part of eighteen parts of the circle *ABC*.

We now make the circle *F* one part of thirty-six parts of the circle *ABC*; the circle *F* will be very much less than the circle *QS*. We make the circle *U* equal to the amount by which the circle *QS* exceeds the circle *F*.

I say that the sum of the lune AEBMA *and the circle* F *is equal to the triangle* ABI, *and that the sum of the lune* BHCKB *and the circle* U *is equal to the sum of the two triangles* BIC *and* IDL.

Proof: The sector *ADBM* is one quarter of the circle, and the sector *ADCB* is one third of the circle. Therefore, the sector *ADBM* is three quarters of the sector *ADCB*. But the square of *AB* is two thirds of the square of *AC*, so the segment *AEB* is two thirds of the segment *ABC*. But the straight line *AI* is two thirds of the straight line *AC*. Therefore, the triangle *ADI* is two thirds of the triangle *ADC*, and the sum of the segment *AEB* and the triangle *ADI* is equal to two thirds of the sector *ADCB*. The sector *ADBM* therefore exceeds the segment *AEB* and the triangle *ADI* by one part of twelve parts of the sector *ADCB*. But the sector *ADCB* is one third of the circle. Therefore, the one part of twelve <parts> of the sector *ADCB* is one part of thirty-six parts of the circle. The sum of the segment *AEB* and the triangle *ADI* and the circle *F* is equal to the sector *ADBM*. Removing the common parts, which are the segment *AMB* and the triangle *ADI*, leaves the lune *AEBMA* and the circle *F* – which is one part of thirty-six parts of the circle *ABC* – whose sum is equal to the triangle *ABI*. But the sum of the two lunes and the circle *QS* is equal to the sum of the two triangles *ABC* and *DIL*, and the sum of the two circles *F* and *U* is equal to the circle *QS*; therefore, the sum of the two lunes and the two circles *F* and *U* is equal to the sum of the two triangles *ABC* and *DIL*.

But as we have shown that the sum of the lune *AEBMA* and the circle *F* is equal to the triangle *ABI*, then the sum of the lune *BHCKB* and the circle *U* is equal to the sum of the two triangles *BIC* and *DIL*.

But, since the circle *QS* is greater than one part of eighteen parts of the circle *ABC*, and the circle *F* is one part of thirty-six parts of the circle *ABC*, the circle *U* will be greater than the circle *F*. The sum of each of the two lunes and a known circle is therefore equal to a known triangle. That is what we wanted to prove.

– **16** – Let us also draw a circle on which lie *A*, *B* and *C*. Let *D* be its centre. We draw a diameter *ADC* and we divide *AD* into two halves at the point *E*. We draw the perpendicular *EB* and we join the straight lines *CB*, *BA* and *BD*.

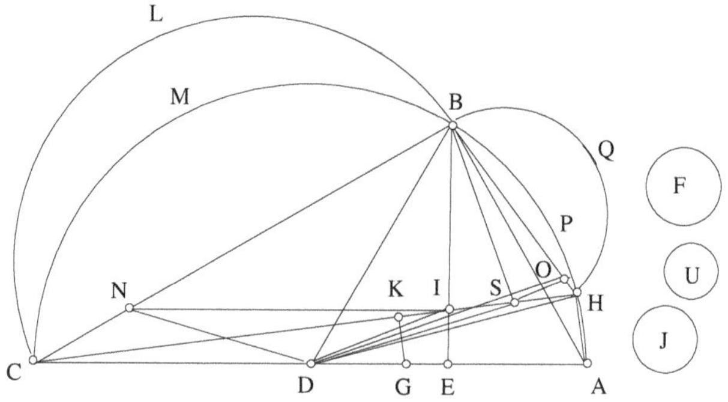

Fig. 3.16

As *AE* is equal to *ED*, and *EB* is perpendicular, *AB* will be equal to *BD*. Therefore, the triangle *ABD* is equilateral, the arc *AB* is one sixth of a circle, and the arc *BC* is one third of a circle. We divide the arc *AB* into two halves, and its half into two halves. Let the arc *AH* be one quarter of the arc *AB*, and let the arc *HB* be three quarters of the arc *AB*. It is therefore one quarter plus one eighth of the arc *BC*, so it is greater than one third of the arc *BC*. But the straight line *AE* is one third of the straight line *EC*. We make the straight line *AG* equal to one quarter plus one eighth of the straight line *GC* and we join the straight lines *DH*, *BH*, *HA* and *HIC*. The angle *BIC* will then be equal to the angle *HBC*, as has been shown in the fourth proposition of this treatise. From the point *G*, we drop the perpendicular *GK* onto the straight line *HC*. This will be parallel to the straight line

AH as the angle *AHC* is a right angle. The point *K* will lie between the two points *I* and *C* as the angle *EIC* is acute. But as *KG* is parallel to *HA*, the ratio of *HK* to *KC* is equal to the ratio of *AG* to *GC*, and the ratio of *AG* to *GC* is equal to the ratio of the arc *HB* to the arc *BC*. Therefore, the ratio of *HK* to *KC* is equal to the ratio of the arc *HB* to the arc *BC*. By composition, the ratio of the straight line *HC* to the straight line *CK* is equal to the ratio of the arc *HBC* to the arc *CB*. By inversion, the ratio of the straight line *KC* to the straight line *CH* will then be equal to the ratio of the arc *BC* to the arc *CBH*. Therefore, the ratio of *IC* to *CH* is greater than the ratio of the arc *BC* to the arc *CBH*. But the ratio of the arc *BC* to the arc *CBH* is equal to the ratio of the angle *BDC* to the angle *CDH*. It is equal to the ratio of the angle *BHC* to the supplement of the angle *HBC* and is equal to the ratio of the sector *BDCM* to the sector *CDHB*. The ratio of *IC* to *CH* is greater than the ratio of the sector *BDCM* to the sector *CDHB*. We draw the straight line *BS* such that the angle *BSH* is equal to the angle *HBC*, and we draw the straight line *SO* parallel to the straight line *DH*. We join *DO*, and we draw the straight line *IN* parallel to the straight line *DC*. We join the straight lines *DN*, *DI* and *DS* and we make the circle *F* equal to the sector *CDHB*. This is possible as the ratio of the sector *CDHB* to the circle *ABC* is a known ratio. We make the ratio of the circle *U* to the circle *F* equal to the ratio of the straight line *SI* to the straight line *HC*, and we make the ratio of the circle *J* to the circle *F* equal to the ratio of the straight line *IK* to the straight line *HC*. On the two straight lines *HB* and *BC*, we construct two segments that are similar to the segment *ABC*. Let them be the segments *HQB* and *BLC*. The sum of the two lunes *HQBPH* and *BLCMB* and the circle *U* is equal to the sum of the two triangles *DOB* and *DBN*, as was shown in the previous proposition. But, as the ratio of *KC* to *CH* is equal to the ratio of the arc *BC* to the arc *CBH*, the ratio of *IK* to *CH* is the amount by which the ratio of *IC* to *CH* exceeds the ratio of the arc *BC* to the arc *CBH*. But the ratio of *KC* to *CH* is equal to the ratio of the sector *BDCM* to the sector *CDHB*, so the ratio of *IK* to *CH* is the amount by which the ratio of *IC* to *CH* exceeds the ratio of the sector *BDCM* to the sector *CDHB*. But the ratio of *IK* to *CH* is equal to the ratio of the circle *J* to the circle *F*, which is equal to the sector *CDHB*. The ratio of *IC* to *CH* is therefore equal to the ratio of the sector *BDCM* plus the circle *J* to the sector *CDHB*. But the ratio of *IC* to *CH* is equal to the ratio of the square of *BC* to the square of *CH*, it is equal to the ratio of the segment *BLC* to the segment *CBH*, and equal to the ratio of the triangle *IDC* to the triangle *CDH*, and to the ratio of the segment *BLC* plus the triangle *IDC* – that is, the triangle *DNC* – to the sector *CDHB*. The ratio of the segment *BLC* plus the triangle *DNC* to the sector *CDHB* is therefore equal to the ratio of the sector *BDCM* plus the circle *J* to the sector *CDHB*. The sum of the segment *BLC* and the triangle *DNC* is therefore equal to that of the sector *BDCM* and the circle *J*. Removing the common parts, which are the segment *BMC* and the triangle *DNC*, leaves the lune *BLCMB* equal to the triangle *BDN* plus the circle *J*.

But, since the sum of the two triangles *DOB* and *DNB* is equal to the sum of the two lunes *HQBPH* and *BLCMB* and the circle *U*, and since the triangle *DNB* is less than the lune *BLCMB* by the circle *J*, the triangle *DOB* exceeds the sum of the lune *HQBPH* and the circle *U* by the circle *J*. The sum of the lune *HQBPH* and the circles *U* and *J* is therefore equal to the triangle *DOB*. That is what we wanted to prove.

– **17** – Let us draw another circle on which lie *A*, *B* and *C*, and draw the chord *AC* which cuts off one third, and the chord *AB* which cuts off one quarter. Let *D* be the centre of the circle. We join the straight lines *AD*, *CD*, *BID* and *BC* and we construct a segment of a circle similar to the segment *ABC* on the straight line *AB*. Let it be the segment *AEB*. We make the circle *K* equal to one part of thirty-six parts of the circle *ABC*. Then the sum of the lune *AEBHA* and the circle *K* will be equal to the triangle *ABI*, as we have shown in the fifteenth proposition of this treatise. We construct a semicircle on the straight line *AB*; let it be *ANB*.

I say firstly that the arc ANB *lies entirely outside the arc* AEB.

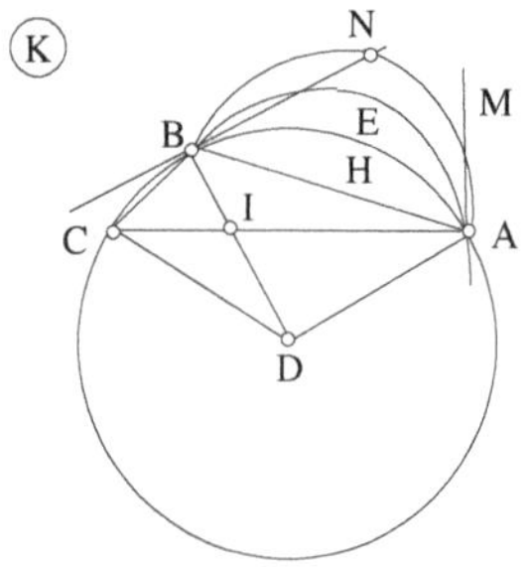

Fig. 3.17.1

Proof: We draw the straight line *AM* tangent to the arc *AEB*. The angle *MAB* will then be two thirds of a right angle, and the straight line *AM* will intersect the arc *AN*. The straight line *AM* is therefore bounded by the two arcs *NA* and *AE*. Similarly, we show that the tangent drawn at the point *B* intersects the arc *BN*. Therefore, the arc *ANB* lies entirely outside the arc *AEB*.

Having proved this, we now say that the lune ANBEA *is equal to the triangle* ADI *plus the circle* K.

Proof: The sum of the lune *AEBHA* and the circle *K* is equal to the triangle *ABI*. We take the triangle *ADI* common. Then, the sum of the lune *AEBHA*, the circle *K* and the triangle *ADI* is equal to the triangle *ADB*. But the lune *ANBHA* is equal to the triangle *ADB*, as we have shown in the ninth proposition of this treatise. The sum of the lune *AEBHA*, the circle *K* and the triangle *ADI* is therefore equal to the lune *ANBHA*. Removing the common lune *AEBHA* leaves the lune *ANBEA* equal to the triangle *ADI* plus the circle *K*.

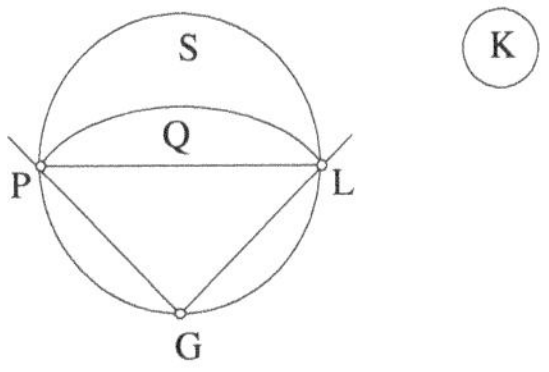

Fig. 3.17.2

We also make the triangle *LGP* a right-angled isosceles triangle equal to the triangle *ADI*. The lune *ANBEA* is therefore equal to the triangle *LGP* plus the circle *K*. With *G* as the centre, and with its distance to *L* and *P*[10] we draw an arc of a circle; let it be *LQP*. We construct a semicircle on the straight line *LP*; let it be *LSP*. Therefore, the lune *LSPQL* is equal to the triangle *LGP*, and the lune *ANBEA* is equal to the lune *LSPQL* plus the circle *K*. That is what we wanted to prove.

<18> Let us draw a circle, on which lie *A*, *B* and *C* and whose centre is at D, and draw the straight line *AC* so as to cut off one third. We divide the arc *ABC* into two halves at the point *B* and we join the straight lines *AB*, *BC*, *AD*, *BD* and *CD*. On the two straight lines *AB* and *BC*, construct two segments that are similar to the segment *ABC*; let them be the segments *AEB* and *BIC*. We draw the two straight lines *BL* and *BM* such that each of the two angles *BLA* and *BMC* is equal to the angle *ABC*. We join the two straight lines *DL* and *DM*. We make a circle *N* equal to one ninth of the circle *ABC*. The sum of the two lunes *AEBHA* and *BICKB*, and the circle *N* is equal to the sum of the two triangles *ABC* and *DLM*, as we have shown in the fourteenth proposition of this treatise. We construct a semicircle on the straight line *AC*; let it be *AFC*.

I say firstly that the arc AFC *lies entirely outside the two arcs* AEB *and* BIC.

[10] Lit.: with the two distances *L* and *P*.

Proof: We draw *AS* tangent to the arc *AEB*. Then the angle *SAB* will be two thirds of a right angle, and the angle *BAC* will be one third of a right angle. Therefore, the angle *SAC* is a right angle, and the straight line *AS* is tangent to the arc *AFC*. But it is tangent to the arc *AEB*, so the arc *AFC* is tangent to the arc *AEB*. Similarly, we show that the arc *AFC* is tangent to the arc *CIB*. Therefore, the arc *AFC* lies entirely outside the two arcs *AEB* and *BIC*.

Having proved this, we now say that the sum of the figure AFCBA *and a known triangle is equal to a known circle.*

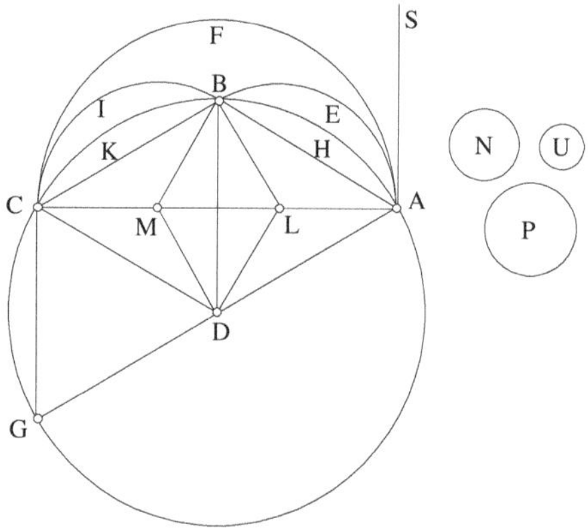

Fig. 3.18

Proof: We produce *AD* as far as *G*, and we join *CG*. The triangle *ACD* will then be equal to the triangle *CDG* and the sector *CDG* will be one sixth of a circle. The triangle *CDG* is therefore less than one sixth of a circle and the triangle *ADC* is less than one sixth of a circle. But the triangle *DLM* is one third of the triangle *ADC*, so the triangle *DLM* is less than one part of eighteen parts of the circle. But the circle *N* is one ninth of the circle. Therefore, the triangle *DLM* is less than one half of the circle *N*. But the sum of the two lunes *AEBHA* and *BICKB* and the circle *N* is equal to the sum of the two triangles *ABC* and *DLM*.[11] The sum of the two lunes and the amount by which the circle *N* exceeds the triangle *DLM* is equal to the triangle *ABC*. We make the circle *U* equal to one part of twenty-four parts of the circle *ABC*. The sum of the two lunes, the amount by which the circle *N* exceeds the triangle *DLM*, and the circle *U* is equal to that of the triangle *ABC* and the circle *U*. But the triangle *ABC* is equal to the triangle *ADC* as the two straight lines *AB* and *BC* are equal to the two straight lines *AD* and

[11] This result is that established in Proposition 14.

CD. The sum of the two lunes, the amount by which the circle *N* exceeds the triangle *DLM*, and the circle *U* is equal to that of the triangle *ADC* and the circle *U*. But the sum of the triangle *ADC* and the circle *U* is equal to the lune *AFCBA*, as we have shown in the thirteenth proposition. The sum of the two lunes *AEBHA* and *BICKB*, the amount by which the circle *N* exceeds the triangle *DLM*, and the circle *U* is equal to the lune *AFCBA*. Removing the two common lunes leaves the figure *AFCIBEA*, bounded by three arcs, which is equal to the amount by which the circle *N* exceeds the triangle *DLM* plus the circle *U*. The sum of the figure *AFCIBEA* and the triangle *DLM* is equal to the sum of the two circles *N* and *U*.

We make the circle *P* equal to the sum of the two circles *N* and *U*. Then the sum of the figure *AFCIBEA* and the triangle *DLM* is equal to the circle *P*. If we construct a right-angled isosceles triangle[12] and then construct a lune on its hypotenuse, as we did in the previous proposition, then this lune will be equal to the triangle *DLM*. The sum of the figure *AFCIBEA* and this lune is therefore equal to the circle *P*. That is what we wanted to prove.

– **19** – Similarly, Euclid has shown in his book *On Division*[13] how to cut off a portion of a known circle between two parallel straight lines such that the ratio of this portion to the whole circle is a known ratio. Here, we shall show what we ourselves make of this.

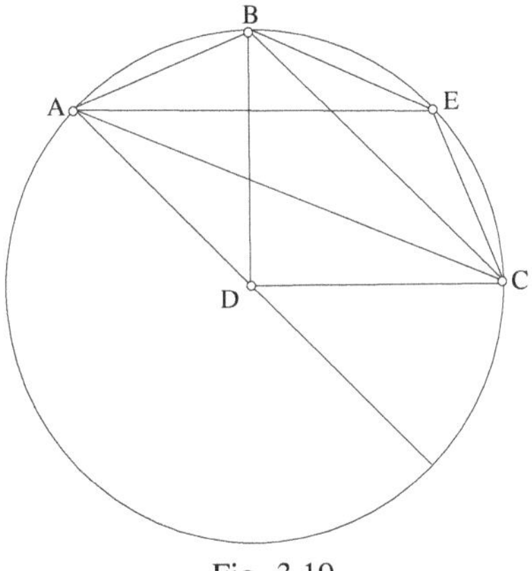

Fig. 3.19

Let there be a circle on which lie *A*, *B* and *C*. Let *D* be its centre. Let the sector *DBC* be equal to one quarter of the circle. We join *BC*, and divide

[12] Right-angled isosceles triangle equivalent to the triangle *DLM*.

[13] Euclid, Proposition 29 of the *Division of Figures*. See R. C. Archibald, *Euclid's Book on Division of Figures*, Cambridge, 1915, pp. 66–7.

the arc *BC* into two halves at the point *E*. We join the two straight lines *BE* and *EC*. We draw *DA* parallel to the straight line *BC*, and join *CA*, *EA* and *BA*. The triangle *ABC* will then be equal to the triangle *BDC*. The triangle *BEC* is common, and therefore the quadrilateral *ABEC* is equal to the quadrilateral *DBEC*. But, since the sector *DBC* is one quarter of a circle, the angle *BDC* is a right angle, and the angle *DBC* is half a right angle. But the angle *BDA* is half a right angle, so the arc *AB* is one eighth of the circle. But the arc *BE* is one eighth of the circle, so the segment *AB* is equal to the segment *BE*. Now, we take the segment *AB* in place of the segment *BE*. The sum of the quadrilateral *ABEC* and the two segments *AB* and *EC* is then equal to the sector *BDCE*. The segment *ABEC* is therefore equal to one quarter of the circle *ABC*.

But, since the arc *EC* is equal to the arc *AB*, the angle *CAE* is equal to the angle *AEB*, and the straight line *AC* is parallel to the straight line *BE*. Therefore, the segment *ABEC* lies between the two parallel straight lines. That is what we wanted to prove.

– **20** – Having proved this, now let us draw a circle on which lie *A*, *B* and *H*. On this circle, we cut off the arc *ABCD* equal to three eighths of the circle, and divide this arc into three eighths. Let these be the arcs *AB*, *BC* and *CD*. We draw the straight line *BC*; it will be parallel to the straight line *AD*, as we have shown in the previous proposition, and the portion *ABCD* which lies between the two parallel lines is one quarter of the circle.

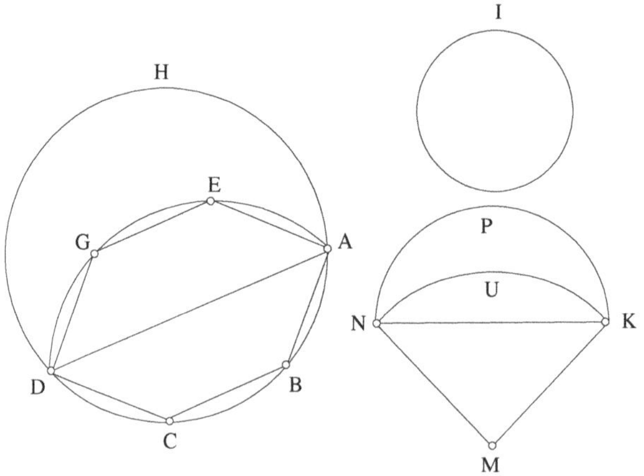

Fig. 3.20

On the straight line *AD*, we construct a portion of a circle on the other side <of *AD*> equal to the portion *ABCD*. Let this portion be *AEGD*, and divide it also into three eighths. Let these be the arcs *AE*, *EG* and *GD*. We join *EG*; it will be parallel to the straight line *AD*. The portion *AEGD* which lies between the two parallel straight lines is one quarter of the circle. The portion *CBAEGDC* which lies between the two straight lines *BC* and *EG*, which are parallel, is one half of the circle *ABC*. The remainder is the lune *AHDGEA* and the segments *BC* and *EG*, whose sum is equal to one half of the circle *ABC*. The sum of the lune *AHDGEA* and the two segments *EG* and *BC* is equal to the portion *CBAEGDC* which lies between the two parallel straight lines *EG* and *BC*. We join the straight lines *AE*, *DG*, *DC* and *BA*. The two segments *AE* and *GD* are equal to the two segments *EG* and *BC*. We remove the two segments *AE* and *GD* from the portion *CBAEGDC*, which lies between the two parallel straight lines *EG* and *BC*. There remains the lune *AHDGEA* equal to the portion *ADCB*, which lies between the two parallel straight lines *AD* and *BC*, which is one quarter of the circle, plus the quadrilateral *AEGD*. Let there be a circle *I*, equal to one quarter of the circle *ABCD*, and a right-angled isosceles triangle *KMN* whose angle *M* is a right angle, and whose two sides *MK* and *MN* are equal. Let this triangle be equal to the quadrilateral *AEGD*. The lune *AHDGEA* is therefore equal to the sum of the circle *I* and the triangle *MKN*. Using the point *M* as the centre, and its distance to the two points *K* and *N*,[14] we draw an arc of a circle; let it be *KUN*. On the straight line *KN*, we construct a semicircle; let it be *KPN*. The lune *KPNUK* is equal to the triangle *MKN*, and the lune *AHDGEA* is therefore equal to the lune *KPNUK* plus the circle *I*. That is what we wanted to prove.

– **21** – This lune, and any other lune in which the sum of the two arcs is a complete circle, has a property that is not shared with other lunes. The parallel straight lines – which lie within the lune and are such that they meet the straight line *AC*[15] at right angles when extended – are all equal. And of them, that which lies at the centre of the lune is equal to that which lies at its extremity.

The proof of this is as follows: we draw a circle on which lie *A*, *B* and *C* and we cut off from this circle any portion that is less than a semicircle; let it be the segment *ADC*. We join *AC*, and on the straight line *AC* we construct a segment *AEC* equal to the segment *ADC*. We divide the straight line *AC* into two halves at the point *G*, we draw the perpendicular *GEB* and extend it to *D*. On the arc *BC*, we assume any point – let it be *M* – and we draw the perpendicular *MQOU*.

I say that the straight line MQ *is equal to the straight line* BE.

[14] Lit.: with the two distances *K* and *N*.

[15] Straight line *AD* in Fig. 3.20.

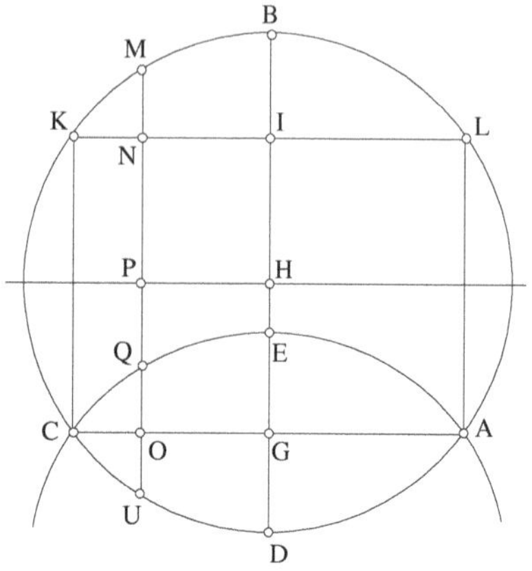

Fig. 3. 21

Proof: *BD* is a diameter that we divide into two halves at the point *H*. *H* is therefore the centre of the circle. From the point *H*, we draw the perpendicular *HP*; it divides *MU* into two halves at the point *P*. Since the segment *ADC* is less than a semicircle, the straight line *BG* is greater than the straight line *GD*. We make *BI* equal to *GD*. There remains *IH* equal to *HG*. From the point *I*, we draw a straight line *LIK* perpendicular to the straight line *BD*. The segment *LBK* is therefore equal to the segment *ADC*. Let the straight line *LK* intersect the straight line *MU* at the point *N*. Then the straight line *PN* is equal to the straight line *PO*. There remains the straight line *MN* equal to the straight line *OU*. We join *AL* and *CK*. They will be equal and parallel, and also equal and parallel to the straight line *IG*. Since *BI* is equal to *GD*, *IG* is the amount by which *BG* exceeds *GD*. Since *MN* is equal to *OU*, *NO* is the amount by which *MO* exceeds *OU*. But *NO* is equal to *IG*, so the amount by which *BG* exceeds *GD* is equal to the amount by which *MO* exceeds *OU*. But *GD* is equal to *GE*, and *OU* is equal to *OQ*, so the amount by which *BG* exceeds *GE* is equal to the amount by which *MO* exceeds *OQ*. The straight line *BE* is therefore equal to the straight line *MQ*. But as each of <the straight lines> *BE* and *MQ* is equal to the amount by which *BG* exceeds *GD*, each of <the straight lines> *BE* and *MQ* is equal to the straight line *IG*. But each of the straight lines *AL* and *CK* is equal to the straight line *IG*, so each of the straight lines *AL* and *CK* is equal to the each of the straight lines *BE* and *MQ*. That is what we wanted to prove.

– **22** – *We also say that if two lunes from two similar segments are constructed on two similar arcs from two circles, then the ratio of one lune to the other lune is equal to the ratio of one circle to the other circle.*

Example: The two lunes *AKBNA* and *DMEOD* lie on two similar arcs *ANB* and *DOE* of the two circles *ABC* and *DEG*, and the two arcs *AKB* and *DME* are similar.

I say that the ratio of one lune to the other lune is equal to the ratio of on circle to the other circle.

Proof: Let the centres of the two circles be *H* and *I*, and join the straight lines *AH*, *AB*, *BH*, *DI*, *DE* and *EI*. Since the two arcs *ANB* and *DOE* are similar, the ratio of the square of *AB* to the square of *DE* is equal to the ratio of the square of the diameter of one circle to the square of the diameter of the other circle; it is equal to the ratio of one circle to the other circle, and equal to the ratio of the sector *AHB* to the sector *DIE*, and equal to the ratio of the triangle *AHB* to the triangle *DIE*, and equal to the ratio of the segment *ANB* to the segment *DOE*, and equal to the ratio of the segment *AKB* to the segment *DME*. The ratio of the segment *AKB* to the segment *DME* is therefore equal to the ratio of the segment *ANB* to the segment *DOE*, and equal to the ratio of the remainder to the remainder. The ratio of the lune *AKBNA* to the lune *DMEOD* is equal to the ratio of the segment *ANB* to the segment *DOE*. But the ratio of the segment *ANB* to the segment *DOE* is equal to the ratio of the circle *ABC* to the circle *DEG*. The ratio of the lune *AKBNA* to the lune *DMEOD* is equal to the ratio of the circle *ABC* to the circle *DEG*. That is what we wanted to prove.

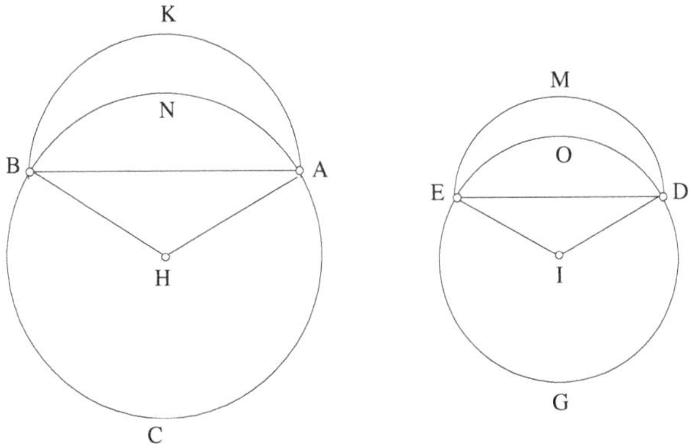

Fig. 3.22

– **23** – Let us draw another circle on which lie *A*, *B* and *C*, and draw the side of a hexagon in this circle; let it be *AB*. On the straight line *AB*, we construct an equilateral triangle such that part of it lies outside the circle. Let it be *ADB*. We make the circle *EGH* equal to three times the circle *ABC*. We also draw the side of the hexagon in this circle; let it be *EG*. On the straight line *EG*, construct a segment such that its circumference is one third of a circle; let it be the segment *EPG*.

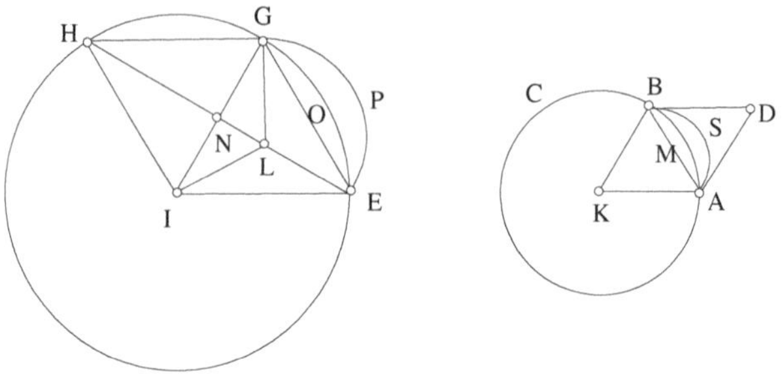

Fig. 3.23

I say that the lune EPGOE *is equal to the figure* ADBMA.

Proof: We mark the centres of the two circles; let them be *K* and *I*. We draw *EH* equal to the side of the triangle.[16] We join the straight lines *AK*, *BK*, *EI*, *GNI*, *HI* and *GH*, and draw the straight line *GL* such that the angle *GLE* is equal to the angle *EGH*. The sum of the lune *EPGOE* and one half of one ninth of the circle *EGH* is equal to the sum of the two triangles *EGN* and *ILN*, as we have proved in the fourteenth proposition. But the triangle *EGN* is equal to the triangle *IHN* as the two triangles *EGH* and *EIH* are equal and the straight line *ING* divides each of them into two halves. The sum of the lune *EPGOE* and one half of one ninth of the circle *EGH* is equal to the triangle *HLI*. But the triangle *HLI* is two thirds of the triangle *EIH*, as the straight line *EL* is one third of the straight line *EH*. But the triangle *EIH* is equal to the triangle *EIG*, as each of them is one half of the quadrilateral *EIHG*. The sum of the lune *EPGOE* and one half of one ninth of the circle *EGH* is equal to two thirds of the triangle *EIG*.

But, as the circle *EGH* is three times the circle *ABC*, the triangle *EIG* is three times the triangle *AKB*. Two thirds of the triangle *EIG* is equal to twice the triangle *ABK*, and the lozenge *ADBK* is twice the triangle *ABK*. Two thirds of the triangle *EIG* is therefore equal to the lozenge *ADBK*. The sum of the lune *EPGOE* and one half of one ninth of the circle *EGH* is equal to the lozenge *ADBK*. But the sector *AKB* is one sixth of the circle *ABC*, and the circle *ABC* is equal to one third of the circle *EGH*. The sector *AKBM* is therefore equal to one half of one ninth of the circle *EGH*. The sum of the lune *EPGOE* and the sector *AKBM* is equal to the lozenge *ADBK*. Removing the common sector leaves the lune *EPGOE* which is equal to the figure *ADBMA*. That is what we wanted to prove.

[16] The inscribed equilateral triangle.

On the straight line *AB*, we construct an arc equal to one third of a circle; let it be *ASB*. The ratio of the lune *EPGOE* to the lune *ASBMA* is therefore equal to the ratio of the circle *EGH* to the circle *ABC*, and the lune *EPGOE* is therefore three times the lune *ASBMA*. The figure *ADMBA* is therefore three times the lune *ASBMA*; consequently, *ADBSA* is twice the lune *ASBMA*.

If we construct a circle equal to twice the circle *ABC*, and if we draw within it the side of a hexagon on which we construct one third of a circle, then the lune so generated will be equal to the figure *ADBSA*.

It is possible to construct many species of lune in the ways that we have described. Where we have made mention of rational lunes, this has been by way of example in order to reveal the universal notion that we have previously shown. The work that we have presented in their regard is sufficient to elucidate what we intended to show.

Let us now conclude this treatise.
The treatise is complete. Praise be to God.
May He receive the thanks that are due to Him.

CHAPTER II

CALCULATION OF VOLUMES OF PARABOLOIDS AND SPHERES AND THE EXHAUSTION METHOD

2.1. INTRODUCTION

This second group of works on infinitesimal mathematics by Ibn al-Haytham deals with the calculation of the volumes of curved solids using the exhaustion method. There are three treatises of differing length by Ibn al-Haytham on this subject, and appearing in the following order:

I. *Treatise on the Measurement of the Paraboloid*

II. *Treatise on the Measurement of the Sphere*

III. *Treatise on the Division of Two Different Magnitudes Mentioned in the First Proposition of the Tenth Book of Euclid.*

As the titles show, Ibn al-Haytham refers to his predecessors: Thābit ibn Qurra and al-Qūhī for the paraboloid; Archimedes, the Banū Mūsā and possibly others for the sphere; and the third treatise had already been the object of some discussion for Ibn Qurra and al-Qūhī. Thus Ibn al-Haytham belongs to this mathematical tradition marked out since al-Kindī and the Banū Mūsā. This Archimedean tradition provided for Ibn al-Haytham problems and methods of research. He was quick to take advantage of the arithmetic methods of his ancient predecessor, Thābit ibn Qurra, and, apparently, of the rediscovery of the method of integral sums by his immediate predecessor al-Qūhī and likely his contemporary Ibn Sahl. The way these two methods were constructed affected the direction in which Ibn al-Haytham would take the study of the infinitesimal, as we shall see later.

The treatises of Ibn al-Haytham represented both the avant-garde and the end of research in this area; whilst shedding new light on its meaning, Ibn al-Haytham's work was also to be the last done in Arabic: no further contributions using the exhaustion method are seen after this, nor indeed was any further research undertaken. This is an area that no historian can fail to investigate, as we now witness a second halt, just as brutal as the first had been, thirteen centuries before. As is customary, we will comment here on the arithmetic of the first two treatises.

2.2. MATHEMATICAL COMMENTARY

2.2.1. *Calculation of volumes of paraboloids*

Ibn al-Haytham's treatise on the volume of the paraboloid comprises an introduction, where the author retraces the history of the problem and where he takes the opportunity to acknowledge the unique contributions of his predecessors Ibn Qurra and al-Qūhī. This is followed by a first section, which is devoted entirely to major arithmetic lemmas necessary to his proofs; then follows a second section, where he studies a paraboloid of revolution, and a third section where he examines other types of paraboloids formed by rotation of a parabola around an ordinate; it concludes with a discussion of the method applied when using infinitesimal determinations to measure areas and volumes. We will take up these chapters one after the other.

2.2.2.1. *Arithmetical lemmas*

Ibn al-Haytham begins his treatise by proving five arithmetical lemmas, four of which deal with the sum of the powers of n natural integers. These four lemmas will be used in establishing the fundamental inequality.

Lemma 1:

$$\sum_{k=1}^{n} k = \frac{n(n+1)}{2} = \frac{n^2}{2} + \frac{n}{2}.$$

Ibn al-Haytham's proof is quasi-general, which is to say, it is established for a specific number, let us say four, and it is then assumed to be valid for any number (in the same way as it is for the specific number). This proof can be rewritten as

$$S_n = 1 + 2 + \ldots + n,$$

$$S_n = n + (n-1) + \ldots + 1,$$

hence

$$S_n = \frac{n(n+1)}{2}.$$

Lemma 2:

$$S_n^{(2)} = \sum_{k=1}^{n} k^2 = \left(\frac{n}{3} + \frac{1}{3}\right)n\left(n + \frac{1}{2}\right) = \frac{1}{3}n^3 + \frac{1}{2}n^2 + \frac{1}{6}n.$$

Ibn al-Haytham proves this lemma by using an archaic form of finite complete induction, a form which was still being used, unchanged, in the seventeenth century. He uses $P_k = (k+1)S_k = S_k^{(2)} + S_k + S_{k-1} + \ldots + S_1$ for this proof, which he proves for the case of $1 \leq k \leq 4$, and in which he expresses the recursive relation. Ibn al-Haytham's calculation is presented in the following manner:

$$(1) \qquad P_1 = 1(1+1) = 1^2 + 1 = S_1^{(2)} + S_1;$$

with the help of (1) it can be proven that

$$(2) \qquad P_2 = (1+2)(2+1) = 2^2 + 1^2 + (1+2) + 1 = S_2^{(2)} + S_2 + S_1.$$

By using (2) it is possible to go to (3):

$$(3) \qquad P_3 = (1+2+3)(3+1) = 3^2 + 2^2 + 1^2(1+2+3) + (1+2) + 1$$
$$= S_3^{(2)} + S_3 + S_2 + S_1.$$

Similarly, following (3), we have

$$(4) \qquad P_4 = (1+2+3+4)(4+1)$$
$$= 4^2 + 3^2 + 2^2 + 1^2 + (1+2+3+4) + (1+2+3) + (1+2) + 1$$
$$= S_4^{(2)} + S_4 + S_3 + S_2 + S_1.$$

This result is true for $k = 1$: $P_1 = (1 + 1)\,1 = 1^2 + 1$. We assume that the order k is true, and we set

$$P_k = (k + 1)\, S_k,$$

which leads to

$$P_k = S_k^{(2)} + S_k + S_{k-1} + \ldots + S_1.$$

We prove that this property is true for the order $k + 1$.

$$P_{k+1} = [(k + 1) + 1]\, S_{k+1} = (k + 1)\, S_{k+1} + S_{k+1},$$

$$P_{k+1} = (S_k + (k + 1))\,(k + 1) + S_{k+1} = P_k + (k + 1)^2 + S_{k+1}$$
$$= S_{k+1}^{(2)} + S_{k+1} + S_k + \ldots + S_1.$$

Having proved this inequality, Ibn al-Haytham proceeds thus: by Lemma 1 we have

$$(n+1)S_n = S_n^{(2)} + \frac{1}{2}S_n^{(2)} + \frac{1}{2}S_n,$$

because

$$S_1 + S_2 + \ldots + S_n = \frac{1}{2}(1\cdot(1+1) + 2\cdot(2+1) + \ldots + n(n+1))$$

$$= \frac{1}{2}\left(1^2 + 2^2 + \ldots + n^2 + 1 + 2 + \ldots + n\right) = \frac{1}{2}\left(S_n^{(2)} + S_n\right);$$

but

$$(n+1)S_n = \left(n + \frac{1}{2}\right)S_n + \frac{1}{2}S_n,$$

hence

$$S_n^{(2)} + \frac{1}{2}S_n^{(2)} = \left(n + \frac{1}{2}\right)S_n$$

and

$$S_n^{(2)} = \frac{2}{3}\left(n + \frac{1}{2}\right)S_n = \frac{1}{3}(n+1)n\left(n + \frac{1}{2}\right).$$

Note that Ibn al-Haytham's proof of this lemma is different from the one given in Archimedes' *On Spirals*. This lemma is proved by a similar method by Abū Kāmil (second half of the ninth century).[1]

Lemma 3:

$$S_n^{(3)} = \sum_{k=1}^{n} k^3 = \left(\frac{n}{4} + \frac{1}{4}\right)n^2(n+1) = \frac{1}{4}n^4 + \frac{1}{2}n^3 + \frac{1}{4}n^2.$$

The proof of this lemma by Ibn al-Haytham can be written as

$$(n+1)S_n^{(2)} = nS_n^{(2)} + S_n^{(2)} = S_n^{(2)} + n^3 + ((n-1)+1)S_{n-1}^{(2)};$$

in the same way, we show that

$$((n-1)+1)S_{n-1}^{(2)} = S_{n-1}^{(2)} + (n-1)^3 + ((n-2)+1)S_{n-2}^{(2)},$$

and so on, down to

$$(n-(n-1)+1)S_{n-(n-1)}^{(2)} = S_1^{(2)} + 1^3.$$

So, from Lemma 2, we get

$$(n+1)S_n^{(2)} = S_n^{(3)} + \sum_{k=1}^{n} S_k^{(2)} = S_n^{(3)} + \frac{1}{3}S_n^{(3)} + \frac{1}{2}S_n^{(2)} + \frac{1}{6}S_n.$$

[1] See *Abū Kāmil: Algèbre et analyse diophantienne*, editing, translation and commentary by R. Rashed, Berlin/New York, 2012.

But

$$(n+1)S_n^{(2)} = \left(n+\frac{1}{2}\right)S_n^{(2)} + \frac{1}{2}S_n^{(2)},$$

therefore

$$\left(n+\frac{1}{2}\right)S_n^{(2)} = S_n^{(3)} + \frac{1}{3}S_n^{(3)} + \frac{1}{6}S_n\,;$$

it follows

$$\frac{3}{4}\left(n+\frac{1}{2}\right)S_n^{(2)} = S_n^{(3)} + \frac{1}{8}S_n.$$

Moreover

$$\frac{3}{4}S_n^{(2)} = \frac{1}{4}(n+1)n\left(n+\frac{1}{2}\right),$$

therefore

$$S_n^{(3)} + \frac{1}{8}S_n = \left(\frac{n}{4}+\frac{1}{4}\right)n\left(n+\frac{1}{2}\right)\left(n+\frac{1}{2}\right) = \frac{1}{4}(n+1)n\left(n+\frac{1}{2}\right)^2.$$

But

$$\frac{1}{2}S_n = \frac{1}{4}(n+1)n,$$

hence

$$S_n^{(3)} = \frac{1}{4}(n+1)n\left[\left(n+\frac{1}{2}\right)^2 - \frac{1}{4}\right] = \frac{1}{4}n^4 + \frac{1}{2}n^3 + \frac{1}{4}n^2.$$

As can be seen, Ibn al-Haytham's proof uses regression and relies on previous lemmas.

Lemma 4:

$$S_n^{(4)} = \sum_{k=1}^{n} k^4 = \left(\frac{n}{5}+\frac{1}{5}\right)n\left(n+\frac{1}{2}\right)\left((n+1)n-\frac{1}{3}\right).$$

We have

$$(n+1)S_n^{(3)} = n^4 + nS_{n-1}^{(3)} + S_n^{(3)};$$

as in the proof of Lemma 3, by using regression, we finally arrive at

$$(n+1)S_n^{(3)} = S_n^{(4)} + \sum_{k=1}^{n} S_k^{(3)};$$

but from the previous lemma, it follows

$$(n+1)S_n^{(3)} = S_n^{(4)} + \frac{1}{4}S_n^{(4)} + \frac{1}{2}S_n^{(3)} + \frac{1}{4}S_n^{(2)},$$

hence

$$\frac{4}{5}(n+1)S_n^{(3)} = S_n^{(4)} + \frac{2}{5}S_n^{(3)} + \frac{1}{5}S_n^{(2)},$$

from which we can deduce

$$\frac{4}{5}(n+1)S_n^{(3)} = S_n^{(4)} + \frac{1}{5}S_n^{(2)},$$

hence

$$S_n^{(4)} = \frac{4}{5}\left(n+\frac{1}{2}\right)S_n^{(3)} - \frac{1}{5}S_n^{(2)}.$$

But from Lemmas 2 and 3, we get

$$\frac{4}{5}S_n^{(3)} = \frac{1}{5}(n+1)n\left(n+\frac{1}{2}\right)$$

and

$$S_n^{(2)} = \frac{1}{3}n(n+1)\left(n+\frac{1}{2}\right),$$

hence

$$S_n^{(4)} = \frac{1}{5}(n+1)n\left(n+\frac{1}{2}\right)\left(n(n+1)-\frac{1}{3}\right).$$

The demonstration of the four previous lemmas proves, by the application of a complete induction (in archaic form) or of regression, the generality of Ibn al-Haytham's method. In fact, his method as it stands is valid for any integer power without adding any supplementary notions. Ibn al-Haytham identifies a general rule for the calculation of the sum of n integers raised to any integer power, which he used in previous cases, and which can be rewritten as

$$(n+1)\sum_{k=1}^{n} k^i = \sum_{k=1}^{n} k^{i+1} + \sum_{p=1}^{n}\left(\sum_{k=1}^{p} k^i\right),$$

in such a way that Ibn al-Haytham would have been able to calculate the sum of powers $i$ths of n as first integers for $i \geq 5$. And the only reason for him stopping at $i = 4$, is because it suited his own aims. Ibn al-Haytham did in fact make fuller use of these powers in his later proofs, specifically the one concerning major inequality (Lemma 5).

Lemma 5:

$$\frac{8}{15}n(n+1)^4 \leq \sum_{k=1}^{n}\left[(n+1)^2-k^2\right]^2 \leq \frac{8}{15}(n+1)(n+1)^4 \leq \sum_{k=0}^{n}\left[(n+1)^2-k^2\right]^2.$$

Ibn al-Haytham's proof of this lemma is very long; it deserves to be looked at again in summary, not only for the text itself, but also to show the scope of arithmetical research being done then by this geometrician.

He first shows the identity for $0 \leq k \leq n$:

(1) $$[2(n+1)^2-k^2]k^2+[(n+1)^2-k^2]^2=(n+1)^4,$$

hence he deduces

(2) $$(n+1)^4-[2(n+1)^2-k^2]k^2=[(n+1)^2-k^2]^2.$$

He also shows

(3) $$2(n+1)^2\sum_{k=1}^{n}k^2-\sum_{k=1}^{n}k^4=\sum_{k=1}^{n}\left[2(n+1)^2-k^2\right]k^2;$$

and by way of summing from (2), he arrives at

(4) $$n(n+1)^4-\sum_{k=1}^{n}\left[2(n+1)^2-k^2\right]k^2=\sum_{k=1}^{n}\left[(n+1)^2-k^2\right]^2.$$

But by Lemma 2, we have

(5) $$\frac{1}{3}n(n+1)\left(n+\frac{1}{2}\right)=\sum_{k=1}^{n}k^2,$$

hence

(6) $$\frac{2}{3}n(n+1)\left(n+\frac{1}{2}\right)(n+1)^2=2(n+1)^2\sum_{k=1}^{n}k^2;$$

and from Lemma 4, we get

(7) $$\frac{1}{5}(n+1)\left(n+\frac{1}{2}\right)n\left[(n+1)n-\frac{1}{3}\right]=\sum_{k=1}^{n}k^4.$$

From (3), (6) and (7), it follows

(8) $$\begin{aligned}A_n&=\sum_{k=1}^{n}\left[2(n+1)^2-k^2\right]k^2\\&=\frac{2}{3}n(n+1)\left(n+\frac{1}{2}\right)(n+1)^2-\frac{1}{5}n(n+1)\left(n+\frac{1}{2}\right)\left[(n+1)n-\frac{1}{3}\right].\end{aligned}$$

Yet $\frac{2}{3}=\frac{7}{15}+\frac{1}{5}$ and $(n+1)^2 = n(n+1)+n+1$, hence

$$\text{(9)}\qquad A_n = n(n+1)\left(n+\frac{1}{2}\right)\left[\frac{7}{15}(n+1)^2+\frac{1}{5}(n+1)+\frac{1}{15}\right] = nH_n.$$

From (4) and (9), we get

$$\text{(10)}\qquad \sum_{k=1}^{n}\left[(n+1)^2-k^2\right]^2 = n(n+1)^4 - nH_n = n\left[(n+1)^4 - H_n\right].$$

But

$$H_n = \frac{7}{15}(n+1)^4 - \frac{7}{30}(n+1)^3 + (n+1)\left(n+\frac{1}{2}\right)\left[\frac{1}{5}(n+1)+\frac{1}{15}\right]$$

and

$$(n+1)^2 = (n+1)\left(n+\frac{1}{2}\right)+\frac{1}{2}(n+1), \qquad \frac{n+1}{2}=\frac{8}{15}\frac{n+1}{2}+\frac{7}{15}\frac{n+1}{2};$$

hence

$$\text{(11)}\qquad K_n = (n+1)^4 - H_n$$

$$= \frac{8}{15}(n+1)^4 + \frac{7}{30}(n+1)^3 - (n+1)\left(n+\frac{1}{2}\right)\left[\frac{1}{5}(n+1)+\frac{1}{15}\right]$$

$$= \frac{8}{15}(n+1)^4 + \frac{7}{30}(n+1)^3 - \frac{(n+1)^3}{5} + \frac{1}{2}\frac{(n+1)^2}{5} - \frac{(n+1)^2}{15} + \frac{n+1}{30}$$

$$= \frac{8}{15}(n+1)^4 + \frac{n+1}{30}\left[(n+1)^2+(n+1)+1\right].$$

However

$$(n+1)^2+(n+1)+1 = \frac{(n+1)^3-1}{n};$$

hence

$$\text{(12)}\qquad nK_n = \frac{8n(n+1)^4}{15} + \frac{n+1}{30}\left[(n+1)^3-1\right].$$

For any natural integer n, then $(n+1)^3 \geq 1$, therefore

$$nK_n \geq \frac{8n(n+1)^4}{15}.$$

We thus verify that

$$\frac{8}{15}n(n+1)^4 \leq \sum_{k=1}^{n}\left[(n+1)^2-k^2\right]^2.$$

Moreover

$$nK_n = \frac{8(n+1)(n+1)^4}{15} - \frac{8(n+1)^4}{15} + \frac{(n+1)(n+1)^3}{30} - \frac{(n+1)}{30},$$

$$= \frac{8(n+1)(n+1)^4}{15} - \frac{(n+1)^4}{2} - \frac{(n+1)}{30};$$

hence for all natural integers n, it is possible to verify that

$$\sum_{k=1}^{n}\left[(n+1)^2-k^2\right]^2 < \frac{8(n+1)(n+1)^4}{15}.$$

We finally have

$$\sum_{k=0}^{n}\left[(n+1)^2-k^2\right]^2 = nK_n + (n+1)^4 = \frac{8(n+1)(n+1)^4}{15} + \frac{(n+1)^4}{2} - \frac{(n+1)}{30}.$$

However, for $n \geq 1$, we get

$$(n+1)^4 > n+1 \text{ and } \frac{(n+1)^4}{2} > \frac{n+1}{30},$$

therefore

$$\sum_{k=0}^{n}\left[(n+1)^2-k^2\right]^2 > \frac{8(n+1)(n+1)^4}{15}$$

and so the inequality is proved.

As can be seen above, proof of this inequality requires calculation of the sum of n first natural integers to the fourth power; as a result, this not only contributes to the understanding of the foregoing, but also to further research into the volume of the second species of paraboloid.

2.2.2.2. *Volume of a paraboloid of revolution*

Ibn al-Haytham proves the following proposition once more:

The volume of a paraboloid of revolution around a diameter is equal to half the volume of the circumscribed cylinder.

He considers three cases, whether the angle ACB is right, acute or obtuse.

First case: Assume that angle $ACB = \frac{\pi}{2}$, let V be the volume of the circumscribed cylinder and v the volume of the paraboloid.

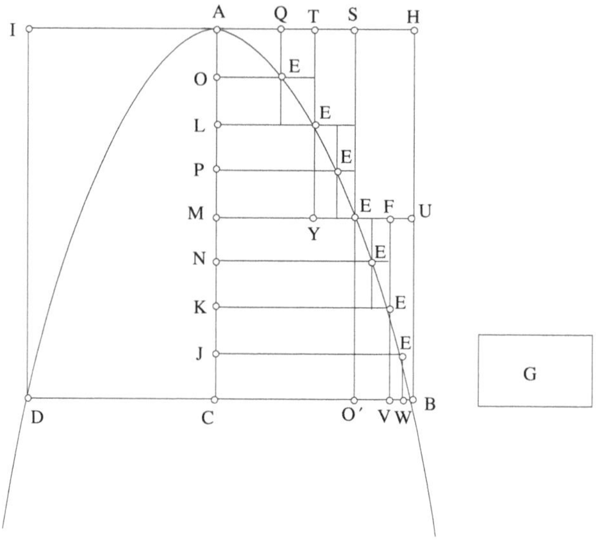

Fig. 2.1.4

• Assume that $v > \frac{1}{2}V$; let $v - \frac{1}{2}V = \varepsilon$.

Let M be the midpoint of AC and draw $MU \parallel BC$ cutting the parabola at E and BH at U. Draw $SEO' \parallel AC$ and cutting BC at O' and AH at S. Let us denote by $[EC]$ the solid generated by rotating surface $MCO'E$, and similarly for other solids. From this

$$(1) \qquad [HE] + [EC] = \frac{1}{2}V \text{ and } [BE] + [AE] = \frac{1}{2}V.$$

We reiterate the construction on point L, the midpoint of AM, then on point K, the midpoint of MC. From this

$$[SE_l] + [ME_l] = \frac{1}{2}[MS] = \frac{1}{2}[AE],$$
$$[UE_k] + [E_kO'] = \frac{1}{2}[UO'] = \frac{1}{2}[BE];$$

therefore

(2) $$[SE_l] + [ME_l] + [UE_k] + [E_kO'] = \frac{1}{2}[AE] + \frac{1}{2}[BE] = \frac{1}{4}V.$$

We reiterate the same construction on points O, P, N, J, the midpoints respectively of AL, LM, MK, KC. Therefore the sum of the eight solids is equal to half (2), that is to say, $\frac{1}{8}V$.

He continues to proceed in the same manner, i.e. removing solids of type (1) and (2) from the circumscribed cylinder. The following would have successively been removed from V:

$$\frac{1}{2}V,\ \frac{1}{2}\left(\frac{1}{2}V\right),\ \frac{1}{2}\left(\frac{1}{2}\left(\frac{1}{2}V\right)\right)$$

and so on. After a finite number of operations, we necessarily come to a remainder smaller than ε, from Lemma X.1 in Euclid's *Elements* (or Ibn al-Haytham's theorem).

Let us assume that the subdivision of the figure corresponds to the step where the remainder is less than ε.

Let V_n be the volume of solids remaining after n steps, therefore $V_n < \varepsilon$ and let v_n be the volume of these solids inside the paraboloid, therefore $v_n < V_n$ and $v_n < \varepsilon$, therefore $v - v_n > \frac{1}{2}V$, according to the hypothesis. But in accordance with the properties of the parabola, we have

$$\frac{AC}{AM} = \frac{CB^2}{EM^2},$$

hence

$$BC^2 = 2\,EM^2.$$

Similarly

$$\frac{BC^2}{AC} = \frac{JE_j^2}{AJ} = \frac{OE_0^2}{AO} = \frac{JE_j^2 + OE_0^2}{AC},$$

hence

$$JE_j^2 + OE_0^2 = BC^2 = 2EM^2.$$

In the same way we can show that

$$KE_k^2 + LE_l^2 = BC^2 = 2EM^2$$

and so on.

If we denote $E_0 = A, E_1, \ldots, E_n = B$, with $n = 2^m$, the points of the parabola corresponding to the points on the axis

$$F_0 = A, \ldots, F_{\frac{n}{2}} = M, \ldots, F_n = C,$$

then it follows

$$\overline{E_i F_i}^2 + \overline{E_{n-i} F_{n-i}}^2 = \overline{BC}^2 = 2\overline{EM}^2 \qquad (0 \le i \le n)$$

and

$$\overline{E_1 F_1}^2 + \ldots + \overline{E_{\frac{n}{2}-1} F_{\frac{n}{2}-1}}^2 + \overline{E_{\frac{n}{2}+1} F_{\frac{n}{2}+1}}^2 + \ldots + \overline{E_{n-1} F_{n-1}}^2 = \frac{1}{2}(n-1)\overline{E_n F_n}^2;$$

therefore

$$\sum_{i=1}^{n-1} \overline{E_i F_i}^2 = \frac{1}{2}(n-1)\overline{E_n F_n}^2.$$

Now let $S_i = \pi \overline{E_i F_i}^2$ $(1 \le i \le n - 1)$ be the areas of the discs of radius $\overline{E_i F_i}$, let S_n be the area of the disc of radius $\overline{E_n F_n} = BC$, then

$$\sum_{i=1}^{n-1} S_i = \frac{1}{2}(n-1)S_n.$$

If W_i is the volumes of cylinders with base S_i and height $h = \frac{1}{n} AC$, and W_n the volume of cylinder of base S_n and height h, then

$$\sum_{i=1}^{n-1} W_i = \frac{1}{2}(n-1)W_n.$$

But

$$\frac{1}{2}(n-1)W_n < \frac{1}{2}V,$$

because $V = n\, W_n$; therefore

$$\sum_{i=1}^{n-1} W_i < \frac{1}{2}V;$$

but

$$\sum_{i=1}^{n-1} W_i = v - v_n > \frac{1}{2}V,$$

which is impossible; therefore

(3) $$v \le \frac{1}{2}V.$$

• Assume now that $v < \frac{1}{2} V$, that is $v + \varepsilon = \frac{1}{2} V$; and proceed as before: successively take away half the volume of the cylinder, then half the remainder, until the remaining volume V_n is less than any given value of ε. Let u_n be part of V_n outside a paraboloid, then $u_n < V_n$. Therefore $u_n < \varepsilon$, hence

$$v + u_n < \frac{1}{2} V;$$

but

$$v + u_n = \sum_{i=1}^{n} W_i;$$

therefore

$$\sum_{i=1}^{n} W_i < \frac{1}{2} V.$$

But we have shown that

$$\sum_{i=1}^{n-1} W_i = \frac{1}{2}(n-1) W_n;$$

yet

$$\sum_{i=1}^{n-1} W_i = \sum_{i=1}^{n} W_i - W_n,$$

therefore

$$\sum_{i=1}^{n} W_i - W_n = \frac{1}{2}(n-1) W_n;$$

hence

$$\sum_{i=1}^{n} W_i - \frac{1}{2} W_n = \frac{n}{2} W_n = \frac{1}{2} V.$$

Therefore

$$\sum_{i=1}^{n} W_i > \frac{1}{2} V,$$

which is impossible; therefore

(4) $$v \geq \frac{1}{2} V;$$

and from (3) and (4), this finally gives

$$v = \frac{1}{2} V.$$

Second case: Assume that angle $ACB < \frac{\pi}{2}$.

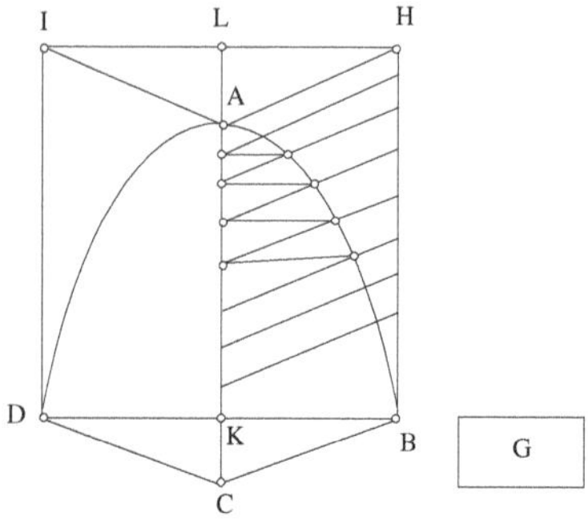

Fig. 2.1.5a

In this case, cones with vertices A' and C' are equal. The conical cylinder is therefore equal to the right cylinder, and is obtained by taking away or adding respectively cones (A') and (C').

- First assume $v > \frac{1}{2}V$.

Subdivision, as used in the first case, is considered; successively taking away half the volume of the cylinder, then half the remaining volume as before, until a solid greater than $\frac{1}{2}V$ is achieved inside the paraboloid; we then show that this solid is less than $\frac{1}{2}V$.

To achieve this, a parabola of diameter AC is drawn on the same figure, giving a paraboloid **P** (second case) and a parabola with axis AC, giving a paraboloid $\mathbf{P}_1$ (first case). The volume of the cylinder circumscribed about $\mathbf{P}_1$ is called V_1. Abscissae at points A, O, L, …, J, C, which divide segment AC, are noted as $x_0, x_1, \ldots, x_n$. Point $E_1(x_i, Y_i)$ of $\mathbf{P}_1$ is associated with every point $E(x_i, y_i)$ of **P**. Perpendicular EE' is dropped from E on AC. Suppose $EE' = z_i$. So we have, for points associated with O and L, for example:

$$\frac{z_1^2}{z_2^2} = \frac{y_1^2}{y_2^2} = \frac{x_1}{x_2} = \frac{Y_1^2}{Y_2^2},$$

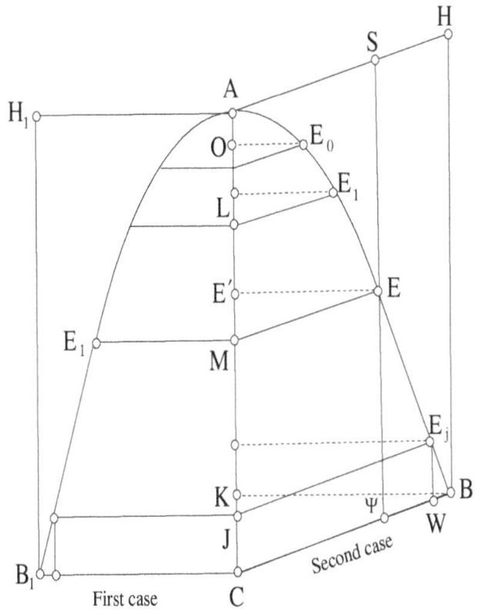

Fig. 2.1.5b

and more generally for $1 \leq i \leq n$

$$(1) \qquad \frac{z_1^2}{Y_1^2} = \frac{z_2^2}{Y_2^2} = \ldots = \frac{z_i^2}{Y_i^2} = \ldots = \frac{BK^2}{CB_1^2} = \frac{V}{V_1} = \frac{\sum_{i=1}^{n} z_i^2}{\sum_{i=1}^{n} Y_i^2}.$$

The solid inscribed in **P** is made up of conical cylinders such as $[CJE_jW]$ whose volume is $\pi\, z^2_{n-1} \cdot h \left(h = CJ = \frac{AC}{n}\right)$, and is associated with the right cylinder of the first case whose volume is $\pi\, Y^2_{n-1} \cdot h$. From (1) we can deduce

$$\frac{\sum_{i=1}^{n} \pi z_i^2 h}{V} = \frac{\sum_{i=1}^{n} \pi Y_i^2 h}{V_1}$$

and if volumes of internal solids inscribed in **P** and $\mathbf{P}_1$ are designated I_n and I_{n1} respectively, then

$$\frac{I_n}{V} = \frac{I_{n1}}{V_1};$$

but in the first case we saw that: $I_{n1} < \frac{1}{2} V_1$; hence

$$I_n < \frac{1}{2} V.$$

The conical solid is therefore less than half the conical cylinder, which contradicts the foregoing; therefore

$$v \leq \frac{1}{2} V.$$

- Assume now $v < \frac{1}{2} V$.

A circumscribed solid smaller than $\frac{1}{2} V$ is constructed in a similar way. As previously, we then show that this solid is larger than $\frac{1}{2} V$, and we finally conclude that $v = \frac{1}{2} V$.

Third case: Assume that angle $ACB > \frac{\pi}{2}$.

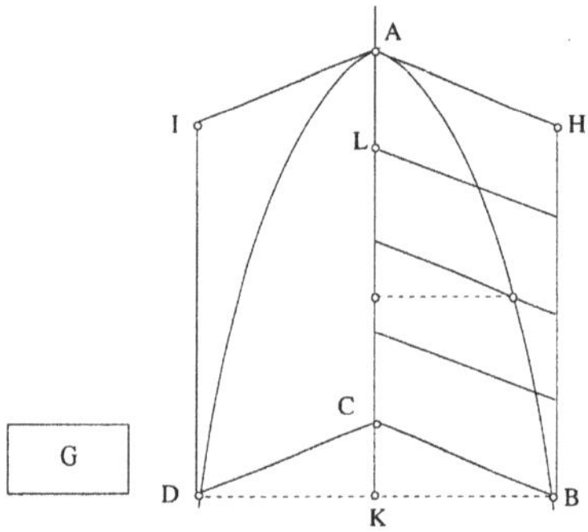

Fig. 2.1.5c

The same method is followed as previously and we show that $v = \frac{1}{2} V$.

The two last cases are linked to the first by an affine transformation, bringing the oblique axes closer to the right-angled axes. Without going into great detail about this transformation, suffice it to say that Ibn al-Haytham links the two last figures to the first, point by point, and preserves the linear relations of these figures.

Ibn al-Haytham then continues with several major corollaries to the proof of the volume of a portion of a paraboloid of revolution. The heights of two portions of a paraboloid are marked h and h', and their respective volumes by v and v'; the areas of two discs of the base of their two associated cylinders marked *S* and *S′* respectively, and the volumes of these two cylinders by *V* and *V′* respectively.

Corollary 1. — *Given a parabola of any diameter* AC, *and ordinate* B_1CB_2, *a paraboloid generated by part* ACB_1 *and a paraboloid generated by part* ACB_2 *have the same volume.*

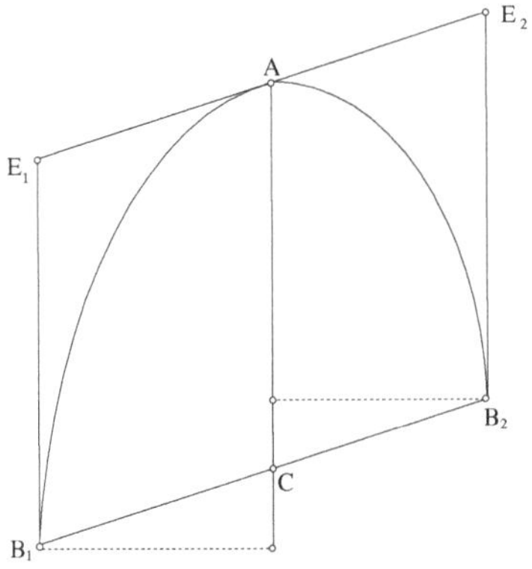

Fig. 2.1.5d

Corollary 2. — *If* S, S′ *and* h, h′ *are respectively the bases and the heights of the right cylinders associated with two paraboloids, and if* S = S′, *then* $\frac{v}{v'} = \frac{h}{h'}$.

Corollary 3. — *If* S ≠ S′ *and* h = h′, *then* $\frac{v}{v'} = \frac{S}{S'}$.

Corollary 4. — *If* S ≠ S′ *and* h ≠ h′, *then* $\frac{v}{v'} = \frac{S}{S'} \cdot \frac{h}{h'}$.

Corollary 5. — *The sum of the volumes of small solids which cross a paraboloid (to step* n *of subdivision and for any high value of* n *we wish) is equal to* $\left(S \cdot \frac{h}{n}\right)$, *that is: for a sufficiently high value of* n, *we have* $J_n - I_n = S \cdot \frac{h}{n}$, *where* J_n *are circumscribed solids and* I_n *are solids inscribed up to step* n *of subdivision.*

Corollary 6. — $I_n = \frac{1}{2}\left[v - S\frac{h}{n}\right]$, *but* $v = \frac{1}{2}V$, *therefore* $v - I_n = \frac{1}{2}S\,\frac{h}{n}$; *and consequently, a paraboloid divides into two halves the sum of those small cylindrical solids which cross it.*

2.2.2.3. *The volume of the second species of paraboloid*

Ibn al-Haytham then determines the volume of a portion of a paraboloid generated by the rotation of a parabola around an ordinate.

Let ABC *be a semiparabola,* BC *its diameter,* AC *its ordinate,* v *the volume of a paraboloid generated by the rotation of* ABC *around* AC*;* V *the volume of the circumscribed cylinder, then* $v = \frac{8}{15}V$.

He considers three cases here, depending on whether angle *ACB* is greater than, smaller than or equal to a right angle.

First case: Assume that $A\hat{C}B = \frac{\pi}{2}$.

Assume first that $v > \frac{8}{15}V$; that is to say $v - \frac{8}{15}V = \varepsilon$.

Let *H* be the midpoint of *AC*, *HS* || *BC*; *HS* cuts the parabola at *M*. Let *QMO* || *AC*; and let us denote by [*U*] the volume generated by the rotation of the surface (*U*); from this

$$[EM] = [MB] \text{ and } [AM] = [MC],$$

hence

$$[EM] + [MC] = \frac{1}{2}V.$$

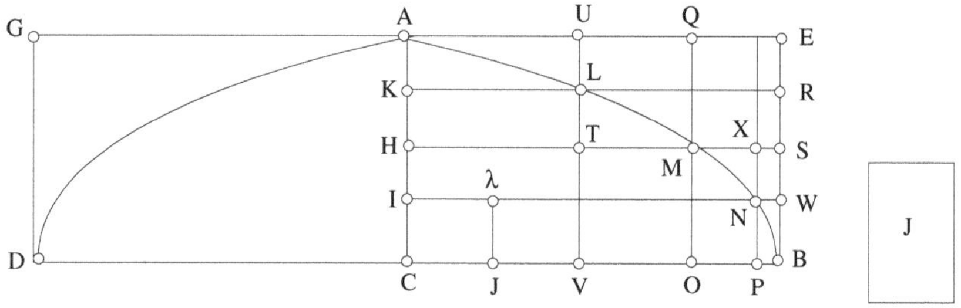

Fig. 2.1.6a

Let *K* be the midpoint of *AH*, *I* be the midpoint of *HC*; then in a similar manner

$$[QL] + [LH] = \frac{1}{2}[AM]$$

and

$$[SN] + [NO] = \frac{1}{2}[BM];$$

hence

$$[SN] + [NO] + [QL] + [LH] = \frac{1}{2}\{[AM] + [BM]\} = \frac{1}{4}V.$$

In this way Ibn al-Haytham first takes a subdivision of AC, in $n = 2^m$ equal parts, and successively subtracts

$$\frac{1}{2}V,\ \frac{1}{2}\left(\frac{1}{2}V\right)$$

and so on. He shows that if points of subdivision are increased sufficiently, then inevitably a remainder which is less than ε is reached.

Assume that we have in fact reached that step, that is

$$[BN] + [NM] + [ML] + [LA] < \varepsilon,$$

or using previous notations

$$V_n < \varepsilon;$$

let v_n be the volume of V_n, inside the paraboloid; it follows

$$v_n < \varepsilon;$$

just as

$$v = \frac{8}{15}V + \varepsilon,$$

therefore

$$v - v_n > \frac{8}{15}V.$$

But $v - v_n$ is equal to the solid whose base is the disc of radius PC and whose vertex is the disc of radius KL. Moreover, because of the properties of the parabola, then

$$\frac{AC^2}{LV^2} = \frac{BC}{BV},\ \frac{LV^2}{MO^2} = \frac{BV}{BO},\ \frac{MO^2}{NP^2} = \frac{BO}{BP};$$

but

$$MO = 2\,NP,\ \ LV = 3\,NP,\ \ AC = 4\,NP.$$

If we set $NP = 1$, then the ratios of NP, MO, LV, AC are equal to the ratios of n first natural integers, and the ratios of BC, BV, BO, BP are equal to the ratios of squares of first natural integers; it follows that the ratios of EA, RL, SM and WN are also equal to the ratios of the squares of first natural integers, as $BP = WN$, $BO = SM$, $BV = RL$, $BC = EA$. But

$$WI = SH = RK = AE$$

and

$$\frac{WN}{SM} = \frac{1^2}{2^2}, \ldots, \frac{RL}{EA} = \frac{3^2}{4^2} = \frac{(n-1)^2}{n^2}.$$

But from Lemma 5, it follows

$$\sum_{k=1}^{n-1}\left(n^2 - k^2\right)^2 \le \frac{8}{15} n \cdot n^4 \le \sum_{k=0}^{n-1}\left(n^2 - k^2\right)^2,$$

which for the corresponding segments gives

$$NI^2 + MH^2 + LK^2 \le \frac{8}{15}\left\{WI^2 + SH^2 + RK^2 + AE^2\right\}$$

and

$$NI^2 + MH^2 + LK^2 + AE^2 \ge \frac{8}{15}\left\{WI^2 + SH^2 + RK^2 + AE^2\right\}.$$

Areas of discs with respective radii the previous segments are marked by S_i, that is

$$S_k = \pi\,(n^2 - k^2)^2;$$

in particular

$$S_0 = \pi\, n^4,$$

therefore

$$\sum_{k=1}^{n-1} S_k \le \frac{8}{15} n S_0 \le \sum_{k=0}^{n-1} S_k.$$

Cylinders of base S_k and of height $h = \dfrac{AC}{n}$ are now marked W_k; we get

$$\sum_{k=1}^{n-1} W_k \le \frac{8}{15} V.$$

However, by construction:

$$\sum_{k=1}^{n-1} W_k = v - v_n;$$

therefore

$$v - v_n < \frac{8}{15} V,$$

which is absurd. Therefore

$$(1) \qquad v \leq \frac{8}{15} V.$$

Assume now that $v < \frac{8}{15} V$, that is $\frac{8}{15} V - v = \varepsilon$, and consider the same subdivision as used at the step where the total of surfaces which surround the parabola is smaller than ε. Let u_n be the volume of V_n, outside the paraboloid, therefore $u_n < \varepsilon$, so $v + u_n < \frac{8}{15} V$.

But solid $v + u_n$ is nothing more than a solid whose base is the disc of radius BC and whose vertex is the disc of radius AU. But we have shown that

$$\sum_{k=0}^{n-1} S_k \geq \frac{8}{15} n S_0;$$

therefore

$$\sum_{k=0}^{n-1} W_k \geq \frac{8}{15} V,$$

which is absurd, since

$$\sum_{k=0}^{n-1} W_k = v + u_n < \frac{8}{15} V.$$

It follows that

$$(2) \qquad v \geq \frac{8}{15} V;$$

and from (1) and (2) we get

$$v = \frac{8}{15} V.$$

Second and third cases: Assume that $A\hat{C}B < \frac{\pi}{2}$ and that $A\hat{C}B > \frac{\pi}{2}$.

Using the two relevant cases, Ibn al-Haytham shows in the same way as before, that for a portion of a paraboloid of revolution:

$$v = \frac{8}{15} V.$$

He also shows that

$$V_n = \frac{1}{2^n} V,$$

with V_n the sum of small cylinders surrounding the parabola. Here

$$\frac{1}{2^n} V = [BI].$$

2.2.2.4. *Study of surrounding solids*

Ibn al-Haytham goes on to investigate the behaviour of surrounding solids, when points of subdivision are increased indefinitely. He raises the problem of the variation of the ratio of two parts which make up infinitesimal solids, that is, parts which are inside and outside the paraboloid. In the case of the first type of paraboloid, these parts had the same volume, but it is not the case here.

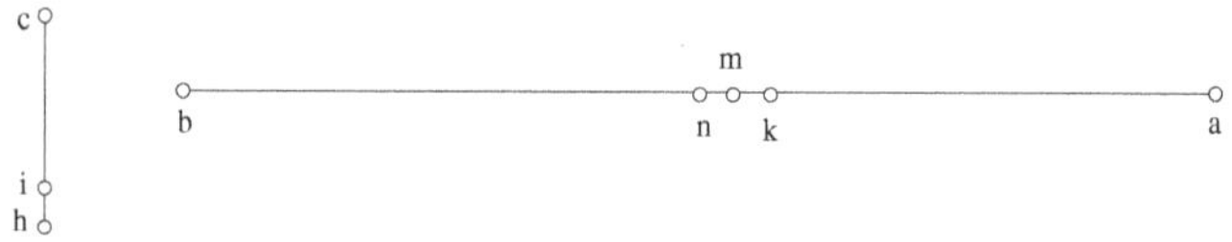

Fig. 2.1.6b

Let ab be the square of the number corresponding to segment AE, ($ab = 2^{2m}$ if AC was divided into 2^m parts)

$$an = \frac{ab}{2}, \quad nk = \frac{1}{30} ab;$$

hence

$$bk = \frac{8}{15} ab.$$

Let $hc = \sqrt{ab}$, $hi = \frac{1}{30}$ and point m such that $\frac{hi}{nm} = \frac{ab}{ch}$; hence

$$ab \cdot nm = \frac{1}{30}ch, \quad ab \cdot kn = \frac{1}{30}ab^2, \quad ab \cdot km = \frac{1}{30}ab^2 - \frac{1}{30}\sqrt{ab}.$$

But from relation (12) in Lemma 5, then

$$\sum_{k=1}^{n-1}\left(n^2 - k^2\right)^2 = \frac{8}{15}(n-1)n^4 + \frac{1}{30}n^4 - \frac{1}{30}n,$$

which is written here for the segments

$$LK^2 + MH^2 + NI^2 = \frac{8}{15}\left(RK^2 + SH^2 + WI^2\right) + \frac{1}{30}ab^2 - \frac{1}{30}\sqrt{ab}$$

$$= \frac{8}{15}\left(RK^2 + SH^2 + WI^2\right) + ab \cdot km;$$

but

$$ab \cdot bk = \frac{8}{15}ab^2,$$

therefore

$$(1) \qquad LK^2 + MH^2 + NI^2 + ab \cdot bm = \frac{8}{15}\left(RK^2 + SH^2 + WI^2 + BC^2\right).$$

Let J be a point on BC (Fig. 2.1.6a) such that

$$(2) \qquad \frac{BC^2}{CJ^2} = \frac{ab}{bm} = \frac{ab^2}{ab \cdot bm},$$

hence

$$CJ^2 = ab \cdot bm.$$

Let L_a be such that $JL_a \parallel CI$; from (1) we have

$$(3) \qquad CJ^2 + NI^2 + MH^2 + LK^2 = \frac{8}{15}\left(RK^2 + SH^2 + WI^2 + BC^2\right).$$

Let discs $S, S_1, \ldots, S_{n-1}$ have respective radii $CJ, NI, \ldots, LK$, and disc S_0 have radius BC, then

$$S + \sum_{k=1}^{n-1} S_k = \frac{8}{15}nS_0.$$

Corresponding cylinders of height $AK = \frac{AC}{n}$ are marked W, $W_1, \ldots,$ W_{n-1}, W_0; from this it follows

$$W + \sum_{k=1}^{n-1} W_i = \frac{8}{15} n W_0 = \frac{8}{15} V.$$

But we have shown that

$$v = \frac{8}{15} V,$$

therefore

$$v = W + \sum_{k=1}^{n-1} W_i ;$$

hence

$$W = v - \sum_{k=1}^{n-1} W_i = v_n,$$

with v_n the sum of parts of small surrounding solids inside the paraboloid.

We have also shown that $V_n = \pi r^2 h$, with V_n the sum of small surrounding solids, $r = BC$, $h = \frac{AC}{n} = IC$. It follows that

$$u_n = V_n - W = u,$$

u_n being the sum of parts of small surrounding solids outside the paraboloid; u_n is therefore equal to the cylinder generated by the rotation of the surface (BL_a). However

$$\frac{u_n}{v_n} = \frac{u}{W} = \frac{BC^2 - JC^2}{JC^2} = \frac{am}{bm},$$

since

$$\frac{BC^2}{JC^2} = \frac{ab}{bm}, \qquad \text{by (2).}$$

Volumes corresponding to the mth subdivision ($n = 2^m$) of u and W are expressed as $u(m)$ and $W(m)$. We show that

$$\frac{u(m+1)}{W(m+1)} > \frac{u(m)}{W(m)}.$$

In fact at step $(m + 1)$, AE corresponds to $(2n)^2$ and ab corresponds to n^2, therefore

$$\frac{AE}{\sqrt{AE}} > \frac{ab}{\sqrt{ab}} = \frac{ab}{ch};$$

but

$$\frac{hi}{nm} = \frac{ab}{ch},$$

therefore

$$\left(\frac{1}{30} / n'm'\right) > \left(\frac{1}{30} / nm\right);$$

$n'm'$ is the correspondent of nm at step $(m + 1)$, hence

$$n'm' < nm$$

and

$$n'b' > nb.$$

At each step of subdivision, it should be noted that AC is equal to the root of the *latus rectum* multiplied by $\sqrt{AE}$ which corresponds to $\sqrt{ab} = \frac{ab}{ch} = \frac{hi}{nm}$.

Passing from subdivision into $n = 2^m$ parts of AC to subdivision into $2n = 2^{m+1}$ parts, $\sqrt{ab}$ becomes $\sqrt{a'b'} = 2\sqrt{ab}$ and nm becomes $n'm' = \frac{nm}{2}$; furthermore nb becomes $n'b' = 4\,nb$. Thus

$$\frac{m'n'}{n'b'} = \frac{1}{8}\frac{mn}{nb};$$

as

$$\frac{mb}{ab} = \frac{mn + nb}{2nb} = \frac{1}{2}\frac{mn}{nb} + \frac{1}{2} \geq \frac{mn}{nb}$$

and

$$\frac{m'b'}{a'b'} = \frac{m'n' + n'b'}{2n'b'} = \frac{1}{2}\frac{m'n'}{n'b'} + \frac{1}{2} = \frac{1}{16}\frac{mn}{nb} + \frac{1}{2}.$$

So

$$\frac{mb}{ab} > \frac{m'b'}{a'b'},$$

from which

$$\frac{a'm'}{m'b'} > \frac{am}{mb},$$

that is

$$\frac{u(m+1)}{W(m+1)} > \frac{u(m)}{W(m)}.$$

Ibn al-Haytham has shown here that as points of subdivision increase, so does the ratio.

2.2.3. *Calculation of the volume of a sphere*

Reminding us that several of his predecessors had already determined the volume of a sphere, Ibn al-Haytham proposes to take up this proof again, to make it shorter and clearer. He does this, using the method he had already applied in the determination of a paraboloid. He starts with arithmetical lemmas to establish the inequalities necessary for determining the volume of a sphere.

Arithmetical lemmas

Ibn al-Haytham starts by restating two lemmas he had already established in his *Treatise on the Measurement of the Paraboloid.* He justifies this by saying that he wants this treatise on the volume of the sphere to be complete in itself. We will briefly go over these lemmas one by one.

Lemma 1:

$$\sum_{k=1}^{n} k = \frac{n^2}{2} + \frac{n}{2}.$$

The proof of this lemma differs from the one given in the previous treatise. This is how it is presented: for any integer n and for any integer $k \leq n$, we have

$$1 + n = k + (n - k + 1),$$

$$2\sum_{k=1}^{n} k = \sum_{k=1}^{n} k + \sum_{k=1}^{n} (n - k + 1) = n(n + 1),$$

hence

$$S_n = \sum_{k=1}^{n} k = \frac{n}{2}(n + 1).$$

Lemma 2:

$$\sum_{k=1}^{n} k^2 = \left(\frac{n}{3} + \frac{1}{3}\right) n \left(n + \frac{1}{2}\right).$$

The proof of this lemma by Ibn al-Haytham is identical to the one given in the previous treatise. In fact, we have

$$(n+1)\,S_n = S_n + n\,S_n = S_n + n^2 + n\,S_{n-1}$$
$$= (S_n + S_{n-1} + \dots + S_1) + (n^2 + (n-1)^2 + \dots + 1^2);$$

but from the previous lemma, we get

$$(n+1)S_n = \sum_{k=1}^{n} k^2 + \frac{1}{2}\sum_{k=1}^{n} k^2 = \frac{1}{2}\sum_{k=1}^{n} k,$$

hence

$$\left(n+\frac{1}{2}\right)S_n = \frac{3}{2}\sum_{k=1}^{n} k^2;$$

therefore

$$S_n^{(2)} = \sum_{k=1}^{n} k^2 = \frac{2}{3}\left(n+\frac{1}{2}\right)\left(\frac{n^2}{2}+\frac{n}{2}\right) = \left(\frac{n}{3}+\frac{1}{3}\right)n\left(n+\frac{1}{2}\right) = \frac{n^3}{3}+\frac{n^2}{2}+\frac{n}{6}.$$

Ibn al-Haytham goes on to show the inequalities.

Lemma 3:

(1)
$$\frac{n^3}{3}+\frac{n^2}{2} < \sum_{k=1}^{n} k^2 \le \frac{n^2}{3}+\frac{2}{3}n^2.$$

These inequalities can be proven immediately by using Lemma 2 and the fact that $\frac{n}{6} \le \frac{n^2}{6}$ because $n \ge 1$.

Let u_1, u_2, …, u_n be an arithmetical progression with ratio u_1, and with first term 0, this also gives

(2)
$$\frac{1}{2}u_n^2 + \frac{1}{3}nu_n^2 < \sum_{k=1}^{n} u_k^2 < \frac{1}{3}nu_n^2 + \frac{2}{3}u_n^2.$$

In fact, we have $u_k = ku_1$ $(1 \le k \le n)$, hence

$$\frac{u_k^2}{u_n^2} = \frac{k^2}{n^2};$$

therefore

$$\frac{1}{u_n^2}\sum_{k=1}^{n} u_k^2 = \frac{1}{n^2}\sum_{k=1}^{n} k^2,$$

hence the result, if Lemma 3 is applied.

In finalizing the arithmetic lemmas and the previous inequalities, Ibn al-Haytham proves the principal theorem of this treatise.

Theorem. — *The volume of a sphere is two thirds the volume of a circumscribed cylinder whose base is the largest disc of the sphere and whose height is equal to the diameter of the sphere.*

Let $AEBG$ be a rectangle, which, by its rotation around AE, forms a cylinder whose base is the largest disc of the sphere and whose height is the radius of the sphere. Using exactly the same rotation, the portion ABE forms a hemisphere, whereas the portion ABC generates a whole sphere. The previous proposition is therefore equivalent to the proposition that: *the volume of a hemisphere generated by rotation of the portion* ABE *is equal to two thirds the volume of a cylinder whose base is the largest disc of the sphere and whose height is the radius of the sphere.*

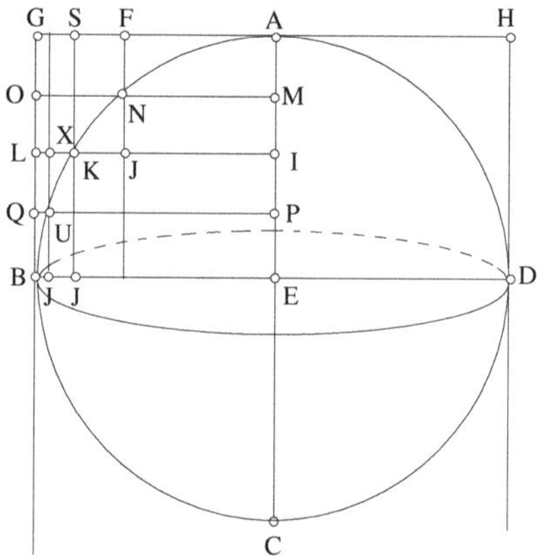

Fig. 2.1.7.

Let v be the volume of a hemisphere, V the volume of the associated cylinder and $[U]$ the volume of the solid U. Ibn al-Haytham proves that

$$v = \frac{2}{3} V.$$

First assume that

$$v > \frac{2}{3} V,$$

that is

$$v = \frac{2}{3} V + \varepsilon \quad (\varepsilon > 0).$$

Cut AE into two halves at I, then cut both AI and IE in half again at M and at P respectively and so on; it follows

$$[EK] + [KG] = \frac{1}{2}\ [AB] = \frac{1}{2} V,$$

$$[NI] + [NS]\ = \frac{1}{2}\ [AK],$$

$$[UJ] + [UL]\ = \frac{1}{2}\ [BK];$$

hence

$$[NI] + [NS] + [UJ] + [UL] = \frac{1}{2}\ [AK] + \frac{1}{2}\ [BK] = \frac{1}{2}\left(\frac{1}{2}V\right).$$

After the first subdivision, the remainder is thus $\frac{1}{2}V$; after the second subdivision, the remainder is $\frac{1}{2}\left(\frac{1}{2}V\right)$ and finally after the nth subdivision, the remainder is $\frac{1}{2^n}V < \varepsilon$, from Lemma X.1 of Euclid's *Elements* (or Ibn al-Haytham's theorem), generalized by Ibn al-Haytham [see *Treatise on the Division of Two Different Magnitudes Mentioned in the First Proposition of the 10th Book of Euclid*]. V_n is the remainder and v_n part of V_n which is inside the sphere. It follows

$$v_n < V_n < \varepsilon,$$

hence

$$v_n < \varepsilon.$$

However

$$v = \frac{2}{3}V + \varepsilon,$$

therefore

$$v - v_n > \frac{2}{3}V;$$

but

$$I_n = v - v_n$$

is the volume of the sum of cylinders of same height. Next, Ibn al-Haytham studies I_n.

Segments EP, EI, EM, EA are terms of an arithmetical progression with a ratio equal to the first term. From Lemma 3, therefore

$$(3) \qquad \frac{1}{2}EA^2 + \frac{1}{3}\left(EB^2 + PQ^2 + IL^2 + MO^2\right) < EP^2 + EI^2 + EM^2 + EA^2$$
$$< \frac{1}{3}\left(EB^2 + PQ^2 + IL^2 + MO^2\right) + \frac{2}{3}EA^2.$$

But

$$EP^2 + PU^2 = EU^2 = R^2 \text{ and } PQ = R;$$

hence

$$EP^2 + PU^2 = PQ^2,$$

$$EI^2 + IK^2 = IL^2,$$

$$EM^2 + MN^2 = MO^2,$$

$$EA^2 = EB^2.$$

Therefore

$$(EP^2 + EI^2 + EM^2 + EA^2) + (PU^2 + IK^2 + MN^2) = PQ^2 + IL^2 + MO^2 + EB^2,$$

hence

$$(4)\ PU^2 + IK^2 + MN^2 = PQ^2 + IL^2 + MO^2 + EB^2 - (EP^2 + EI^2 + EM^2 + EA^2).$$

If we take (3) into account, we have

$$PU^2 + IK^2 + MN^2 < \frac{2}{3}\left(EB^2 + PQ^2 + IL^2 + MO^2\right) - \frac{1}{2}EA^2,$$

a fortiori

$$PU^2 + IK^2 + MN^2 < \frac{2}{3}\left(EB^2 + PQ^2 + IL^2 + MO^2\right),$$

but

$$I_n = \pi\,(PU^2 + IK^2 + MN^2)\,EP;$$

hence

$$I_n < \frac{2}{3}\pi\,(EB^2 + PQ^2 + IL^2 + MO^2)\,EP.$$

Therefore

$$I_n < \frac{2}{3}V,$$

which is impossible and consequently

$$v \leq \frac{2}{3}V.$$

Assume now that

$$v < \frac{2}{3}V,$$

that is

$$v + \varepsilon = \frac{2}{3}V \qquad (\varepsilon > 0).$$

Let u_n be a part of V_n outside the sphere and assume that the figure represents the step where subdivisions determine $u_n < \varepsilon$. It follows

$$v + u_n < \frac{2}{3}V.$$

C_n is marked as $C_n = v + u_n$; C_n is therefore the volume of the sum of cylinders of same height. According to (3) and (4), we have

$$PU^2 + IK^2 + MN^2 > \frac{2}{3}\left(EB^2 + PQ^2 + IL^2 + MO^2\right) - \frac{2}{3}EA^2$$

and

$$PU^2 + IK^2 + MN^2 + EA^2 > \frac{2}{3}\left(EB^2 + PQ^2 + IL^2 + MO^2\right) + \frac{1}{3}EA^2;$$

a fortiori, we have

$$PU^2 + IK^2 + MN^2 + EA^2 > \frac{2}{3}\left(EB^2 + PQ^2 + IL^2 + MO^2\right).$$

But

$$C_n = \pi\,(EB^2 + PU^2 + IK^2 + MN^2) \cdot EP;$$

hence

$$C_n > \frac{2}{3}\,V,$$

which is impossible. Therefore

$$v = \frac{2}{3}\,V$$

and the proposition is proved.

2.3. *Translated texts*

Al-Ḥasan ibn al-Haytham

2.3.1. *On the Measurement of the Paraboloid*

2.3.2. *On the Measurement of the Sphere*

2.3.3. *On the Division of Two Different Magnitudes as Mentioned in the First Proposition of the Tenth Book of Euclid's Elements*

In the Name of God, the Forgiving, the Merciful
Glory to God in the Highest

TREATISE BY AL-ḤASAN IBN AL-ḤASAN IBN AL-HAYTHAM

On the Measurement of the Paraboloid

<*Introduction*>

For every discourse and for every writing, there is a reason that prompted the speaker or writer to say or write that which he said or wrote. We have made a careful study of the book by Abū al-Ḥasan Thābit ibn Qurra[1] *On the Measurement of the Paraboloid*, and we have concluded that he followed a course without any plan, forcing him to take a path through his explanation that was both long and laboriously difficult.

We then obtained a copy of a treatise by Abū Sahl Wayjan ibn Rustum al-Qūhī *On the Measurement of the Paraboloid*. We found it to be uncluttered and concise, but we noted that the author states that his reason for writing the treatise arose from his study of the book by Abū al-Ḥasan Thābit ibn Qurra on the measurement of this solid, the difficulties he found there, and his rejection of the method proposed. However, we find that the treatise by Abū Sahl, albeit less cluttered and easier to follow, includes a proof of only one of the two species of paraboloid.

There are two species of paraboloid, as we shall discover later. One of these is accessible and easy, the other is difficult and arduous. We find that Abū Sahl has limited his treatise to the measurement of <paraboloids belonging to> the accessible species, and that he makes no mention of the second species.

Having found the characteristics on which we have just commented in both these treatises, we therefore felt compelled to write this treatise. We intend in this treatise to provide a full discussion of the measurement of the two species of this solid, and to treat exhaustively all the concepts related to their measurement. However, in all that we shall mention and explain, we shall confine ourselves to the shortest possible paths by which the subject

[1] See Volume I, Chapter II.

may be approached, together with the most concise methods for determining the proofs, while still dealing fully with the concepts involved.

Let us now begin this study. May God help and support the accomplishment of that which pleases Him.

Given any plane figure, in the plane of which we draw a given straight line, fixed so that its position does not vary, and if the plane figure is rotated about the line until it returns to its original position, then that rotation will generate a compact body.

Given any portion of a parabola, in the plane of which we draw a given straight line, fixed so that its position does not vary, and if the portion of a parabola is rotated about the line until it returns to its original position, then that rotation will generate a compact body. The body generated in this way is known as a paraboloid.

Any given straight line in the plane of a parabola is either parallel to the diameter of the portion in whose plane it lies, or is the diameter itself, or it meets the diameter, either in their original positions, or in their extensions. If it is parallel to the diameter, then it is also a diameter. If it meets the diameter, then it meets the section at two points, and if it meets the section at two points, then it is an ordinate to one of the diameters of the section, as has been shown by the eminent Apollonius in his work on the *Conics*.

All the possible given straight lines in the plane of one of the portions of a parabola belong to one of two species; the diameters and the ordinates. If this is the case, then all the paraboloids generated by the movement of the parabola around one of the given straight lines in its plane may also be divided into two species; the solids generated by the movement of the section around its diameters, and the solids generated by the movement of the section around its ordinates. We now seek to measure <paraboloids in> each of these two species, and we begin by introducing a number of lemmas.

One of the two species, that is those <paraboloids> generated by the movement of the section around its diameters, does not require any of these lemmas. We have already mentioned in the introduction to this treatise that this species is simple and easy to deal with. But the second species, that is those <solids> generated by the movement of the section around its ordinates, and which is the more difficult of the two, requires these arithmetic lemmas.

Here is one of the lemmas: If we take any quantity of numbers beginning with one and increasing from one by one, and if we take half of the greatest of these and half of unity, which is the first of these, and if we add them and multiply their sum by the last number – which is the greatest of the numbers – then we obtain the sum of all the numbers.

If we take one third of the greatest of the successive numbers and one third of unity, and if we add them together and then multiply their sum by the last number, which is the greatest of the numbers, and if we then add one half of unity to the greatest number and then multiply the result by the result of the first multiplication, then the result of this multiplication is the sum of the squares of the numbers.

If we take one quarter of the greatest of the successive numbers, and if we add one quarter of unity and then multiply by the greatest number, and if we add one to the greatest number and then multiply the result by the greatest number, and if we finally multiply this product by the product from the first multiplication, then the result is the sum of the cubes of the successive numbers.

If we take one fifth of the greatest of the successive numbers, and if we add one fifth of unity to it, and if we multiply this sum by the greatest number, and if we then add one half of unity to the greatest number and then multiply the result by that obtained from the first multiplication, saving this product, and if we then add one to the greatest number and then multiply the result by the greatest number, and if we subtract one third of unity from the product, and if we finally multiply the remainder by what we saved earlier, then the result of all these is the sum of the square-squares of the successive numbers.

We shall begin by proving all these lemmas.

<*Lemmas*>

<1> Let the numbers *AB*, *CD*, *EG* and *HI* be successive numbers. Let *AB* be equal to one, and let the others increase from one by one.

I say that, if we take one half of HI, *and if we add one half of unity and multiply the sum by the number* HI, *then the result is the sum of the numbers* AB, CD, EG *and* HI.

Proof: We associate another <sequence of> successive numbers with these numbers, beginning with one and increasing from one by one. Arrange these in the reverse order to that in which the original numbers are arranged, and let them be *KH*, *LE*, *NC* and *MA*. Let *KH* be unity and let the other numbers increase from one by one. As *HI* exceeds *EG* by one unity, and as *KH* is equal to one, then *KI* exceeds *EG* by two. *LE* is equal to two. Therefore, *LG* is equal to *KI*. As *HI* exceeds *CD* by two, *KI* must exceed *CD* by three. But *NC* is equal to three. *ND* is therefore equal to *KI*. Similarly, we can also show that *MB* is equal to *KI*. All the numbers *MB*, *ND*, *LG* and *KI* are therefore equal. And the successive numbers beginning with one and increasing from one by one are equal in number to the number of unities contained in the last number. Therefore, the multiplicity of the numbers *AB*, *CD*, *EG* and *HI* is equal to the number of unities contained in *HI*. And the multiplicity of the numbers *AB*, *CD*, *EG* and *HI* is equal to the multiplicity of the numbers *MB*, *ND*, *LG* and *KI*. The multiplicity of the numbers *MB*, *ND*, *LG* and *KI*, which are equal, is therefore equal to the number of unities contained in *HI*. If the number *KI* is multiplied by the unities in *HI*, the result of the multiplication is the sum of the numbers *MB*, *ND*, *LG* and *KI*. The numbers *AB*, *CD*, *EG* and *HI* are successive, beginning with one and increasing from one by one. The numbers *KH*, *LE*, *NC* and *MA* are also successive, beginning with one and increasing from one by one, and the multiplicity of these numbers is equal to that of the previous numbers. They are therefore equal to them. The sum of all <these numbers> is therefore twice the sum of the numbers *AB*, *CD*, *EG* and *HI*. The sum of these numbers is therefore one half of the sum of the numbers *MB*, *ND*, *LG* and *KI*, and <the product of> *KI* and the unities in *HI* is equal to the sum of these numbers. The product of one half of *KI* and *HI* is therefore the sum of the numbers *AB*, *CD*, *EG* and *HI*. But *IK* is equal to the number *HI* – which is the last of the successive numbers – plus *KH*, which is one. One half of *KI* is therefore equal to one half of *HI* plus one half of unity.

The same can be shown for all the successive numbers beginning with one, regardless of their multiplicity.

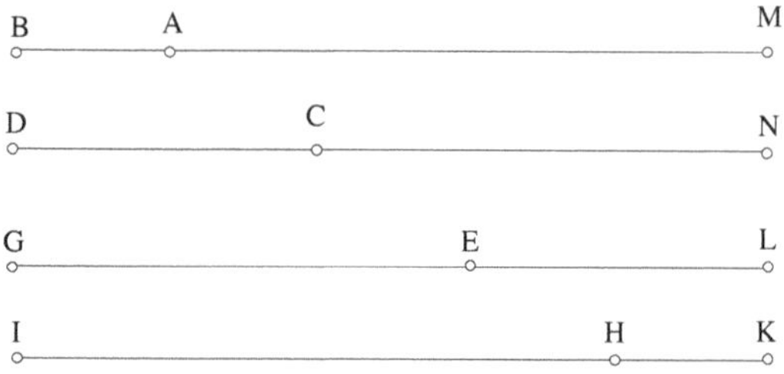

Fig. II.1.1

Therefore, if we take one half of the greatest of the successive numbers beginning with one and increasing from one by one, and if add one half of unity to it, and if we multiply this by the greatest of the numbers, then the result of this multiplication will be the sum of <all> the successive numbers beginning with one. That is what we wanted to prove.

This proof clearly shows that the sum of the successive numbers is one half of the square of the greatest number, plus one half of the number itself. The product of this latter number by its half is equal to one half of its square, and its product with one half of one is equal to one half of the number itself.

<**2**[2]> Similarly, let *AB*, *BC*, *CD* and *DE* be successive numbers in the position shown in the figure. By that, I mean the figure in the second proposition. Let *BG*, *CH*, *DI* and *EK* also be successive numbers beginning with one. *AB* is therefore equal to *BG*, *BC* is equal to *CH*, *CD* is equal to *DI*, and *DE* is equal to *EK*. Add one to each of the numbers *BG*, *CH*, *DI* and *EK*, and let these be the unities *PG*, *NH*, *MI* and *LK*. The product of *AB* and *BP* is therefore equal to the product of *AB* and *BG* plus the product of *AB* and *GP*. The product of *AB* and *BG* is equal to the square of *BG*, and the product of *AB* and *GP* is equal to *AB* itself, as *GP* is equal to one. And the product of *AC* and *CN* is equal to the product of *AC* and *CH* plus the product of *AC* and *HN*. As for the product of *AC* and *HN*, it is equal to *AC* itself, as *HN* is equal to one. The product of *AC* and *CH* is equal to the product of *BC* and *CH* plus the product of *AB* and *CH*. And the product of *BC* and *CH* is equal to the square of *CH*, as *BC* is equal to *CH*. Hence, the product of *AC* and *CN* is equal to *AC* itself plus the square of *CH*, plus the product of *AB* and *CH*. But the product of *AB* and *CH* is equal to the product of *AB* and *BP*, as *BP* is equal to *CH*. *BP* is equal to *BC*, as *BP* exceeds *BG*, which is equal to *BA*, by one unity; *BC* exceeds *AB* by one unity, and *BC* is equal to *CH*. *BP* is therefore equal to *CH*.

We have already shown that the product of *AB* and *BP* is equal to the square of *BG* plus *AB* itself. The product of *AC* and *CN* is therefore equal to the square of *BG* plus the square of *CH* plus *AB* itself plus *AC* itself.

Similarly, the product of *AD* and *DM* is equal to the product of *AD* and *DI*, plus the product[3] of *AD* and *IM*. The product of *AD* and *IM* is equal to *AD* itself, as *IM* is equal to one. And the product of *AD* and *DI* is equal to the product of *CD* and *DI*, plus the product of *AC* and *DI*. But the product of *CD* and *DI* is equal to the square of *DI*, and the product of *AC* and *DI* is equal to the product of *AC* and *CN*, as *DI* is equal to *CN*. *CN* exceeds *CH* – which is equal to *CB* – by one unity. Therefore, *CN* is equal to *CD*. But *CD* is equal to *DI*; therefore, *NC* is equal to *DI*. Therefore, the product of *AD* and *DM* is equal to *AD* itself plus the square of *DI*, plus the product of *AC* and *CN*. But we have already shown that the product of *AC* and *CN* is equal to the square of *HC* plus the square of *BG* plus *AC* itself plus *AB* itself. The product of *AD* and *DM* is therefore equal to the square of *DI* plus the square of *CH* plus the square of *BG* plus *AD* itself plus *AC* itself plus *AB* itself.

[2] See the statement for the previous lemma.
[3] We have sometimes inserted 'product' for the purposes of translation.

Using a similar method, we can also show that the product of *AE* and *EL* is equal to *AE* itself, plus the square of *EK*, plus the product of *AD* and *DM*. But we have already shown that the product of *AD* and *DM* is equal to the square of *DI* plus the square of *CH* plus the square of *BG* plus *AD* itself plus *AC* itself plus *AB* itself. The product of *AE* and *AL* is therefore equal to the square of *EK* plus the square of *DI* plus the square of *CH* plus the square of *BG* plus *AE* itself plus *AD* itself plus *AC* itself plus *AB* itself. But *AE* itself is equal to the sum of the successive numbers beginning with one and increasing from one by one, the last of which is *DE*, which is equal to *EK*. *AE* is therefore equal to one half of the square of *KE*, plus one half of *KE*, as we have shown in the corollary[4] to the first proposition. Similarly, *AD* is equal to one half of the square of *DI*, plus one half of *DI*, *AC* is equal to one half of the square of *CH*, plus one half of *CH*, and *AB* is equal to one half of the square of *BG*, plus one half of *BG*. The product of *AE* and *EL* is therefore equal to the sum of the squares of *EK*, *DI*, *CH*, *BG*, plus the halves of their squares, plus the halves of the numbers themselves. But the sum of the halves of the numbers *EK*, *DI*, *CH* and *BG* is equal to one half of *AE*, as *AE* is the sum of these numbers. The product of *AE* and *EL* is therefore equal to the squares of the successive numbers, the last of which is *KE*, plus the halves of their squares, plus one half of *AE*.

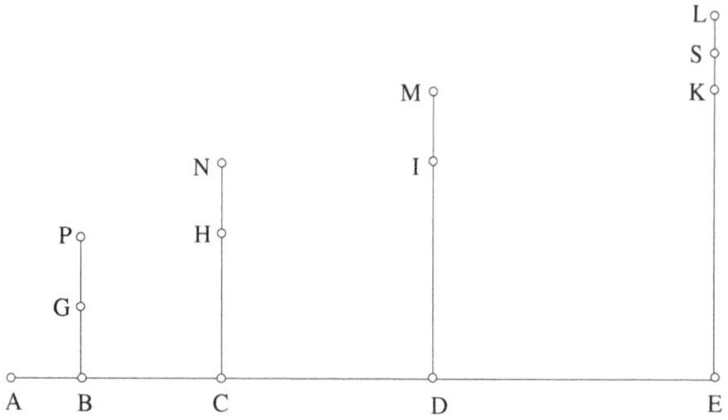

Fig. II.1.2

We divide *LK* into two halves at the point *S*. The product of *AE* and *EL* is then equal to the product of *AE* and *ES* plus the product of *AE* and *SL*. The product of *AE* and *SL* is equal to one half of *AE*, as *SL* is equal to one half of one. But the product of *AE* and *EL* is equal to the squares of the successive numbers, plus the halves of their squares, plus one half of *AE*. It follows that the product of *AE* and *ES* is equal to the squares of the successive numbers, the last of which is *KE*, plus the halves of their squares.

[4] *ʿaqib, corollarium,* πόρισμα.

The product of two thirds of *AE* and *ES* is therefore equal to the sum of the squares of the successive numbers, the last of which is *KE*. But we have already shown in the first proposition that the product of one half of *LE* – which is the last number plus unity – and *KE* is equal to the whole of *AE*. The product of two thirds of one half of *LE* – which is one third of *LE* – and *KE*, is therefore equal to two thirds of *AE*. If we take one third of *LE*, which is one third of *KE* – the greatest of the numbers – plus one third of unity, and if we multiply that by *KE* – which is the greatest of the numbers – and if we then multiply the product by *ES* – which is the greatest of the numbers plus one half of unity – then the result of this multiplication is equal to the sum of the squares of *EK*, *DI*, *CH* and *BG*, which are the successive numbers beginning with one and increasing from one by one. That is what we wanted to prove.

It follows clearly from this proof that the sum of the squares of the successive numbers is equal to one third of the cube of the greatest of them, plus one half of its square, plus one sixth of the number itself. The product of one third of *LE* and *EK* is equal to one third of the square of *EK*, plus one third of *EK*. If this is then multiplied by *ES*, the product of one third of the square of *EK* and *ES* is then equal to one third of the cube of *EK*, plus one sixth of the square of *EK*, as *KS* is equal to one half of one. And the product of one third of *EK* and *ES* is equal to one third of the square of *EK*, plus one sixth of *EK* itself. The product of one third of *LE* and *EK*, then the product of that result and *SE*, is therefore equal to one third of the cube of *EK*, plus one half of its square, plus one sixth of *EK* itself.

– **3** – Similarly, let the numbers *AB*, *BC*, *CD* and *DE* be successive square numbers. Let *AB* be unity – that is, the square of unity – *BC* the square of two, *CD* the square of three, and *DE* the square of four. Let the numbers *BG*, *CH*, *DI* and *EK* be the successive numbers themselves. Hence, *BG* is equal to one, *CH* to two, *DI* to three, and *EK* to four. The product of *DE* and *EK* is therefore equal to the cube of *EK*, and the product of *CD* and *DI* is equal to the cube of *DI*. The same applies to the rest. Now we add one to each of the successive numbers, as shown in the figure. The product of *AE* and *EL* is therefore equal to the product of *AE* and *EK*, plus the product of *AE* and *KL*. The product of *AE* and *KL* is equal to *AE* itself, as *KL* is equal to one. The product of *AE* and *EK* is equal to the product of *DE* and *EK*, plus the product of *AD* and *EK*. The product of *DE* and *EK* is the cube of *EK*, as *DE* is the square of *EK*. And the product of *AD* and *EK* is equal to the product of *AD* and *DM*, as *DM* is equal to *EK*, as we have shown earlier. The product of *AE* and *EL* is therefore equal to *AE* itself, plus the cube of *EK*, plus the product of *AD* and *DM*.

By following a similar proof, it can be shown that the product of *AD* and *DM* is equal to *AD* itself, plus the cube of *DI*, plus the product of *AC* and *CN*. The product of *AC* and *CN* is equal to *AC* itself, plus the cube of *CH*, plus the product of *AB* and *BP*, and the product of *AB* and *BP* is equal to *AB* itself, plus the cube of *BG*. The product of *AE* and *EL* is therefore equal to the cube of *EK* plus the cube of *DI* plus the cube of *CH* plus the cube of *BG* plus *AE* itself plus *AD* itself plus *AC* itself plus *AB* itself. But *AE* is the sum of the successive squares; *AE* is therefore equal to one third of the cube of *EK*, plus one half of its square, plus one sixth of *EK* itself, as we have shown earlier. Similarly, *AD* is equal to one third of the cube of *DI*, plus one half of its square, plus one sixth of *DI* itself, and *AC* is equal to one third of the cube of *CH*, plus one half of its square, plus one sixth of *CH* itself. Similarly, *AB* is equal to one third of the cube of *BG*, plus one half of its square, plus one sixth of *BG* itself, as the unity possesses such a property. The product of *AE* and *EL* is therefore equal to the sum of the cubes of the successive numbers, the last of which is *EK*, one third of their cubes,[5] the halves of their squares, and the sixths of the numbers themselves.

The product of *AE* and *EL* is equal to the product of *AE* and *ES*, plus the product of *AE* and *SL*. But the product of *AE* and *SL* is equal to one half of *AE*, as *SL* is equal to one half of unity. And one half of *AE* is equal to the <sum of the> halves of the squares of all the successive numbers, the last of which is *EK*. There remains the product of *AE* and *ES* equal to the cubes of all these numbers, plus the thirds of their cubes, plus the sixths of the numbers themselves. But *AE* results from taking the product of one third of *LE* and *EK*, which is then multiplied by *ES*.[6] The product of three quarters of one third of *LE* – that is one quarter of *LE* – and *EK*, which is then multiplied by *ES*, is therefore equal to three quarters of *AE*. If three quarters of *AE* is multiplied by *ES*, the result is the sum of the cubes of the successive numbers, plus the eighths of the numbers themselves, as if the whole of *AE* is multiplied by *ES*, the result is the sum of the cubes of these numbers, plus the thirds of their cubes, plus the sixths of the numbers themselves. Therefore, if we take one quarter of *LE* – which is equal to one quarter of *EK* plus one quarter of unity – and if that is multiplied by *EK*, and if that product is then multiplied by *ES*, and if the result is finally multiplied again by *ES*, then the result is the sum of the cubes of the numbers *EK*, *DI*, *CH* and *BG*, plus one eighth of the sum of these numbers. But the product of one quarter of *LE* and *EK*, multiplied by *ES*, and then

[5] One third of the sum of their cubes.

[6] See Proposition 2.

multiplied again by *ES*, is equal to the product of one quarter of *LE* and *EK*, multiplied by the square of *ES*. If there are three numbers, then the product of the first and the second, multiplied by the third, is equal to the product of the third and the second, multiplied by the first. The result of the multiplication of one quarter of *LE* by *EK* is a certain number, *ES* is a second number, and *ES* is also a third number. Therefore, one quarter of *LE* is multiplied by *EK*, and if that product is then multiplied by the square of *ES*, then the result is the sum of the cubes of the numbers *EK*, *DI*, *CH* and *BG*, plus one eighth of the sum of these numbers. But we have shown[7] that the product of one half of *LE* and *EK* is equal to the sum of these numbers. The product of one quarter of *LE* and *EK* is therefore equal to one half of the sum of these numbers. And the product of this half and one quarter of unity is equal to one eighth of the sum of these numbers. Therefore, if the product of one quarter of *LE* and *EK*, which is equal to one half of the sum of these numbers, is multiplied by the square of *ES*, then the result is the sum of the cubes of the successive numbers plus one eighth of their sum. If one quarter of unity is subtracted from the square of *ES*, and if the remainder is multiplied by the product of one quarter of *LE* and *EK*, that is one half of the sum of the numbers, then the result is the sum of the cubes of the successive numbers, and no more. But the square of *ES* is equal to the product of *LE* and *EK*, plus the square of *KS*, as may be shown by multiplying each of these numbers by the others. But the square of *KS* is equal to one quarter of unity, as *KS* is equal to one half of unity. If one quarter of unity is subtracted from the square of *ES*, then the remainder is the product of *LE* and *EK*. Therefore, if one quarter of *LE* is multiplied by *EK*, and if that product is then multiplied by the product of *LE* and *EK*, then the result is the sum of the cubes of the numbers *EK*, *DI*, *CH* and *BG*.

If we take one quarter of the greatest of a group of successive numbers beginning with one and increasing from one by one, regardless of their number, and if we add one quarter of unity, and if the result is multiplied by the greatest number, and if that product is multiplied by the product of the greatest number and the number that is one greater than it, we obtain from all these the sum of the cubes of the successive numbers beginning with one. That is what we wanted to prove.

It follows clearly from this proof that the sum of the cubes of the successive numbers is equal to one quarter of the square-square of the greatest of them, plus one half of its cube, plus one quarter of its square.

The product of one quarter of *LE* and *EK* is equal to one quarter of the square of *EK*, plus one quarter of *EK* itself, as one quarter of *LE* is equal to

[7] See Proposition 1.

one quarter of *EK* plus one quarter of unity. And the product of one quarter of *EK* and *EK* is equal to one quarter of the square of *EK*. The product of one quarter of unity and *EK* is equal to one quarter of *EK* itself. But the product of *LE* and *EK* is equal to the square of *EK* plus *EK* itself. The product of the square of *EK* and one quarter of the square of *EK* is equal to one quarter of the square-square of *EK*. The product of *EK* itself and one quarter of the square of *EK* is equal to one quarter of the cube of *EK*. The product of the square of *EK* and one quarter of *EK* itself is equal to one quarter of the cube of *EK*. And the product of *EK* itself and one quarter of *EK* is equal to one quarter of the square of *EK*. It follows that the result of multiplying one quarter of the square of *EK* plus one quarter of *EK* itself by the product of *LE* and *EK* is equal to one quarter of the square-square of *EK*, plus one half of the cube of *EK*, plus one quarter of the square of *EK*. The sum of the cubes of the successive numbers is therefore equal to one quarter of the square-square of the greatest of them, plus one half of its cube, plus one quarter of its square.

<4> Similarly, let the numbers *AB*, *BC*, *CD* and *DE* be successive cubic numbers, and let *BG*, *CH*, *DI* and *EK* be the successive numbers themselves. The product of *DE* and *EK* is then equal to the square-square of *EK*, the product of *CD* and *DI* is equal to the square-square of *DI*, the product of *BC* and *CH* is equal to the square-square of *CH*, and the product of *AB*, which is unity, and *BG*, which is also unity, is equal to the square-square of unity. We add to unity all these numbers beginning with one, as shown in the figure. The product of *AE* and *EL* is therefore equal to the product of *AE* and *EK*, plus the product of *AE* and *KL*. The product of *AE* and *KL* is equal to *AE* itself[8] and the product of *AE* and *EK* is equal to the product of *DE* and *EK*, plus the product of *AD* and *EK*. But the product of *DE* and *EK* is the square-square of *EK*, as *DE* is the cube of *EK*, and the product of *AD* and *EK* is equal to the product of *AD* and *DM*, as *DM* is equal to *EK*. The product of *AE* and *EL* is therefore equal to *AE* itself, plus the square-square of *EK*, plus the product of *AD* and *DM*. Therefore, the product of *AD* and *DM* is equal to *AD* itself plus the square-square of *DI*, plus the product of *AC* and *CN*. And similarly for the rest, given that this may be proved as shown previously. The product of *AE* and *EL* is therefore equal to the square-squares of *EK*, *DI*, *CH* and *BG*, plus the numbers *AE*, *AD*, *AC* and *AB* themselves. But we have shown[9] that *AE* is equal to one quarter of the square-square of *EK*, plus one half of the cube *EK*, plus one quarter of the square of *EK*, as *AE* is the sum of the cubes of the successive

[8] As $KL = 1$.
[9] Proposition 3.

numbers, the greatest of which is *EK*. Similarly, *AD* is equal to one quarter of the square-square of *DI*, plus one half of its cube, plus one quarter of its square. And similarly, *AC* is equal to one quarter of the square-square of *CH*, plus one half of its cube, plus one quarter of its square. *AB*, which is unity, is equal to one quarter of the square-square of *BG*, plus one half of its cube, plus one quarter of its square. The product of *AE* and *EL* is therefore equal to the <sum of the> square-squares of all the successive numbers, the greatest of which is *EK*, plus the quarters of their square-squares, plus the halves of their cubes, plus the quarters of their squares. Therefore, if four fifths of *AE* is multiplied by *EL*, the result is the square-squares of the successive numbers, plus two fifths of their cubes, plus one fifth of their squares. The product of four fifths of *AE* and *SL*, which is one half of one, is equal to two fifths of *AE*, which is equal to the sum of the cubes of these successive numbers. The product of four fifths of *AE* and *ES* remains equal to the <sum of the> square-squares of the successive numbers plus one fifth of their squares. But *AE* is the result of multiplying one quarter of *LE* by *EK*, which product is then multiplied by the product of *LE* and *EK*.[10] Therefore, if four fifths of one quarter of *LE* – which is equal to one fifth of *LE* – is multiplied by *EK*, and that product is then multiplied by the product of *LE* and *EK*, the result is equal to four fifths of *AE*. If this is multiplied by *ES*, the result is the sum of the square-squares of the successive numbers, plus one fifth of their squares. Therefore, if one fifth of *LE* is multiplied by *EK*, and that product is then multiplied by the product of *LE* and *EK*, and if that result is then multiplied by *ES*, the final result is equal to the square-squares of the successive numbers, plus one fifth of their squares. If the product of the numbers is commuted,[11] it remains the same. Therefore, if one fifth of *LE* is multiplied by *EK*, and that product is then multiplied by *ES*, and if that result is then multiplied by the product of *LE* and *EK*, the final result is equal to the square-squares of the successive numbers, plus one fifth of their squares.

But[12] the product of one third of *LE* and *EK*, subsequently multiplied by *ES*, is the sum of the squares of the successive numbers, the greatest of which is *EK*. And the product of one fifth of *LE* and *EK*, subsequently multiplied by *ES*, is therefore equal to three fifths of the squares of the successive numbers, as one fifth is equal to three fifths of one third. The product of three fifths of the squares of the successive numbers, the last of which is *EK*, and the product of *LE* and *EK* is therefore equal to the <sum of the> square-squares of the successive numbers plus one fifth of their

[10] See Proposition 3.
[11] Lit.: moving them forwards and then backwards.
[12] Proposition 2.

squares. But the product of one third of one and three fifths of the squares of the successive numbers is equal to one fifth of their squares. Therefore, if one third of one is subtracted from the product of *LE* and *EK*, and if the remainder is multiplied by three fifths of the squares of these successive numbers, then the result will be equal to the square-squares of these numbers alone. The product of one fifth of *LE* and *EK*, subsequently multiplied by *ES*, and then multiplied again by the product of *LE* and *EK*, from which one third of unity has been subtracted, is therefore the sum of the square-squares of these numbers.

If we take one fifth of the greatest of the successive numbers beginning with one and increasing from one by one, plus one fifth of unity, and if the result is multiplied by the greatest number, and if we then multiply this product by the greatest number plus one half of unity, and if we retain that result, and if we then add one to the greatest number and multiply that result by the greatest number, and if we subtract only one third of unity from that product, and if we multiply the remainder by the result we retained earlier, then the final result of all that is the sum of the square-squares of the successive numbers. That is what we wanted to prove.

<5> Similarly, let the numbers *AB*, *CD*, *EG*, *HI* and *KL* be the squares of successive numbers. Let each of <the numbers> *MB*, *ND*, *PG* and *OI* be equal to *KL*.

I say that the sum of the squares of AM, CN, EP *and* HO *is less than one third plus one fifth of the sum of the squares of* MB, ND, PG, OI *and* KL, *and greater than one third plus one fifth of the sum of the squares of* MB, ND, PG *and* OI, *and that the sum of the squares of* AM, CN, EP, HO *and* KL *is greater than one third plus one fifth of <the sum of> the squares of* MB, ND, PG, OI *and* KL.

Proof: Let *SB* be equal to twice *MB*, *SD* equal to twice *ND*, *SG* equal to twice *PG*, and *SI* equal to twice *OI*. The product of *SH* and *HI*, plus the square of *HO*, is therefore equal to the square of *OI*, and the product of *SE* and *EG*, plus the square of *EP*, is therefore equal to the square of *PG*. The same applies to the subsequent numbers. If the product of *SH* and *HI* is subtracted from the square of *OI*, the remainder is equal to the square of *HO*, and the same applies to the subsequent numbers. But, if the square of *HI* is subtracted from the product of *SI* and *IH*, the remainder is equal to the product of *SH* and *HI*. Similarly, if the square of *GE* is subtracted from the product of *SG* and *GE*, the remainder is equal to the product of *SE* and *EG*. If the square of *DC* is subtracted from the product of *SD* and *DC*, the remainder is equal to the product of *SC* and *CD*. And, if the square of *BA* is subtracted from the product of *SB* and *AB*, the remainder is equal to the product of *SA* and *AB*. But the product of *SI* and *IH*, plus the product of *SG* and *GE*, plus the product of *SD* and *DC*, plus the product of *SB* and *BA*, is equal to the product of *SI* and the sum of *IH*, *GE*, *DC* and *BA*, which is the sum of the squares of the successive numbers. The square of *IH*, the square of *GE*, the square of *DC*, and the square of *BA* are the square-squares of the successive numbers. If twice *OI*, that is twice *KL*, is multiplied by the sum of the squares of the successive numbers, the last of which is the square number *HI*, and if the square-squares of the successive numbers, the last of which is *HI*, is subtracted from the result obtained, then the remainder is equal to the sum of the product of *SH* and *HI*, the product of *SE* and *EG*, the product of *SC* and *CD*, and the product of *SA* and *AB*. And if this remainder is subtracted from the sum of the squares of *OI*, *PG*, *ND* and *MB*, which are equal, the remainder is equal to the sum of the squares of *HO*, *EP*, *CN* and *AM*.

Let *UQ* be the side of the square *KL*, and let *UJ* be unity. *JQ* is therefore the side of the square *HI*. Divide *UJ* into two halves at the point *V*. As *JQ* is the side of the square *HI*, *JQ* is therefore the last of the successive numbers whose squares are *AB*, *CD*, *EG* and *HI*. *JU* is equal to unity. The product of one third of *UQ* and *QJ*, then the product of that result and *QV*, is equal to the sum of *AB*, *CD*, *EG* and *HI*, which are the successive squares.[13] Therefore, if one third of *UQ* is multiplied by *QJ*, and if the result is then multiplied by *QV*, and that result is then multiplied by twice *KL*, the final result is the product of twice *KL* and the sum of *AB*, *CD*, *EG* and *HI*. But the product of one third of *UQ* and *QJ*, then the product of that result and *QV*, then the product of that result and twice *KL*, is equal to the product of *UQ* and *QJ*, then the product of that result and *QV*, then the product of that result and one third of twice *KL*, which is equal to two thirds of *KL*. The product of *UQ* and *QJ*, then the product of that result and *QV*, and the product of that result and two thirds of *KL*, is therefore equal to the product of twice *KL* and the sum of *AB*, *CD*, *EG* and *HI*, which are the successive squares. We have already shown[14] that the product of one fifth of *UQ* and *QV*, multiplied by *QJ*, and again multiplied by the product of *UQ* and *QJ* from which is subtracted one third of unity, is equal to <the sum of> the square-squares of the successive numbers. It is therefore equal to the sum of the squares of *AB*, *CD*, *EG* and *HI*, which are the squares of the successive numbers. Let *LY* be the product of *UQ* and *QJ*, and let *YZ* be one third of unity. The product of one fifth of *UQ* and *QV*, then the product of that result and *QJ*, then the product of that result and *LZ*, is equal to the sum of the squares of *AB*, *CD*, *EG* and *HI*. If the product of each of the

[13] Proposition 2.
[14] Proposition 4.

numbers and the others is commuted,[15] it remains the same. The product of *UQ* and *QV*, then the product of that result and *QJ*, then the product of that result and *LZ*, is therefore equal to the sum of the squares of *AB*, *CD*, *EG* and *HI*. If *UQ* is multiplied by *QV*, then that result is multiplied by *QJ*, and that result is multiplied by one fifth of *LZ*, and if the product of *UQ* and *QV*, then the product of that result and *QJ*, then the product of that result and two thirds of *KL* is subtracted from the first result, then the remainder is equal to the product of *SH* and *HI*, plus the product of *SE* and *EG*, plus the product of *SC* and *CD*, plus the product of *SA* and *AB*. But if we subtract the product of *UQ* and *QV*, then that result is multiplied by *QJ*, and that result is multiplied by one fifth of *LZ*, from the product of *UQ* and *QV*, then the product of that result and *QJ*, then the product of that result and two thirds of *KL*, then the remainder is equal to the product of *UQ* and *QV*, then the product of that result and *QJ*, then the product of that result and one fifth plus one sixth plus one tenth of *LZ*, then the product of that result and two thirds of *KZ*.

Let *LT* be equal to the product of *UQ* and *QV*. There remains *TK* equal to one half of *UQ*, as *KL* is equal to the square of *UQ*. Therefore, *KL* is equal to the product of *UQ* and *QV*, plus <the product of> *UQ* and *UV*. The product of *UQ* and *UV* is equal to one half of *UQ*, as *UV* is equal to one half of one. Therefore, *TY* is also equal to *TK*, as *YK* is equal to *UQ*, and *TK* is the product of *UQ* and *UJ*, which is one. The product of *UQ* and *QV*, then the product of that result and *QJ*, then the product of that result and one fifth plus one sixth plus one tenth of *LZ*, plus two thirds of *KZ*, is equal to the product of *LT* and one fifth plus one sixth plus one tenth of *LZ*, plus two thirds of *KZ*, then the product of that result and *QJ*. But, as *LT* is equal to the product of *UQ* and *QV*, and two thirds of *KZ* is equal to one fifth plus one sixth plus one tenth of *KZ*, plus one fifth of *KZ* as well, the product of *LT* and one fifth plus one sixth plus one tenth of *LZ*, plus one fifth plus one sixth plus one tenth of *KZ* – this sum being equal to one fifth plus one sixth plus one tenth of *LK* – plus <the product of *LT*> and one fifth of *KZ*, then the product of that result and *QJ*, is equal to the product of *SH* and *HI*, plus the product of *SE* and *EG*, plus the product of *SC* and *CD*, plus the product of *SA* and *AB*. But the product of *LT* and one fifth plus one sixth plus one tenth of *LK* is equal to the product of *LK* and one fifth plus one sixth plus one tenth of *LT*, and the product of *LT* and one fifth of *KZ* is equal to the product of *LT* and two fifths of *KT* plus two fifths of one sixth of unity, as *KT* is one half of *KY*, and one sixth is equal to one half of *YZ*. The product of *KL* and one fifth plus one sixth plus one tenth of *LT*, plus the product of *LT* and two fifths of *KT* plus two fifths of one sixth of unity, that is two

[15] Lit.: moving them forwards and then backwards.

thirds of one tenth of unity, then the product of that result and *QJ*, is equal to the sum of the products of *SH* and *HI*, *SE* and *EG*, *SC* and *CD*, and *SA* and *AB*. But, as the numbers *AB*, *CD*, *EG*, *HI* and *KL* are the squares of the successive numbers, and *UQ* is the side of *KL*, *UQ* is the last of the successive numbers whose precedents are the squares. The number of unities in *UQ* is therefore equal to the number of these numbers, and the number of these successive numbers is equal to the number of their squares. The number of *AB*, *CD*, *EG*, *HI* and *KL* is therefore the number of unities in *UQ*, and *UJ* is equal to one. *QJ* contains the same number of unities as the number of *AB*, *CD*, *EG* and *HI*. The number of these numbers is equal to the number of *MB*, *ND*, *PG* and *OI*, which are equal and equal to *KL*. Therefore, if the square of *KL* is multiplied by the unities in *QJ*, the result is equal to the sum of the squares of the numbers *OI*, *PG*, *ND* and *MB*. But we have shown that, if *KL* is multiplied by one fifth plus one sixth plus one tenth of *LT*, and if this product is added to the product of *LT* and two fifths of *KT* plus two thirds of one tenth of unity, and if the result obtained is multiplied by *QJ*, then the final result is the sum of the products of *SH* and *HI*, *SE* and *EG*, *SC* and *CD*, and *SA* and *AB*. Therefore, if we subtract the product of *KL* and one fifth plus one sixth plus one tenth of *LT*, plus the product of *LT* and two tenths of *KT*, plus two thirds of one tenth of unity, from the square of *KL*, and if the remainder is multiplied by *QJ*, then the result is that which remains of <the sum of> the squares of *OI*, *PG*, *ND* and *MB*, which is <the sum of> the squares of *MA*, *NC*, *PE* and *OH*.[16] But, if we take the square of *KL* and subtract the product of *KL* and one fifth plus one sixth plus one tenth of *LT*, plus the product of *LT* and two fifths of *KT*, plus two thirds of one tenth of unity, then the remainder is the product of *KL* and one third plus one fifth of *LT*, plus the product of *KL* and the whole of *KT*, from which is subtracted the product of *LT* and two fifths of *KT* plus two thirds of one tenth of unity. But the whole of *KT* is equal to one third plus one fifth of *KT* plus one fifth plus one sixth plus one tenth of *KT*. That which remains of the square of *KL* is therefore the product of *KL* and one third plus one fifth of *KL* plus one fifth plus one sixth plus one tenth of *KT*, from which is subtracted the product of *LT* and two fifths of *KT* plus two thirds of one tenth of unity. Therefore, if *KL* is multiplied by one third plus one fifth of *KL* plus one fifth plus one sixth plus one tenth of *KT*, from which is subtracted the product of *LT* and two fifths of *KT* plus two thirds of one tenth of unity, and if the remainder is multiplied by *QJ*, then the result is the sum of the squares of *MA*, *NC*, *PE* and *OH*.

[16] In other words, the result is:
$OI^2 + PG^2 + ND^2 + MB^2 - [(SH \cdot HI) + (SE \cdot EG) + (SC \cdot CD) + (SA \cdot AB)] = MA^2 + NC^2 + PE^2 + OH^2$.

We set the ratio of *LK* to *KT* equal to the ratio of *TK* to *KO′*.[17] The ratio of *KL* to *LT* is therefore equal to the ratio of *KT* to *TO′*. The product of *LT* and *TK* is therefore equal to the product of *KL* and *TO′*, and the product of *LT* and two fifths of *KT* is equal to the product of *KL* and two fifths of *O′T*. But, as the ratio of *LK* to *KT* is equal to the ratio of *TK* to *KO′*, the product of *LK* and *KO′* is equal to the square of *KT*. *KT* is one half of *UQ*, as has been shown earlier, and its square is therefore one quarter of the square of *UQ*. But *KL* is the square of *UQ*, and the square of *KT* is therefore one quarter of *KL*. The product of *KL* and *KO′* is then equal to one quarter of *KL*, and *KO′* is therefore equal to one quarter of unity.

We set *O′Z* equal to one sixth of unity.[18] The product of *LT* and two thirds of one tenth of unity is therefore equal to the product of *LT* and two fifths of *O′Z*. We set the ratio of *O′Z* to *ZW* equal to the ratio of *TK* to *KO′*,[19] which is equal to the ratio of *LK* to *KT*. The ratio of *KL* to *LT* is therefore equal to the ratio of *ZO′* to *O′W*. The product of *LT* and *O′Z* is therefore equal to the product of *KL* and *WO′*, and the product of *LT* and two fifths of *O′Z* is equal to the product of *KL* and two fifths of *WO′*. The product of *LT* and two fifths of *KT*, multiplied by two thirds of one tenth of unity, is therefore equal to the product of *KL* and two fifths of *WT*.

We set *TI′* equal to six sevenths of *TW*.[20] The ratio of *WT* to *TI′* is then equal to the ratio of one fifth plus one sixth plus one tenth, that is 14 over 30, to two fifths, that is 12 over 30. The product of *KL* and two fifths of *WT* is therefore equal to the product of *KL* and one fifth plus one sixth plus one tenth of *TI′*. The product of *LT* and two fifths of *KT*, multiplied by two thirds of one tenth of unity, is therefore equal to the product of *KL* and one fifth plus one sixth plus one tenth of *TI′*. If the product of *KL* and one fifth plus one sixth plus one tenth of *TI′* is subtracted from the product of *KL* and one fifth plus one sixth plus one tenth of *KT*, the remainder is equal to the product of *KL* and one fifth plus one sixth plus one tenth of *KI′*. If the product of *KL* and one fifth plus one sixth plus one tenth of *LT*, plus the product of *LT* and two fifths of *KT* multiplied by two thirds of a tenth of unity, is subtracted from the square of *KL*, the remainder is equal to the product of *KL* and one third plus one fifth of *KL*, multiplied by one fifth plus one sixth plus one tenth of *KI′*. If this is then multiplied by *QJ*, the result is the sum of the squares of *MA*, *NC*, *PE* and *OH*. But the product of *KL* and

[17] The point *O′* is chosen such that this equality holds.

[18] The letter *Z* used here designates a point that is different from the point *Z* referred to earlier.

[19] The point *W* is chosen so as to satisfy this ratio.

[20] The point *I′* is chosen such that the equality $TI' = \frac{6}{7}TW$ holds.

one third plus one fifth of *KL* is equal to one third plus one fifth of the square of *KL*. If this is then multiplied by *QJ*, the result is equal to one third plus one fifth of the sum of the squares of *OI*, *PG*, *ND* and *MB*, as the number of unities in *QJ* is equal to the number of these numbers. The squares of *MA*, *NC*, *PE* and *OH* are equal to one third plus one fifth of the squares of *MB*, *ND*, *PG* and *OI*, plus the product of *KL* and one fifth plus one sixth plus one tenth of *KI'*, multiplied by *QJ*. But the product of *KL* and one fifth plus one sixth plus one tenth of *KI'*, multiplied by *QJ*, is equal to the product of one fifth plus one sixth plus one tenth of *KI'* and *QJ*, multiplied by *KL*. But one fifth plus one sixth plus one tenth of *KI'* is equal to one fifth plus one sixth plus one tenth of *I'W* plus one fifth plus one sixth plus one tenth of *WZ*, plus one fifth plus one sixth plus one tenth of *ZK*. *I'W* is therefore equal to one seventh of *WT*, as *TI'* is equal to six sevenths of *WT*. And one fifth plus one sixth plus one tenth of one seventh is equal to one seventh of one fifth plus one sixth plus one tenth, that is fourteen parts out of thirty parts. The seventh is therefore equal to two <parts> out of thirty, that is two thirds of one tenth. One fifth plus one sixth plus one tenth of *I'W* is therefore equal to two thirds of one tenth of *TW*. We take two thirds of one tenth of *KW* and we add them to this result. Of the one fifth plus one sixth plus one tenth of *KW*, there remains two fifths <of *KW*>. Therefore, two thirds of one tenth of *TW*, plus two thirds of one tenth of *KW*, equals two thirds of one tenth of *KT*. Therefore, one fifth plus one sixth plus one tenth of *KI'* is equal to two thirds of one tenth of *KT* plus two fifths of *KW*. But two thirds of one tenth of *KT* is equal to one third of one tenth of *UQ*, as *KT* is one half of *UQ*. If one third of one tenth of *UQ* is multiplied by *QJ*, the result is one third of one tenth of *LY*, as the product of *UQ* and *QJ* is *LY*. The product of one fifth plus one sixth plus one tenth of *KI'* and *QJ* is therefore equal to one third of one tenth of *LY*, plus the product of two fifths of *KW* and *QJ*. But *KZ* is equal to one half of one sixth of unity, as *KO'* is equal to one quarter of unity and *O'Z* is equal to one sixth of unity. Two fifths of *KZ* is therefore equal to one third of one tenth of unity. Therefore, if this is multiplied by *QJ*, the result is equal to one third of one tenth of *QJ*, which is equal to *KY* less one unity, as *KY* is equal to *UQ*. Therefore, if one third of one tenth of *QJ* is added to one third of one tenth of *LY*, the result is one third of one tenth of *KL* less one third of one tenth of unity. The product of one fifth plus one sixth plus one tenth of *KI'* and *QJ* is therefore equal to one third of one tenth of *KL*, less one third of one tenth of unity plus the product of two fifths of *ZW* and *QJ*. But, if one third of one tenth of *KL* less one third of one tenth of unity is multiplied by *KL*, the result is equal to one third of one tenth of the square of *KL* less one third of one tenth of *KL*, as the product of one third of one tenth of unity and *KL* is equal to one third of one tenth of *KL*. The product of *KL* and one fifth plus one sixth plus one tenth of *KI′*, multiplied by *QJ*, is therefore equal to one third of one tenth of the square of *KL*, less one third of one tenth of *KL*, plus the product of *KL* and two fifths of *ZW*, multiplied by *QJ*. But we have assumed the ratio of *O′Z* to *ZW* as being equal to the ratio of *TK* to *KO′*, which is equal to the ratio of *LK* to *KT*. The ratio of *KL* to *KT* is therefore equal to the ratio of *O′Z* to *ZW*. The product of *LK* and *ZW* is therefore equal to the product of *KT* and *O′Z*. But the product of *KT* and *O′Z* is equal to one sixth of *KT*, as *O′Z* is equal to one sixth of unity. The product of *KL* and *ZW* is therefore equal to one sixth of *KT*. The product of *KL* and two fifths of *ZW* is therefore equal to two fifths of one sixth of *KT*, which is equal to two thirds of one tenth of *KT*, which is equal to one third of one tenth of *UQ*, as *KT* is equal to one half of *UQ*. Therefore, if one third of one tenth of *UQ* is multiplied by *QJ*, the result is one third of one tenth of *LY*, as the product of *UQ* and *QJ* is *LY*. The product of *KL* and two fifths of *ZW*, multiplied by *QJ*, is equal to one third of one tenth of *LY*. The product of *KL* and one fifth plus one sixth plus one tenth of *KI′*, multiplied by *QJ*, is equal to one third of one tenth of the square of *KL*, plus one third of one tenth of *LY*, less one third of one tenth of *KL*. But one third of one tenth of *KL* is equal to one third of one tenth of *LY*, plus one third of one tenth of *KY*. If that which is added is taken away from that which is subtracted, there remains one third of one tenth of *KL* and one third of one tenth of *KY*, which is equal to *UQ*. The product of *KL* and one fifth plus one sixth plus one tenth of *KI′*, multiplied by *QJ*, is therefore equal to one third of one tenth of the square of *KL*, less one third of one tenth of *UQ*, which is the side of *KL*. *UQ* is equal to a whole number of unities, as it is the last of the successive numbers. But *KL* is the square of *UQ*, and therefore *KL* is greater than *UQ*. One third of one tenth of *UQ* is therefore less than one third of one tenth of the square of *KL*. It is also less than one third of one tenth of *KL* itself, as *KL* is also equal to a whole number of unities. *KL* is therefore a multiple of *UQ*.

But we have already shown that the sum of the squares of *MA*, *NC*, *PE* and *OH* is equal to one third plus one fifth of the sum of the squares of *MB*, *ND*, *PG* and *OI*, plus the product of *KL* and one fifth plus one sixth plus one tenth of *KI′*, then multiplied by *QJ*. The sum of the squares of *MA*, *NC*, *PE* and *OH* is therefore equal to one third plus one fifth of the sum of the squares of *MB*, *ND*, *PG* and *OI*, plus one third of one tenth of the square of *KL*, less one third of one tenth of the side of *KL*. But one third of one tenth of the square of *KL*, less one third of one tenth of its side, is less than one third plus one fifth of the square of *KL*, by one half of the square of *KL* plus one third of one tenth of its side, as if one third of a tenth is subtracted from one third plus one fifth, the remainder is one half. The squares of *MA*, *NC*, *PE* and *OH* are less than one third plus one fifth of the sum of the squares of *MB*, *ND*, *PG*, *OI* and *KL*, by one half of the square of *KL* plus one third of one tenth of its side. Therefore, if the squares of *MA*, *NC*, *PE* and *OH* are added to one half of the square of *KL* plus one third of one tenth of the side of *KL*, the result is equal to one third plus one fifth of the squares of *BM*, *ND*, *PG*, *OI* and *KL*. And if the squares of *MA*, *NC*, *PE* and *OH* are added to the whole square of *KL*, the result is greater than one third plus one fifth of the squares of *MB*, *ND*, *PG*, *OI* and *KL*, less one third of one tenth of the side of *KL*.

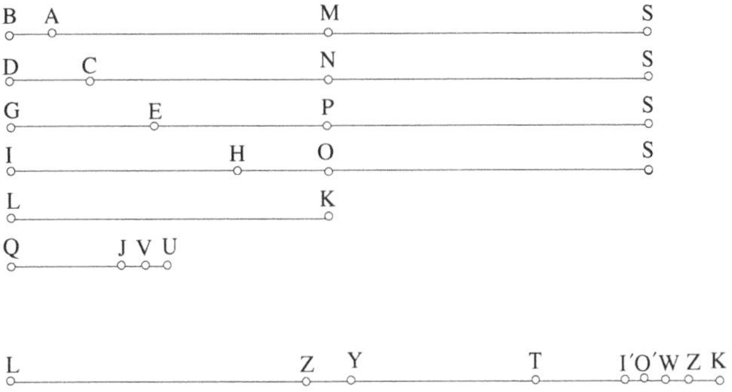

Fig. II.1.3*

But one half of the square of *KL* is greater than one third of one tenth of *UQ*. The sum of the squares of *MA*, *NC*, *PE*, *OH* and *KL* is therefore greater than one third plus one fifth of the sum of the squares of *MB*, *ND*, *PG*, *OI* and *KL*.

From all we have mentioned, we have therefore shown that the sum of the squares of *MA*, *NC*, *PE* and *OH* is less than one third plus one fifth of the squares of *MB*, *ND*, *PG*, *OI* and *KL* and greater than one third plus one fifth of the squares of *MB*, *ND*, *PG* and *OI*; and that the squares of *MA*, *NC*, *PE*, *OH* and *KL* are greater than one third plus one fifth of the squares of *MB*, *ND*, *PG*, *OI* and *KL*. That is what we wanted to prove.

From this proof it can clearly be seen that, given any number of equal straight lines, and given a set of successive square numbers beginning with one and equal in number to these straight lines, and if a magnitude is subtracted from the first straight line such that the ratio of the entire straight line to this magnitude is equal to the ratio of the greatest square to unity, in this case the ratio of *MB* to *BA*, and if a magnitude is subtracted from the next straight line such that the ratio of the straight line to this magnitude is

* For the figure relating to the line *LK*, see Note 18.

equal to the ratio of the greatest square to the square following unity, in this case the ratio of *ND* to *DC*, and if a magnitude is subtracted from the next straight line such that the ratio of the straight line to this magnitude is equal to the ratio of the greatest square to the third square, in this case the ratio of *PG* to *GE*, and if this process is continued for all the equal straight lines until there only remains the indivisible unity straight line corresponding to the greatest square, then the sum of the squares of the straight lines remaining from the divided lines once the straight lines corresponding to the squares had been subtracted, is less than one third plus one fifth of the sum of the squares of the initial straight lines[21] including the square of the indivisible straight line. And the sum of the squares of the straight lines remaining from the initial straight lines, including the square of the indivisible straight line, is greater than one third plus one fifth of the sum of the squares of the initial straight lines, including the square of the indivisible straight line.

If the ratio of the initial straight lines to their parts is equal to the ratio of the numbers *MB*, *ND*, *PG* and *OI* to the numbers *BA*, *DC*, *GE* and *HI*, then the ratio of the initial straight lines to their remainders is equal to the ratio of the numbers *BM*, *DN*, *GP* and *IO* to the numbers *MA*, *NC*, *PE* and *OH*. The ratio of the squares of the remaining parts of these straight lines to the squares of these lines themselves is equal to the ratio of the squares of the numbers corresponding to the numbers *MA*, *NC*, *PE* and *OH* to the squares of the numbers corresponding to the numbers *MB*, *ND*, *PG*, *OI* and *KL*. If successive parts are subtracted from a set of equal straight lines, and if there remains an indivisible straight line, and if this indivisible straight line, together with the parts that have been subtracted from the initial straight lines, are in the ratio of the successive square numbers beginning with one, then the sum of the squares of the remainders of the initial straight lines is less than one third plus one fifth of the sum of the squares of the initial straight lines, including the square of the indivisible straight line, and the sum of the squares of the remainder, including the square of the straight line that was not divided, is greater than one third plus one fifth of the sum of the squares of the initial straight lines, including the straight line that was not divided. That is what we wanted to prove.

<*The paraboloid: first species*>

Now that these lemmas have been proved, let us move on to the measurement of the paraboloid.

[21] Lit.: the divided straight lines.

Let AB be a portion of a parabola, and let AC be its diameter, A its vertex, and the straight line BC its ordinate, extended as far as its two extremities.[22] Let the angle ACB be a right angle in the first figure, an acute angle in the second, and an obtuse angle in the third. Let us fix the diameter AC in its position and unchanging. Now, let us rotate the section ABC about the diameter AC until it returns to its original position, such that it generates the solid ABD.

I say that the solid ABD *is equal to one half of the right cylinder, the half diameter of the base of which is the perpendicular dropped from the point* B *onto the diameter* AC, *and whose height is the diameter* AC.

We drop a perpendicular from the point B onto the diameter AC. In the first figure, this is the straight line BC, which is an ordinate as the angle ACB is a right angle by hypothesis. In the two other figures, let the perpendicular be BK. From the point B, we draw a straight line BH parallel to AC in the plane of the section ABC. We make BH equal to CA, and we join AH. This will be parallel to the straight line BC. In the second and third figures, we produce the perpendicular HL from the point H. We imagine the surface $ACBH$ in the first figure being rotated about the straight line AC until it returns to its original position. Moving it in this way generates a right cylinder, and the two lines BC and HA generate two parallel circles, which are the two bases of the cylinder. The straight line AC is the axis of the cylinder. In the second figure, we imagine the surface $HLCB$ being rotated about the straight line LC. The surface $HLKB$ then generates a right cylinder, and the two triangles BKC and HAL generate two right cones. In the third figure, we imagine the surface $HAKB$ being rotated about the straight line AK. The surface $HLKB$ then generates a right cylinder, and the two triangles BKC and HAL generate two right cones. In all three figures, let the right cylinder be $BHID$.

I say that the solid ABD *in each of the three figures is equal to one half of the cylinder* BHID.

Proof: If this solid is not equal to one half of the cylinder, then it must be either greater than or less than one half of the cylinder.

Let us first assume that the paraboloid is greater than one half of the cylinder $BHID$. Let it exceed this half by the solid G. In the first figure, we divide the diameter AC into two halves at the point M. Draw the ordinate ME, and we extend it until it meets the straight line HB. Let this meeting point be the point U. We draw a straight line parallel to the straight line AC through the point E; let it be SEO'. As AM is equal to MC, SE will be equal to EO', the surface HE will be equal to the surface EB, and the surface AE will be equal to the surface EC. If the surface $AHBC$ is rotated about the

[22] This doubtless refers to the extremities of the entire portion.

straight line *AC*, generating the cylinder *HBDI*, then the surface *SC* will generate a cylinder, the surface *HO′* will generate a cylindrical body[23] circumscribed around the cylinder generated by the surface *SC*, and the straight line *MU* will generate a circle which cuts the cylinder generated by the surface *SC* into two halves, and which also cuts the cylindrical body generated by the surface *HO′* into two halves. The body generated by the rotation of the surface *HE* added to the cylinder generated by the rotation of the surface *EC* is equal to one half of the large cylinder generated by the rotation of the surface *HC*.

Similarly, we now divide the straight line *AM* into two halves at the point *L*. We draw an ordinate *LE* through the point *L* and we extend it until it meets the straight line *HB*. We draw a straight line *TY* parallel to the diameter *AM* through the point *E* on the straight line *EL*. <The sum of> the two bodies generated by the rotation of the two surfaces *SE* and *ME* is equal to one half of the cylinder generated by the rotation of the surface *SM*.

Similarly, we now divide the straight line *MC* into two halves at the point *K*. We draw an ordinate *KE* through the point *K* and we extend this until it meets the straight line *BH*. We draw a straight line *FEV* parallel to the straight line *MC* through the point *E* on the straight line *EK*. The body generated by the rotation of the two surfaces *UE* and *EO′* is equal to one half of the cylindrical body generated by the rotation of the surface *UO′*, as the surface *UE* is equal to one half of the surface *FB*, and the surface *EO′* is equal to one half of the surface *FO′*. The sum of the four solids generated by the rotation of the surfaces *UE*, *EO′*, *SE* and *EM* is therefore equal to one half of the two solids generated by the rotation of the two surfaces *BE* and *EA*. And these are the two bodies that remain from the cylinder once the two bodies generated by the rotation of the two surfaces *HE* and *EC* have been removed.

Similarly, we divide each of the straight lines *AL*, *LM*, *MK* and *KC* into two halves at the points *O*, *P*, *N* and *J*. We draw the ordinates *OE*, *PE*, *NE* and *JE* through these points and we extend them until they meet the straight line *HB*. We draw straight lines parallel to the diameter through the points *E*. In this way, the remaining surfaces have been divided in half, and in half again. The solids generated by the rotation of these half-surfaces are equal to one half of what remains of the cylinder following the first two divisions. If this is done, the greatest cylinder is reduced by half, then the remainder reduced by half, and then the remainder reduced by half again. If this is done, that which remains of the greatest cylinder is a magnitude less

[23] Lit.: circular body. We shall continue to translate this expression in this way from now on.

than the magnitude *G* as, if a magnitude is reduced by half, and then the remainder is reduced by half if we do that twice, then that <quantity> removed from the magnitude is greater than half of it. If this remainder is also reduced by half, and that remainder also reduced by half twice, then the <quantity> removed is greater than half of it. And if a magnitude is reduced by half, and the remainder reduced by half, and if we continue to proceed in this way, then the <quantity> removed from the magnitude is greater than half of it, and the <quantity> removed from the remainder is greater than half of it, as in each division, the parts removed are greater than half. But the cylinder is greater than the magnitude *G*. Therefore, if the cylinder is reduced by half, and the remainder reduced by half, as shown in the figure, and if we continue to proceed in this way, then there necessarily remains a magnitude less than the magnitude *G*. Let us continue the division until this point is reached. But that which remains of the cylinder when it has been divided in this way, are the cylindrical bodies through the middles of which the surface of the paraboloid passes, and the points *E* are at <the vertices of> their angles. The cylindrical bodies whose angles lie at the points *E*, when added together, are less than the magnitude *G*. The portions of these cylindrical bodies that lie within the paraboloid are therefore very much less than the magnitude *G*.

If it is so, that which remains of the paraboloid, once those parts of the cylindrical bodies lying within it have been removed, is greater than one half of the cylinder *BHID*, as this paraboloid exceeded one half of this cylinder by the magnitude *G*. But that which remains of the paraboloid, once those parts of the cylinder lying within it have been removed, is the solid[24] that divides the circles generated by the rotation of the ordinates. Its base is the circle whose half diameter is *WC*, its vertex is the circle whose half-diameter is *OE*, and its circular angles are limited by the circles generated by the rotation of the point *E*. This solid* is therefore greater than one half of the cylinder *BHID*.

But the section *AB* is a parabola. The ratio of *CA* to *AM* is therefore equal to the ratio of the square of *BC* to the square of *EM*. *CA* is twice *AM*. Therefore the square of *BC* is twice the square of *EM*. But *BC* is equal to *UM*, and the square of *UM* is therefore equal to twice the square of *ME*. Similarly, the ratio of the square of *BC* to the square of *EJ* is equal to the

[24] The original Arabic term is *manshūr*, the generally accepted translation of which is 'prism'. This word, which has the same root as the verb *nashara* (to saw), as does the Greek word πρῖσμα and the verb πρίζω (to saw), cannot however be translated as the English word 'prism' with the risk of confusion. We have therefore chosen to use the more general term, 'solid'. This term will be followed by an asterisk in the remainder of the text.

ratio of *CA* to *AJ*. Therefore, by separation, the ratio of the difference between the square of *BC* and the square of *EJ* to the square of *EJ* is equal to the ratio of *CJ* to *JA*. The ratio of the square of *EO* to the square of *EJ* is equal to the ratio of *OA* to *AJ*. But *OA* is equal to *CJ*, and the ratio of the square of *EO* to the square of *EJ* is therefore equal to the ratio of the difference between the square of *BC* and the square of *EJ* to the square of *EJ*. The difference between the square of *BC* and the square of *EJ* is therefore equal to the square of *EO*. The square of *EJ* plus the square of *EO* is therefore equal to the square of *BC*. Consequently, their sum is equal to twice the square of *EM*. Similarly, the ratio of the square of *BC* to the square of *EK* is equal to the ratio of *CA* to *AK*. By separation, the ratio of the difference between the square of *BC* and the square of *EK* to the square of *EK* is equal to the ratio of *CK* to *KA*. But the ratio of the square of *EL* to the square of *EK* is equal to the ratio of *AL*, which is equal to *CK*, to *AK*. The difference between the square of *BC* and the square of *EK* is therefore equal to the square of *EL*. The square of *EK* plus the square of *EL* is therefore equal to the square of *EM*. The same process may be applied for the squares of *EN* and *EP*.

The sum of the squares of the straight lines *EJ*, *EK*, *EN*, *EP*, *EL* and *EO* is therefore a multiple of the square of *EM*, the multiple being equal to the number of these straight lines. For each pair of these squares, <the sum> is equal to twice the square of *EM*. But the square of *EM* is equal to one half of the square of *BC*. The sum of the squares of these straight lines is therefore equal to one half of the sum of the squares of the straight lines passing through the points *O*, *L*, *P*, *N*, *K* and *J* which cut the surface *HC*, and each of which is equal to the straight line *BC*. But the square of *EM* is also equal to one half of the square of *UM*. The sum of the squares of the straight lines *EO*, *EL*, *EP*, *EM*, *EN*, *EK* and *EJ* is therefore equal to one half of the squares of the straight lines equal to the straight line *BC* and passing through the points *O*, *L*, *P*, *M*, *N*, *K* and *J*. The same applies to their doubles which cut the surface *BI*, that is, the sum of the squares of the straight lines which cut the section – and which are each twice the straight lines *EO*, *EL*, *EP*, *EM*, *EN*, *EK* and *EJ* – is equal to one half of the sum of the squares of the straight lines which cut the parallelogram *BI* and which pass through the points *O*, *L*, *P*, *M*, *N*, *K* and *J*, each of which is equal to the straight line *BD*. The same applies to the circles whose diameters pass through these points. Using one of the equal parts of the diameter *AC* as the common height, that is, the straight line *AO*, gives a set of cylinders whose bases are the circles that cut the paraboloid, and whose diameters are the ordinates, and whose height is the straight line *AO*. The sum of these cylinders is therefore equal to one half of the cylinders whose bases are the circles which cut the large cylinder, and whose height is the straight line *AO*. The cylinders whose height is the straight line *AO* are the same as the cylinders whose heights are the straight lines *OL*, *LP*, *PE*, *EN*, *NK*, *KJ* and *JC*, as these are the equal straight lines. The sum of the cylinders whose heights are these straight lines, and whose bases are the circles that cut the paraboloid, is equal to the solid* whose base is the circle whose half-diameter is the straight line *WC* and whose vertex is the circle whose half-diameter is *EO*. But the cylinders whose heights are the straight lines *OL*, *LP*, *PM*, *MN*, *NK*, *KJ* and *JC*, and whose bases are the circles that cut the large cylinder, are equal to the cylinder whose base is the circle whose half-diameter is *BC*, and whose height is the straight line *OC*. The solid* inscribed within the paraboloid is therefore equal to one half of the cylinder whose height is the straight line *OC* and whose base is the base of the large cylinder. It is therefore less than one half of the large cylinder *BHID*. But we have shown that this solid* is greater than one half of this cylinder, which is impossible.

This impossibility arises from the hypothesis that the paraboloid is greater than one half of the cylinder. The paraboloid is therefore not greater than one half of the cylinder.

I say that neither is it less than one half of the cylinder.

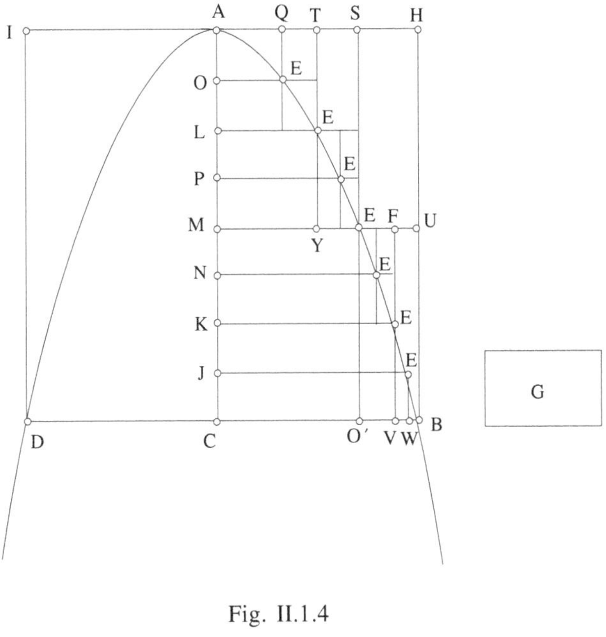

Fig. II.1.4

If this were possible, let it be less than one half of the cylinder, in other words, let it be equal to one half of the cylinder less a solid of magnitude *G*.

We take away one half of the cylinder, then we remove half from the reminder, and so on in the same way as before, until there remains just the sum of the cylindrical bodies through the middles of which the surface of the paraboloid passes, and which is less than the solid *G*. That part of these cylindrical bodies which lies outside the paraboloid must therefore be very much less than the magnitude *G*. But the paraboloid plus the magnitude *G* is equal to one half of the cylinder *BHID*. The paraboloid plus that part of the cylindrical bodies that lies outside the paraboloid is less than one half of the cylinder. But the paraboloid plus that part of the cylindrical bodies that lies outside the paraboloid is equal to the solid* whose base is the base of the cylinder, and whose vertex is the circle whose half-diameter is *QA*. The solid* whose base is the base of the cylinder and whose vertex is the circle whose half-diameter is *QA* is therefore less than one half of the cylinder.

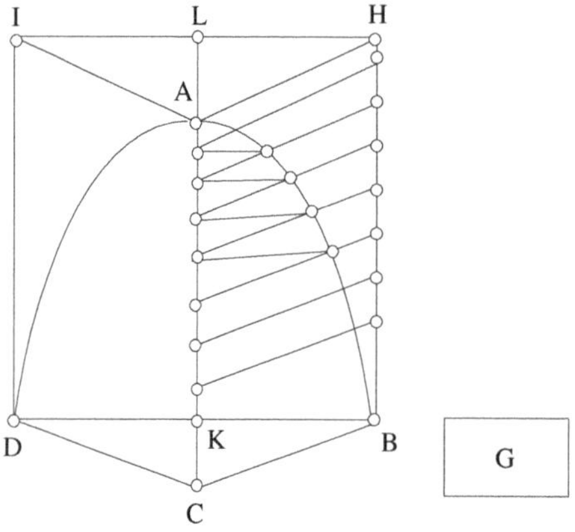

Fig. II.1.5a

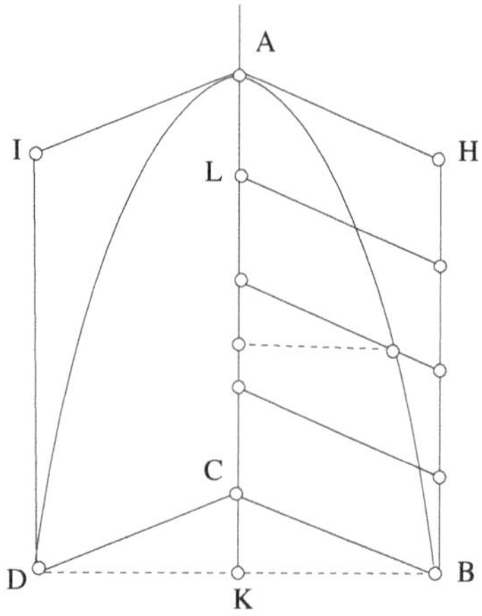

Fig. II.1.5b

We have shown that the solid* inscribed within the paraboloid is equal to one half of the cylinder whose height is *CO* and whose base is the circle whose half-diameter is *BC*. But the solid* inscribed within the paraboloid is equal to the solid* circumscribed around the paraboloid whose base is the circle whose half-diameter is the straight line *EJ* and whose vertex is the circle whose half-diameter is *QA*, as *EJ* is equal to *WC*, *QA* is equal to *EO*, and the height *AJ* is equal to the height *CO*. But the cylinder whose height is *CO* is equal to the cylinder whose height is *AJ*. The solid* circumscribed around the paraboloid whose base is the circle whose half-diameter is *EJ* is therefore equal to one half of the cylinder whose height is *AJ*. If this solid* is added to one half of the cylinder whose base is the circle whose half-diameter is *BC* and whose height is *JC*, then the sum is equal to one half of the cylinder *BHID*. Therefore, if this solid* circumscribed around the paraboloid whose base is the circle whose half-diameter is *EJ* is added to the whole cylinder whose height is *JC* and whose base is the circle whose half-diameter is *BC*, then the sum is greater than one half of the cylinder *BHID*. But if the solid* circumscribed around the paraboloid whose base is the circle whose half-diameter is *EJ* and whose height is *AJ* is added to the cylinder whose base is the circle whose half-diameter is *BC* and whose height is the straight line *CJ*, then the solid* so obtained will be the solid circumscribed around the paraboloid whose base is the base of the large cylinder, that is the cylinder *BHID*, and whose vertex is the circle whose half-diameter is *QA*. The solid* is therefore greater than one half of the cylinder *BHID*. But we have shown that this solid* is less than one half of this cylinder, which is impossible.

The paraboloid is therefore not less than one half of the cylinder, nor greater than one half of the cylinder. It is therefore equal to one half of this cylinder.

The paraboloid in the second figure has a conical base and the circumscribed cylinder is also conical. The cone generated by the triangle *BCK* is equal to the cone generated by the triangle *HLA*. If the cone whose vertex is at the point *C* is subtracted from the conical cylinder, and if the cone whose vertex is at the point *A* is added, then the result is a right cylinder equal to the conical cylinder. If we assume that the paraboloid is greater than one half of the cylinder, and if the conical cylinder is divided in the same way as shown in the first figure, then the cylinder is reduced by half, and then the remainder is reduced by half, and then that remainder also reduced by half, then the result[25] is the solid* inscribed within the paraboloid that is greater than one half of the cylinder, as has already been shown in the first figure. This solid* is conical. As in the first figure, it can be shown that this solid* is less than one half of the conical cylinder. Thus, if perpendiculars are drawn onto the diameter from the vertices of the ordinates, the ratio of these perpendiculars, each to the others, is equal to the ratio of the ordinates, each to the others. And the ratio of the ordinates in this figure, each to the others, is equal to the ratio of the ordinates in the first figure, each to the others. The ratios of the perpendiculars from the vertices of the ordinates to the diameter in this figure, each to the others, are equal to the ratios of the ordinates in the first figure, each to the others. If these perpendiculars are extended until they reach the straight line *BH*, then the ratios of <those parts of> the perpendiculars that lie within the section to <the external parts of> these perpendiculars, terminating at the straight line *BH*, are equal to the ratios of the ordinates to their <external parts>,

[25] Lit.: there remains.

terminating at the straight line *BH*. In the second figure, the ratios of the ordinates to the <external> parts terminating on the straight line *BH* are equal to the ratios of the ordinates in the first figure to the <external> parts terminating on the straight line *BH*. The ratios of the <parts of the> perpendiculars which lie within the section in the second figure to the <external> parts terminating on the straight line *BH* are therefore equal to the ratios of the ordinates in the first figure to the external portions terminating on the straight line *BH*. The ratios of the circles, each to the others, whose half-diameters are the <parts of the> perpendiculars that lie within the section in the second figure are therefore equal to the ratios of the circles, each to the others, whose half-diameters are the ordinates in the first figure. The ratios of the right cylindrical bodies in the second figure to the right cylinder in the same figure are therefore equal to the ratios of the cylindrical bodies in the first figure to their cylinder. And the ratio of the right solid* inscribed within the second figure to the right cylinder is equal to the ratio of the solid* in the first figure to the cylinder in the same figure. But the solid* in the first figure is less than one half of the <right> cylinder. The right solid* in the second figure is therefore less than one half of the right cylinder. The right cylinder is equal to the conical cylinder, and the right solid* is equal to the conical solid, as each of the right cylindrical bodies is equal to its homologue in the conical cylindrical bodies. This may be shown in the same way as for the right and conical cylinders. It therefore necessarily follows that the conical solid* is less than one half of the conical cylinder.

Similarly, if we assume that the paraboloid is less than one half of the cylinder, then the circumscribed solid* is less than one half of the conical cylinder. As before, it can be shown that the solid* circumscribed around the paraboloid is greater than one half of the conical cylinder. Using the same proof as was used in the case of the first figure, it necessarily follows that the paraboloid in the second figure is equal to one half of the conical cylinder. But the conical cylinder is equal to the right cylinder. The paraboloid in the second figure is therefore equal to one half of the right cylinder.

We continue with the same proof in the third figure. The two cones and the perpendiculars in the third figure are in the same situation as the two cones and the perpendiculars in the second figure.

The paraboloid generated by the rotation of the section *ABC* around the diameter *AC* is therefore, in all three figures, equal to one half of the cylinder whose base is the circle whose half-diameter is the perpendicular dropped from the point *B* onto the diameter *AC*, and whose height is equal to the diameter *AC*. That is what we wanted to prove.

<Corollaries to the first species>

For any parabola whose diameter encloses different angles with its ordinates, the paraboloid generated by the acute angle part is equal to the paraboloid generated by the obtuse angle part.

Thus, their two right cylinders are equal, as the axis of each of the two cylinders is equal to the diameter of the section, and the half-diameter of the base of each of the two cylinders is equal to the perpendicular dropped from the extremity of the ordinate onto the diameter. But the two perpendiculars drawn from the two extremities of the ordinate to the diameter are equal, as the diameter divides the ordinate into two halves. The two right cylinders are therefore equal, and each of the two paraboloids[26] is equal to one half of its cylinder. The two paraboloids generated by the two parts of the section are therefore equal.

The same applies to the parabola whose diameter is an axis. Let this axis be equal to the diameter of another section, <which cuts the ordinate> at two different angles, and let the ordinate <at the extremity> of the axis, and which is the base of the section, be equal to each of the perpendiculars drawn from the two extremities of the two ordinates in the section with two different angles. The paraboloid generated by the rotation of this section about its axis is therefore equal to each of the two paraboloids generated by the rotation of each of the two sections of the parabola, with two different angles, around its diameter.

It can be shown, from that which we have already described, that if the bases of their two cylinders are equal, then the ratio of one paraboloid to another paraboloid is equal to the ratio of the height of one to the height of the other, as the ratio of one paraboloid to another paraboloid is equal to the ratio of the cylinder of one to the cylinder of the other.

If the bases of their cylinders are different and their heights equal, then the ratio of one to the other is equal to the ratio of the bases.

If both the heights and the bases are different, the ratio of one to the other is then compounded of both the ratio of the heights and the ratio of the bases. For all paraboloids of this species, the heights are the diameters of the sections from which the paraboloids were generated.

It can clearly be seen from the previous proof that the sum of the cylindrical bodies through the middles of which the surface of the paraboloid

[26] Lit.: solids. In a context such as this, we have chosen to translate the term as 'paraboloids'.

passes is equal to the cylinder whose base is the base of the large cylinder and whose height is the straight line *CJ*.

We have thus shown that the sum of the two cylindrical bodies generated by the rotation of the two surfaces *SEM* and *BUE* is equal to one half of the large cylinder. The cylinder generated by the rotation of the surface *BM* is equal to one half of the large cylinder. The sum of the two cylindrical bodies is therefore equal to the cylinder generated by the rotation of the surface *BM*. The two cylindrical bodies generated by the rotation of the two surfaces *TEL* and *EYE* are equal to one half of the cylindrical body generated by the rotation of the surface *SEM*. The two cylindrical bodies generated by the rotation of the two surfaces *EFE* and *BVE* are equal to one half of the cylindrical body generated by the rotation of the surface *UO'*. The four cylindrical bodies generated by the rotation of the surfaces *TEL*, *EYE*, *EFE* and *BVE* are equal to one half of one half of the large cylinder. But the cylinder generated by the rotation of the surface *BK* is equal to one half of one half of the large cylinder. Therefore the four cylindrical bodies that we have thus defined are equal to the cylinder generated by the rotation of the surface *BK*.

Similarly, we can also show that each of the four cylindrical bodies that we have defined is divided into half and into half again by the two cylindrical bodies lying within it, and through the middles of which the surface of the paraboloid passes. The sum of all the small cylindrical bodies through the middles of which the paraboloid passes is therefore equal to one half of the four cylindrical bodies that we have defined. The cylinder generated by the rotation of the surface *BJ* is equal to one half of the cylinder generated by the rotation of the surface *BK*, which we have shown to be equal to the four cylindrical bodies. The final small cylindrical bodies through the middles of which the surface of the paraboloid passes are therefore equal to the cylinder whose base is the base of the large cylinder, and whose height is the straight line *CJ*.

Similarly, it can be shown that if the cylinder is divided indefinitely into cylindrical bodies that are smaller than these cylindrical bodies, then their sum is equal to the small cylinder whose base is the base of the large cylinder, and whose height is a single part of the diameter. That is what we wanted to prove.

We have also shown that the solid* inscribed within the paraboloid whose base is the circle whose half-diameter is *WC*, and whose vertex is the circle whose half-diameter is *EO*, is equal to one half of the cylinder whose base is the base of the large cylinder and whose height is the straight line *CO*, which is equal to the straight line *JA*. We have shown that the paraboloid is equal to one half of the large cylinder. The amount by which the paraboloid exceeds the solid* that is inscribed within it is therefore equal to one half of the cylinder whose base is the base of the large cylinder and whose height is the straight line *CJ*. But the amount by which the paraboloid exceeds the solid* that is inscribed within it is that part of the small cylindrical bodies through the middles of which the surface of the paraboloid passes that lies within the paraboloid. That part of these cylindrical bodies that lies within the paraboloid is equal to one half of the cylinder whose base is the base of the large cylinder and whose height is the straight line *CJ*.

But we have shown that the sum of these cylindrical bodies is equal to the cylinder whose base is the base of the large cylinder and whose height is the straight line *CJ*. The surface of the paraboloid therefore divides all the small cylindrical bodies through the middles of which it passes into half. That is what we wanted to prove.

This conclusion[27] follows necessarily for the paraboloids whose base is the circle whose half-diameter is the straight line *EJ*, for the paraboloids whose half diameter is *EK*, and for all other paraboloids.

We have therefore shown that the surface of the paraboloid divides each of the small cylindrical bodies into two halves.[28]

That which we have just proved is the measurement of one of the two species of paraboloids, those generated by the rotation of the section around its diameter.

<The paraboloid: second species>

Let us now turn to the second species; the paraboloids generated by the rotation of the section around its ordinate.

Let *ABC* be a parabola, *BC* its diameter, and *AC* its ordinate. Let the angle *ACB* be a right angle. From the point *B*, let us draw a straight line parallel to the straight line *AC*, namely *BE*. We draw the straight line *AE* parallel to the straight line *CB*. We fix the straight line *AC* so that its position does not change. We rotate the rectangle *ACBE* around the straight line *AC*. The rotation of the surface *AB* generates a circular cylinder, the half-diameter of whose base is the straight line *BC*, namely *BG*. The section *BAC* generates a paraboloid whose base is the circle whose half-diameter is the straight line *BC*, which is *BAD*.

[27] Lit.: notion.
[28] Lit.: into two halves, and then into two halves.

I say that the paraboloid BAD *is equal to one third plus one fifth of the cylinder* ED.

Proof: If it is not equal to one third plus one fifth of the cylinder, then it must be either greater than one third plus one fifth of the cylinder or less than one third plus one fifth of the cylinder.

Let it be first greater than one third plus one fifth of the cylinder, and let the amount by which it exceeds one third plus one fifth of the cylinder be the solid *J*. We divide *AC* into two halves at the point *H*, and we draw the straight line *HMS* parallel to the straight line *BC*. Through the point *M*, we draw the straight line *QMO*, parallel to the two straight lines *BE* and *AC*. As the straight line *QM* is equal to the straight line *MO* – *AH* being equal to *HC* – the surface *EM* is equal to the surface *MB*, and the surface *AM* is equal to the surface *MC*. If the surface *BA* is rotated about the straight line AC until it returns to its original position, then the two cylindrical bodies generated by the rotation of the two surfaces *AM* and *MC* will be equal, and the two cylindrical bodies generated by the rotation of the two surfaces *EM* and *MB* will also be equal. The sum of the two cylindrical bodies generated by the rotation of the two surfaces *ME* and *MC* will therefore be equal to one half of the cylinder *BG*.

We also divide the straight line *AH* into two halves at the point *K*. From the point *K*, we draw a straight line parallel to the two straight lines *HS* and *AE*, namely the straight line *KLR*. Through the point *L*, we draw the straight line parallel to the two straight lines *AC* and *EB*, namely the straight line *ULTV*. We also divide the straight line *HC* into two halves at the point *I*. From the point *I*, we draw a straight line parallel to the two straight lines *CB* and *HS*, namely the straight line *INW*. Through the point *N*, we draw the straight line *XNP*, parallel to the two straight lines *TV* and *RS*. As before, it can be shown that the two cylindrical bodies generated by the rotation of the two surfaces *QL* and *LH* are equal to one half of the cylindrical body generated by the rotation of the surface *AM*. Similarly, it can be shown that the two cylindrical bodies generated by the rotation of the two surfaces *SN* and *NO* are equal to one half of the cylindrical body generated by the rotation of the surface *SO*. The sum of the four cylindrical bodies generated by the rotation of the surfaces *SN*, *NO*, *QL* and *LH* is therefore equal to one half of the two cylindrical bodies generated by the rotation of the two surfaces *BM* and *MA*. But, if the two cylindrical bodies generated by the rotation of the two surfaces *EM* and *MC* – which are equal to one half of the cylinder – are subtracted from the entire cylinder *BG*, then the remainder is the two cylindrical bodies generated by the rotation of the two surfaces *BM* and *MA*. And if the four cylindrical bodies generated by the rotation of the surfaces *QL*, *LH*, *SN* and *NO* are subtracted from the two cylindrical bodies generated by the rotation of the two surfaces *BM* and *MA*, and which are equal to one half of these latter two cylindrical bodies, then the remainder is the cylindrical bodies generated by the rotation of the surfaces *BN*, *NM*, *ML* and *LA*. If each of the parts of the straight line *AC* is divided into two halves, and if straight lines are drawn from the points[29] of division parallel to the straight line *BC*, and if straight lines are drawn through the points[30] of intersection of these straight lines and the section *AB*[31] of the straight lines parallel to the straight line *AC*, then the cylindrical bodies resulting from the rotation of the surfaces will, two by two, generate half of the cylindrical body in which they are located, as we have shown earlier.

If there are two different magnitudes, and if one[32] is halved, and the remainder is also halved, and if we continue to proceed in this way, then the remainder must necessarily be a magnitude that is less than the smaller magnitude <of the two magnitudes>, as we have shown in the previous proposition. If the cylinder *BG* is divided in this manner, the remainder must eventually be less than the magnitude *J*. Let the division be continued to this limit, and let the remainder of the cylinder *BG* be equal to the cylindrical bodies generated by the rotation of the surfaces *BN*, *NM*, *ML* and *LA*. These cylindrical bodies are therefore less than the magnitude *J*. But that portion of these cylindrical bodies that lies within the paraboloid is less than these cylindrical bodies. That part of these cylindrical bodies that lies within the paraboloid is therefore very much less than the solid *J*. So if the paraboloid *BAD* exceeds the solid *J* by one third plus one fifth of the cylinder *BG*, and that part of the small cylindrical bodies that lies within the paraboloid is less than the solid *J*, then the remainder of the paraboloid once those parts that lie within it have been removed is greater than one third plus one fifth of the cylinder. But the remainder of the paraboloid once those parts of the small cylindrical bodies that lie within it have been removed is the solid* whose base is the circle whose half-diameter is *PC* and whose vertex is the circle whose half-diameter is *LK*. This solid* is therefore greater than one third plus one fifth of the cylinder *BG*.

As the section *ABC* is a parabola with diameter *BC* and ordinate *AC*, the square of the straight line *AC* is equal to the product of *BC* and the *latus rectum*. And as the straight lines *LV*, *MO* and *NP* are parallel to the straight line *AC*, these straight lines are also ordinates. The square of *LV* is therefore equal to the product of *BV* and the *latus rectum*, the square of *MO* is equal

[29] Lit.: positions (*mawāḍi'*).
[30] Lit.: positions.
[31] Lit.: which lie between these straight lines and the section *AB*.
[32] i.e. the greater of the two.

to the product of *BO* and the *latus rectum*, and the square of *NP* is equal to the product of *BP* and the *latus rectum*. The ratio of the square of *AC* to the square of *LV* is therefore equal to the ratio of *CB* to *BV*, the ratio of the square of *LV* to the square of *MO* is equal to the ratio of *VB* to *BO*, and the ratio of the square of *MO* to the square of *NP* is equal to the ratio of *OB* to *BP*. The ratio of the straight lines *BC*, *BV*, *BO* and *BP*, each to the others, is therefore equal to the ratio of the squares of the straight lines *AC*, *LV*, *MO* and *NP*, each to the others. As the straight line *NP* is equal to the straight line *CI*, the straight line *MO* is equal to the straight line *CH*, and the straight line *HC* is equal to twice the straight line *CI*, then *MO* is equal to twice *NP*. And, as the parts *AK*, *KH*, *HI* and *IC* are equal, *KC* is equal to three times *CI*, and the straight line *LV* is therefore equal to three times *NP*. Similarly, *AC* is equal to four times *CI*, and therefore *AC* is equal to four times *NP*. Therefore, depending on the magnitude needed to make the straight line *NP* equal to one, *MO* is two, *LV* is three, and *AC* is four. The ratios of the straight lines *NP*, *MO*, *LV*, *AC*, each to the others, are therefore equal to the ratios of the successive numbers beginning with one and increasing from one by one, each to the others. Similarly, if these straight lines were greater in number than they are, they would all be in the ratios of the successive numbers. It is for this reason that the ratios of the squares of the straight lines *NP*, *MO*, *LV* and *AC*, each to the others, are equal to the ratios of the squares of the successive numbers, each to the others. But the ratios of the squares of the straight lines *NP*, *MO*, *LV* and *AC*, each to the others, are equal to the ratios of the straight lines *BP*, *BO*, *BV* and *BC*, each to the others. The ratios of the straight lines *BP*, *BO*, *BV* and *BC*, each to the others, are therefore equal to the ratios of the successive numbers beginning with one and increasing from one by one, each to the others. But the straight line *BP* is equal to *WN*, *BO* is equal to *SM*, *BV* is equal to *RL*, and *BC* is equal to *EA*. The straight lines *WN*, *SM*, *RL* and *EA* are therefore in the ratios of the successive square numbers beginning with one, each to the others, and the straight lines *WI*, *SH*, *RK* and *EA* are equal.

We have shown in the lemmas that we have introduced that, given a set of equal straight lines from which straight lines are separated, and if there remains one which is indivisible,[33] and if the ratios of these separated straight lines to the straight line that has not been divided are successively the ratios of the successive square numbers beginning with one, then the sum of the squares of the parts of these straight lines that remain is less than one third plus one fifth of the sum of the squares of all the straight lines that were equal to one another, and equal to the greatest straight line, and the sum of the squares of the remaining parts, plus the square of the straight line

[33] See Lemma 5.

that was not divided, is greater than one third plus one fifth of the sum of the squares of all the straight lines that are equal to one another. The squares of the straight lines *NI*, *MH* and *LK* are therefore less than one third plus one fifth of the squares of the straight lines *WI*, *SH*, *RK* and *AE*, and the squares of the straight lines *NI*, *MH*, *LK* and *AE* are greater than one third plus one fifth of the squares of the straight lines *WI*, *SH*, *RK* and *AE*.

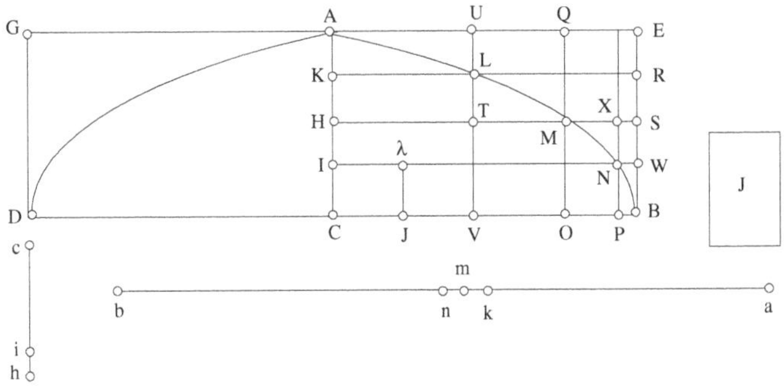

Fig. II.1.6

The ratios of the squares of the straight lines, each to the others, are thus equal to the ratios of the circles whose half-diameters are these straight lines, each to the others. The circles whose half-diameters are the straight lines *NI*, *MH* and *LK* are less than one third plus one fifth of the circles whose half-diameters are *WI*, *SH*, *RK* and *AE*. And the circles whose half-diameters are the straight lines *NI*, *MH*, *LK* and *EA* are greater than one third plus one fifth of the circles whose half-diameters are the straight lines *WI*, *SH*, *RK* and *EA*. We make the straight line *AK* as the common height. The small cylinders whose bases are the circles whose half-diameters are the straight lines *NI*, *MH* and *LK*, and whose height is equal to the straight line *AK*, are less than one third plus one fifth of the cylinders whose bases are the circles whose half-diameters are the straight lines *WI*, *SH*, *RK* and *EA*, and whose height is equal to the straight line *AK*. And the cylinders whose bases are the circles whose half-diameters are the straight lines *NI*, *MH* and *LK*, and whose height is equal to the straight line *AK*, form a solid* whose base is the circle whose half-diameter is the straight line *CP*, equal to the straight line *NI*, and whose vertex is the circle whose half-diameter is the straight line *LK*. Each of the heights *KH*, *HI* and *IC* is in fact equal to the straight line *AK*. The cylinders whose bases are the circles whose half-diameters are the straight lines *WI*, *SH*, *RK* and *EA*, and whose height is equal to the straight line *AK* are equal to the cylinder whose base is the circle whose half-diameter is the straight line *EA* and whose height is the straight line *AC*, that is the cylinder *BG*. The solid* whose base is the circle whose half-diameter is the straight line *PC* and whose vertex is the circle whose half-diameter is *LK* is therefore less than one third plus one fifth of the cylinder *BG*.

This solid* is the solid* inscribed within the paraboloid, and we have already shown that it is greater than one third plus one fifth of the cylinder *BG*. Therefore, this is absurd. Therefore, the paraboloid is not greater than one third plus one fifth of the cylinder.

And I also say that neither is it less than one third plus one fifth of the cylinder.

If this were possible, then let this solid be less than one third plus one fifth of the cylinder, and let it be less than one third plus one fifth of the cylinder by the solid magnitude *J*. We subtract the cylindrical bodies from the cylinder, as we did before. The remainder is the cylindrical bodies generated by the rotation of the surfaces *BN*, *NM*, *ML* and *LA*, which are less than the solid *J*. The parts of these cylindrical bodies that lie outside the paraboloid, and which surround it, are very much less than the solid *J*.

The paraboloid plus these parts is therefore less than one third plus one fifth of the cylinder. But the paraboloid plus these parts forms the solid* whose base is the circle whose half-diameter is the straight line *BC* and whose vertex is the circle whose half-diameter is the straight line *AU*. This solid* is therefore less than one third plus one fifth of the cylinder *BG*.

But we have shown that the circles whose half-diameters are the straight lines *NI*, *MH*, *LK* and *EA* are greater than one third plus one fifth of the circles whose half-diameters are the straight lines *WI*, *SH*, *RK* and *EA*. We set *AK* the common height. We take *BC* in place of *EA*, as they are equal. The small cylinders whose bases are the circles whose half-diameters are the straight lines *BC*, *NI*, *MH* and *LK*, and whose height is equal to the straight line *AK*, are therefore greater than one third plus one fifth of the cylinders whose bases are the circles whose half-diameters are the straight lines *BC*, *WI*, *SH* and *RK*, and whose height is equal to the straight line *AK*. But the cylinders whose bases are the circles whose half-diameters are the straight lines *BC*, *NI*, *MH* and *LK*, and whose height is equal to the straight line *AK*, are the cylinders generated by the rotation of the surfaces *BI*, *NH*, *MK* and *LA*. The sum of the cylinders generated by the rotation of these surfaces is equal to the solid* whose base is the circle whose half-diameter is *BC*, and whose vertex is the circle whose half-diameter is *UA*. But the cylinders whose bases are the circles whose half-diameters are the straight lines *BC*, *WI*, *SH* and *RK*, and whose height is equal to the straight line *AK*, are the cylinders generated by the rotation of the surfaces *BI*, *WH*, *SK* and *RA*. The sum of these cylinders is the cylinder generated by the rotation of the surface *BA* – the sum of the surfaces that we have mentioned is equal to the surface *BA* – and this cylinder is the cylinder *BG*. The solid* whose base is the circle whose half-diameter is the straight line *BC* and whose vertex is the circle whose half-diameter is *UA* is therefore greater than one third plus one fifth of the cylinder *BG*.

But we have shown that this solid* is less than one third plus one fifth of the cylinder *BG*, which is absurd and cannot be. Therefore, the paraboloid *BAD* is not less than one third plus one fifth of the cylinder *BG*.

But we have also shown that it is not greater than one third plus one fifth of this cylinder. The paraboloid *BAD* is therefore equal to one third plus one fifth of the cylinder *BG*. That is what we wanted to prove.

If now the angle *ACB* is either acute or obtuse, then we proceed for this section in the same way as in the second and third figures of the previous proposition. We can then show, as in this proposition, that the paraboloid is equal to one third plus one fifth of the right cylinder whose base is the circle whose half diameter is the perpendicular dropped from the extremity of the diameter onto the ordinate, and whose height is equal to the ordinate. That is what we wanted to prove.

<Corollaries to the second species>

As in the previous proposition, it can be shown that the sum of the small cylindrical bodies through the middles of which the surface of the paraboloid passes is equal to the cylindrical body generated by the rotation of the surface *BI*.

The ratio of the small cylindrical bodies to the cylinder is equal to one half of one half of one half.[34] The same applies to the cylindrical body generated by the rotation of the surface *BI*. Each time that the cylindrical bodies through the middles of which the surface of the solid passes, are divided, the cylindrical body generated by the rotation of the surface *BI* is divided into two halves. The cylindrical bodies through the middles of which the surface of the paraboloid passes are therefore equal to the cylindrical body generated by the rotation of the surface *BI*.

[34] Lit.: is equal to the ratio of the half and of the half of the half. The meaning is $1/2^n$.

<The small cylindrical bodies>

We set *AB* the square number corresponding to the straight line *EA*, as the straight lines *WN*, *SM*, *RL* and *EA* are in the ratios of the successive square numbers beginning with one. We divide *AB* into two halves at the point *N*, and we set *NK* equal to one third of one tenth of *AB*. Then, *BK* is equal to one third plus one fifth of *AB*. Let *CH* be the side of the number *AB*, which is a square. We set *HI* be one third of one tenth of unity, and we set the ratio of *HI* to *NM* equal to the ratio of *AB* to *CH*. The product of *AB* and *NM* is equal to the product of *CH* and *HI*. But the product of *CH* and *HI* is equal to one third of one tenth of *CH*, as *HI* is one third of one tenth of unity. The product of *AB* and *MN* is therefore equal to one third of one tenth of *CH*, and the product of *AB* and *KN* is equal to one third of one tenth of the square of *AB*. The product of *AB* and *KM* is therefore equal to one third of one tenth of the square of *AB*, less one third of one tenth of the side of *AB*. We have shown in the arithmetical lemmas[35] that we have introduced that the sum of the squares of the numbers corresponding to the straight lines *LK*, *MH* and *NI* exceeds one third plus one fifth of the squares of the numbers corresponding to the straight lines *RK*, *SH* and *WI* by one third of one tenth of the square of the number corresponding to the straight line *AE*, less one third of one tenth of the side of this number. The squares of the numbers corresponding to the straight lines *LK*, *MH* and *NI* therefore exceed one third plus one fifth of the squares of the numbers corresponding to the straight lines *RK*, *SH* and *WI* by the product of *AB* and *KM*. But the product of *AB* and *BK* is equal to one third plus one fifth of the square of *AB*. The squares of the numbers corresponding to the straight lines *LK*, *MH* and *NI*, plus the product of *AB* and *BM*, are equal to one third plus one fifth of the squares of the numbers corresponding to the straight lines *RK*, *SH*, *WI* and *BC*.

We set the ratio of the square of *BC* to the square of *CJ* equal to the ratio of *AB* to *BM*. But the ratio of *AB* to *BM* is equal to the ratio of the square of *AB* to the product of *AB* and *BM*. The ratio of the square of *BC* to the square of *CJ* is therefore equal to the ratio of the square of *AB* to the product of *AB* and *BM*. Therefore, the square of *CJ* is equal to the product of *AB* and *BM*. We draw the straight line JL_a parallel to the straight line *IC*. As the product of *AB* and *BM*, plus the squares of the numbers corresponding to the straight lines *LK*, *MH* and *NI*, is equal to one third plus one fifth of the squares of the numbers corresponding to the straight lines *RK*, *SH*, *WI* and *BC*, the squares of the straight lines *LK*, *MH*, *NI* and *JC* are therefore equal to one third plus one fifth of the squares of the straight

[35] See Lemma 5.

lines *RK*, *SH*, *WI* and *BC*. But the circles whose half-diameters are these straight lines are also in this ratio, and the cylindrical bodies whose bases are these circles, and whose heights are the straight lines *AK*, *KH*, *HI* and *IC*, are also in this ratio. The solid* inscribed within the paraboloid – whose vertex is the circle whose half-diameter is *LK* and whose base is the circle whose half-diameter is *PC* – plus the cylindrical body generated by the rotation of the surface *JI* are therefore equal to one third plus one fifth of the cylindrical bodies whose bases are the circles whose half-diameters are the straight lines *RK*, *SH*, *WI* and *BC*, and whose heights are the straight lines *AK*, *KH*, *HI* and *IC*. But these cylindrical bodies are the cylinder *BG*. The solid* inscribed within the paraboloid plus the cylindrical body generated by the rotation of the surface *JI* is therefore equal to one third plus one fifth of the cylinder *BG*. But the paraboloid is equal to one third plus one fifth of the cylinder *BG*. The solid* inscribed within the paraboloid plus the cylindrical body generated by the rotation of the surface *JI* is therefore equal to the paraboloid. The cylindrical body generated by the rotation of the surface *JI* is therefore equal to those parts of the small cylindrical bodies through the middles of which the surface of the paraboloid passes that lie within the paraboloid.

But we have shown that the sum of all the small cylindrical bodies is equal to the cylindrical body generated by the rotation of the surface *BI*. Those parts of the small cylindrical bodies through the middles of which the surface of the paraboloid passes, and which lie outside it and surround it, are therefore equal to the cylindrical body generated by the rotation of the surface BL_a. But the ratio of the external parts of these cylindrical bodies to their internal parts is equal to the ratio of the cylindrical body generated by the rotation of the surface BL_a to the cylindrical body generated by the rotation of the surface *JI*. The ratio of these two cylindrical bodies, each to the other, is equal to the ratio of their bases, each to the other. But the ratio of their bases, each to the other, is equal to the ratio of the difference between the square of *BC* and the square of *CJ* to the square of *CJ*, and the ratio of the difference between the square of *BC* and the square of *CJ* to the square of *CJ* is equal to the ratio of *AM* to *MB*, as the ratio of the square of *BC* to the square of *CJ* is equal to the ratio of *AB* to *BM*. The ratio of the parts of the small cylindrical bodies that lie outside the paraboloid to those parts that lie inside the paraboloid is therefore equal to the ratio of the number *AM* to the number *MB*.

This ratio follows necessarily for each of the cylindrical bodies, as we have shown in the previous proposition. And from this ratio it necessarily follows that, the smaller the small cylindrical body, the greater the amount by which the ratio of its external part to its internal part exceeds the ratio of the external part to the internal part of the next largest cylindrical body. As the small cylindrical bodies become smaller, the straight lines that are homologous to the straight lines *LK*, *MH*, *NI* and *CB* become larger. Therefore, the straight lines that are homologous to the straight lines *WN*, *SM*, *RL* and *EA* also become larger. The square number corresponding to the straight line *AE* is therefore greater than the number *AB*. Its ratio to its side is therefore greater than the ratio of *AB* to *CH*, as the further one of the successive square numbers is from unity, the greater is the ratio between the number and its side. The ratio of one third of one tenth of unity, which is equal to *HI*, to the number corresponding to the number *NM* is therefore greater than the ratio of *HI* to *NM*. The number corresponding to the number *NM* is therefore less than *NM*, and one half of the square number corresponding to the number *NB* is greater than *NB*. The ratio of *MN* to *NB* is therefore greater than the ratio of the number corresponding to *NM* to the number corresponding to *NB*, in the case of the greatest square number corresponding to the number *AB*. By composition, the ratio of *MB* to *BA* is greater than the ratio of the number corresponding to *NM* to the number corresponding to *NB*. But the ratio of *MB* to *BA* is equal to the ratio of one half of this number to the whole of the number. The ratio of *MB* to *BA* is therefore greater than the ratio of the number corresponding to the number *MB*, in the case of the greatest square number, to this greater square number. By inversion, the ratio of this number, which is the greatest square number, to the part of it that corresponds to the number *BM* is greater than the ratio of *AB* to *BM*. By separation, the ratio of the number corresponding to the number *AM* to the number corresponding to the number *MB* is therefore greater than the ratio of *AM* to *MB*. The ratio of the external parts of the smaller cylindrical bodies to their internal parts is therefore greater than the ratio of the external parts of the cylindrical bodies that are greater than them to their internal parts. That is what we wanted to prove.

It also necessarily follows that, for any parabola in this species which has two different enclosed angles between the ordinate and the diameter, the paraboloid generated by the part with the acute angle is equal to the paraboloid generated by the part with the obtuse angle, as their two cylinders are equal. The two heights of the two cylinders are equal to the two ordinates, the two ordinates are equal, and the half-diameter of the base of each of the two cylinders is the perpendicular dropped from the extremity of the diameter onto the ordinate, and which is the same perpendicular. The two paraboloids resulting from these two parts are therefore equal.

Similarly, the paraboloid obtained from the section whose diameter is equal to the perpendicular dropped from the extremity of the diameter onto the ordinate, and whose ordinate is equal to the ordinate of the section with two unequal angles, is equal to each of the two paraboloids generated by these two sections with two unequal angles.

The ratios of the paraboloids in this species, each to the others, are the same as those shown for the first species.

<Proof by reductio ad absurdum>

But it may be that the proof by *reductio ad absurdum*, if it looks like the proof of these two propositions, raises difficulties for many people. Perhaps some of them, who do not take the examination far enough, may believe that, if it is assumed that the paraboloid is equal to a part of the cylinder that is not one third plus one fifth, in this species, or one half, in the first species, then a proof identical to that which we have put forward for the two propositions would also be successful. This being the case, we must therefore show the reason (*'illa*) that underlies the success of the proof, and which produced what is sought, together with the reason why a paraboloid generated by the rotation of the parabola about its ordinate is equal to one third plus one fifth <of the cylinder>, while the paraboloid generated by the rotation of the parabola about its diameter is equal to one half <of the cylinder>.

We say that the reason that enables us to show that the paraboloid generated by the rotation of the section about its ordinate is equal to one third plus one fifth <of the cylinder> is that any solid* inscribed within the paraboloid – in accordance with the property that we have explained in the proof – is less than one third plus one fifth of the cylinder, and that any solid* circumscribed around the paraboloid, in accordance with the property that we have also explained in the proof, is greater than one third plus one fifth of the cylinder. We can also show that, for any other assumed part which is not one third plus one fifth, there must exist several solids* inscribed within the paraboloid in accordance with the previous property, together with several solids circumscribed around it, such that those inscribed and those circumscribed are together either greater than this part or less than this part, and there exists no part such that any solid* inscribed within the paraboloid is less than it, and such that any solid* circumscribed around the paraboloid is greater than it, with the single exception of one third plus one fifth <of the cylinder>. It is this notion that has produced the proof. In the discussion above, and in that which follows, the word 'part' (*juz'*) is intended to mean 'portion' (*ba'ḍ*).[36] It now remains for us to show the truth of that which we have just mentioned by means of a proof.

Let us assume any part less than one third plus one fifth of the cylinder. I then say that there exist many solids* inscribed within the paraboloid, each of which is greater than this part. The difference between the assumed part, which is less than one third plus one fifth of the cylinder, and one third plus one fifth of the cylinder, is any magnitude. Therefore, if the cylinder is divided into cylindrical bodies – in two halves – and its half into two halves, and if we continue to proceed in this way, then it necessarily follows that the remainder of the cylinder will become a magnitude less than this difference. But the remainder of the cylinder following this division is the set of small cylindrical bodies through the middles of which the surface of the paraboloid passes, and the sum of these cylindrical bodies is equal to the cylindrical body that is homologous to the cylindrical body generated by the rotation of the surface *BI*. The cylindrical body that is homologous to the cylindrical body generated by the rotation of the surface *BI* is therefore less than this difference. The cylindrical bodies generated by the rotation of the surface that is homologous to the surface *JI* are very much less than[37] this difference. The assumed part plus the cylindrical body generated by the rotation of the surface that is homologous to the surface *JI* is less than one third plus one fifth <of the cylinder>. We have shown that the solid* inscribed within the paraboloid plus the cylindrical body generated by the rotation of the surface that is homologous to the surface *JI* is one third plus one fifth of the cylinder. Therefore, the solid* plus the cylindrical body generated by the rotation of the surface that is homologous to the surface *JI* is greater than this part plus the same cylindrical body. Hence, the solid* inscribed within the paraboloid is greater than this part. If the small cylindrical bodies are then also divided into two halves, again and again, each of the remainders of the cylinder will be less than the preceding remainder. Each of the solids* inscribed within the paraboloid will be very much greater than this part. It is clear from this explanation that, for all magnitudes assumed to be less than one third plus one fifth <of the cylinder>, there exist may solids* inscribed within the paraboloid, each of which is greater than that part.

In the same way, let us now assume any part greater than one third plus one fifth <of the cylinder>. Then there will be a difference between it and one third plus one fifth <of the cylinder>. Therefore, if the cylinder is divided – into cylindrical bodies – in two halves, and its half into two halves, and if we continue to proceed in this way, then a remainder will be obtained

[36] i.e. portion of the magnitude.
[37] Have a sum that is very much less.

that is less than this difference. The remainder of the cylinder is the small cylindrical bodies through the middles of which the surface of the paraboloid passes, and <the sum of> these cylindrical bodies is equal to the cylindrical body that is homologous to the cylindrical body generated by the rotation of the surface BI. Therefore the cylindrical body generated by the rotation of the surface that is homologous to the surface BI is less than this difference. One third plus one fifth of the cylinder plus the cylindrical body generated by the rotation of the surface that is homologous to the surface BI is less than this part. One third plus one fifth <of the cylinder> plus the cylindrical body generated by the rotation of the surface that is homologous to the surface BL_a is very much less than this part. But one third plus one fifth, plus the cylindrical body generated by the rotation of the surface that is homologous to the surface BL_a, is the solid* circumscribed around the paraboloid, as the solid* circumscribed around the paraboloid exceeds one third plus one fifth <of the cylinder> by the cylindrical body generated by the rotation of the surface that is homologous to the surface BL_a. The solid* circumscribed around the cylinder is therefore less than this assumed part, which is greater than one third plus one fifth. If the small cylindrical bodies are also divided into halves, each of the solids* generated and circumscribed around the paraboloid will be very much less than this part.

For any part less than one third plus one fifth of the cylinder, there exist many solids* inscribed within the paraboloid, each of which is greater than this part. Each of the solids* circumscribed around the paraboloid and associated with these <inscribed> solids* is also greater than this part, as it is greater than the solid* inscribed within the paraboloid.

For any part greater than one third plus one fifth of the cylinder, there exist many solids* circumscribed around the paraboloid, each of which is less than this part. Each of the solids* inscribed within the paraboloid and associated with these <circumscribed> solids* is also less than this part, as it is less than the solid* circumscribed around the paraboloid.

And for any assumed part other than one third plus one fifth <of the cylinder>, there exist many solids* inscribed within the paraboloid and many solids* circumscribed around the paraboloid, such that those inscribed and circumscribed are both greater than this part and less than this part.

We have previously shown that any solid* inscribed with the paraboloid is less than one third plus one fifth of the cylinder, and that any solid* circumscribed around the paraboloid is greater than one third plus one fifth of the cylinder. It clearly follows from this explanation that no part of the cylinder – that is, no magnitude that is a portion of the cylinder – other than one third plus one fifth of the cylinder, can be less than any solid* inscribed within the paraboloid or greater than any solid* circumscribed around the paraboloid. But the paraboloid is a portion of the cylinder, and any solid* inscribed <within this paraboloid> is less than it, and any solid* circumscribed around it is greater than it. Therefore, if the paraboloid is a portion of the cylinder, and if any solid* inscribed <within this paraboloid> is less than it, and if any solid* circumscribed around it is greater than it, and if no other portion of the cylinder other than one third plus one fifth is such that any solid* inscribed within the paraboloid is less than it, and such that any solid* circumscribed around the paraboloid is greater than it, then it necessarily follows that the paraboloid is one third plus one fifth <of the cylinder>.

This then is the reason why it is necessary that the paraboloid generated by the rotation of the parabola about its ordinate is equal to one third plus one fifth of the cylinder, and why it is not possible for this solid to be other than one third plus one fifth <of the cylinder>. This reason is that any solid* inscribed within the paraboloid is less than one third plus one fifth of the cylinder, and any solid* circumscribed around the paraboloid is greater than one third plus one fifth of the cylinder.

Using the same method in relation to the first species, it can be shown that the reason why it is necessary that the paraboloid generated by the rotation of the parabola about its axis is one half of the cylinder, and why any solid* inscribed within this paraboloid is less than one half of the cylinder, and why any solid* circumscribed around the paraboloid is greater than one half of the cylinder. It is this reason that has given rise to the proof. The method for proving this is the same method that has been described in relation to the second species. We have described the proof for the second species as this proof is more difficult and less obvious. It was this difficulty and obscurity that made it necessary for us to explain it, reveal the reason, and compare the first and second species.

If a notion may be established by a proof by *reductio ad absurdum* – by removing from the magnitude its half and a half from its half – or more than its half – until an impossibility occurs, then the reason[38] which produced the proof is similar to the reason that we have described in this proposition.

We have established the measurement of two species of paraboloids, we have revealed the reasons behind these proofs, and we have treated exhaustively <the measurement> of paraboloids. Let us now conclude our treatise.

The book is completed. Praise be to God, Lord of worlds. May the blessing and salvation of God be upon the prophet Muḥammad and all his People.

[38] Lit.: its cause.

In the Name of God, the Forgiving, the Merciful

TREATISE BY AL-ḤASAN IBN AL-ḤASAN IBN AL-HAYTHAM

On the Measurement of the Sphere

One may arrive at many geometric notions after having followed different routes, and one may successfully prove them using many different methods. Mathematicians have always acted <such that> one may express himself about a notion already considered by another or may achieve a result already achieved by his predecessors, provided that he has found a route that has been taken by none other than himself, and which has not been followed by any of his predecessors. Many mathematicians have spoken on the measurement of the sphere and have proved the magnitude[1] of this measurement. Each of those who have spoken on this subject have followed a path different from those taken by the others.

When their sayings on this notion became available to us and we knew of their proofs, we devoted some thought to the measurement of the sphere. We asked ourselves whether it was possible to achieve the same result using a different method from that adopted by the author. When we examined the question carefully, a method for measuring the sphere was made known to us that is shorter and more concise than any of the methods used by our predecessors, while the proof is clearer and the formulation more obvious than before. This justifies our speaking of the measurement of the sphere despite the fact that many mathematicians[2] have already spoken of it.

<Arithmetical lemmas>

<**1**> In order to do this, let us introduce a simple arithmetical lemma, which will make it easy to understand our objective. This lemma is the following: if we take the successive numbers beginning with one and

[1] This is the volume of the sphere and 'magnitude' is a literal translation of 'quantity'.

[2] Lit.: people of this art. He certainly refers to Archimedes and the Banū Mūsā.

increasing from one by one, if we then take one third of the greatest number and one third of the unit and add them together, if we multiply the result by the greatest number, then we add the greatest number to half of one and we multiply the result by the product previously obtained. The result is the sum of the squares of these numbers.

We have proved this lemma by means of a certain proof in our book *On the Measurement of the Paraboloid*. We repeat the proof here so that this treatise shall require no other.

Let the numbers *AB*, *BC*, *CD* and *DE* be successive numbers beginning with one and increasing from one by one.

I say that if one third of DE *plus one third of the unit is multiplied by the number* DE, *and if* DE *is then added to one half of the unit, and if this is multiplied by the result of the first multiplication, then the result is the sum of the squares of* AB, BC, CD *and* DE.

Proof: Let *BG* be equal to *BA*, *CH* equal to *CB*, *ID* equal to *DC*, and *KE* equal to *ED*. Let each of <the numbers> *GP*, *HN*, *IM* and *KL* be equal to one.

We say first of all that one half of the square of *DE* plus one half of *DE* is the sum of the numbers *AB*, *BC*, *CD* and *DE*, which is *AE*.

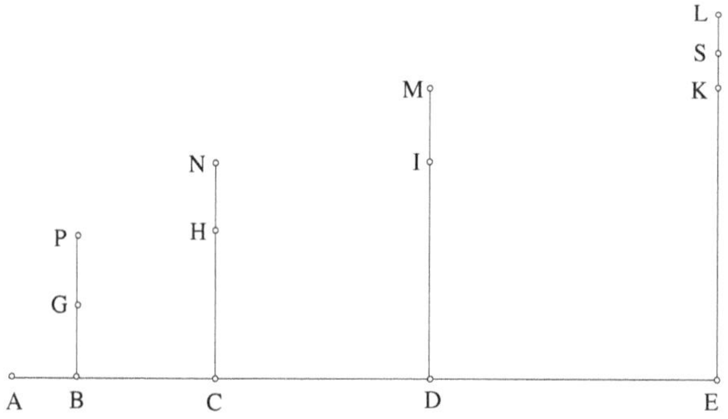

Fig. II.2.1

This is because *CB* is one greater than *BA*, and *CD* is one less than *DE*. Therefore, *AB* plus *DE* is equal to *CB* plus *CD*. Similarly, if the number of numbers is greater than these, then the sum of the two extreme numbers is equal to the sum of the two numbers which follow them, and the sum of those two numbers is equal to the sum of those which follow them, and so on. Therefore, if the number of numbers is odd, then their mean term is one half of the sum[3] of the two extremes, as it is one half of the sum of the two

[3] We have sometimes inserted 'sum' for the purposes of translation.

numbers which surround it,[4] and this is because it is one greater than the number which precedes it and one less than the number which follows it. It is therefore half of the sum of the two numbers which surround it. It follows that the sum of the numbers *AB*, *BC*, *CD* and *DE*, which is the number *AE*, is the multiple of the sum of the two numbers *AB* and *DE*. This multiple is one half of the number of numbers *AB*, *BC*, *CD* and *DE*. But the number of numbers is the number of units contained in the final number, as the first of the numbers is one and they increase from one by one. The number *AE* is therefore a multiple of the sum of the two numbers *AB* and *DE*, and this multiple is one half of the number of units in *DE*. If the number *AB* plus *DE* is multiplied by one half of units in *DE*, then the result of the multiplication is the whole of the number *AE*. The product of one half of *DE* and *DE* is half of the square of *DE*, and the product of one half of *DE* and *AB* is one half of *DE*, as *AB* is the unit. The product of one half of *DE* and the number *AB* plus *DE* is therefore equal to one half of the square of *DE* plus one half of *DE*. The number *AE* is therefore equal to one half of the square of *DE* plus one half of *DE*.

Similarly, the product of *AE* and *EL* is equal to the product of *AE* and *KL* plus the product of *AE* and *EK*. But the product of *AE* and *KL* is *AE*, as *KL* is the unit, and the product of *AE* and *EK* is equal to the product of *DE* and *EK* plus the product of *AD* and *EK*. But the product of *DE* and *EK* is equal to the square of *EK*, as *DE* is equal to *EK*, and the product of *AD* and *EK* is equal to the product of *AD* and *DM*, as *DM* is equal to *EK*. This is because *EK* is one greater than *DI* and *MD* is one greater then *DI*. *MD* is therefore equal to *EK*. The product of *AE* and *EL* is therefore equal to *AE* itself plus the square of *EK*, plus the product of *AD* and *DM*. But the product of *AD* and *DM* is equal to the product of *AD* and *IM* plus the product of *AD* and *DI*. The product of *AD* and *IM* is equal to *AD* itself, as *IM* is the unit. The product of *AD* and *DI* is equal to the product of *CD* and *DI* plus the product of *AC* and *DI*. The product of *CD* and *DI* is the square of *DI*, as *CD* is equal to *DI*. The product of *AC* and *DI* is equal to the product of *AC* and *CN*, as *CN* is equal to *DI*. The product of *AE* and *EL* is therefore equal to *AE* itself, plus *AD* itself, plus the square of *EK*, plus the square of *DI*, plus the product of *AC* and *CN*. But the product of *AC* and *CN* is equal to *AC* itself, plus the square of *CH*, plus the product of *AB* and *BC*, as this can be proved in the same way as it was for the numbers *EL* and *DM*. The product of *AB* and *BP* is equal to *AB* itself, plus the square of *BG*, as *BG* is equal to *BA* and *GP* is the unit.

The product of *AE* and *EL* is equal to *AE* itself, plus *AD* itself, plus *AC* itself, plus *AB* itself, plus the square of *EK*, plus the square of *DI*, plus the

[4] Lit.: which are to one side of it and the other.

square of *CH*, plus the square of *BG*. But *AE* itself is one half of the square of *DE* plus one half of *DE*. Similarly, *AD* is one half of the square of *CD* plus one half of *CD*; similarly, *AC* is one half of the square of *BC* plus one half of *BC*, and *AB*, which is the unit, is one half of the square of *AB* plus one half of *AB*. The numbers *AB*, *BC*, *CD* and *DE*, which are the successive numbers, are equal to the numbers *BG*, *CH*, *DI* and *EK*. The product of *AE* and *EL* is therefore equal to the sum of the squares of *BG*, *CH*, *DI* and *EK*, plus the halves of their squares, plus the halves of themselves.

Divide *LK* into two halves at the point *S*. The product of *AE* and *EL* is then equal to the product of *AE* and *ES* plus the product of *AE* and *SL*. The product of *AE* and *SL* is equal to one half of *AE* as *SL* is one half of the unit. But the product of *AE* and *EL* was the sum of the squares of the successive numbers, plus the halves of their squares, plus the halves of themselves. The product of *AE* and *ES* is therefore the sum of the squares of the successive numbers, the last of which is *EK*, plus the halves of their squares only. The product of two thirds of *AE* and *ES* is therefore equal to the sum of the squares of *BG*, *CH*, *DI* and *EK* only. But we have already shown that the product of one half of *DE* and the sum of *AB* and *DE* is the whole of *AE*, that *DE* is equal to *EK*, and that *AB* is equal to *KL*. The product of one half of *EK* and *EL* is therefore equal to *AE*. The product of two thirds of one half of *EK* – i.e. one third of *EK* – and *EL* is therefore equal to two thirds of *AE*. The product of two thirds of *AE* and *ES* is the sum of the squares of *BG*, *CH*, *DI* and *EK*. If one third of *EK* is then multiplied by *EL*, since this was obtained from *SE*, the result is equal to the sum of the squares of *BG*, *CH*, *DI* and *EK*. But the product of one third of *EK* and *EL* is equal to the product of one third of *EL* and *EK*. *EL* is equal to *EK* plus one. If one third of *EK* plus one third of the unit is multiplied by *EK*, since this was obtained from *ES* – which is equal to *EK* plus one half of the unit – then the result is the sum of the squares of *BG*, *CH*, *DI* and *EK*, which are the successive numbers beginning with one and increasing from one by one, the last of which is *EK*. That is what we wanted to prove.

<2> Similarly, the product of one third of *EK* and *EK* is equal to one third of the square of *EK*, and the product of one third of the unit and *EK* is equal to one third of *EK*. The product of one third of *EK* plus one third of the unit and *EK* is therefore equal to one third of the square of *EK* plus one third of *EK*. If one third of the square of *EK* plus one third of *EK* is multiplied by *ES*, then the result will be the sum of the squares of *BG*, *CH*, *DI* and *EK*.

The product of one third of the square of *EK* and *ES* is the product of one third of the square of *EK* and *EK* and *KS*. The product of one third of the square of *EK* and *EK* is equal to one third of the sum of the equal squares, each one of which is equal to the square of *EK* and whose number is equal to the number of units in *EK*, as the product of one third of the square of *EK* and *EK* is equal to a certain number of times one third of the square of *EK*, this number being the number of units in *EK* and each time being one third of the square of *EK*. The product of one third of *EK* and *EK* is equal to one third of the square of *EK*. The product of one third of the square of *EK* plus one third of *EK* and *EK* is therefore equal to one third of the sum of the equal squares, each of which is equal to the square of *EK*, and whose number is the number of units in *EK*, to which is added one third of the square of *EK*. But the product of one third of the square of *EK* and *KS* is equal to one sixth of the square of *EK*, as *KS* is equal to one half of a unit. The product of one third of *EK* and *KS* is equal to one sixth of *EK*. The product of one third of the square of *EK* plus one third of *EK* and *ES* is therefore equal to one third of the sum of the equal squares, each of which is equal to the square of *EK*, and whose number is equal to the number of units in *EK*, to which is added one third of the square of *EK* plus one sixth of the square of *EK* plus one sixth of *EK*. But one third of the square of *EK* plus one sixth of the square of *EK* is equal to one half of the square of *EK*. A sixth of *EK* is less than one sixth of the square of *EK*, as for any number greater than one, one sixth of that number is less than one sixth of its square as, if the number itself is greater than one, then it will be less than its square. Half of the square of *EK* plus one sixth of *EK* is less than two thirds of the square of *EK*. The product of one third of the square of *EK* plus one third of *EK* and *ES* therefore exceeds one third of the sum of the equal squares, each of which is equal to the square of *EK*, and whose number is equal to the number of units in *EK*, by less than two thirds of the square of *EK* and by more than one half of the square of *EK*. But the product of one third of the square of *EK* plus one third of *EK* and *ES* is equal to the sum of the squares of the numbers *BG*, *CH*, *DI* and *EK*. The number of units in *EK* is the number of numbers *BG*, *CH*, *DI* and *EK*. The sum of the squares of the numbers *BG*, *CH*, *DI* and *EK* therefore exceeds one third of the sum of the equal squares, each of which is equal to the square of *EK*, and whose number is equal to the number of numbers *BG*, *CH*, *DI* and *EK*, by less than two thirds of the square of *EK* and by more than one half of the square of *EK*.

<*Corollary*>

Similarly, let *BG*, *CH*, *DI* and *EK* be straight lines increasing by an equal amount, each increase being equal to the straight line *BG*. These straight lines will be successive multiples of the straight line *BG*, in the same way as the succession of numbers, each of which is one greater than its predecessor. The ratios of these straight lines, each to the other, are equal to the ratios of the successive numbers beginning with one and each being one greater than its predecessor, each to the other. The ratios of the squares of these straight lines, each to the other, will be equal to the ratios of the squares of the successive numbers each to the other. As each straight line is divided into equal parts, its square is therefore a multiple of the square of one of its parts, the number of which is equal to the number of multiples of the square of one, i.e. one, contained by the square of the homonymous number of the parts of this straight line. The ratio of the sum of the squares of the straight lines *BG*, *CH*, *DI* and *EK* to the square of *EK* is equal to the ratio of the sum of the squares of the successive numbers beginning with one and increasing from one to one, and the number of numbers being equal to the number of straight lines, to the square of the greatest of the numbers, equivalent to the straight line *EK*. But the sum of the squares of the successive numbers beginning with one and increasing from one to one exceeds one third of the sum of the equal squares, each being equal to the square of the greatest number and whose number is equal to the number of successive numbers, by less than two thirds of the square of the greatest number and by more than one half of its square. The squares of the straight lines *BG*, *CH*, *DI* and *EK* exceed one third of the sum of the equal squares, each of which is equal to the square of *EK* and whose number is equal to the number of straight lines *BG*, *CH*, *DI* and *EK*, by less than two thirds of the square of *EK*, and by more than one half of its square.

<Theorem>

Having proved that, we now say: Any sphere is equal to two thirds of the circular cylinder whose base is the greatest circle to be found in the sphere, and whose height is equal to the diameter of the sphere.

Example: Let the sphere be *ABCD*, and let its centre be *E*.

I say that it is two thirds of the cylinder whose base is the greatest circle to be found in the sphere and whose height is the diameter of the sphere.

Let there be a plane cutting the sphere and passing through its centre, i.e. the point *E*. This plane generates a circle which is one of the great circles found within the sphere. Let this be the circle *ABCD*. We draw two diameters in this circle that cross at right angles. Let these two diameters be *AEC* and *BED*. From the point *B*, we draw a straight line parallel to the straight line *EA*, and let this straight line be *BG*. From the point *A*, we draw a straight line parallel to the straight line *EB*, and let this straight line be *AG*. The surface *AEBG* therefore has parallel sides and right angles.

If we fix the straight line *AE* and we rotate the surface *AEBG* about the straight line *AE* until it returns to its original position, then the surface *AEBG* generates a circular cylinder whose base is the circle whose half-diameter is the straight line *EB*, which is also the half-diameter of the sphere, and whose height is the straight line *EA*, which is also the half-diameter of the sphere. The circle whose half-diameter is the half-diameter of the sphere is the greatest circle that can be found within the sphere. The cylinder generated by the rotation of the surface *BA* about the straight line *EA* has a base which is the greatest circle to be found within the sphere, and a height equal to the half-diameter of the sphere. Let this cylinder be *BH*. If the surface *BA* rotates about the straight line *EA*, then the sector *ABE* will also rotate about the straight line *EA*. If the sector *ABE* rotates about the straight line *EA*, its rotation will generate a hemisphere whose base is the circle whose half-diameter is the straight line *BE*, as one half of the circle *ABCD* – in which lies *ABC*, and whose diameter is *AC* – if it rotates about the diameter *AC* until it returns to its <original> position, generates the sphere *ABCD* as it rotates. The rotation of the straight line *EB* generates a circle which divides the sphere into two halves. If then the surface *BA* rotates about the straight line *EA*, its rotation will generate a cylinder whose base is the greatest circle to be found within the sphere *ABCD*, and whose height is the straight line *EA*, which is the half-diameter of the sphere *ABCD*, and the rotation of the section *ABE* will generate the hemisphere *ABCD*.

We say that one half of the sphere generated by the rotation of the sector ABE *is equal to two thirds of the cylinder generated by the rotation of the surface* BA, *which is the cylinder* BH.

Proof: It is impossible for this to be otherwise. If it were possible, let the hemisphere be not equal to two thirds of the cylinder *BH*. If the hemisphere is not equal to two thirds of the cylinder *BH*, then it must be either greater than two thirds of the cylinder or less than two thirds of the cylinder.

First, let the hemisphere be greater than two thirds of the cylinder, and let the amount by which the hemisphere exceeds two thirds of the cylinder be a magnitude *T*.

We divide *AE* into two halves at the point *I* and we draw a straight line through the point *I* parallel to the straight line *EB*; let it be *IK*. *IK* is therefore perpendicular to the straight line *AE*. We extend *IK* as far as *L*. *IL* will then be equal to the straight line *EB*. We draw a straight line through the point *K* parallel to the two straight lines *EA* and *BG*; let it be *SKJ*. Then, *SK* is equal to *KJ*, as *AI* is equal to *IE*. The surface *KE* is then equal to the surface *KA*, and the surface *KB* is equal to the surface *KG*. If the surface *BA* rotates about the straight line *EA*, then the two surfaces *EK* and *KA* generate two equal cylinders, and the two surfaces *KB* and *KG* generate two equal cylindrical bodies surrounding the two equal cylinders. The sum of the cylinder generated by the rotation of the surface *KE* and the cylindrical body generated by the rotation of the surface *KG* is equal to one half of the cylinder *BH*. Similarly, if we divide *AI* into two halves at the point *M* and we draw a straight line through the point *M* parallel to the straight line *EB* – let it be *MN* – then *MN* is perpendicular to the straight line *AE*. We extend *MN* as far as *O*. Then, *MO* is equal to the straight line *EB*. We draw a straight line through the point *N* parallel to the two straight lines *KS* and *LG*; let it be *FNJ*. Then *FN* is equal to *NJ*, the surface *NI* is equal to the surface *NA*, and the surface *NK* is equal to the surface *NS*. If the surface *BA* then rotates about the straight line *EA*, the surface *KA* rotates, the two surfaces *NI* and *NA* generate two equal cylinders, and the two surfaces *NK* and *NS* generate two equal cylindrical bodies. The sum of the cylinder generated by the rotation of the surface *NI* and the cylindrical body generated by the rotation of the surface *NS* is equal to one half of the cylinder generated by the rotation of the surface *KA*.

Similarly, if we divide the straight line *IE* into two halves at the point *P*, and we draw a straight line through the point *P* parallel to the straight line *EB* – let it be *PU* – then *PU* is perpendicular to *AE*. We extend *PU* as far as *Q*. Then, *PQ* is equal to the straight line *EB*. We draw a straight line through the point *U* parallel to the two straight lines *EI* and *BL*; let it be *XUJ*. Then, *XU* is equal to *UJ*, the surface *UK* is equal to the surface *UJ*, and the surface *UB* is equal to the surface *UL*. If the surface *BA* rotates about the straight line *EA*, the surface *BK* rotates and, in its rotation, generates *a cylindrical body.[5] The rotation of the two surfaces *UK* and *UJ* generates two equal cylindrical bodies, and the rotation of the two surfaces *UB* and *UL* generates two equal cylindrical bodies. The cylindrical body generated by the rotation of the surface *UJ* plus the cylindrical body generated by the rotation of the surface *UL* are therefore equal to one half of the cylindrical body generated by the rotation of the surface *KB*. The cylinder generated by the rotation of the surface *NI* plus the cylindrical body

[5] Lit.: cylindrical-circular (*mudawwara mustadīra*).

It is clear that the copyist of MS [C] made his copy from a copy whose folios were not in order, or from a copy transcribed from another copy whose folios were not in order. The order of the paragraphs in [C] is as follows: *cylindrical body ... half-: fol. 117ᵛ–118ʳ (*Mathématiques infinitésimales*, vol. II, pp. 310–14); +diameter ... is also: fol. 116ᵛ–117ᵛ (*ibid.*, pp. 314–18); #less than ... less than: fol. 118ʳ (*ibid.*, p. 318); which shows that the order of the folios has been altered.

generated by the rotation of the surface *NS* plus the two cylindrical bodies generated by the rotation of the two surfaces *UJ* and *UL* are equal to one half of the cylinder generated by the rotation of the surface *KA* plus one half of the cylindrical body generated by the rotation of the surface *KB*. But we have shown that the cylinder generated by the rotation of the surface *KE* plus the cylindrical body generated by the rotation of the surface *KG* are equal to one half of the cylinder *BH*.

If this is so, we have taken half away from the cylinder *BH*, and we have take half away from the remainder. If we divide each of the straight lines *AM*, *MI*, *IP* and *PE* into two halves, and if we draw straight lines through the dividing points parallel to the straight line *EB*, and if we draw straight lines through the dividing points on the arc *AB* parallel to the straight line *AE*, then each of the surfaces *BU*, *UK*, *KN* and *NA* will be divided into four parts such that the sum of two opposing surfaces is equal to one half of the surface from which it is taken, and such that the cylindrical bodies generated by the rotation of these surfaces are half of the cylindrical bodies generated by the rotation of the surfaces *BU*, *UK*, *KN* and *NA*. And if we continue to proceed in this way, we will have taken half away from the cylinder *BH*, and half away from the remainder.

If we have two different magnitudes and half is taken away from the larger of these, and half from the remainder, and so on, there must eventually remain a magnitude that is smaller than the smaller of the two magnitudes, as if one half is taken from the magnitude, and then half from the remainder, twice, then the amount taken from the magnitude will be greater than one half of it. If one half is then taken from the magnitude and one half from the remainder several times, then the sum of the two quantities removed is greater than the half.[6] If we have two different magnitudes and half is taken away from the larger of these, and half from the remainder, and if we continue to proceed in this way, then there must necessarily remain a magnitude that is less than the smaller of the two magnitudes. The cylinder *BH* and the magnitude *T* are two different magnitudes, the greatest of which is the cylinder *BH*. If one half is then removed from the cylinder *BH*, and one half is taken from the remainder, and one half is taken from the remainder, and so on as we have described, and if we continue to proceed in this way, then there must necessarily remain a magnitude that is less than the magnitude *T*.

But, if one half is removed from the cylinder *BH*, and one half is taken from the remainder, and one half is taken from the remainder, in the way that we have described, then that which remains of the cylinder is <the sum of> the cylindrical bodies generated by the surfaces *BU*, *UK*, *KN* and *NA*,

[6] He means by this the half of the magnitude whose two quantities are removed.

and their homologues, through the middles of which the surface of the sphere passes.

Let the parts resulting from the division of the cylinder in the way that we have described, and which are less than the magnitude *T*, be the cylindrical bodies generated by the rotation of the surfaces *BU*, *UK*, *KN* and *NA*. The portion of these cylindrical bodies that lies within the hemisphere is very much less than magnitude *T*. But the hemisphere exceeds two thirds of the cylinder *BH* by a magnitude *T*. That portion of the hemisphere that remains once the parts of the cylindrical bodies that lie within it have been removed is greater than two thirds of the cylinder *BH*. But, that portion of the hemisphere that remains once the parts of the cylindrical bodies that lie within it have been removed is the solid[7] which lies within the hemisphere whose base is the circle with the half-diameter *BE*, and whose vertex is the circle whose half-*diameter+ is *NM*. This solid is greater than two thirds of the cylinder *BH*.

Similarly, the straight lines *AM*, *MI*, *IP* and *PE* are equal. Each of the straight lines *EP*, *EI*, *EM* and *EA* therefore exceeds its successor by a straight line equal to *EP*. The ratios of the straight lines *EP*, *EI*, *EM* and *EA*, each to the others, are therefore equal to the ratios of the successive numbers beginning with one and increasing from one by one. The sum of the squares of the straight lines *EP*, *EI*, *EM* and *EA* exceeds one third of the sum of the equal squares, each of which is equal to the square *EA* and whose number is equal to the number of the straight lines *EP*, *EI*, *EM* and *EA* by less than two thirds of the square of *EA*, as we have shown in the lemma. The number of straight lines *EP*, *EI*, *EM* and *EA* is equal to the number of <points> of separation *P*, *I*, *M* and *A*. The number of <points> of separation *P*, *I*, *M* and *A* is equal to the number of <points> of separation *E*, *P*, *I* and *M*, taking *E* in place of *A*. The number of <points> of separation *E*, *P*, *I* and *M* is the number of straight lines *EB*, *PQ*, *IL* and *MO*. The straight lines *EB*, *PQ*, *IL* and *MO* are equal, each of them is equal to the straight line *EB*, and *EB* is equal to the straight line *EA*. The squares of the straight lines *EP*, *EI*, *EM* and *EA* therefore exceed one third of the squares of the straight lines *EB*, *PQ*, *IL* and *MO* by less than two thirds of the square of *EA*. But the square of *EP* plus the product of *CP* and *PA* is equal to the square of *EA*, and the product of *CP* and *PA* is equal to the square of *PU*. The square of *EP* plus the square of *PU* is therefore equal to the square of *EA*, which is equal to the square of *PQ*. Similarly, the square of *EI* plus the square of *IK* is equal to the square of *EA*, which is equal to the square of *IL*, and the square of *EM* plus the square of *MN* is equal to the square of *EA*, which is equal to the square of *MO*. But the square of *EA* is equal to

[7] See *On the Measurement of the Paraboloid*, note 24, p. 199.

the square of *EB*. The sum of the squares of *EP*, *EI*, *EM* and *EA*, and the squares of *PU*, *IK* and *MN* is therefore equal to the sum of the squares of *EB*, *PQ*, *IL* and *MO*. But the squares of *EP*, *EI*, *EM* and *EA* exceed one third of the squares of *EB*, *PQ*, *IL* and *MO* by less than two thirds of the square of *EA*. It remains that the sum of the squares of *PU*, *IK* and *MN* is less than two thirds of the sum of the squares of *EB*, *PQ*, *IL* and *MO* by less than two thirds of the square of *EA*.[8]

The circles[9] whose half-diameters are the straight lines *PU*, *IK* and *MN* are less than two thirds of the circles whose half-diameters are the straight lines *EB*, *PQ*, *IL* and *MO*. But the ratio of the circles to the circles is equal to the ratio of the cylinders of which these circles are the bases, each to the others, providing that the heights of the cylinders are equal. The sum of the cylinders whose bases are the circles whose half-diameters are the straight lines *PU*, *IK* and *MN*, and whose heights are the straight lines *EP*, *PI* and *IM* is therefore less than two thirds of the sum of the cylinders whose bases are the circles whose half-diameters are the straight lines *EB*, *PQ*, *IL* and *MO*, and whose heights are the straight lines *EP*, *PI*, *IM* and *MA*, which are equal. But the sum of the cylinders whose bases are the circles whose half-diameters are the straight lines *PU*, *IK* and *MN*, and whose heights are the straight lines *EP*, *PI* and *IM* is the solid whose base is the circle whose half-diameter is *BE*, and whose vertex is the circle whose half-diameter is *MN*, which lies within the hemisphere. But the sum of the cylinders whose bases are the circles whose half-diameters are the straight lines *EB*, *PQ*, *IL* and *MO*, and whose heights are the straight lines *EP*, *PI*, *IM* and *MA*, is the cylinder *BH*.

The solid that lies within the hemisphere is therefore less than two thirds of the cylinder *BH*.

But we have just shown that this solid is greater than two thirds of the cylinder *BH*. This is impossible.

This impossibility follows from our hypothesis that the hemisphere is greater than two thirds of the cylinder *BH*. Therefore, the hemisphere is not greater than two thirds of the cylinder *BH*.

I say that neither is the hemisphere + less# than two thirds of the cylinder BH.

If this were possible, let it be less# than two thirds of the cylinder, and let the difference between the hemisphere and two thirds of the cylinder be

[8] By inadvertence, Ibn al-Haytham proceeds by upper bounding of this subtractive sum $EP^2 + EI^2 + EM^2 + EA^2$ instead of lower bounding of it, and obtains the upper bounding of $PU^2 + IK^2 + MN^2$. This inadvertence is corrected in the Mathematical commentary, p. 172.

[9] This is the sum of the circles.

the magnitude *T*. The magnitude *T* will therefore be less than the cylinder *BH*.

If one half is taken away from the cylinder *BH*, and one half is taken from the remainder, and one half is taken from the remainder, as we have described earlier, then there must eventually remain a magnitude that is less than the magnitude *T*. But that which remains of the cylinder after it has been divided up, as we have described, is <the sum of> the cylindrical bodies generated by the rotation of the surfaces *BU*, *UK*, *KN* and *NA* and their homologues, through the middles of which the surface of the sphere passes. Let the division continue until one arrives at a magnitude less than the magnitude *T*, and let this be the cylindrical bodies generated by the rotation of the surfaces *BU*, *UK*, *KN* and *NA*. The sum of the parts of the cylindrical bodies that lie outside the hemisphere is therefore very much less than the magnitude *T*. But the hemisphere plus the magnitude *T* is equal to two thirds of the cylinder *BH*. The hemisphere plus those parts of the cylindrical bodies that lie outside it is therefore very much less than two thirds of the cylinder *BH*. But the hemisphere plus those parts of the cylindrical bodies that lie outside it is equal to the solid whose base is the circle whose half-diameter is the straight line *EB*, and whose vertex is the circle whose half-diameter is the straight line *AF*, and which surrounds the hemisphere. This solid is therefore less than two thirds of the cylinder *BH*.

But we have already shown that the sum of the squares of the straight lines *PU*, *IK* and *MN* is less than two thirds of the sum of the squares of the straight lines *EB*, *PQ*, *IL* and *MO* by less than two thirds of the square of *EA*. If we add the entire square of *EB*, which is equal to the square of *EA*, to the squares of the straight lines *PU*, *IK* and *MN*, then the sum of the squares of the straight lines *EB*, *PU*, *IK* and *MN* is greater than two thirds of the squares of the straight lines *EB*, *PQ*, *IL* and *MO*. The sum of the circles whose half-diameters are the straight lines *EB*, *PU*, *IK* and *MN* is therefore greater than two thirds of the sum of the circles whose half-diameters are the straight lines *EB*, *PQ*, *IL* and *MO*, and the sum of the cylinders whose bases are the circles whose half-diameters are the straight lines *EB*, *PU*, *IK* and *MN*, and whose heights are the straight lines *EP*, *PI*, *IM* and *MA*, which are equal, is greater than two thirds of the sum of the cylinders whose bases are the circles whose half-diameters are the straight lines *EB*, *PQ*, *IL* and *MO*, and whose heights are the straight lines *EP*, *PI*, *IM* and *MA*. But the cylinders whose bases are the circles whose half-diameters are the straight lines *EB*, *PU*, *IK* and *MN*, and whose heights are the straight lines *EP*, *PI*, *IM* and *MA*, are equal to the solid whose base is the circle whose half-diameter is the straight line *EB*, and whose vertex is the circle whose half-diameter is *AF*, which is the solid that surrounds the hemisphere. And the cylinders whose bases are the circles whose half-diameters are the straight lines *EB*, *PQ*, *IL* and *MO*, and whose heights are the straight lines *EP*, *PI*, *IM* and *MA*, are equal to the cylinder *BH*. The solid that surrounds the hemisphere is therefore greater than two thirds of the cylinder *BH*.

But we have already shown that this solid is less than two thirds of the cylinder *BH*. This is impossible.

This impossibility follows from our hypothesis that the hemisphere is less than two thirds of the cylinder *BH*. Therefore, the hemisphere is not less than two thirds of the cylinder *BH*. But we have shown that it is not greater than two thirds of the cylinder *BH*. Therefore, if the hemisphere is not greater than two thirds of the cylinder *BH*, neither is it less than this two thirds, then it must be equal to two thirds of the cylinder *BH*. The whole sphere is twice the hemisphere, and the cylinder whose base is the circle whose half-diameter is the straight line *EB*, and therefore whose height is the straight line *AC*, which is the diameter of the sphere and twice the straight line *AE*, is equal to twice the cylinder *BH*. The sphere *ABCD* is therefore equal to two thirds of the cylinder whose base is the greatest circle that can be found on the sphere, and whose height is equal to the diameter of the sphere. That is what we wanted to prove.

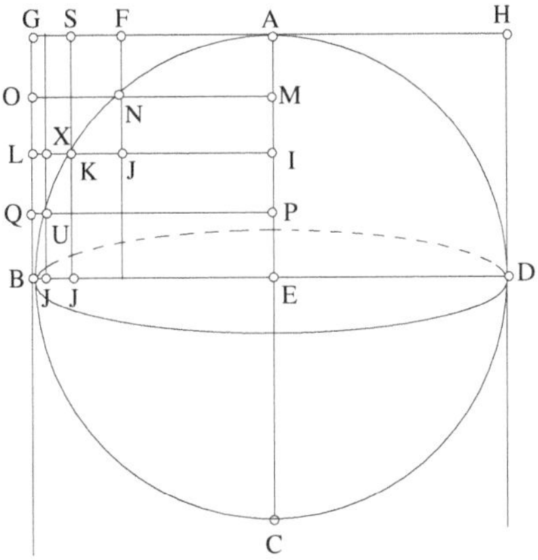

Fig. II.2.2

The treatise *On the Measurement of the Sphere* is complete.

In the Name of God, the Forgiving, the Merciful

TREATISE BY AL-ḤASAN IBN AL-ḤASAN IBN AL-HAYTHAM

On the Division of Two Different Magnitudes as Mentioned in the First Proposition of the Tenth Book of Euclid's *Elements*

Many mathematicians believe that the first proposition in the tenth book of Euclid's *Elements* is a particular case and can only be true in the manner mentioned by Euclid, i.e. that if there are two magnitudes, and more than one half is taken away from the greater of these, and more than one half is then taken from the remainder, and if we continue to proceed in this way, then there will remain a magnitude that is less than the smaller of the two original magnitudes.[1]

The reality is, however, different from that which these people believe. If Euclid limited his discussion to this particular notion, that the <magnitude> taken away is greater than one half, then it is because that was the result that he used in his work, and he confined himsef to this notion because that was what he needed.

For some geometrical notions that we had determined, we were faced with the need to take one half away from the greater of the two different magnitudes, then one half from the remainder, and half again from the remainder, and so on until the division resulted in a magnitude that was less than the smaller of the two original magnitudes. In this way, we were able to determine the notion that we needed. We then proceeded to examine this

[1] Ibn al-Haytham quotes almost verbatim the text of the Arabic translation of the *Elements* said to be by Isḥāq-Thābit (ms. Tehran, Malik 3433, folios not numbered):

إذا كان مقداران موضوعان غير متساويين، وفصل من أعظمهما أكثر من نصفه، ومما يبقى أكثر من نصفه، وفعل ذلك دائماً، فإنه سيبقى منه مقدار ما أقل من المقدار الأصغر الموضوع.

Note that the Arab translator has translated the Greek καὶ τοῦτο ἀεὶ γίγνηται as *wa-fa'ala thālika dā'iman* which quite literally means 'and this has always been done'. The expression ἀεί has been translated by *dā'iman*, a Koranic expression which indicates the permanence of an action or of a thing through time, as well as the indefinite repetition of an action. In the Koran there is *alladhīna hum 'alā ṣalātihim dā'imūn*. The choice of translation is therefore precise, for, as in Greek, the Arabic expression can be written in a variety of ways, all of which use the expression 'always', or an equivalent in other languages – 'if we always act in the same way', 'if we continue to proceed in this way'...

notion attentively, and we have meditated upon it. We then found that it was universal, and that it was one of the properties of proportions, i.e. that if the ratio of the <magnitude> taken away to the larger of the two magnitudes is set to be any ratio whatsoever, and if all the <magnitudes> taken away are in the same ratio, then the division necessarily reaches a magnitude that is less than the smaller of the two original magnitudes. We have therefore decided to reveal this notion and make it evident, so that all who have need of it may make use of it, and to show the error of the belief mentioned earlier that this is a particular notion. We have taken this up, and we have arrived at a proof which shows the universality of this notion, which is moreover extremely brief and concise. It is as follows:

If there are two different magnitudes, and if, from the greater of these two magnitudes, a magnitude is taken away whose ratio to the larger of the two original magnitudes is equal to any given ratio, such that this ratio will be a ratio of the least to the greatest, and if from the remainder, a magnitude is taken away whose ratio to the remainder is the same ratio, and if from the remainder, a magnitude is taken away whose ratio to the remainder is the same ratio, and if we continue to proceed in this way, then the division will eventually result in a magnitude that is less than the smaller of the two original magnitudes.

Example: Let the two magnitudes be *AB* and *CD*, let *AB* be greater than *CD* and let the ratio of *EG* to *GH* be known.

I say that if we separate from the magnitude AB *a magnitude whose ratio to this magnitude is equal to the ratio of* EG *to* GH, *and if we separate from what remains a magnitude whose ratio to the remaining is this ratio, the division will lead to what remains of* AB *a magnitude less than the magnitude* CD.

Proof: We set the ratio of *IC* to *CD* equal to the ratio of *GE* to *EH*; then we multiply the magnitude *IC* until it reaches to a magnitude greater than the magnitude *AB*. Let *KL*, *LM* and *MN*, these multiples,[2] and let *KN* be greater than *AB*; we set the ratio of *FI* to *ID* equal to the ratio of *IC* to *CD*, we set the ratio of *QF* to *FD* equal to the ratio of *FI* to *ID* and we continue to proceed in this way until the magnitudes in proportion, added to the magnitude *CD*, become in an equal number to the number of the multiples[3] which are in *KN*. Let these magnitudes added to the magnitude *CD* – whose number is equal to the number of multiples[4] which are in *KN* – be the magnitudes *QF*, *FI* and *IC*.

[2] i.e. the magnitudes whose sum is the multiple of *KN*.
[3] *Ibid*.
[4] *Ibid*.

Since the ratio of *FI* to *ID* is equal to the ratio of *IC* to *CD*, if then we permute, the ratio of *FI* to *IC* is equal to the ratio of *ID* to *DC*. But *ID* is greater than *DC*. The magnitude *FI* is therefore greater than the magnitude *IC*. Likewise, we show that *QF* is greater than *FI*. The number of the magnitudes *QF*, *FI*, *IC* is therefore equal to the number of the magnitudes *KL*, *LM*, *MN*, and the magnitudes *KL*, *LM*, *MN* are equal, each of them being equal to the magnitude *IC*; and the magnitudes *QF*, *FI*, *IC* are different, as the smallest of them is *IC*. Therefore, the whole *QC* is greater than the whole *KN*; the magnitude *QD* is thus much greater than the magnitude *KN*. But *KN* is greater than *AB*; then the magnitude *QD* is greater than *AB*, and the magnitude *AB* is less than the magnitude *QD*. But the division of *AB* is following the ratios of the parts of the magnitude *QD*; let the division be in *S*, *O*, *U*. The ratio of *AS* to *SB* is therefore equal to the ratio of *QF* to *FD*; but the ratio of *QF* to *FD* is equal to the ratio of *GE* to *EH*, then the ratio of *AS* to *SB* is equal to the ratio of *GE* to *EH*. Similarly, the ratio of *SO* to *OB* is equal to the ratio of *FI* to *ID*, which is equal to the ratio of *GE* to *EH*; and likewise the ratio of *OU* to *UB* is equal to the ratio of *IC* to *CD*, which is equal to the ratio of *GE* to *EH*. The ratio of *SA* to *SB* is therefore equal to the ratio of *EG* to *GH*, the ratio of *OS* to *OB* is equal to the ratio of *EG* to *GH*, the ratio of *UO* to *UB* is equal to the ratio of *EG* to *GH*, the ratio of *AB* to *BS* is therefore equal to the ratio of *QD* to *DF*, the ratio of *SB* to *BO* is equal to the ratio of *FD* to *DI*, and the ratio of *OB* to *BU* is equal to the ratio of *ID* to *DC*.

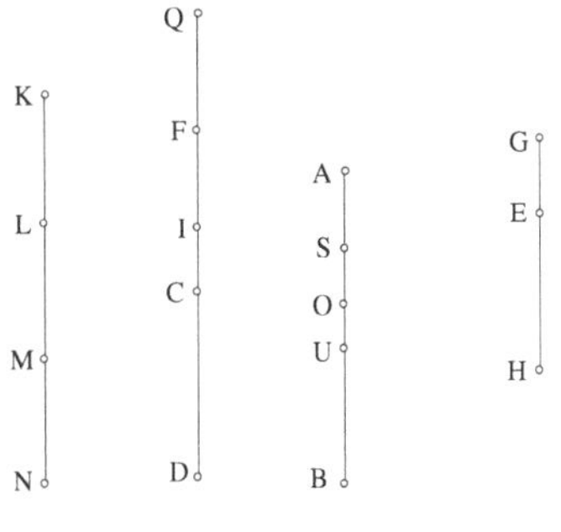

Fig. III.3.1

Since the magnitudes *AB*, *SB*, *OB*, *UB* are following the ratios of the magnitudes *QD*, *FD*, *ID*, *CD*, they will be in a ratio of equality: the ratio of *AB* to *BU* is equal to the ratio of *QD* to *CD*. If we permute, the ratio of *AB* to *QD* is equal to the ratio of *UB* to *CD*. But we have shown that *AB* is smaller than *QD*; the magnitude *BU* is therefore smaller than the magnitude *CD*, which is the smallest one.

We have then separated from the magnitude *AB* a magnitude whose ratio to *AB* is equal to the ratio of *EG* to *GH*, from what remains a magnitude whose ratio to this remainder is the same ratio, and from the remainder a magnitude whose ratio to this remainder is the same ratio. The ratio has led to a magnitude smaller than the magnitude *CD*, the smallest, namely, the magnitude *UB*. This is what we wanted to prove.

The treatise is complete. Glory be to God, and may Grace be with Him. Blessed be our Lord Muḥammad and His people, and may peace be upon them.

Reading and comparison have both arrived at this term.

CHAPTER III

THE PROBLEMS OF ISOPERIMETRIC AND ISEPIPHANIC FIGURES AND THE STUDY OF THE SOLID ANGLE

3.1. INTRODUCTION

The third area of infinitesimal mathematics studied by Ibn al-Haytham concerns isoperimetric and isepiphanic figures: he shows that, out of all the plane figures with a given perimeter, the disc always has the largest area; and, similarly, as far as space is concerned, that out of all the solids with a given surface area, the sphere has the greatest volume. This age-old problem – evident in the writings of Greek astronomers and mathematicians translated into Arabic – had attracted scholars relatively early; al-Kindī was interested in it as early as the middle of the ninth century;[1] a century later, al-Khāzin among others took it up again. And, although he stands alone, Ibn al-Haytham is very much a part of this Graeco-Arabic tradition. However, the only reason he returned to this problem was in order to completely remodel the way in which it was studied, as can be seen immediately from the introduction to the treatise. He says without any ambiguity that he is not satisfied with the positing of this problem. He writes:

> Mathematicians have made mention of this notion and have made use of it. However, they have provided no formal proof of this notion, nor has any convincing argument of its truth come down to us from them.[2]

Did Ibn al-Haytham ignore here the works of his predecessors? We shall discuss this matter later. For the time being, we will only note the promise that Ibn al-Haytham made in terms of presenting a 'universal proof' to establish its extremal properties. While this commitment is fulfilled by him in the case of the disc, it remains however unaccomplished in the difficult case of figures in space. Nonetheless, such a shortcoming is far from being purely

[1] We know from *al-Fihrist* of al-Nadīm that al-Kindī wrote a book: *The sphere is the biggest of the solid figures and the circle the biggest of the plane figures* (ed. R. Tajaddud, Tehran, 1971, p. 316).

[2] Cf. later.

negative in character, since, reversely, it constituted a genuine inventiveness in another sector of mathematics. Having offered a sketchy account of Ibn al-Haytham's undertaking in this regard, we shall engage in their detailed mathematical analysis and commentary.

Following on from the isoperimetric problem, he comes to the isepiphanic problem and tries to demonstrate the fifth proposition as follows:

1) of two regular polyhedra with similar faces and same total area, the one with the greater number of faces has the greater volume.

2) of two regular polyhedra with similar faces, inscribed in the same sphere, the one with the greater number of faces has the greater area and the greater volume.

To prove this proposition, Ibn al-Haytham establishes five lemmas (Lemmas 6–10). And these actually deal with inequalities of ratios of solid angles and ratios of areas. This, as far as we are aware, is the first important and extensive application of the solid angle and, therefore, the first substantial study of some of its properties. The method used by Ibn al-Haytham is just as important as the inequalities themselves: he applies a combination of conical projection and infinitesimal determinations in his work on sections of a pyramid. He did encounter some problems proving these difficult lemmas, however this did not affect the final outcome, and in fact, he was able to establish this fifth proposition for all cases, but his method applies only for the tetrahedron, the octahedron and the icosahedron, since the number of faces of a regular polyhedron with square or pentagonal faces is fixed at six or twelve. The first part of Ibn al-Haytham's proposition maintains that if a regular tetrahedron, octahedron or icosahedron each have the same area, then their volumes increase in the following order: tetra, octa, icosahedron. The second part of the proposition goes on to say that if a regular tetrahedron, octahedron or icosahedron are inscribed in a same sphere, their volumes also increase in this order.

It is therefore impossible to approach the question of the sphere by using an infinite series of polyhedra inscribed therein.

It must be admitted that such an inadvertence on the part of Ibn al-Haytham (who knew Euclid's *Elements* better than anyone) is rather disconcerting. How could he not have noticed that his polyhedra were the same as Euclid's? Or that they are finite in number? However, this should in no way overshadow the richness and depth of this treatise, especially Ibn al-Haytham's work on the solid angle and infinitesimal mathematics.

As for the text itself, which is cited by ancient biobibliographers, we can testify to its undoubted authenticity. If more proof were needed, this treatise is also cited by Ibn al-Haytham in two other works – *On Place* and *The Resolution of Doubts on the Almagest*.[3]

But the list of Ibn al-Haytham's writings drawn up by Ibn Abī Uṣaybiʿa contains a treatise entitled *The Greatest Line Lying in a Segment of Circle* (*Fī aʿẓam al-khuṭūṭ allatī taqaʿ fī qiṭʿat al-dāʾira*). We know nothing, even indirectly, of the content of this treatise, which also studies an extremal property. From the title, and in the context of the mathematics of the time, it is possible that the subject might have been a comparison between various convex curves in the segment of a circle, considering the length of each curve as the upper bounding of inscribed polygons; and to bring as such the comparison between curves to bear on that of polygons. This conjecture, if true, allows us to dwell on Ibn al-Haytham's intentions. Perhaps he wanted to expand on Archimedes' famous postulate as introduced in *The Sphere and Cylinder*:

> Of other lines in a plane and having the same extremities, [any two] such are unequal whenever both are concave in the same direction and one of them is either wholly included between the other and the straight line which has the same extremities with it, or is partly included by, and is partly common with, the other; and that [line] which is included is the lesser [of the two].[4]

Former bibliographers make no mention of any other title which might have dealt with the isoperimetric problem or connected problems or even those general topics which would later become part of the calculation of variations. Ibn al-Haytham himself makes no reference to any other contribution in the works which have come down to us; therefore, at this time, we are only able to analyse the treatise on isoperimetrics: this we intend to do, and in some detail.

It only remains for us to note with regret the absence of any writings on the centre of gravity and the *qarasṭūn* (research on infinitesimal mechanics).

[3] Cf. Introduction, p. 36.

[4] Archimedes, *The Sphere and the Cylinder*, in *The Works of Archimedes*, ed. T. L. Heath, New York, 1953, p. 4. This text had been translated into Arabic and so available to Ibn al-Haytham (Istanbul, Süleymaniye, Fātiḥ 3414, fol. 7ᵛ):

وأما الخطوط الأخر التي في سطح ونهاياتُها واحدة، فإنها مختلفة؛ وأسمّي بهذه الأسماء الخطوط التي انحناؤها في جهة واحدة، وهذه الخطوط إما أن يكون كلٌّ واحد منها يشتمل على الذي يليه حتى يكون الخط المستقيم الذي يصل بين نهاياتها مشتركًا لها كلها أو يكون الخط يشتمل على بعض الخط الذي يليه ويكون باقيه مشتركًا، والخط الذي يشتمل عليه الخط هو أصغر منها.

3.2. MATHEMATICAL COMMENTARY

Proposition 1. — *If a circle and a regular polygon have the same perimeter, the area of the circle is greater than that of the polygon.*

Let (Γ) be a circle, r its radius, $2p_1$ its perimeter and A_1 its area and let there be a regular polygon with n sides, with perimeter $2p_2$ and area A_2.
If $2p_1 = 2p_2 = 2p$, then $A_1 > A_2$.

Proof: The bisectors of the angles of a regular polygon are concurrent at a point I. Triangles with vertex I and with one side of a polygon as a base are isosceles and equal, with h as common height. The circle with centre I and radius h is tangent to all the sides of the polygon. Let this circle be (Γ') and its perimeter $2p'$. Let EG be a side of the polygon, let $IK \perp GE$ and let L and M be the intersections of straight lines IE and IG with (Γ')

$$\frac{h \cdot EG}{2} = \text{area } (IEG) = s,$$

$$\frac{h \cdot \widehat{ML}}{2} = \text{area sect. } (IMKL) = s',$$

(according to Archimedes' *Measurement of a Circle*).

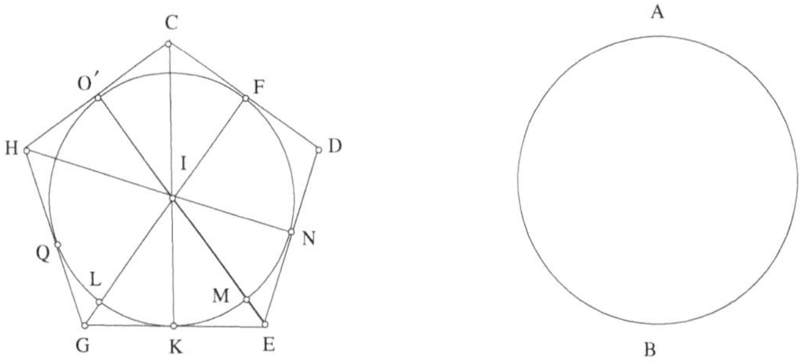

Fig. 3.1

Then $s > s'$, hence $EG > \widehat{ML}$; it can be deduced $n \cdot EG > n \cdot \widehat{ML}$, that is $2p_2 > 2p'$, or $2p > 2p'$. It follows $r > h$ and $p \cdot r > p \cdot h$. However the area of the circle is $A_1 = p \cdot r$ and the area of the polygon is $A_2 = p \cdot IK = p \cdot h$; from this $A_1 > A_2$.

Proposition 2. — *Of two regular polygons with the same perimeter, the one with the greater number of sides has the greater area.*

Let P_1 and P_2 be two regular polygons with same perimeter $2p$. Let n_1 be the number of sides of P_1 and let A_1 be its area, n_2 the number of sides of P_2 and A_2 its area. If $n_1 < n_2$, then $A_1 < A_2$.

Let DE be a side of P_1 and LM a side of P_2.

So $2p = n_1 \cdot DE = n_2 \cdot LM$, hence $DE > LM$ since $n_2 > n_1$. If P and U are respectively the midpoints of DE and of LM, then

$$PE > UM \text{ and } \frac{PE}{UM} = \frac{n_2}{n_1},$$

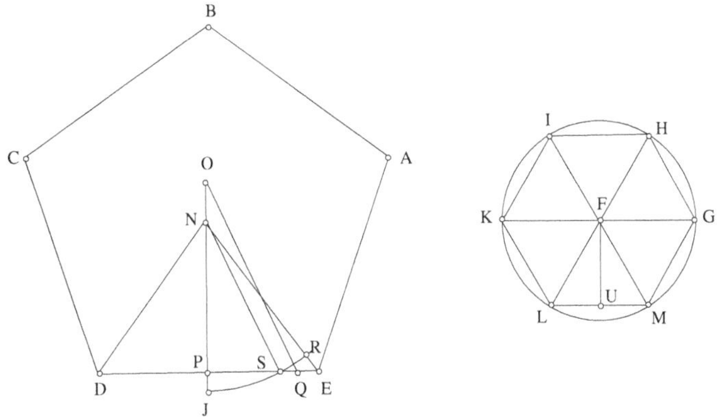

Fig. 3.2

with N and F as centres of polygons P_1 and P_2 respectively, therefore

$$P\hat{N}E = \frac{2\pi}{2n_1} = \frac{\pi}{n_1}, \text{ similarly } U\hat{F}M = \frac{\pi}{n_2};$$

from which we deduce

$$\frac{P\hat{N}E}{U\hat{F}M} = \frac{n_2}{n_1},$$

hence

$$P\hat{N}E > U\hat{F}M \text{ and } \frac{P\hat{N}E}{U\hat{F}M} = \frac{PE}{UM}.$$

Let S be a point on DE such that $P\hat{N}S = U\hat{F}M$, then $\dfrac{P\hat{N}E}{P\hat{N}S} = \dfrac{PE}{UM}$.

Circle (N, NS) cuts NP at J and NE at R. Therefore

$$\frac{E\hat{N}P}{S\hat{N}P} = \frac{\text{area sect.}(RNJ)}{\text{area sect.}(SNJ)} = \frac{EP}{MU},$$

$$\text{area tr.}(SNE) > \text{area sect.}(SNR) \text{ and area tr.}(SNP) < \text{area sect.}(SNJ),$$

hence

$$\frac{\text{tr.}(SNE)}{\text{tr.}(SNP)} > \frac{\text{sect.}(SNR)}{\text{sect.}(SNJ)} \Rightarrow \frac{\text{tr.}(PNE)}{\text{tr.}(SNP)} > \frac{\text{sect.}(RNJ)}{\text{sect.}(SNJ)}.$$

However

$$\frac{\text{tr.}(PNE)}{\text{tr.}(SNP)} = \frac{PE}{PS},$$

hence

$$\frac{PE}{PS} > \frac{PE}{MU} \text{ and } PS < MU.$$

Right-angled triangles PNS and UFM are similar, as $P\hat{N}S = U\hat{F}M$; and as $PS < MU$, then $NP < FU$; however $A_1 = p \cdot NP$ and $A_2 = p \cdot FU$, therefore $A_2 > A_1$.

Proposition 3. — *Of two regular polygons inscribed in the same circle, the one with the greater number of sides has the greater perimeter and the greater area.*

Lemma. — *Let* $\widehat{AB}$ *and* $\widehat{BC}$ *be two arcs such that* $\widehat{AB} > \widehat{BC}$ *and* $\widehat{AB} + \widehat{BC} \leq \frac{2}{3}$ *circle, therefore* $\frac{\widehat{AB}}{\widehat{BC}} > \frac{AB}{BC}$.

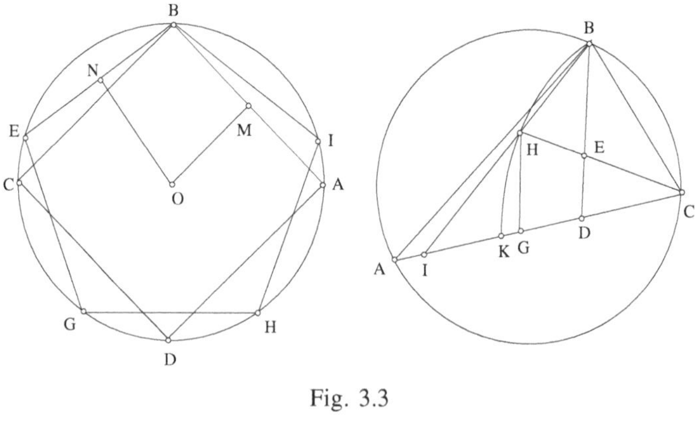

Fig. 3.3

If $\widehat{AB} > \widehat{BC}$ and $\widehat{AB} + \widehat{BC} \leq \frac{2}{3}$ circle, then $\widehat{BC} < \frac{1}{3}$ circle and $\widehat{AC} \geq \frac{1}{3}$ circle, hence $\widehat{BC} < \widehat{AC}$ and $B\hat{A}C < A\hat{B}C$. This allows the construction of $C\hat{B}D = B\hat{A}C$, with D on segment AC.

ABC and *BDC* are two similar triangles, hence

$$(1) \qquad \frac{AC}{BC} = \frac{BC}{CD} = \frac{AB}{BD};$$

as $AC > BC$ and $AB > BC$, we can deduce $BC > CD$ and $BD > CD$.

We construct $D\hat{C}E = C\hat{B}E = B\hat{A}C$ inside $A\hat{C}B$.

CDE and *BDC* are two similar triangles, hence

$$(2) \qquad \frac{CB}{CE} = \frac{CD}{DE} = \frac{BD}{DC}.$$

Circle (C, CB) cuts CE at H and CA at K; we draw a straight line HG parallel to BD from point H; BH cuts AC at I.

HGC and *EDC* are two similar triangles, hence

$$(3) \qquad \frac{HG}{ED} = \frac{GC}{DC} = \frac{HC}{EC} = \frac{CB}{EC}.$$

(2) and (3) give $\frac{CD}{DE} = \frac{HG}{ED}$, hence $HG = CD$.

HG and BD are parallel, hence $\frac{BI}{IH} = \frac{BD}{HG}$, from which we can deduce $\frac{BI}{IH} = \frac{BD}{CD}$.

From (1), $\frac{BD}{DC} = \frac{AB}{BC}$, hence $\frac{BI}{IH} = \frac{AB}{BC}$.

Then area sect.(CBH) > area tr.(CBH) and area sect.(CHK) < area tr.(CHI), hence

$$\frac{\text{sect.}(CBH)}{\text{sect.}(CHK)} > \frac{\text{tr.}(CBH)}{\text{tr.}(CHI)}.$$

We can deduce

$$\frac{\widehat{BH}}{\widehat{HK}} > \frac{BH}{HI} \quad \text{and} \quad \frac{B\hat{C}H}{H\hat{C}I} > \frac{BH}{HI},$$

hence

$$\frac{B\hat{C}H + H\hat{C}I}{H\hat{C}I} > \frac{BH + HI}{HI}.$$

And from this

$$\frac{B\hat{C}A}{B\hat{A}C} > \frac{AB}{BC},$$

therefore

$$\frac{\widehat{AB}}{\widehat{BC}} > \frac{AB}{BC}.$$

Comment: In the study of regular polygons inscribed in a circle, the greatest possible arc corresponds to the side of an equilateral triangle, which justifies the hypothesis introduced here by Ibn al-Haytham – $\widehat{AB} + \widehat{BC} \leq \frac{2}{3}$circle; it is also used in construction of point D in the argument.

Let us note that if the radian measures of arcs $\widehat{AB}$ and $\widehat{BC}$ are 2α and 2β, the result established is simply $\frac{\alpha}{\beta} > \frac{\sin\alpha}{\sin\beta}$, for $\frac{\pi}{2} > \alpha > \beta$.

Proof of the theorem: An equilateral triangle is a regular convex polygon with the smallest number of sides.

For any regular polygon with more than three sides, the arc subtended on the circle circumscribed by one of the sides is less than one third of the circle.

Let $ABCD$ be a square with perimeter C_1 and area A_1 and $BEGHI$ be a pentagon with perimeter C_2 and area A_2. $\widehat{AB} + \widehat{BE} < \frac{2}{3}$ circle, therefore $\frac{\widehat{AB}}{\widehat{BE}} > \frac{AB}{BE}$, hence

(1) $$\frac{\widehat{AB}}{\widehat{BI}} > \frac{AB}{BI}$$ from the lemma.

If C is the perimeter of a circumscribed circle, then

$$\frac{\widehat{AB}}{C} = \frac{AB}{C_1} \text{ and } \frac{\widehat{BI}}{C} = \frac{BI}{C_2},$$

hence by division

$$\frac{\widehat{AB}}{\widehat{BI}} = \frac{AB}{BI} \cdot \frac{C_2}{C_1},$$

hence

$$\frac{AB}{BI} \cdot \frac{C_2}{C_1} > \frac{AB}{BI},$$

which implies $C_2 > C_1$.

Then $A_1 = \frac{1}{2}C_1 \cdot OM$ and $A_2 = \frac{1}{2}C_2 \cdot ON$, with $OM < ON$ ($AB > BE$ $\Rightarrow OM < ON$), hence $A_2 > A_1$.

Let us note that the argument is independent of the nature of the regular polygon.

Proposition 4. — *If a sphere and a regular polyhedron inscribed in a sphere have the same area, then the volume of the sphere is greater than that of the polyhedron.*

Lemmas:

1) Let a sphere have radius R, volume V_S and area A_S and a right cylinder of radius R and height $2R$ and volume V_C, then $V_S = \frac{2}{3} V_C$ (Archimedes). Let s be the area of a great circle of the sphere equal to the base of the cylinder.

$V_C = s \cdot \text{height} = s \cdot 2R$, therefore $V_S = \frac{2}{3}\, s \cdot 2R = \left(1+\frac{1}{3}\right) s \cdot R$, but $\left(1+\frac{1}{3}\right) s = \frac{1}{3} A_S$, hence

(1) $$V_S = \frac{1}{3} A_S \cdot R = \frac{4}{3}\pi R^3.$$

2) Let a regular polyhedron be inscribed in a sphere. With each face of the polyhedron is associated a regular pyramid whose vertex is centre B of the sphere. In this way we define a solid angle of vertex B, a spherical surface area and a section of the sphere.

Let A be the area of a sphere, s be the area of the spherical surface, v be the volume of the spherical section, V be the volume of the sphere, α be the solid angle, $\frac{\pi}{2}$ be the solid right angle. Then

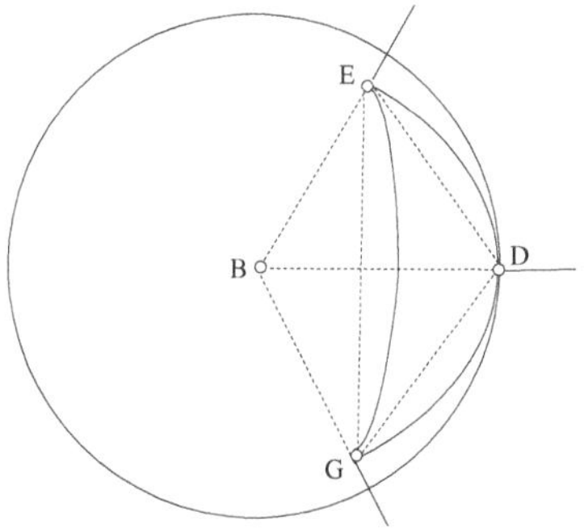

Fig. 3.4

(2) $$\frac{v}{V} = \frac{s}{A} = \frac{\alpha}{4\pi}.$$

Let us note that each of these ratios is equal to $\frac{1}{n}$, if n is the number of faces of the polyhedron.

The sum of solid angles in the centre of the sphere is in fact eight solid right angles, since the three sets of perpendicular lines through one point determine eight equal solid angles; each one of them is a solid right angle.

From (2) and (1), we can deduce $v = \frac{1}{3}s \cdot R$.

Proof of Proposition 4: This proof does not include the nature of the polyhedron.

Let (Π) be the plane of one of the faces; let G, D and E be three vertices of this face; let B be the centre of the circumscribed sphere, then $BG = BD = BE$. If $BC \perp (\Pi)$, then $CG = CD = CE$, and the face is inscribed in circle (C, CD). For all the faces in equal polygons, the circles defined in this way are equal and B is equidistant from the planes of all the faces. Therefore sphere (B, BC) is inscribed in the polyhedron.

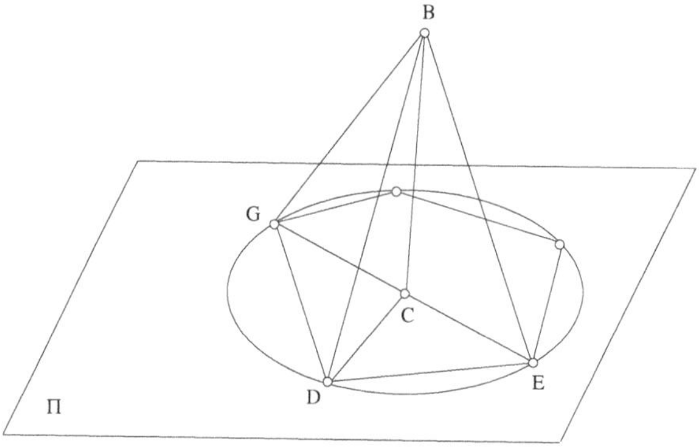

Fig. 3.5

Let P be a pyramid with apex B and base the face GDE of the polyhedron. Let v_1 be its volume, s_1 the area of its base, V_1 the volume of the polyhedron and S_1 its total area, then $V_1 = nv_1$ and $S_1 = ns_1$, if n is the number of faces of the polyhedron. Pyramid P determines a spherical section in sphere (B, BC). Let v_2 be its volume, s_2 the area of corresponding spherical surface, S_2 the area of the sphere and V_2 its volume, then $S_2 = ns_2$ and $V_2 = nv_2$ and $v_2 < v_1$. We have $v_1 = \frac{1}{3} s_1 \cdot BC$ and, from the lemma,

$v_2 = \frac{1}{3} s_2 \cdot BC$; as $v_1 > v_2$, then $s_1 > s_2$, hence $S_1 > S_2$. However S_1 is also the area of sphere A, then

$$S_1 \text{ (area of } A) > S_2 \text{ (area of } (B, BC)),$$

therefore the radius of sphere A is greater than BC.

Volume of polyhedron B: $V_1 = \frac{1}{3} S_1 \cdot BC$

Volume of sphere A: $V = \frac{1}{3} S_1 \cdot R$.

$R > BC \Rightarrow V > V_1$.

Proposition 5:

5a. — *Of two regular polyhedra with similar faces and with the same total area, the one with the greater number of faces has the greater volume.*

5b. — *Of two regular polyhedra with similar faces, inscribed in the same sphere, the one with the greater number of faces has the greater area and the greater volume.*

Preliminary. — Let A be the centre of a sphere, and consider pyramids P_1 $(A, BCDE)$ and P_2 (A, HFG). P_1 with a solid angle α_1 intercepts a spherical surface s_1 and delimits a section of the sphere of volume v_1; and similarly α_2, s_2 and v_2 in P_2. Then

$$\frac{\alpha_1}{\alpha_2} = \frac{s_1}{s_2} = \frac{v_1}{v_2}.$$

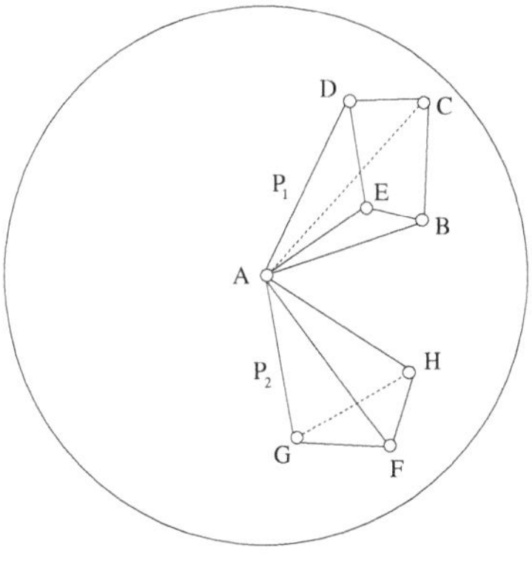

Fig. 3.6

At n times pyramid P_1 is associated with a portion of the sphere whose volume is nv_1, the intercepted spherical surface ns_1 and the solid angle $n\alpha_1$, and similarly for P_2.

If $nv_1 > nv_2$, then $n\alpha_1 > n\alpha_2$ and $ns_1 > ns_2$.
If $nv_1 < nv_2$, then $n\alpha_1 < n\alpha_2$ and $ns_1 < ns_2$.
If $nv_1 = nv_2$, then $n\alpha_1 = n\alpha_2$ and $ns_1 = ns_2$.

If $n\alpha_1 > n\alpha_2$, then $ns_1 > ns_2$ and $nv_1 > nv_2$.
If $n\alpha_1 < n\alpha_2$, then $ns_1 < ns_2$ and $nv_1 < nv_2$.
If $n\alpha_1 = n\alpha_2$, then $ns_1 = ns_2$ and $nv_1 = nv_2$.

Let us note that the explanations given here by Ibn al-Haytham do not constitute a proof of the stated property:

$$\frac{\alpha_1}{\alpha_2} = \frac{s_1}{s_2} = \frac{v_1}{v_2}.$$

In Lemma 2 of Proposition 4, he considers a regular polyhedron inscribed in a sphere, a polyhedron decomposable of n regular pyramids; for each pyramid, then $\frac{v}{V} = \frac{s}{A} = \frac{\alpha}{4\pi}$ (see later).

If, in this Preliminary, P_1 and P_2 come from two regular polyhedra with n_1 and n_2 faces respectively, then

$$\frac{v_1}{V} = \frac{s_1}{A} = \frac{\alpha_1}{4\pi} = \frac{1}{n_1} \quad \text{and} \quad \frac{v_2}{V} = \frac{s_2}{A} = \frac{\alpha_2}{4\pi} = \frac{1}{n_2},$$

hence

$$\frac{v_1}{v_2} = \frac{s_1}{s_2} = \frac{\alpha_1}{\alpha_2}.$$

But Ibn al-Haytham does not give details of the nature of pyramids P_1 and P_2.

Lemma 6. — *Let* ABCD *be a pyramid such that* $A\hat{B}C \geq \frac{\pi}{2}$ *and* $A\hat{B}D \geq \frac{\pi}{2}$; *if* E *is a point on* BD *such that* $A\hat{E}C \geq \frac{\pi}{2}$ *or* $A\hat{C}E \geq \frac{\pi}{2}$, *then*

$$\frac{\text{area } (DBC)}{\text{area } (EBC)} > \frac{\text{solid angle } (A, BDC)}{\text{solid angle } (A, EBC)}.$$

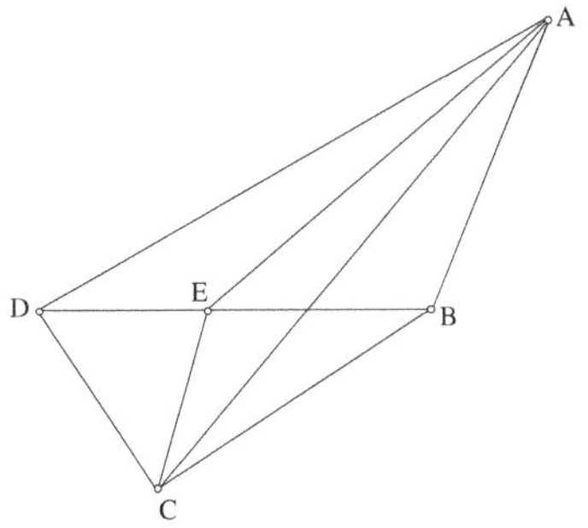

Fig. 3.7

Let Σ be a sphere with centre A and radius AB, which cuts AC at H, AD at I, AE at L, such that $AB = AH = AI = AL$. Therefore arc BH is in plane (BAC), arc BLI in plane (BAD), arc HI in plane (ACD) and arc HGL in plane (ACE). Straight line BL is in plane (BAD) and cuts AD at K (since angle ABL is acute and angle BAD is acute). Arc LGH is on Σ, therefore K is outside Σ, and $AK > AI$. The conical surface with vertex B defined by arc LGH cuts plane (ADC) following an arc KFH, 266s this plane at F outside sphere Σ; arc KFH, except for point H, is outside sphere Σ. Therefore the section of sphere $AILGH$ is inside solid $AKFHGL$, and is limited by planes and part of the conical surface, since part GF of the generating line is outside Σ and the section of sphere $ALHB$ is greater than solid $ALHB$, itself limited by planes and another part of the conical surface, since portion BG of the generating line of the cone is inside Σ.

$$\left.\begin{array}{l}\text{sect.}\ (A, ILH) < \text{sol.}\ (A, KFHGL)\\ \text{sect.}\ (A, LHB) > \text{sol.}\ (A, HGLB)\end{array}\right\} \Rightarrow \frac{\text{sect.}\ (A, ILH)}{\text{sect.}\ (A, LHB)} < \frac{\text{sol.}\ (A, KFHGL)}{\text{sol.}\ (A, LGHB)}.$$

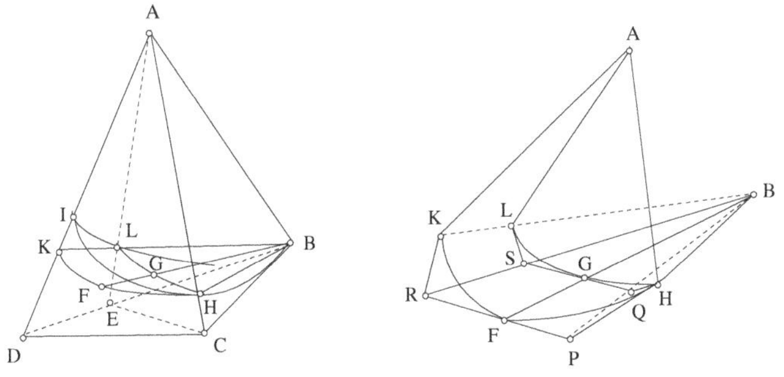

Fig. 3.8

By composition,

$$(*)\qquad \frac{\text{sect. }(A, IHB)}{\text{sect. }(A, LHB)} < \frac{\text{sol. }(B, AKFH)}{\text{sol. }(B, AHGL)}.$$

In this way, Ibn al-Haytham introduces a proposition into the proof of Lemma 6, stated as

$$(**)\qquad \frac{\text{area tr. }(AEC)}{\text{area sect. }(ALGH)} \leq \frac{\text{area tr. }(ADC)}{\text{area sect. }(AKFH)}.$$

In other words, how does the conical projection of centre B of plane (AEC) onto plane (ADC) increase some ratios of star areas in relation to A?

Ibn al-Haytham's proof of this proposition relies on the *apagogic method* to compare areas of triangles with apex A.

Let us take up the steps of this proof. We assume that

$$(1)\qquad \frac{\text{area }(AEC)}{\text{area sect. }(ALGH)} > \frac{\text{area }(ADC)}{\text{area sect. }(AKFH)};$$

area L_a (4th proportional) exists such that

$$(2)\qquad \frac{\text{area }(AEC)}{L_a} = \frac{\text{area }(ADC)}{\text{area sect. }(AKFH)}.$$

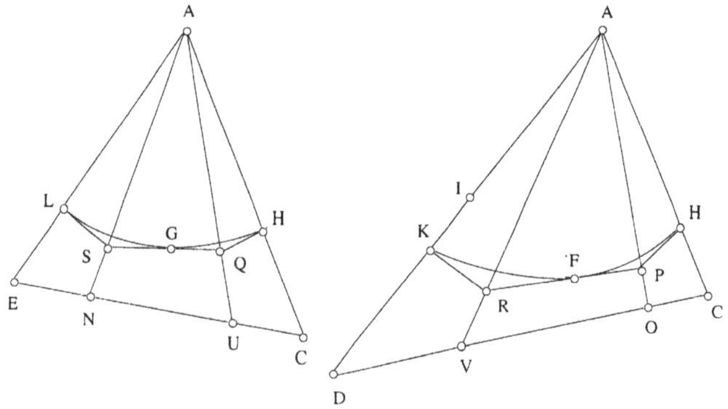

Fig. 3.9

Hypothesis (1) can therefore be written as $L_a >$ area sect. ($ALGH$), and therefore polygon $LSQH$ circumscribed about the arc of circle LGH exists such that

(3) $\text{area}\ (ALSQH) < L_a$.

Ibn al-Haytham puts forward a case with a polygon having three sides *LS*, *SQ*, *QH* tangent to the arc of the circle at points *L*, *G* and *H* respectively. Polygon *LSQH* is projected onto plane *ADC* as polygon *KRPH* with sides *KR*, *RP* and *PH* tangent to the projection *KFH* (conical arc) of the arc of circle *LGH* at points *K*, *F* and *H* respectively.

Let us note that Ibn al-Haytham is perfectly aware of the fact that conical projection maintains the points of contact; this reminds us of the properties of a plane tangent to a cone proved by Ibn Sahl.[5]

The statement is thus reduced to the inequality

$$(4) \qquad \frac{\text{area}\ (ADC)}{\text{area sect.}\ (AKRPH)} > \frac{\text{area}\ (AEC)}{\text{area sect.}\ (ALSQH)}.$$

In fact, the second ratio of (4) is, according to (3), greater than

$$\frac{\text{area}\ (AEC)}{L_a} = \frac{\text{area}\ (ADC)}{\text{area sect.}\ (AKFH)} \qquad (2),$$

from which we deduce that

$$\text{area}\ (AKRPH) < \text{area sect.}\ (AKFH),$$

which is absurd since the polygon is circumscribed about the arc of the curve.

Ibn al-Haytham then confirms that inequality (4) results from the following inequalities:

$$(5) \qquad \frac{\text{area}\ (AEN)}{\text{area}\ (ALS)} < \frac{\text{area}\ (ADV)}{\text{area}\ (AKR)};$$

$$\frac{\text{area}\ (ANU)}{\text{area}\ (ASQ)} < \frac{\text{area}\ (AVO)}{\text{area}\ (ARP)};$$

$$\frac{\text{area}\ (AUC)}{\text{area}\ (AQH)} < \frac{\text{area}\ (AOC)}{\text{area}\ (APH)}.$$

[5] R. Rashed, *Geometry and Dioptrics in Classical Islam*, London, 2005, pp. 46–54.

To demonstrate the inequalities in (5), Ibn al-Haytham constructs triangles AZW, AWJ and AJC in plane ADC such that

(5′) $$\frac{\text{area }(AZW)}{\text{area }(AKR)} = \frac{\text{area }(AEN)}{\text{area }(ALS)};$$

$$\frac{\text{area }(AWJ)}{\text{area }(APR)} = \frac{\text{area }(ANU)}{\text{area }(ASQ)};$$

$$\frac{\text{area }(AJC)}{\text{area }(APH)} = \frac{\text{area }(AUC)}{\text{area }(AQH)}.$$

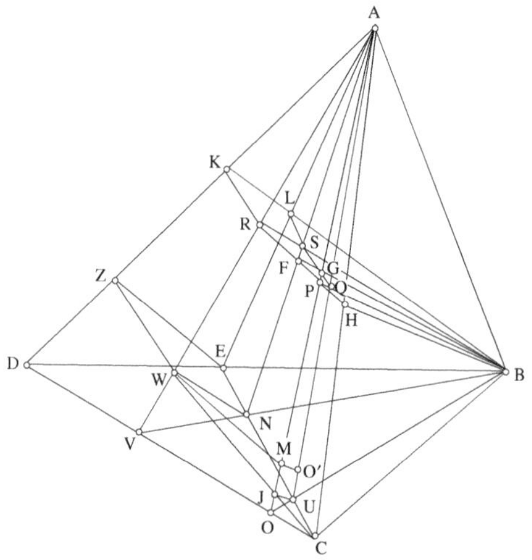

Fig. 3.10

Let us assume first that angle AEC is a right angle; LS is parallel to EC, hence

$$\frac{AE}{AL} = \frac{AN}{AS} = k.$$

In plane ABD, we draw a straight line parallel to BK from E; it cuts AD at Z, so

$$\frac{AE}{AL} = \frac{AZ}{AK} = k.$$

In plane ABV, we draw a straight line parallel to BS from N; it cuts AV at W, so

$$\frac{AN}{AS} = \frac{AW}{AR} = k.$$

Therefore $\frac{WA}{AR} = \frac{AZ}{AK}$, and it follows that WZ is parallel to RK. We can deduce

(a) $$\frac{\text{area } (AWZ)}{\text{area } (ARK)} = \frac{\text{area } (ANE)}{\text{area } (ASL)} = k^2.$$

The line drawn from N parallel to SQ cuts AU at O' between Q and U, since angle ANU is obtuse and angle ASQ is acute. The line drawn from O' parallel to PQ cuts AO at M, and therefore

$$\frac{AM}{AP} = \frac{AO'}{AQ} = \frac{AN}{AS} = \frac{AW}{AR}.$$

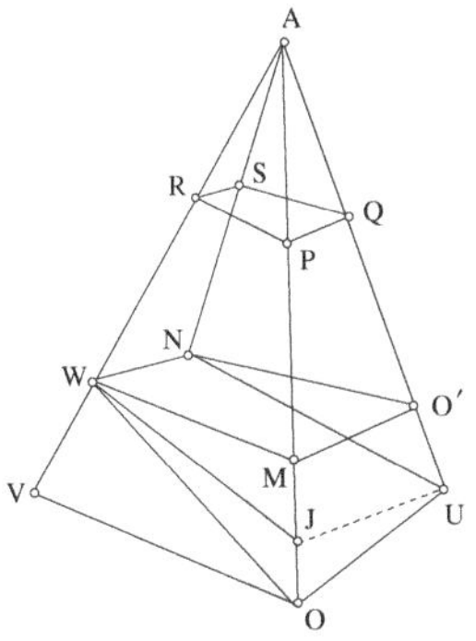

Fig. 3.11

From this we deduce that MW is parallel to PR and that

(6) $$\frac{\text{area } (AMW)}{\text{area } (APR)} = \frac{\text{area } (ANO')}{\text{area } (ASQ)}.$$

The line drawn from U parallel to QP, cuts AO at J between M and O, since UJ is parallel to MO'. Then $\frac{AU}{AO'} = \frac{AJ}{AM}$. But

$$\frac{\text{area } (ANU)}{\text{area } (ANO')} = \frac{AU}{AO'} \text{ and } \frac{\text{area } (AWJ)}{\text{area } (AWM)} = \frac{AJ}{AM},$$

hence

(7) $$\frac{\text{area }(AWJ)}{\text{area }(AWM)} = \frac{\text{area }(ANU)}{\text{area }(ANO')}.$$

From (6) and (7), we get

(b) $$\frac{\text{area }(AWJ)}{\text{area }(APR)} = \frac{\text{area }(ANU)}{\text{area }(ASQ)}.$$

Perpendicular $O'I'$ to AC is drawn from O'; then $O'I'$ is parallel to QH and point I' is between H and C; we deduce

$$\frac{MA}{AP} = \frac{O'A}{AQ} = \frac{I'A}{AH},$$

hence $I'M \parallel PH$ and it follows

$$\frac{\text{area }(AMI')}{\text{area }(APH)} = \frac{\text{area }(AO'I')}{\text{area }(AQH)}.$$

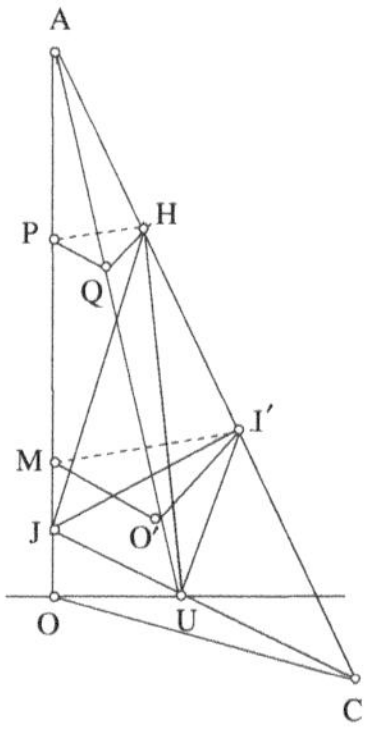

Fig. 3.12

Moreover

$$\frac{\text{area }(AJI')}{\text{area }(AMI')} = \frac{AJ}{AM} = \frac{AU}{AO'} = \frac{\text{area }(AUI')}{\text{area }(AO'I')};$$

therefore

(8) $$\frac{\text{area }(AJI')}{\text{area }(APH)} = \frac{\text{area }(AUI')}{\text{area }(AQH)}.$$

but

$$\frac{\text{area}\,(AUC)}{\text{area}\,(AUI')} = \frac{AC}{AI'} = \frac{\text{area}\,(AJC)}{\text{area}\,(AJI')},$$

hence

$$\frac{\text{area}\,(AJC)}{\text{area}\,(AJI')} = \frac{\text{area}\,(AUC)}{\text{area}\,(AUI')}. \tag{9}$$

Multiplying member by member in (8) and (9) gives

$$\frac{\text{area}\,(AJC)}{\text{area}\,(APH)} = \frac{\text{area}\,(AUC)}{\text{area}\,(AQH)}. \tag{c}$$

Since (a), (b), (c) constitute the equalities in (5′) it is possible to construct triangles AZW, AWJ and AJC with the required properties.

Later Ibn al-Haytham assumes angle AEC to be obtuse and proceeds as follows.

It is possible to construct $A\hat{E}\nu = \frac{\pi}{2}$, with ν on AN and with $\nu\mu$ drawn parallel to BSR from point ν, so $\nu\mu \parallel NW$. Therefore

$$\frac{AN}{A\nu} = \frac{AW}{A\mu},$$

hence

$$\frac{\text{area}\,(AZW)}{\text{area}\,(AZ\mu)} = \frac{\text{area}\,(AEN)}{\text{area}\,(AE\nu)}.$$

Moreover

$$\frac{A\mu}{AR} = \frac{A\nu}{AS} = \frac{AE}{AL} = \frac{AZ}{AK} \quad (\text{since } E\nu \parallel LS \text{ and } EZ \parallel LK);$$

from this we deduce $Z\mu \parallel KR$, therefore

$$\frac{\text{area}\,(AZE)}{\text{area}\,(AKL)} = \frac{\text{area}\,(AZ\mu)}{\text{area}\,(AKR)} = \frac{\text{area}\,(AE\nu)}{\text{area}\,(ALS)},$$

and from the previous relation, we have

$$\frac{\text{area}\,(AZW)}{\text{area}\,(AKR)} = \frac{\text{area}\,(AEN)}{\text{area}\,(ALS)};$$

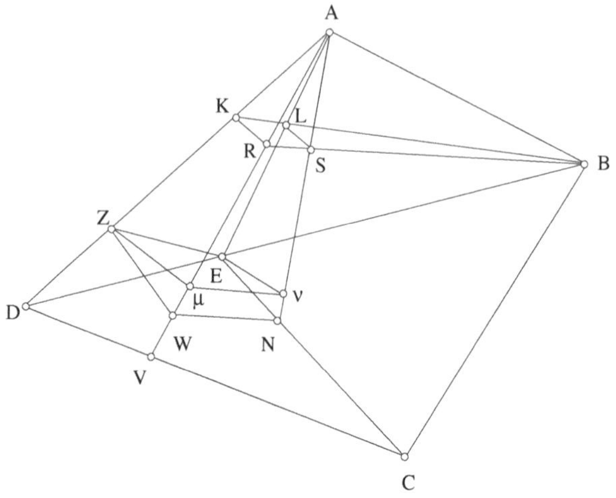

Fig. 3.13

however area (ADV) > area (AZW), hence

$$\frac{\text{area }(ADV)}{\text{area }(AKR)} > \frac{\text{area }(AEN)}{\text{area }(ALS)}.$$

We draw a line parallel to SQ from point ν. We continue as in the first case. The method is the same if the hypothesis is $A\hat{C}E \geq \frac{\pi}{2}$.

Unfortunately, Ibn al-Haytham's statement that (4) follows from (5) is not true for all cases.

Let us examine the relation between the two types of inequalities, and let us suppose that

$$\lambda_1 = \frac{\text{area }(AEN)}{\text{area }(ALS)},\ \lambda'_1 = \frac{\text{area }(ADV)}{\text{area }(AKR)};\ \lambda_2 = \frac{\text{area }(ANU)}{\text{area }(ASQ)},\ \lambda'_2 = \frac{\text{area }(AVO)}{\text{area }(ARP)};$$

$$\lambda_3 = \frac{\text{area }(AUC)}{\text{area }(AQH)},\ \lambda'_3 = \frac{\text{area }(AOC)}{\text{area }(APH)},$$

so that (5) may be written as: $\lambda_1 < \lambda'_1$, $\lambda_2 < \lambda'_2$, $\lambda_3 < \lambda'_3$.

Therefore

$$\begin{aligned}\text{area }(AEC) &= \text{area }(AEN) + \text{area }(ANU) + \text{area }(AUC)\\ &= \lambda_1 \text{ area }(ALS) + \lambda_2 \text{ area }(ASQ) + \lambda_3 \text{ area }(AQH);\end{aligned}$$

similarly

$$\text{area }(ADC) = \lambda'_1 \text{ area }(AKR) + \lambda'_2 \text{ area }(ARP) + \lambda'_3 \text{ area }(APH).$$

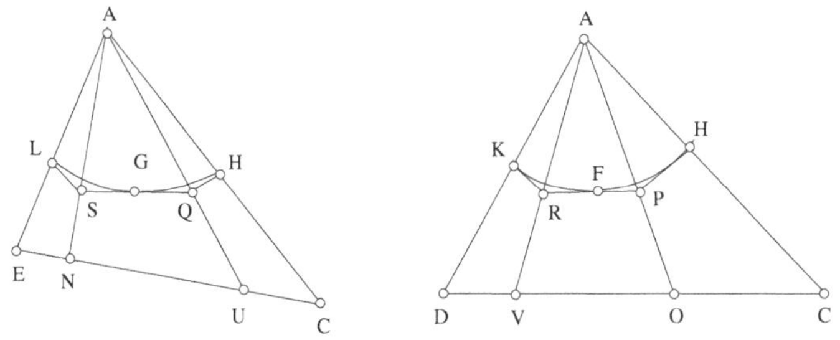

Fig. 3.14

Suppose

$$\mu_1 = \frac{\text{area }(ALS)}{\text{area }(ALSQH)},\ \mu_2 = \frac{\text{area }(ASQ)}{\text{area }(ALSQH)},\ \mu_3 = \frac{\text{area }(AQH)}{\text{area }(ALSQH)},$$

$$\mu'_1 = \frac{\text{area }(AKR)}{\text{area }(AKRPH)},\ \mu'_2 = \frac{\text{area }(ARP)}{\text{area }(AKRPH)},\ \mu'_3 = \frac{\text{area }(APH)}{\text{area }(AKRPH)}.$$

So that the two members of (4) may be written respectively

$$\lambda_1 \mu_1 + \lambda_2 \mu_2 + \lambda_3 \mu_3 \quad \text{and} \quad \lambda'_1 \mu'_1 + \lambda'_2 \mu'_2 + \lambda'_3 \mu'_3.$$

From (5) we have

$$(10) \qquad \lambda'_1 \mu'_1 + \lambda'_2 \mu'_2 + \lambda'_3 \mu'_3 > \lambda_1 \mu'_1 + \lambda_2 \mu'_2 + \lambda_3 \mu'_3 \\ = (\lambda_1 - \lambda_2)\, \mu'_1 + (\lambda_2 - \lambda_3)\,(\mu'_1 + \mu'_2) + \lambda_3$$

where the last equality comes from $\mu'_1 + \mu'_2 + \mu'_3 = 1$. Since we have

$$\lambda_1 \mu_1 + \lambda_2 \mu_2 + \lambda_3 \mu_3 = (\lambda_1 - \lambda_2)\, \mu_1 + (\lambda_2 - \lambda_3)\,(\mu_1 + \mu_2) + \lambda_3,$$

it is sufficient in order to prove (4) to establish that

(α) $\qquad \lambda_1 > \lambda_2 > \lambda_3$

(β) $\qquad \mu_1 < \mu'_1,\ \mu_1 + \mu_2 < \mu'_1 + \mu'_2 \ \text{ or } \ \mu_3 > \mu'_3.$

Inequalities (β) are true for all cases, whereas for inequalities (α) this is only true when angle $A\hat{C}E$ is a right angle or obtuse.

Let us first establish (β), that is

$$\frac{\text{area }(ALS)}{\text{area }(AKR)} < \frac{\text{area }(ALSQH)}{\text{area }(AKRPH)} < \frac{\text{area }(AQH)}{\text{area }(APH)}.$$

This aims to show that the ratio of the area of a triangle with apex A, in plane AEC, to the area of its conical projection with apex B in plane ADC increases by E towards C. Let us consider two adjacent triangles AMM_1 and AM_1M_2 in plane AEC and their projections $AM'M'_1$, $AM'_1M'_2$ in plane ADC.

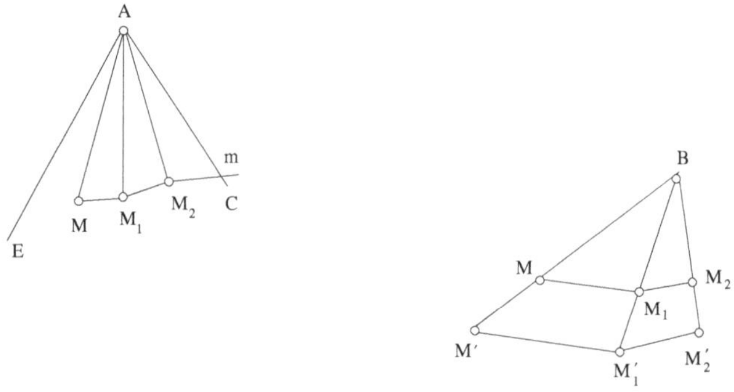

Fig. 3.15

Describing volumes of pyramids $ABMM_1$ and $ABM'M'_1$, using two different methods, we can see that

$$\frac{\delta'}{\delta}\frac{\text{area }(AM'M_1')}{\text{area }(AMM_1)} = \frac{\text{area }(BM'M_1')}{\text{area }(BMM_1)},$$

where δ and δ' designate the distances from B to planes AEC and ADC respectively. We thus have

(11) $$\frac{\text{area }(AMM_1)}{\text{area }(AM'M_1')} = \frac{\delta'}{\delta}\frac{\text{area }(BMM_1)}{\text{area }(BM'M_1')}$$

and similarly

$$\frac{\text{area }(AM_1M_2)}{\text{area }(AM_1'M_2')} = \frac{\delta'}{\delta}\frac{\text{area }(BM_1M_2)}{\text{area }(BM_1'M_2')}.$$

It is therefore necessary to show that

$$\frac{\text{area }(BMM_1)}{\text{area }(BM'M_1')} < \frac{\text{area }(BM_1M_2)}{\text{area }(BM_1'M_2')}.$$

The value of the first member is $\frac{BM \cdot BM_1}{BM' \cdot BM_1'}$, and the second $\frac{BM_1 \cdot BM_2}{BM_1' \cdot BM_2'}$, so we obtain the inequality $\frac{BM}{BM'} < \frac{BM_2}{BM_2'}$.

To evaluate $\rho = \frac{BM}{BM'}$, we write $\overrightarrow{BM} = \rho\overrightarrow{BM'}$ and we denote $\vec{u}$ as the unit vector perpendicular to plane ADC, oriented so that $\overrightarrow{BA} \cdot \vec{u} = \delta'$. Then

$$\overrightarrow{BM} \cdot \vec{u} = \rho\overrightarrow{BM'} \cdot \vec{u} = \rho\delta';$$

or, by introducing orthogonal projection m from M onto AC:

$$\rho\delta' = \overrightarrow{Bm} \cdot \vec{u} + \overrightarrow{mM} \cdot \vec{u} = \delta' + \overrightarrow{mM} \cdot \vec{u} = \delta' - mM \cdot \sin\varphi$$

lettering the angle of planes AEC and ADC $\left(0 < \varphi < \frac{\pi}{2}\right)$ as φ. Therefore

$$(12) \qquad \rho = 1 - \frac{mM}{\delta'} \sin\varphi.$$

Similarly

$$\rho_2\text{:} \ \frac{BM_2}{BM'_2} = 1 - \frac{m_2M_2}{\delta'} \sin\varphi,$$

and $\rho < \rho_2$ can also be written as $mM > m_2M_2$. Inequalities (β) mean that M is further away from AC than M_2.

In the case which interests us, the distances from L, S and Q to AC do decrease, which establishes inequalities (β).

Let us prove inequalities in (α) from the hypothesis where angle $A\hat{C}E$ is a right angle or an obtuse angle

$$(\alpha) \qquad \frac{\text{area }(AEN)}{\text{area }(ALS)} > \frac{\text{area }(ANU)}{\text{area }(ASQ)} > \frac{\text{area }(AUC)}{\text{area }(AQH)}.$$

Let us take again the adjacent triangles AMM_1 and AM_1M_2, both with apex A; let us produce their sides from A to the meeting points N, N_1, N_2 on the straight line EC, namely AN, AN_1, AN_2, with the aim of establishing

$$\frac{\text{area }(ANN_1)}{\text{area }(AMM_1)} > \frac{\text{area }(AN_1N_2)}{\text{area }(AM_1M_2)}.$$

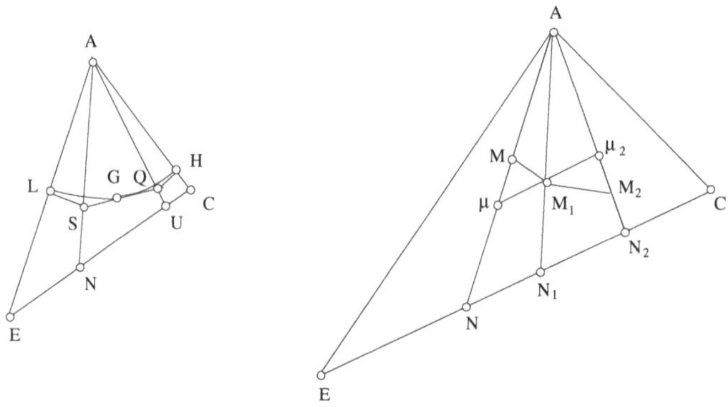

Fig. 3.16

We draw line $\mu\mu_2$ through M_1 parallel to EC, which meets AM and AM_2 at μ and μ_2 respectively. Then

$$\frac{\text{area }(ANN_1)}{\text{area }(AMM_1)} = \frac{\text{area }(ANN_1)}{\text{area }(A\mu M_1)} \cdot \frac{\text{area }(A\mu M_1)}{\text{area }(AMM_1)} = \left(\frac{AN_1}{AM_1}\right)^2 \cdot \frac{\text{area }(A\mu M_1)}{\text{area }(AMM_1)},$$

and similarly

$$\frac{\text{area }(AN_1N_2)}{\text{area }(AM_1M_2)} = \left(\frac{AN_1}{AM_1}\right)^2 \cdot \frac{\text{area }(AM_1\mu_2)}{\text{area }(AM_1M_2)}.$$

The aim is therefore to show that

$$\frac{\text{area }(A\mu M_1)}{\text{area }(AMM_1)} > \frac{\text{area }(AM_1\mu_2)}{\text{area }(AM_1M_2)}.$$

As angles AN_1N and AN_2N are obtuse in our hypothesis, then $A\hat{M}_1\mu > A\hat{M}_1M$ and $A\hat{\mu}_2M_1 > A\hat{M}_2M_1$ if angles AM_1M and AM_2M_1 are acute, then μ is between M and N, and μ_2 between A and M_2; so

$$\frac{\text{area }(A\mu M_1)}{\text{area }(AMM_1)} > 1 > \frac{\text{area }(AM_1\mu_2)}{\text{area }(AM_1M_2)}.$$

In this case, angles ASL, AQS are acute and angles ALS, AHQ are right angles.

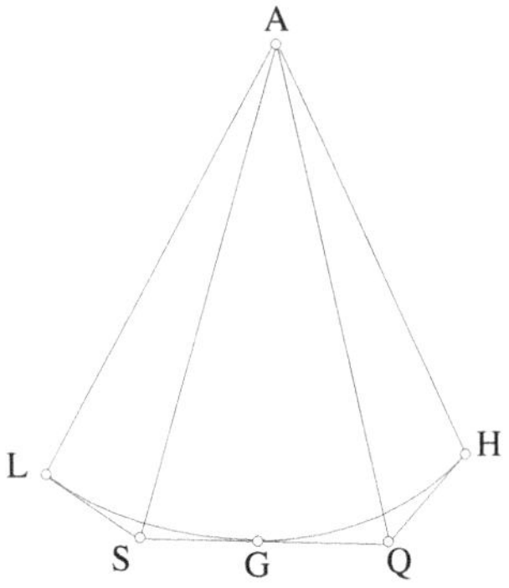

Fig. 3.17

We have thus established Ibn al-Haytham's property (inequality (4)) in the case where angle ACE is a right angle or an obtuse angle.

But what is the real meaning of Ibn al-Haytham's method and what ideas did he have at the back of his mind? In order to answer these questions, it will be necessary to go back over his proof using a different formal language, that of integral calculus.

If AMM_1 is an infinitesimal triangle in plane AEC with area $\frac{1}{2}\, r^2 d\theta$ ($r = AM$, $d\theta = M\hat{A}M_1$), its ratio to projected triangle $AM'M'_1$ in plane ADC, from (11), is

$$\frac{\delta'}{\delta}\,\frac{\text{area}\,(BMM_1)}{\text{area}\,(BM'M'_1)} = \frac{\delta'}{\delta}\left(\frac{BM}{BM'}\right)^2$$

(to infinitesimals of the nearest higher order) and

$$\frac{BM}{BM'} = 1 - \frac{mM}{\delta'}\sin\varphi = 1 - \frac{r}{\delta'}\sin\varphi\sin\theta$$

from (12), and with $\theta = C\hat{A}M$ $\left(\text{between } 0 \text{ and } \frac{\pi}{2}\right)$.

The element of projected area $AM'M'_1$ is therefore

$$\frac{\delta}{2\delta'}\,\frac{r^2\, d\theta}{(1 - r/\delta' \sin\varphi\sin\theta)^2}.$$

Let us now consider, in plane AEC, star areas with apex A, defined by the respective integrals

I $\quad 0 \le \theta \le \theta_1, \quad (0 < \theta_1 < \frac{\pi}{2}), \qquad 0 \le r \le R(\theta)$

II $\quad 0 \le \theta \le \theta_1, \quad 0 \le r \le \lambda(\theta)\, R(\theta)$

where the ratio λ of radius vectors remains greater than 1, so that area II contains area I. Let us also consider the projections I′ and II′ of I and II in plane *ADC*. The aim would be to establish that $\frac{\text{II}}{\text{I}} < \frac{\text{II}'}{\text{I}'}$, that is

$$(13) \qquad \frac{\int_0^{\theta_1} \lambda^2 R^2 d\theta}{\int_0^{\theta_1} R^2 d\theta} < \frac{\int_0^{\theta_1} \frac{\lambda^2 R^2 d\theta}{\left(1 - \frac{\lambda R}{\delta'} \sin\varphi \sin\theta\right)^2}}{\int_0^{\theta_1} \frac{R^2 d\theta}{\left(1 - \frac{R}{\delta'} \sin\varphi \sin\theta\right)^2}}.$$

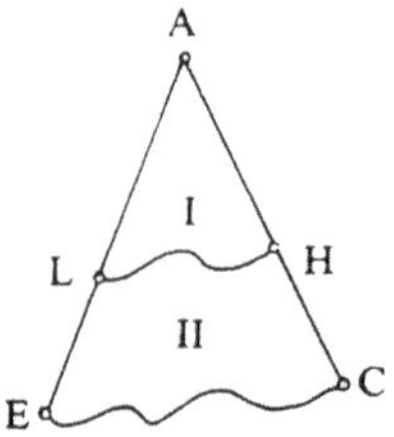

Fig. 3.18

In the corresponding Ibn al-Haytham's inequalities in (5), we have on the elementary areas

$$\frac{\lambda^2 R^2 d\theta}{R^2 d\theta} = \lambda^2 < \frac{\frac{\lambda^2 R^2 d\theta}{\left(1 - \frac{\lambda R}{\delta'} \sin\varphi \sin\theta\right)^2}}{\frac{R^2 d\theta}{\left(1 - \frac{R}{\delta'} \sin\varphi \sin\theta\right)^2}} = \lambda^2 \left(\frac{1 - \frac{R}{\delta'} \sin\varphi \sin\theta}{1 - \frac{\lambda R}{\delta'} \sin\varphi \sin\theta} \right)^2$$

since $\frac{\lambda R}{\delta'} \sin\varphi \sin\theta > \frac{R}{\delta'} \sin\varphi \sin\theta$.

The second member of (13) becomes the minor by replacing its numerator with

$$\int_0^{\theta_1} \frac{\lambda^2 R^2 d\theta}{\left(1-\frac{R}{\delta'}\sin\varphi\sin\theta\right)^2};$$

this integral corresponds to Ibn al-Haytham's area *AZVWJ*.

Let us put

$$f(\theta)=\int_0^{\theta} R^2(\omega)d\omega,\ g(\theta)=\int_0^{\theta}\frac{R^2(\omega)d\omega}{\left(1-\frac{R(\omega)}{\delta'}\sin\varphi\sin\omega\right)^2} \text{ and } h(\theta)=\lambda(\theta)^2.$$

The aim is to show

$$(13') \qquad \frac{1}{f(\theta_1)}\int_0^{\theta_1} h(\theta)\,df(\theta) < \frac{1}{g(\theta_1)}\int_0^{\theta_1} h(\theta)\,dg(\theta),$$

using integration by parts (a stage which corresponds to transformation (10) of the above proof), this inequality becomes

$$h(\theta_1)-\frac{1}{f(\theta_1)}\int_0^{\theta_1} f(\theta)\,dh(\theta) < h(\theta_1)-\frac{1}{g(\theta_1)}\int_0^{\theta_1} g(\theta)\,dh(\theta),$$

or

$$\frac{1}{f(\theta_1)}\int_0^{\theta_1} f(\theta)\,dh(\theta) > \frac{1}{g(\theta_1)}\int_0^{\theta_1} g(\theta)\,dh(\theta).$$

Let $\gamma=\frac{g}{f}$; the second member is written

$$\frac{1}{\gamma(\theta_1)f(\theta_1)}\int_0^{\theta_1} \gamma(\theta)\,f(\theta)\,dh(\theta),$$

and we have to show that

$$\int_0^{\theta_1} f(\theta)\,dh(\theta) > \frac{1}{\gamma(\theta_1)}\int_0^{\theta_1} \gamma(\theta)\,f(\theta)\,dh(\theta).$$

This inequality is guaranteed if we assume

(α) h is the increasing function of θ (not a constant)

(β) γ is the increasing function of θ (not a constant).

However

$$\gamma(\theta)=\frac{1}{\int_0^\theta R^2(\omega)d\omega}\int_0^\theta \frac{R^2(\omega)d\omega}{\left(1-\frac{R(\omega)}{\delta'}\sin\varphi\sin\omega\right)^2}$$

is increasing if the same holds for $\frac{1}{\left(1-\frac{R(\theta)}{\delta'}\sin\varphi\sin\theta\right)^2}$, that is to say if $R(\theta)\sin(\theta)$ is increasing.

We have thus proved the inequality (13) using the hypothesis that $\lambda(\theta)$ and $R(\theta)\sin\theta$ (distance from the floating point on the boundary of area I to AC) are both increasing functions of θ. In Ibn al-Haytham's case, R is constant and equal to AB and $\lambda(\theta)=\frac{AN}{R}$ is increasing when angle ACE is a right angle or an obtuse angle. In this case, areas II = AEC and II′ = ADC are elementary (triangles) whereas area I equals $\frac{1}{2}R^2\theta_1$.

The inequality (13) can therefore be simplified as

$$\frac{\text{area}\,(AEC)}{\frac{1}{2}R^2\theta_1}<\frac{\text{area}\,(ADC)}{\frac{\delta}{2\delta'}R^2\int_0^{\theta_1}\frac{d\theta}{\left(1-\frac{R}{\delta'}\sin\varphi\sin\theta\right)^2}},$$

or

$$(14)\qquad \frac{1}{\theta_1}\int_0^{\theta_1}\frac{d\theta}{\left(1-\frac{R}{\delta'}\sin\varphi\sin\theta\right)^2}<\frac{\delta'\,\text{area}\,(ADC)}{\delta\,\text{area}\,(AEC)}=\frac{\text{area}\,(BDC)}{\text{area}\,(BEC)}=\frac{BD}{BE}.$$

This is guaranteed if $\left(\frac{BK}{BL}\right)^2=\frac{1}{\left(1-\frac{R}{\delta'}\sin\varphi\sin\theta_1\right)^2}\leq\frac{BD}{BE}$,

that is, if

$$(15)\qquad \left(\frac{BL}{BK}\right)^2\geq\frac{BE}{BD},$$

an inequality which only involves the points of triangle ABD. This reduction obviously escaped Ibn al-Haytham.

Taking orthogonal axes from point B, with BA as the y axis, $(0, a)$ as coordinates of A and (m, n) those of D, with angle ABD either a right angle or obtuse, then $m < 0$, $n \leq 0$ (assuming $a > 0$ and D to the left of AB)

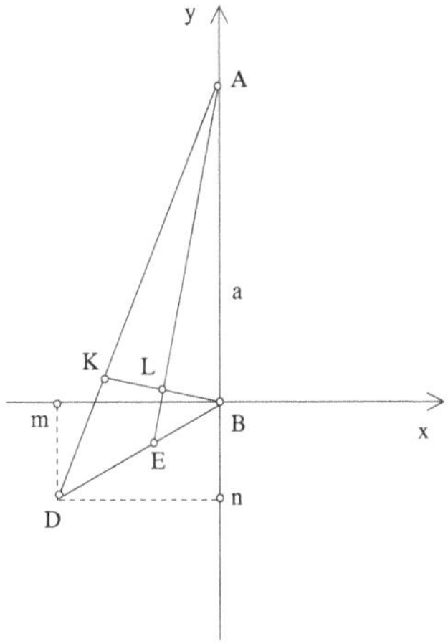

Fig. 3.19

The coordinates (x, y) of E are linked by $y = \frac{n}{m}x$, and equations of AD and AE are respectively $Y - a = \frac{n-a}{m}X$ and $Y - a = \frac{y-a}{x}X$. The abscissa of L equals $\frac{ax}{AE}$, therefore its ordinate is $a + \frac{y-a}{x}\frac{ax}{AE} = \frac{a}{AE}(AE + y - a)$. The equation of BL is therefore written as $Y = (AE + y - a)\frac{X}{x}$ and the abscissa of K is given as

$$(AE + y - a)\frac{X}{x} = a + \frac{n-a}{m}X,$$

that is to say $X = \dfrac{ax}{AE - a\left(1 - \frac{x}{m}\right)}$.

So

$$(15') \qquad \frac{BL}{BK} = 1 - \frac{a}{AE}\left(1 - \frac{x}{m}\right)$$

and condition (14) becomes

$$1-\frac{2a}{AE}\left(1-\frac{x}{m}\right)+\frac{a^2}{AE^2}\left(1-\frac{x}{m}\right)^2 \geq \frac{x}{m},$$

by simplifying by $1-\frac{x}{m}$ (which is strictly positive since E is between B and D), we have

$$1-\frac{2a}{AE}+\frac{a^2}{AE^2}\left(1-\frac{x}{m}\right) \geq 0,$$

that is to say

$$\left(x^2+(y-a)^2+a^2\left(1-\frac{x}{m}\right)\right)^2 \geq 4a^2\left(x^2+(y-a)^2\right)$$

or again

$$(16) \qquad \left(x^2+y^2-\frac{a^2x}{m}-2ay\right)^2-\frac{4a^4x}{m} \geq 0.$$

This inequality means that point E is not inside a quartic curve Γ tangent at B to AB (a and m being fixed). If we suppose $\frac{a}{m}=\alpha$, $\frac{x}{m}=\xi$, $\frac{y}{m}=\eta$, the equation of Γ becomes

$$(17) \qquad (\xi^2+\eta^2-\alpha^2\xi-2\alpha\eta)^2-4\alpha^4\xi=0$$

where there only remains the parameter α (negative).

For $\alpha = -4$, it is Pascal's *limaçon* of polar equation $\rho = 8(\cos\theta + \sqrt{2})$ with double point $(4, -4)$; for other values of α, the genus of Γ is 1.

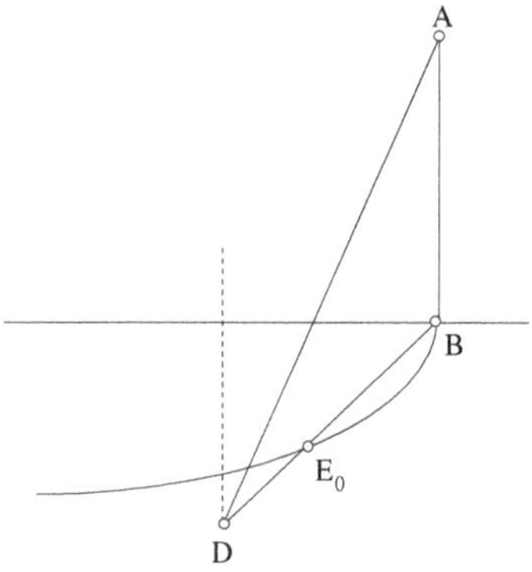

Fig. 3.20

If triangle ABD is given, then condition (14) is verified when E belongs to segment E_0D, and where E_0 denotes the distinct point of intersection of BD with Γ, if in fact this point exists. One can easily see that E_0 still exists if $0 > \frac{a}{m} \geq -\frac{\sqrt{3}}{3}$; but if $\frac{a}{m} < -\frac{\sqrt{3}}{3}$, then angle ABD has to be bigger than a minimum value $\arctan\frac{\sqrt{4\alpha^2 - 1} - \alpha}{1 - 3\alpha^2}$. If E is on E_0D, then inequality (**) stated by Ibn al-Haytham is proven, whatever the other elements of the pyramid $ABCD$; but if E is on BE_0, further discussion is necessary using values of $\theta_1 = C\hat{A}E$.

The first member of (14) can be calculated in an elementary way. Then

$$\frac{R}{\delta'}\sin\varphi\sin\theta_1 = 1 - \frac{BL}{BK} = \frac{KL}{BK} = \frac{a}{AE}\left(1 - \frac{x}{m}\right) = \beta,$$

(cf. (15′)); therefore $\frac{R}{\delta'}\sin\varphi = \frac{\beta}{\sin\theta_1}$ and the expression under examination is written as

$$I = \frac{1}{\theta_1}\int_0^{\theta_1}\frac{d\theta}{\left(1 - \beta\frac{\sin\theta}{\sin\theta_1}\right)^2} = \frac{1}{\theta_1}\int_0^{t_1}\frac{2\left(1 + t^2\right)dt}{\left(t^2 - 2vt + 1\right)^2},$$

by saying $\tan\frac{\theta}{2} = t$, $\tan\frac{\theta_1}{2} = t_1$, $v = \frac{\beta}{\sin\theta_1}$.

If $v < 1$, let it be $a^2\left(1 - \frac{x}{m}\right)^2 < [(x^2 + (y - a)^2]\sin^2\theta_1$.

$$\xi^2(\sin^2\theta_1 - \alpha^2) + \eta^2\sin^2\theta_1 + 2\alpha^2\xi - 2\alpha\eta\sin^2\theta_1 - \alpha^2\cos^2\theta_1 > 0, \tag{18}$$

that is, if point (ξ, η) is outside a conic **C**, then we put $v = \sin\theta_0$ $\left(0 < \theta_0 < \frac{\pi}{2}\right)$, or $\sin\theta_0\sin\theta_1 = \beta$, and we use the change of variable $t = u\cos\theta_0 + \sin\theta_0$. From this we obtain

$$I = \frac{2}{\theta_1\cos^3\theta_0}\left(\arctan\frac{t_1 - \sin\theta_0}{\cos\theta_0} + \theta_0\right) + \frac{\sin\theta_1}{\theta_1}\frac{\beta}{\sin^2\theta_1 - \beta^2}\left(1 - \frac{\cos\theta_1}{1 - \beta}\right). \tag{19}$$

If on the contrary $v > 1$, that is if (ξ, η) is inside **C**, then we put $\frac{1}{v} = \sin\theta_0 = \frac{\sin\theta_1}{\beta}$ and from this we obtain

$$(20) \qquad I = \frac{\tan^3\theta_0}{\theta_1} \log \frac{\sin\left(\theta_0 - \frac{\theta_1}{2}\right) - \sin\frac{\theta_1}{2}}{\sin\left(\theta_0 + \frac{\theta_1}{2}\right) - \sin\frac{\theta_1}{2}} + \frac{\sin\theta_1}{\theta_1} \frac{\beta}{\beta^2 - \sin^2\theta_1}\left(\frac{\cos\theta_1}{1 - \beta} - 1\right).$$

If in the last case, $v = 1$, that is if (ξ, η) is on **C**, then we get

$$(21) \qquad I = \frac{1}{3}\left(1 + \frac{2}{1 - \sin\theta_1} + \cos\theta_1 \frac{3 - 2\sin\theta_1}{(1 - \sin\theta_1)^2}\right).$$

Let us calculate (19) and (20) numerically with a few examples to show that Ibn al-Haytham's inequality (**) does not always hold true.

Example 1. — Let $a = 4$, $m = -1$, $n = -\frac{1}{2}$, therefore $BD = \frac{\sqrt{5}}{2} = 1.11803399\ldots$ and $AD = \frac{\sqrt{85}}{2} = 4.60977223$. If $\frac{BE}{BD} = \frac{x}{m} = \frac{1}{10}$, then $x = -\frac{1}{10}$, $y = -\frac{1}{20}$; $BE = \frac{\sqrt{5}}{20} = 0.11803399$, $AE = \frac{\sqrt{6565}}{20} = 4.05123438$ and $\beta = \frac{72}{\sqrt{6565}} = 0.888618051$. It is possible to take $\theta_1 = \frac{5\pi}{12}$ to give $\sin\theta_1 = 0.965925826 > \beta$, so that in the case where $v < 1$, then $\sin\theta_0 = 0.919965102$, hence $\theta_0 = 66°55'15'',53$. The calculation of (19) gives the value $14.1533141 > \frac{BD}{BE} = 10$ for I. It is possible to construct a pyramid taking angle $AEC = \frac{\pi}{2}$, which gives

$$AC = \frac{AE}{\sin\frac{\pi}{12}} = \frac{\sqrt{6565\,(2 + \sqrt{3})}}{10} = 15.6527677$$

and

$$EC = AE\tan\frac{5\pi}{12} = \left(\frac{2+\sqrt{3}}{20}\right)\sqrt{6565} = 15.1194125.$$

Then, by choosing BC between $EC - BE = 15.00137851$ and $EC + BE = 15.23744649$, for example $BC = 15.1$, therefore we get

$$\cos\ A\hat{B}C = \frac{AB^2 + BC^2 - AC^2}{2AB \cdot BC} = -\ 0.1076087378,$$

$$A\hat{B}C = 96°10'38'',96$$

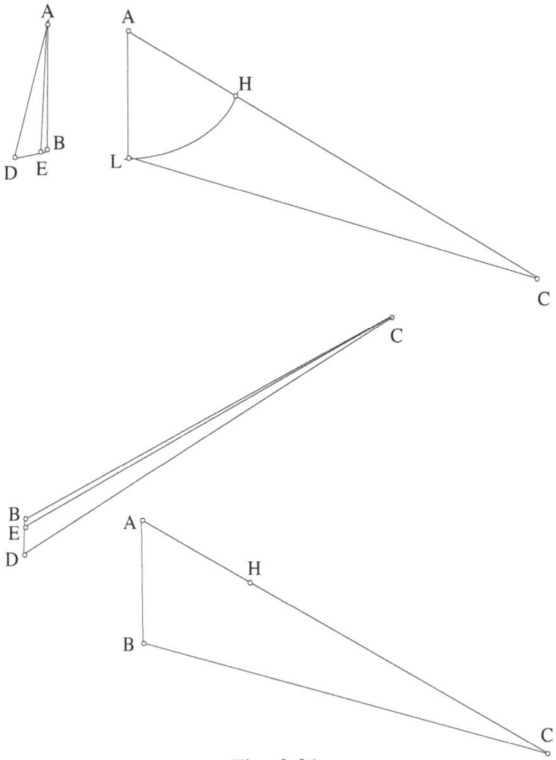

Fig. 3.21

and

$$DC^2 = BD^2 + BC^2 - 2BD \cdot BC \cos D\hat{B}C = BD^2 + BC^2 - \frac{BD}{BE}\left(BC^2 + BE^2 - EC^2\right)$$
$$= ED\left(BD - \frac{BC^2}{BE}\right) + \frac{EC^2 \cdot BD}{BE} = \left(1 - \frac{x}{m}\right)\left(BD^2 - BC^2\frac{m}{x}\right) + EC^2\frac{m}{x}$$
$$= 235.001355452;$$

hence $DC = 15.3297539$.

For this pyramid $\frac{\text{area}(ABC)}{\text{area}(\widehat{ALH})} = 2.92464268$ whilst $\frac{\text{area}(ADC)}{\text{area}(\widehat{AKH})} =$ 2.06640131, which is contrary to Ibn al-Haytham's inequality (**) above.

Example 2. — Let $a = 4$, $m = -1$, $n = -\frac{1}{5}$. Therefore $BD = \frac{\sqrt{26}}{5} =$ 1.0198039 and $AD = \frac{\sqrt{466}}{5} = 4.317406629$.

If $\frac{BE}{BD} = \frac{x}{m} = \frac{1}{10}$, then $x = -\frac{1}{10}$, $y = -\frac{1}{50}$, $BE = \frac{\sqrt{26}}{50} = 0.10198039$,

$$AE = \frac{\sqrt{40426}}{50} = 4.02124359 \text{ and } \beta = \frac{180}{\sqrt{40426}} \cong 0.895245443.$$

We may take $\theta_1 = \frac{\pi}{3}$, which gives $\sin\theta_1 = \frac{\sqrt{3}}{2} = 0.866025404 < \beta$, to be in the case where $v > 1$; we have $\sin\theta_0 = \frac{\sin\theta_1}{\beta} = 0.967360863$, hence $\theta_0 = 75°19'15'',72$. The calculation of (20) gives the value for I as $10.9012463 > \frac{BD}{BE} = 10$. We may construct a pyramid by taking angle $AEC = \frac{7\pi}{12}$, which gives

$$AC = AE\frac{\sin\frac{7\pi}{12}}{\sin\frac{\pi}{12}} = 15.00748538$$

and

$$EC = AE\frac{\sin\frac{\pi}{3}}{\sin\frac{\pi}{12}} = 13.45534329.$$

Then choosing BC between $EC - BE = 13.3533629$ and $EC + BE = 13.55732368$, for example $BC = 13.4$; therefore we get

$$\cos A\hat{B}C = -0.276722178,\ A\hat{B}C = 106°37'52'',14,$$
$$DC^2 = 195.35863136,$$

hence

$$DC = 13.9770752076.$$

For this pyramid $\frac{\text{area}(AEC)}{\text{area}(\widehat{ALH})} = 3.11925113383$, whereas $\frac{\text{area}(ADC)}{\text{area}(\widehat{AKH})} =$ 2.86137112032, which is again contrary to Ibn al-Haytham's inequality (**).

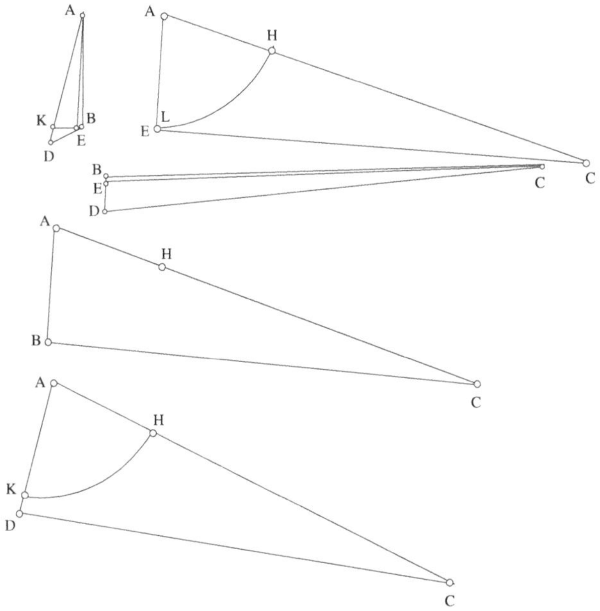

Fig. 3.22

So in the case where $A\hat{C}E$ is a right angle or an obtuse angle, we have

$$\frac{\text{area tr. }(AEC)}{\text{area sect. }(AHGL)} \leq \frac{\text{area tr. }(ADC)}{\text{area sect. }(AHFK)},$$

but we have seen that, in the case where $A\hat{E}C$ is a right angle or an obtuse angle, this property is only true when E is sufficiently far away from B.

At the end of this (incomplete) proof, Ibn al-Haytham returns to Lemma 6

$$\frac{\text{area }(DBC)}{\text{area }(EBC)} > \frac{\text{solid angle }(A, BCD)}{\text{solid angle }(A, EBC)},$$

that is

$$\frac{\text{pyr. }(ABCD)}{\text{pyr. }(ABCE)} > \frac{\text{sect. sph. }(A, BCD)}{\text{sect. sph. }(A, BCE)},$$

as the second member is increased by

$$\frac{\text{curv. pyr. }(BAHFK)}{\text{circ. pyr. }(BAHGL)} \quad \text{(see (*))};$$

it is sufficient to show that

$$\frac{\text{pyr. }(BACD)}{\text{pyr. }(BAEC)} \geq \frac{\text{curv. pyr. }(BAHFK)}{\text{circ. pyr. }(BAHGL)}.$$

However, the inequality (**) means that

$$\frac{\text{pyr. }(BAEC)}{\text{circ. pyr. }(BAHGL)} \leq \frac{\text{pyr. }(BACD)}{\text{curv. pyr. }(BAHFK)}.$$

We now examine the validity of Lemma 6, which is at the heart of Ibn al-Haytham's treatise, by using a direct analytical method. We choose orthogonal axes from origin *A*, with point *B* on *Ax* and point *D* in plane *Axy*, and we assume their orientation to be such that coordinates of *B* and *D* are positive.

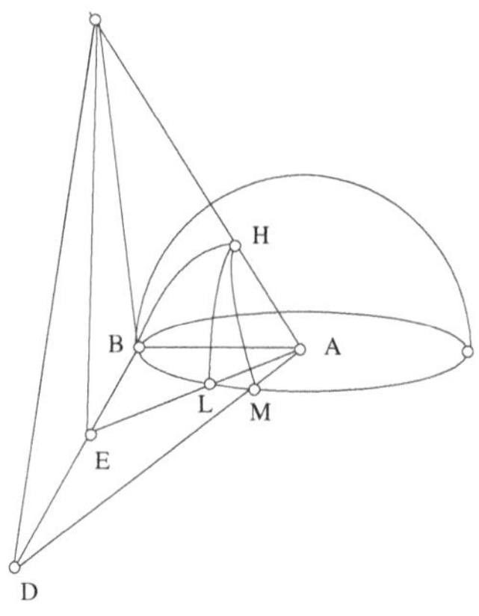

Fig. 3.23

Let λ and μ be lines of longitude from points *L* and *M*, where straight lines *AE* and *AD* respectively meet the sphere with centre *A* and radius $a = AB$.

The coordinates of points *B*, *L*, *M*, *E* and *D* are:

B: $(a, 0, 0)$; L: $(a \cos \lambda, a \sin \lambda, 0)$; M: $(a \cos \mu, a \sin \mu, 0)$;
E: $(e \cos \lambda, e \sin \lambda, 0)$; D: $(d \cos \mu, d \sin \mu, 0)$

with $e = AE$ and $d = AD$. By writing that points *B*, *E*, and *D* are aligned, we find that

$$e = a\left(1 - \frac{1}{\rho}\right)\frac{\sin\mu}{\sin(\mu - \lambda)} \quad \text{and} \quad d = a(\rho - 1)\frac{\sin\lambda}{\sin(\mu - \lambda)}$$

where $\rho = \dfrac{BD}{BE} > 1$.

The condition where angle ABD is a right angle or an obtuse angle is written as $d\cos\mu \geq a$, which can be expressed as $\mu < \frac{\pi}{2}$ and

$$(22) \qquad \rho \geq \frac{\tan\mu}{\tan\lambda}$$

(with $0 < \lambda < \mu$).

If θ and φ designate the co-latitude and the longitude of point H where AC pierces the sphere, the Cartesian coordinates of H and C are written

H: $(a\sin\theta\cos\varphi, a\sin\theta\sin\varphi, a\cos\theta)$;
$C = (c\sin\theta\cos\varphi, c\sin\theta\sin\varphi, c\cos\theta)$.

Angle ABC being superior or equal to a right angle, we have

$$(23) \qquad c\sin\theta\cos\varphi \geq a$$

which requires that $|\varphi| < \frac{\pi}{2}$ and $c \geq \dfrac{a}{\sin\theta\cos\varphi}$.

One of the angles AEC or ACE is a right angle or an obtuse angle:

$$(24) \qquad A\hat{E}C \geq \frac{\pi}{2} \Leftrightarrow e \leq c\sin\theta\cos(\lambda - \varphi)$$

$$(24') \qquad A\hat{C}E \geq \frac{\pi}{2} \Leftrightarrow c \leq e\sin\theta\cos(\lambda - \varphi).$$

These conditions demand that $|\lambda - \varphi| < \frac{\pi}{2}$; (24) requires that c be large enough, but (24′) is not compatible with (23), except if

$$\frac{a}{\sin\theta\cos\varphi} \leq e\sin\theta\cos(\lambda - \varphi) \text{ or } \cos(2\varphi - \lambda) \geq \frac{2a}{e\sin^2\theta} - \cos\lambda.$$

The expression $\dfrac{2a}{e\sin^2\theta} - \cos\lambda$ is a decreasing function of ρ which tends to the limit $\dfrac{2\sin(\mu-\lambda)}{\sin\mu\sin^2\theta} - \cos\lambda$ for an infinite value of ρ.

We may choose ρ and φ in accordance with (23) and (24′) if

$$\frac{2\sin(\mu-\lambda)}{\sin\mu\sin^2\theta} - \cos\lambda \leq 1 \quad \text{or} \quad \tan\mu\left(\cos^2\theta - \tan^2\frac{\lambda}{2}\right) \leq 2\tan\frac{\lambda}{2}.$$

The figure can therefore be constructed with $A\hat{C}E \geq \dfrac{\pi}{2}$ in one or the other of the following two cases:

(*i*) $\quad \tan\dfrac{\lambda}{2} \geq \cos\theta = \tan\dfrac{\lambda_0}{2}$, that is $\lambda \geq \lambda_0$;

(*ii*) $\quad \lambda < \lambda_0$ and $\tan\mu \leq \dfrac{2\tan\dfrac{\lambda}{2}}{\cos^2\theta - \tan^2\dfrac{\lambda}{2}} = \tan\lambda \, \dfrac{1+\cos\lambda_0}{1-\dfrac{\cos\lambda_0}{\cos\lambda}} = \tan\mu_0$

that is $\lambda < \mu \leq \mu_0$.

Solid angles (A, BCE) and (A, BCD) are measured by the areas of spherical triangles BHL and BHM, by expressing the area of an infinitesimal spherical triangle HNN' with N and N' between B and M, and with respective longitudes v and $v + dv$.

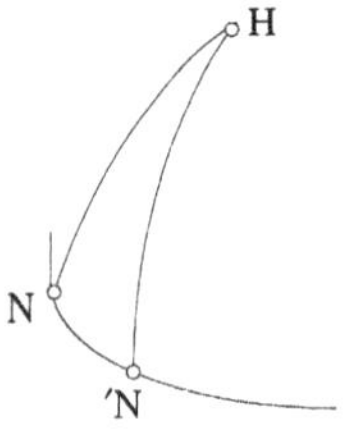

Fig. 3.24

This area $a^2\, d\sigma = a^2\,(\hat{H} + \hat{N} + \hat{N}' - \pi)$ is calculated using S. Lhuillier's formula

$$\frac{1}{4}d\sigma \approx \tan\frac{d\sigma}{4} = \sqrt{\tan\frac{p}{2a}\tan\frac{p-\widehat{NN'}}{2a}\tan\frac{p-\widehat{HN}}{2a}\tan\frac{p-\widehat{HN'}}{2a}}$$

where we have set

$$p = \frac{1}{2}\left(\widehat{HN}+\widehat{HN'}+\widehat{NN'}\right) \cong \widehat{HN}+\frac{a}{2}dv.$$

We get

$$p-\widehat{NN'} \cong \widehat{HN}-\frac{a}{2}dv;\; p-\widehat{HN} = \frac{1}{2}\left(\widehat{HN'}-\widehat{HN}+\widehat{NN'}\right) \cong \frac{a}{2}\left(d\frac{\widehat{HN}}{a}+dv\right)$$

and

$$p-\widehat{HN'} \cong \frac{1}{2}\left(\widehat{HN}-\widehat{HN'}+\widehat{NN'}\right) = \frac{a}{2}\left(-d\frac{\widehat{HN}}{a}+dv\right)$$

so

$$\tan\frac{p}{2a}\tan\frac{p-\widehat{NN'}}{2a}\tan\frac{p-\widehat{HN}}{2a}\tan\frac{p-\widehat{HN'}}{2a} \approx \tan^2\frac{\widehat{HN}}{2a}\cdot\frac{1}{16}\left(dv^2-\left(d\frac{\widehat{HN}}{2a}\right)^2\right).$$

However $\cos\frac{\widehat{HN}}{a} = \sin\theta\cos(v-\varphi)$ gives

$$d\frac{\widehat{HN}}{a} = \frac{\sin\theta\sin(v-\varphi)}{\sin\frac{\widehat{HN}}{a}}dv;$$

and it follows that

$$\begin{aligned} dv^2-\left(d\frac{\widehat{HN}}{a}\right)^2 &= \frac{dv^2}{\sin^2\frac{\widehat{HN}}{a}}\left(1-\sin^2\theta\cos^2(v-\varphi)-\sin^2\theta\sin^2(v-\varphi)\right) \\ &= \frac{dv^2\cos^2\theta}{\sin^2\frac{\widehat{HN}}{a}}. \end{aligned}$$

Finally

$$(25) \qquad d\sigma = \left(\frac{\tan\frac{\widehat{HM}}{2a}}{\sin\frac{\widehat{HM}}{a}}\right)dv\cos\theta = \frac{dv\cos\theta}{1+\sin\theta\cos(v-\varphi)}.$$

The solid angles under consideration are integrals of this differential element between 0 and λ and between 0 and μ respectively. The inequality of Ibn al-Haytham's Lemma 6 is therefore written as

$$(26) \qquad \int_0^{\mu} \frac{dv}{1+\sin\theta\cos(v-\varphi)} < \rho \int_0^{\lambda} \frac{dv}{1+\sin\theta\cos(v-\varphi)}.$$

Because of inequality (22), relation (26) is true if

$$\frac{1}{\tan\lambda}\int_0^{\lambda} \frac{dv}{1+\sin\theta\cos(v-\varphi)}$$

is a decreasing function of λ. The derivative of this expression is written as

$$-\frac{1+\tan^2\lambda}{\tan^2\lambda}\int_0^{\lambda} \frac{dv}{1+\sin\theta\cos(v-\varphi)} + \frac{1}{\tan\lambda}\frac{1}{1+\sin\theta\cos(\lambda-\varphi)}$$

and is negative if

$$(27) \qquad \frac{\sin 2\lambda}{2}\frac{1}{1+\sin\theta\cos(\lambda-\varphi)} \leq \int_0^{\lambda} \frac{dv}{1+\sin\theta\cos(v-\varphi)}.$$

The two members of (27) are null for $\lambda = 0$, therefore the inequality is proven as long as

$$\frac{\cos 2\lambda}{1+\sin\theta\cos(\lambda-\varphi)} + \frac{\sin 2\lambda}{2}\frac{d}{d\lambda}\left(\frac{1}{1+\sin\theta\cos(\lambda-\varphi)}\right) \leq \frac{1}{1+\sin\theta\cos(\lambda-\varphi)},$$

that is

$$\frac{d}{d\lambda}\log\frac{1}{1+\sin\theta\cos(\lambda-\varphi)} \leq 2\tan\lambda = \frac{d}{d\lambda}\log\frac{1}{\cos^2\lambda},$$

which again means that $\dfrac{\cos^2\lambda}{1+\sin\theta\cos(\lambda-\varphi)}$ is a decreasing function of λ. The derivative of this last expression is written

$$-\frac{\cos\lambda}{2\,(1+\sin\theta\cos(\lambda-\varphi))^2}(3\sin\theta\sin\varphi + \sin(2\lambda-\varphi)\sin\theta + 4\sin\lambda);$$

it is negative if $\sin 2\lambda\cos\varphi + 2\,(1+\sin^2\lambda)\sin\varphi \geq -4\dfrac{\sin\lambda}{\sin\theta}$.

With λ, θ as givens, this shows that the point (cos φ, sin φ) of the unit circle is above the straight line of equation

$$x \sin \lambda \cos \lambda + y\,(1 + \sin^2 \lambda) = -2\,\frac{\sin \lambda}{\sin \theta};$$

for $\lambda = 0$, this straight line becomes the axis of $x : y = 0$, and the corresponding condition is $\varphi \geq 0$.

Therefore for $\varphi \geq 0$, the inequality (27) is satisfied for whatever value of λ, between 0 and $\frac{\pi}{2}$, and from this, the inequality (26) on solid angles can be deduced.

In summary, Lemma 6, independent of the sub-lemma, is true if it is assumed that the longitude φ of *C* is positive, that is with points *C* and *D* on the same side of the perpendicular plane at *ABD* passing through *AB*; this condition escaped Ibn al-Haytham.

But, if angle φ is negative, the conclusion of Lemma 6 is not always valid; the difference between the two members of (27) begins decreasing from 0 to a minimum negative value, then increases and becomes positive from a value $\lambda_0\,(\theta, \varphi)$ of λ which is given by

$$\text{(28)} \qquad \arctan\left(\tau \tan \frac{\lambda_0 - \varphi}{2}\right) + \arctan\left(\tau \tan \frac{\varphi}{2}\right) = \frac{\sin 2\lambda_0}{4}\,\frac{\cos\theta}{1 + \sin\theta \cos(\lambda_0 - \varphi)}$$

$$\left(-\frac{\pi}{2} < \varphi < 0 < \theta < \frac{\pi}{2};\ \ \tau = \tan\left(\frac{\pi}{4} - \frac{\theta}{2}\right)\right).$$

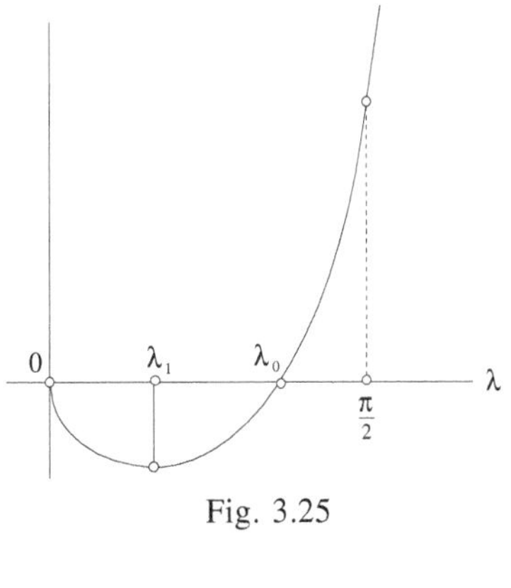

Fig. 3.25

In fact, the value of this difference for $\lambda = \frac{\pi}{2}$ is

$$\int_0^{\frac{\pi}{2}} \frac{dv}{1+\sin\theta\cos(v-\varphi)} \geq 0.$$

If $\lambda_0 \leq \lambda \leq \mu$, the inequality of Lemma 6 is still true; but if $\lambda < \lambda_0$, as indicated in the example, then it is not necessarily true.

Example: Let $\varphi = -87°$, $\tau = \frac{1}{10}$, and let $\theta = 78°34'43'',72$; then $\lambda_0 = 88°32'7'',42$. By taking $\lambda = 1°$ and $\mu = 21°$, we have $\frac{\tan\mu}{\tan\lambda} = 21.99155584$ and we may choose $\rho = 22$. If $a = 4 = AB$, then

$$AD = d = 4 \cdot 21 \cdot \frac{\sin 1°}{\sin 20°} = 4.286303511,$$

$$AE = e = 4 \cdot \frac{21}{22} \cdot \frac{\sin 21°}{\sin 20°} = 4.000682462.$$

Then go on to choose $AC = c > \frac{4}{\sin\theta\cos\varphi} = \frac{4 \cdot 101}{99\cos 87°} = 77.9733165$ and $AC = c \geq \frac{e}{\sin\theta\cos(\lambda-\varphi)} = 116.9502347.$

For example for $c = 117$, solid angles have the following respective values:

$$2\left(\text{arc tan}\frac{\tan 54°}{10} - \text{arc tan}\frac{\tan 43°30'}{10}\right) = 4°51'54'',58$$

and

$$2\left(\text{arc tan}\frac{\tan 44°}{10} - \text{arc tan}\frac{\tan 43°30'}{10}\right) = 11'23'',46;$$

their ratio is $25.62634243 > \rho = 22 = \frac{\text{area }(BCD)}{\text{area }(BCE)}$ and Lemma 6 is not satisfied.

In the case $BD = 1.536074643$, $BE = 0.069821575$,
$EC = 116.9315223$, $BC = 116.8630976$, $DC = 118.3688195$,

$A\hat{B}D = 90°03'36'',07$ $A\hat{B}C = 90°58'53'',82$

$A\hat{E}C = 90°00'03''$ $A\hat{D}C = 70°23'26'',07$

$\theta_1 = C\hat{A}E = 88°02'22'',63$ or 1.536581233 radians, and

$v = \sin\theta_0 = 0.954941529$ or $\theta_0 = 1.26946271$ radians in the integral form which gives the curvilinear area *AKH*. We find

$$\int_0^{\theta_1} \frac{d\theta}{(1 - v\sin\theta)^2} = 102.766964$$

for this integral, hence

$$\frac{\text{area }(ADC)}{\text{area }(AKH)} = 6.259142771 \text{ and } \frac{\text{area }(AEC)}{\text{area }(ALH)} = 19.02787015$$

approximately three times greater, contrary to the sub-lemma.

Comment: The validity of the sub-lemma relies solely on the positions of points A, B, D, E and angle $\theta_1 = C\hat{A}E$. Whatever the values of these parameters, and, by implication, whether this sub-lemma is valid or not, it is possible to complete the pyramid by choosing point C in such a way that angle AEC is superior or equal to a right angle, angle φ is positive, and the Lemma 6 inequality is therefore satisfied, even if the sub-lemma is not.

In fact, assume as givens $AB = a$, negative coordinates x, y of E along axes from origin B and angle θ_1, then $AE = e = \sqrt{x^2 + (y-a)^2}$. For $AC = c$ and $BC = b$ as givens, coordinates of C along axes from origin A are

$$\frac{a^2 + c^2 - b^2}{2a}, \quad \frac{e\cos\theta_1}{|x|} + \frac{a-y}{x}\,\frac{a^2 + c^2 - b^2}{2a},$$

hence

$$\tan\varphi = \frac{1}{|x|}\left(\frac{2aec\,\cos\theta_1}{a^2 + c^2 - b^2} - a + y\right);$$

angle φ is positive if $a^2 + c^2 - b^2 \leq \dfrac{2\,aec\cos\theta_1}{a - y}$, that is

$$b^2 \geq a^2 + c^2 - \frac{2\,aec\,\cos\theta_1}{a - y}.$$

The following inequalities must also be stated:

(*i*) $c \geq \dfrac{e}{\cos\,\theta_1} \quad \left(A\hat{E}C \geq \dfrac{\pi}{2}\right)$

(*ii*) $b \leq BE + EC$

(*iii*) $c^2 \geq a^2 + b^2 \quad \left(A\hat{B}C \geq \dfrac{\pi}{2}\right).$

Inequalities (*ii*) and (*iii*) define a non-empty interval for b if

$$a^2 + c^2 - \frac{2\,aec\,\cos\,\theta_1}{a - y} \leq c^2 - a^2$$

and

$$\leq BE^2 + EC^2 + 2BE \cdot EC$$

$$= 2x^2 + 2y^2 - 2ay + a^2 + c^2 - 2\,ec\,\cos\,\theta_1 + 2\sqrt{(x^2+y^2)(e^2+c^2-2ec\cos\theta_1)}\,.$$

The first inequality is written $a \leq ec\dfrac{\cos\,\theta_1}{a - y}$ or $c \geq a\,\dfrac{a - y}{e\,\cos\,\theta_1}$ and it is satisfied, because $e \geq a\dfrac{a-y}{e}$.

The second inequality becomes

$$0 \leq x^2 + y^2 - ay + \frac{ecy\,\cos\,\theta_1}{a-y} + \sqrt{(x^2+y^2)(e^2+c^2-2ec\cos\theta_1)}\,,$$

which is true if it is assumed that

(*iv*) $\dfrac{ec\mid y\mid\cos\,\theta_1}{a - y} \leq x^2 + y^2 - ay.$

Inequalities (*i*) and (*ii*) determine a non-empty interval for c if

$$e^2\mid y\mid \leq (x^2 + y^2 - ay)\,(a - y)$$

or

$$(x^2 + y^2 - ay)(a - y) + y\,(x^2 + y^2 - 2ay + a^2) = ax^2 \geq 0$$

which condition always remains satisfied.

So in the above example 1, it is sufficient to take $BC = 15.13$ instead of 15.1 in order for angle φ to become positive ($\varphi = 22°27'23'',17$); Lemma 6 is therefore true, and the sub-lemma is false.

In example 2, we have $\varphi = -46°29'38'',6$ and $\theta = 48°15'03'',31$, which gives $\lambda_0 = 19°30'21'',92$ greater than λ and μ. Solid angles are $37'46'',92$ and $6°10'18'',94$ respectively and their ratio is $9.801378082 < \rho = 10$; Lemma 6 is therefore satisfied – this comes from the fact that $\rho = 10 > \frac{\tan \mu}{\tan \lambda} = 9.571428571$.

If we take $\rho = 9.6$ with the same values of λ, μ, φ and θ, then Lemma 6 is no longer true.

Ibn al-Haytham's hypotheses have been examined to confirm the inequality in Lemma 6 and they have been shown to be insufficient. Ibn al-Haytham deduced the following lemmas from Lemma 6, and so it is to be expected that these lemmas are not true either. Albeit, this is not the case at all – these lemmas are all true. Let us look at them one by one.

Lemma 7. — *Let* ABCD *be a pyramid such that* AB *is perpendicular to plane* BCD *and angle* BCD $\geq \frac{\pi}{2}$. *If* E $\in$ [CD], *then*

$$\frac{\text{area }(DBC)}{\text{area }(EBC)} > \frac{\text{solid angle }(A, BDC)}{\text{solid angle }(A, EBC)}.$$

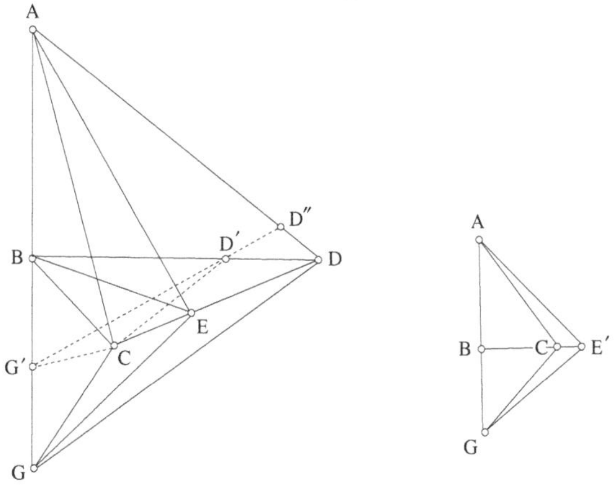

Fig. 3.26

In plane ABE, let us draw a perpendicular from E to AE. This cuts AB at G. Since angle $BCE \geq \frac{\pi}{2}$, then $BE > BC$, and since $AB \perp$ plane (BCD), then $AE > AC$. If BC is produced up to E' such that $BE' = BE$, then $A\hat{E}'G = A\hat{E}G = \frac{\pi}{2}$, hence $A\hat{C}G$ is an obtuse angle.

Let us show how $A\hat{C}D \geq \frac{\pi}{2}$:

• if $B\hat{C}D = \frac{\pi}{2}$, then $CB \perp CD$; on the other hand, CD is orthogonal to AB, therefore $CD \perp$ plane (ABC), and it follows that $CD \perp AC$, that is $A\hat{C}D = \frac{\pi}{2}$.

• if $B\hat{C}D > \frac{\pi}{2}$, then CD' can be drawn inside angle BCD such that $D'C \perp BC$, therefore $CD' \perp$ plane (ACB) and $AC \perp CD'$, therefore $A\hat{C}D' = \frac{\pi}{2}$ and $D' \in\ ]BD[$.

$CG' \perp AC$, $G' \in\ ]BG[$ can be drawn in plane ACG. Therefore plane $CD'G'$ is perpendicular to AC, cutting plane ABD along the straight line $G'D'$ which meets AD between A and D at D'', then $A\hat{C}D'' = \frac{\pi}{2}$, therefore $A\hat{C}D > \frac{\pi}{2}$. So, pyramid $AGCD$ does satisfy the conditions of Lemma 6, in the case where angle AEG is a right angle, which we have shown to be doubtful.

Ibn al-Haytham continues the argument as follows:

$$\frac{\text{area }(GCD)}{\text{area }(GCE)} > \frac{\text{solid angle }(A, GCD)}{\text{solid angle }(A, GCE)};$$

but

$$\frac{\text{area }(GCD)}{\text{area }(GCE)} = \frac{CD}{CE} = \frac{\text{area }(DBC)}{\text{area }(EBC)}$$

and

$$\text{solid angle }(A, GDC) = \text{solid angle }(A, BCD)$$

$$\text{solid angle }(A, GCE) = \text{solid angle }(A, BCE),$$

hence the result

$$\frac{\text{area }(DBC)}{\text{area }(EBC)} > \frac{\text{solid angle }(A, BCD)}{\text{solid angle }(A, BCE)}.$$

However, by relying on analytical methods once again, it is possible to show that this lemma is true: using orthogonal axes from origin A with C on Ax and D in plane Axy, and assuming the orientation to be such that coordinates of C and D are positive. If coordinates of points of the figure are written as

C: $(c, 0, 0)$; E: $(e \cos \lambda, e \sin \lambda, 0)$;
D: $(d \cos \mu, d \sin \mu, 0)$ (with $0 < \lambda < \mu < \pi$).

We have $\frac{CD}{CE} = \rho > 1$ with C, E, D in alignment, which gives

$$e = c\left(1 - \frac{1}{\rho}\right)\frac{\sin \mu}{\sin(\mu - \lambda)} \quad \text{and} \quad d = c(\rho - 1)\frac{\sin \lambda}{\sin(\mu - \lambda)}.$$

Points B and H have as coordinates

H: $(c \sin \theta \cos \varphi, c \sin \theta \sin \varphi, c \cos \theta)$
B: $(b \sin \theta \cos \varphi, b \sin \theta \sin \varphi, b \cos \theta)$

with $0 \leq \theta < \frac{\pi}{2}$. By expressing AB as perpendicular to plane BCD it can be seen that $b = AB$ is the projection of $c = AC$ on AH, and $b = c \sin \theta \cos \varphi$; this condition requires that $|\varphi| < \frac{\pi}{2}$. In the same way AB is the projection of AD on AH, and $b = d \sin \theta \cos (\mu - \varphi)$, hence

$$c \cos \varphi = d \cos(\mu - \varphi) = c(\rho - 1)\frac{\sin \lambda}{\sin(\mu - \lambda)}\cos(\mu - \varphi);$$

and from this

$$\rho = \frac{\sin \mu \cos(\lambda - \varphi)}{\sin \lambda \cos(\mu - \varphi)}.$$

Angle BCD is now expressed as greater than or equal to a right angle, the orthogonal projection of $A\hat{C}D$. The condition is therefore equivalent to $A\hat{C}D \geq \frac{\pi}{2}$, that is

$$c \leq d \cos \mu = c (\rho - 1) \frac{\sin \lambda \cos \mu}{\sin (\mu - \lambda)} = c \frac{\cos \varphi \cos \mu}{\cos (\mu - \varphi)}.$$

This requires that $\mu < \frac{\pi}{2}$ (then $\rho > 1$ and $0 < \lambda < \mu$) and that $|\mu - \varphi| < \frac{\pi}{2}$; the condition can be written as $\frac{\sin\mu \sin\varphi}{\cos(\mu - \varphi)} \leq 0$, or $-\frac{\pi}{2} < \varphi \leq 0$.

In this way Ibn al-Haytham's lemma is not assured *a priori*. However, the inequality of the lemma is explicitly written as

$$\frac{\int_0^\mu \frac{dv}{1+\sin\theta\cos(v-\varphi)}}{\int_0^\lambda \frac{dv}{1+\sin\theta\cos(v-\varphi)}} \leq \rho = \frac{\sin\mu\cos(\lambda-\varphi)}{\sin\lambda\cos(\mu-\varphi)};$$

which signifies that

$$\frac{\cos(\lambda-\varphi)}{\sin\lambda}\int_0^\lambda \frac{dv}{1+\sin\theta\cos(v-\varphi)}$$

is a decreasing function of λ. Let us calculate its derivative

$$\left(-\frac{\sin(\lambda-\varphi)}{\sin\lambda}-\frac{\cos(\lambda-\varphi)\cos\lambda}{\sin^2\lambda}\right)\int_0^\lambda \frac{dv}{1+\sin\theta\cos(v-\varphi)}+\frac{\cos(\lambda-\varphi)}{\sin\lambda(1+\sin\theta\cos(\lambda-\varphi))}$$

$$=\frac{1}{\sin^2\lambda}\left(\frac{\cos(\lambda-\varphi)\sin\lambda}{1+\sin\theta\cos(\lambda-\varphi)}\right)-\cos\varphi\int_0^\lambda \frac{dv}{1+\sin\theta\cos(v-\varphi)}.$$

This derivative is negative if

$$\frac{\cos(\lambda-\varphi)\sin\lambda}{1+\sin\theta\cos(\lambda-\varphi)} \leq \cos\theta\int_0^\lambda \frac{dv}{1+\sin\theta\cos(v-\varphi)},$$

which inequality can be satisfied for $\lambda = 0$.

The inequality persists as long as the derivative of the first member is less than the derivative of the second:

$$\frac{\cos(2\lambda-\varphi)}{1+\sin\theta\cos(\lambda-\varphi)}+\frac{\cos(\lambda-\varphi)\sin\lambda\sin\theta\sin(\lambda-\varphi)}{(1+\sin\theta\cos(\lambda-\varphi))^2} \leq \frac{\cos\varphi}{1+\sin\theta\cos(\lambda-\varphi)},$$

or

$$\cos(2\lambda-\varphi)+\cos^2(\lambda-\varphi)\cos\lambda\sin\theta \leq \cos\varphi+\sin\theta\cos(\lambda-\varphi)\cos\varphi$$

$$0 \leq 2\sin\lambda\sin(\lambda-\varphi)+\sin\theta\cos(\lambda-\varphi)\sin\lambda\sin(\lambda-\varphi)$$
$$=\sin\lambda\sin(\lambda-\varphi)(2+\sin\theta\cos(\lambda-\varphi))$$

which is satisfied as soon as $\varphi \leq \lambda$. However, here $\varphi \leq 0 < \lambda$; the lemma is therefore established.

Lemma 8. — *Let* ABCD *be a pyramid such that* AB *is perpendicular to plane* (CBD) *and that* BC = BD; *if* EG ∥ CD, *then*

$$\frac{\text{area }(CDB)}{\text{area }(EBG)} > \frac{\text{solid angle }(A, BCD)}{\text{solid angle }(A, BEG)}.$$

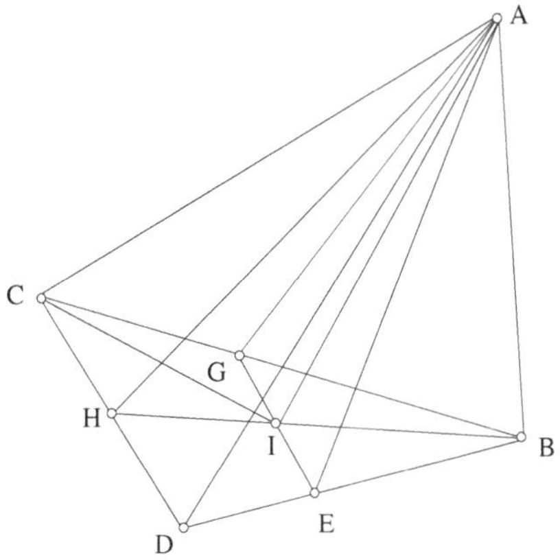

Fig. 3.27

Since $EG \parallel CD$, triangle BGE is isosceles. If I is the midpoint of EG, then $BI \perp EG$ and BI cuts DC at H, the midpoint of CD.

$AB \perp (CBD) \Rightarrow (ABC) \perp (CBD)$ and $(ABH) \perp (CBD)$. But $BH = (ABH) \cap (BCD)$ and the straight line GI in plane (BCD) is perpendicular to BH, therefore it is perpendicular to plane (ABH), therefore $A\hat{I}G = \frac{\pi}{2}$, in the same way $A\hat{H}C = \frac{\pi}{2}$.

We know therefore that $A\hat{I}H$ and $A\hat{I}C$ are obtuse and that $B\hat{H}C = \frac{\pi}{2}$, and Lemma 6 can be applied with $A\hat{I}C$ as an obtuse angle; again this is not a straightforward case, but it is consistent, since points C and H are on the same side of the plane which is perpendicular to ABH passing through AB (CH is parallel to this plane)

$$\frac{\text{area }(BCH)}{\text{area }(BCI)} > \frac{\text{solid angle }(A, BCH)}{\text{solid angle }(A, BCI)}.$$

The same Lemma 6 with angle $A\hat{I}G$ as a right angle (true in all cases, with $A\hat{I}G$ playing the role of angle $A\hat{C}E$ in Lemma 6) gives

$$\frac{\text{area }(CBI)}{\text{area }(IBG)} > \frac{\text{solid angle }(A, BCI)}{\text{solid angle }(A, BIG)}.$$

Multiplying member by member, we obtain

$$\frac{\text{area }(HBC)}{\text{area }(IBG)} > \frac{\text{solid angle }(A, BCH)}{\text{solid angle }(A, BIG)}.$$

By multiplying each term of the ratios by two, it can be deduced that

$$\frac{\text{area }(DBC)}{\text{area }(BEG)} > \frac{\text{solid angle }(A, BCD)}{\text{solid angle }(A, BEG)},$$

moreover, a direct integral calculus would prove this independently of Lemma 6.

Lemma 9. — *Let there be two regular pyramids with the same vertex* A, *whose bases are similar, regular but unequal polygons, inscribed in a sphere with centre* A.

Let P_1 be the pyramid with the largest base and P_2 the other pyramid, then

$$\frac{\text{solid angle } A \text{ of } P_1}{\text{solid angle } A \text{ of } P_2} > \frac{\text{base of } P_1}{\text{base of } P_2}.$$

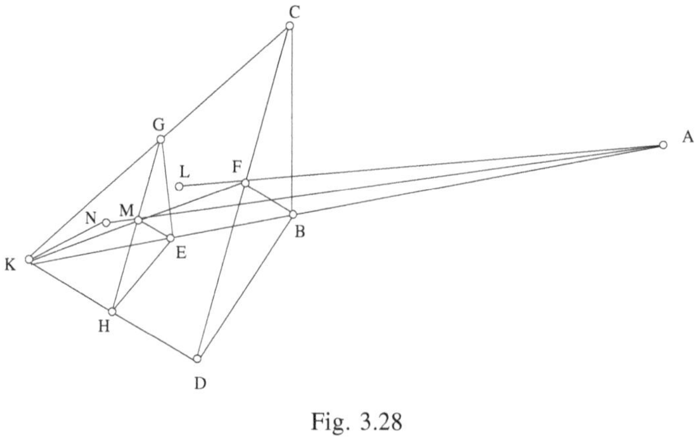

Fig. 3.28

Let (B, BC) be a circle circumscribed about polygon with base P_1 and AB perpendicular to the plane of this circle. Let the base polygon be split into equal isosceles triangles by radii which terminate in its vertices, and let BCD be one of these triangles.

The pyramid P_1 is hence decomposed into pyramids that are equal to one another, and let $ABCD$ be one of them.

One does not limit the problem by presupposing that the plane of base of P_2 is parallel to plane BCD, and that circle (E, EG) circumscribed about the second base is smaller than the circle (B, BC), and therefore $AE > AB$.

As in P_1, P_2 is divided into pyramids equal to each other, and let $AEGH$ be one of them. Assume $EG \parallel BC$ and $EH \parallel BD$, EGH is similar to BCD, therefore GH is parallel to CD.

If $BF \perp CD$, F is the midpoint of CD and if $EM \perp GH$, M is the midpoint of GH and therefore $EM < BF$ since $\frac{BF}{EM} = \frac{BC}{EG}$.

Straight line AF cuts the sphere at L, therefore L is the midpoint of arc CD in plane ACD. In the same way straight line AM cuts the sphere at N, and N is the midpoint of arc GH in plane AGH.

Straight line FM cuts straight line AB, since $FB \parallel EM$ and $FB > EM$. Let K be the point of intersection. Therefore $\frac{BK}{KE} = \frac{BF}{EM} = \frac{BC}{EG}$; however BC and EG are parallel, hence GC passes through K, and in the same way straight line DH passes through K.

Comment: K is the centre of homothety for two circles with centre B and E. Let K be this homothety:

$K: E \to B$
$G \to C$
$H \to D$
$M \to F$

If we draw from M the parallel to the straight line FA, it cuts straight line AB at point O between points E and A.

Since $OM \parallel FA$ and $GH \parallel CD$, then planes HOG and ACD are parallel and cut plane AKC following straight lines OG and AC, then $OG \parallel AC$. Similarly, they cut plane AKD, following straight lines OH and AD, therefore $OH \parallel AD$, and as a result $G\hat{O}H = C\hat{A}D$.

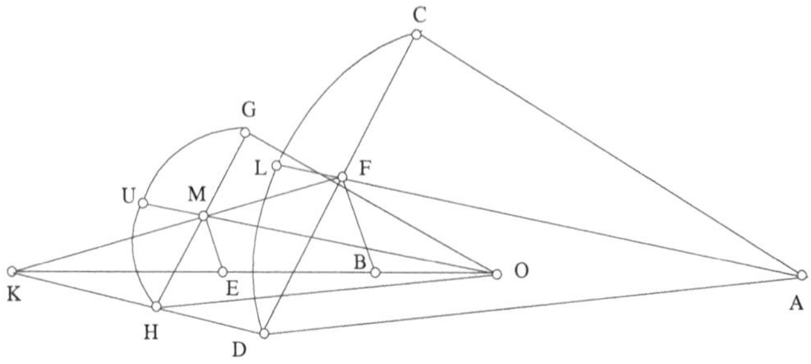

Fig. 3.29

In the homothety K, the point A has for image the point O, the sphere (A, AC) has for image the sphere (O, OG), and the plane (ACD) has for image the plane (OGH). Therefore intersections $(A, AC) \cap (ACD)$ and $(O, OG) \cap (OGH)$ are homothetic circles, arcs $\overset{\frown}{CLD}$ and $\overset{\frown}{GUH}$ are homothetic, and the same holds for sectors $(A, \overset{\frown}{CLD})$ and $(O, \overset{\frown}{GUH})$:

$$\frac{\text{sect. } (A, CLD)}{\text{sect. } (O, GUH)} = \frac{AC^2}{OG^2} = \frac{\text{tr. } (ACD)}{\text{tr. } (OGH)} = \frac{\text{segm. } (CLD)}{\text{segm. } (GUH)}$$

(with U the midpoint of $\overset{\frown}{GH}$); from this we deduce

$$\frac{\text{pyr. } (KACD)}{\text{pyr. } (KOGH)} = \frac{\text{circ. pyr. } (KCLD)}{\text{circ. pyr. } (KGUH)}$$

since on the one hand $(KACD)$ and $(KCLD)$ and on the other hand $(KOGH)$ and $(KGUH)$ are of the same height. But

$$\frac{\text{pyr. } (KACD)}{\text{pyr. } (KOGH)} > \frac{\text{pyr. } (KACD)}{\text{pyr. } (KAGH)},$$

therefore

$$\frac{\text{circ. pyr. } (KCLD)}{\text{circ. pyr. } (KGUH)} > \frac{\text{pyr. } (KACD)}{\text{pyr. } (KAGH)}.$$

Plane $AKLF$ contains points L, M, N, U since it is the intervening plane between CD and GH, and its intersection with the sphere is a great circle which passes through L, N and I (I on AK).

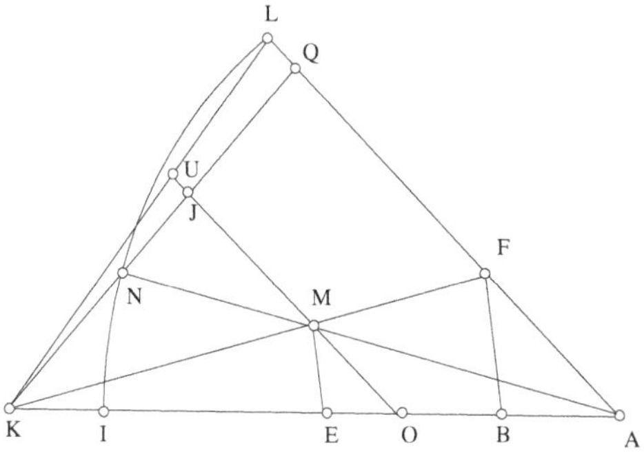

Fig. 3.30

In the previous homothety $\frac{AC}{OG} = \frac{AF}{OM} = \frac{FK}{KM}$, but $AC = AL$ and $OG = OU$, hence $\frac{AL}{OU} = \frac{AF}{OM} = \frac{AL - AF}{OU - OM} = \frac{LF}{MU}$; however, $LF \parallel MU$ and $\frac{KF}{KM} = \frac{LF}{MU}$, and it follows that straight line LU passes through the centre of homothety K. Straight line KN is in plane $AKLF$, and cuts straight line FL at point Q. It also cuts the plane of arc GUH whose centre is O, at a point J on straight line OU; from which we deduce that points K, N, J, Q are aligned.

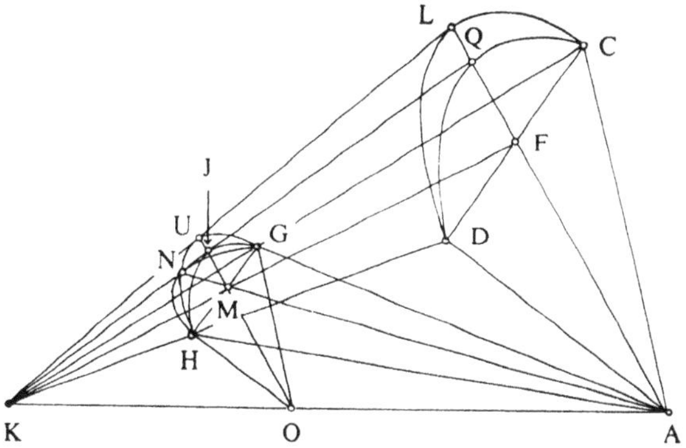

Fig. 3.31

In a circular pyramid let K be the vertex and $AHNG$ be the base of the sector. If the pyramid is elongated, it cuts plane ACD, following a line which includes points C, Q and D. The pyramid also cuts the plane of sector $OHGU$ following a line which passes through points G, J and H. Arcs GUH and GIH are homologous, in the homothety with centre K, to arcs CLD and CQD respectively. Therefore

$$\frac{\text{circ. pyr. }(KCLD)}{\text{circ. pyr. }(KGUH)} = \frac{\text{circ. pyr. }(KCQD)}{\text{circ. pyr. }(KGJH)} \ (= k^3),$$

if k is the ratio of homothety $k = \frac{AC}{OG}$.

But

$$\frac{\text{circ. pyr. }(KCLD)}{\text{circ. pyr. }(KGUH)} > \frac{\text{pyr. }(KACD)}{\text{pyr. }(KAGH)},$$

hence

$$\frac{\text{circ. pyr. }(KCQD)}{\text{circ. pyr. }(KGJH)} > \frac{\text{pyr. }(KACD)}{\text{pyr. }(KAGH)}.$$

From this we deduce

$$\frac{\text{circ. pyr. }(KACQD)}{\text{solid }(KAGJH)} > \frac{\text{pyr. }(KACD)}{\text{pyr. }(KAGH)} \quad \left[\text{since } \frac{a}{b} > \frac{c}{d} \Rightarrow \frac{a+c}{b+d} > \frac{c}{d}\right].$$

Note that our reference to the circular pyramid ($KACQD$) is a misuse of the term since the arc CQD is an arc of a conic section, not an arc of a circle; it would be more correct to call the figure a curvilinear pyramid. We have allowed ourselves to continue to misuse this term, because it does not risk causing confusion. The circular pyramid ($KAGNH$) is inside the solid ($KAGJH$), so

$$\frac{\text{circ. pyr. }(KACQD)}{\text{cir.pyr. }(KAGNH)} > \frac{\text{circ. pyr. }(KACQD)}{\text{solid }(KAGJH)} > \frac{\text{pyr. }(KACD)}{\text{pyr. }(KAGH)}.$$

The portion of pyramid $KACQD$ contained between planes $ACQD$ and $AHNG$ is inside the portion of the sphere bounded by circular sectors $ACLD$ and $AGNH$, therefore solid ($ACLDHNG$) > solid ($ACQDHNG$).

The spherical section which is bounded by the straight line AI and the sector $AGNH$ is inside the circular pyramid $KAGNH$, therefore section ($AIGNH$) < circular pyramid ($KAGNH$), hence

$$\frac{\text{solid }(ACLDHNG)}{\text{section }(AIGNH)} > \frac{\text{solid }(ACQDHNG)}{\text{pyr. circ. }(KAGNH)}$$

by composition

$$\left[\frac{a}{b} > \frac{c}{d} \Rightarrow \frac{a+b}{b} > \frac{c+d}{d}\right]$$

$$\frac{\text{spherical sect. with solid angle } (A, BCD)}{\text{spherical sect. with solid angle } (A, EGH)} > \frac{\text{circ. pyr. } (KACQD)}{\text{pyr. circ. } (KAGNH)} > \frac{\text{pyr. } (KACD)}{\text{pyr. } (KAGH)};$$

hence

$$\frac{\text{solid angle } (A, BCD)}{\text{solid angle } (A, EGH)} > \frac{\text{tr. } (KCD)}{\text{tr. } (KGH)};$$

but

$$\frac{\text{tr. } (KCD)}{\text{tr. } (KGH)} = \left(\frac{KC}{KG}\right)^2 = \left(\frac{BC}{EG}\right)^2 = \frac{\text{tr. } (BCD)}{\text{tr. } (EGH)},$$

hence

$$\frac{\text{solid angle } (A, BCD)}{\text{solid angle } (A, EGH)} > \frac{\text{tr. } (BCD)}{\text{tr. } (EGH)}.$$

If n is the number of faces of each of the two pyramids under study, then by multiplying by n each term of the ratios of this inequality, we obtain

$$\frac{\text{solid angle of } P_1}{\text{solid angle of } P_2} > \frac{\text{base of } P_1}{\text{base of } P_2}.$$

Comment: Lemma 9 of Ibn al-Haytham's treatise on figures with equal perimeters, solids with equal areas and the study of the solid angle is particularly important. The proof he proposes is long and complicated: it involves the introduction of no fewer than seventeen distinct points, thirty-five different straight lines, nine distinct planes, eighteen three-dimensional figures and eight different curved lines.

Let us examine the validity of Lemma 9 by using a direct analytical method. Lemma 9 can be written

$$\frac{\text{solid angle } A \text{ of } P_1}{\text{solid angle } A \text{ of } P_2} > \frac{\text{base of } P_1}{\text{base of } P_2}$$

where P_1 and P_2 are two pyramids with vertex A.

This same lemma can be rewritten

$$\frac{\text{spherical area } ICD}{\text{plane area } BCD} > \frac{\text{spherical area } IGH}{\text{plane area } EGH};$$

So we need to prove that the ratio $\frac{\text{spherical area } ICD}{\text{plane area } BCD}$ is a decreasing function of the distance AB, since our assumptions mean that $AB < AE$.

Let the radius of the sphere be r and let us put $AB = u$. The radius BC is equal to $\sqrt{r^2-u^2}$, so the coordinates of C are $x=\sqrt{r^2-u^2}\cdot\cos\theta$, $y=\sqrt{r^2-u^2}\cdot\sin\theta$, $z = u$ where θ is half the angle CBD, let $\theta = C\hat{B}F$. It follows that the plane area BCD is equal to $(r^2 - u^2)\sin\theta\cos\theta$.

The area ICD, which is part of the surface of a sphere, is the product of r^2 by the difference between a plane angle and the sum of the angles $CID = 2\theta$, $ICD = \varphi$, $CDI = \varphi$ (a result called Albert Girard's theorem). So the curved area is

$$2r^2\left(\theta+\varphi-\frac{\pi}{2}\right)=2r^2\psi$$

where

$$\psi=\theta+\varphi-\frac{\pi}{2}.$$

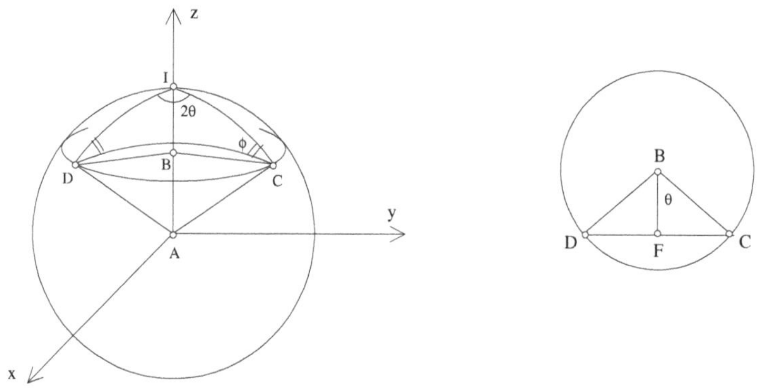

Fig. 3.32

Let us calculate φ, the dihedral angle between the planes ACI and ACD. The equations of these planes are respectively: $x\sin\theta - y\cos\theta = 0$ and $ux - z\cos\theta\sqrt{r^2-u^2} = 0$, so

$$\cos\varphi=\frac{u\sin\theta}{\sqrt{u^2+\left(r^2-u^2\right)\cos^2\theta}}=\frac{u\sin\theta}{\sqrt{u^2\sin^2\theta+r^2\cos^2\theta}}.$$

We also have

$$\sin\varphi=\frac{r\cos\theta}{\sqrt{u^2\sin^2\theta+r^2\cos^2\theta}}$$

and

$$\sin\psi = -\cos(\theta+\varphi) = \sin\theta\sin\varphi - \cos\theta\cos\varphi = \frac{(r-u)\sin\theta\cos\theta}{\sqrt{u^2\sin^2\theta + r^2\cos^2\theta}}.$$

The value of the spherical area *ICD* is thus

$$2r^2\sin\psi.\frac{\psi}{\sin\psi} = \frac{2r^2(r-u)\sin\theta\cos\theta}{\sqrt{u^2\sin^2\theta + r^2\cos^2\theta}}.\frac{\psi}{\sin\psi}$$

and the value of the ratio $\frac{\text{spherical area } ICD}{\text{plane area } BCD}$ is

$$\frac{2r^2}{(r+u)\sqrt{u^2\sin^2\theta + r^2\cos^2\theta}}.\frac{\psi}{\sin\psi}.$$

The first factor is a decreasing function of u and the second an increasing function $\psi \in \left]0, \frac{\pi}{2}\right]$. Now $\psi = \theta + \varphi - \frac{\pi}{2}$ is an increasing function of φ and $\sin\varphi = \frac{r\cos\theta}{\sqrt{u^2\sin^2\theta + r^2\cos^2\theta}}$ is a decreasing function of u; thus $\frac{\psi}{\sin\psi}$ is a decreasing function of u, which proves the lemma.

This proof makes a direct comparison between the areas, whereas Ibn al-Haytham's proof compares volumes. However, our proof shows Ibn al-Haytham's basic idea, which is to measure a solid angle by the spherical area of that which it subtends.

Lemma 10. — *Let* P_1 *and* P_2 *be two regular pyramids with the same vertex* A *and with* n_1 *and* n_2 *faces respectively, whose bases are regular polygons* B_1 *and* B_2 *inscribed in a same sphere with centre* A, *with areas* s_1 *and* s_2 *respectively.*

If $n_1 > n_2$ and $s_1 < s_2$, then $\frac{\text{solid angle } A \text{ of } P_2}{\text{solid angle } A \text{ of } P_1} > \frac{s_2}{s_1}$.

Let r_1 and r_2 be radii of circles circumscribed about B_1 and B_2. To show that $r_1 < r_2$, Ibn al-Haytham uses a polygon B with n_2 sides, area s_1, and radius of the circumscribed circle r. However B is similar to B_2 and has an area smaller than B_2, therefore $r < r_2$. But B and B_1 have the same area and B has less sides than B_1, hence $r_1 < r$. It follows that $r_1 < r_2$. From this it can be deduced that $h_1 > h_2$, if h_1 and h_2 are the respective heights of the pyramids.

The method continues as before. Let A be the centre of the sphere, AB perpendicular to the plane of the larger base and BCD one of its triangles,

AE perpendicular to the plane of the smaller base and EGH one of its triangles, then $AE > AB$. It can be assumed without limiting the problem that points A, E and B are aligned and that $EG \parallel BC$, $G\hat{E}H \left(= \frac{2\pi}{n_1}\right)$ is smaller than $C\hat{B}D \left(= \frac{2\pi}{n_2}\right)$, therefore EH is not parallel to BD.

Draw $EK \parallel BD$, then $G\hat{E}K = C\hat{B}D$. Let N, M and F be the respective midpoints of chords DC, GH and GK, then $BN \perp CD$, $EM \perp GH$, $EF \perp GK$, pyramids $ABCD$, $AEGH$ and $AEGK$ are divided into two equal pyramids by planes ABN, AEM and AEF respectively. Triangles EGK and BCD are similar, therefore

(1) $$\frac{\text{solid angle pyr. } (A, BCD)}{\text{solid angle pyr. } (A, EGK)} > \frac{\text{area } (BCD)}{\text{area } (EGK)} \text{ (Lemma 9).}$$

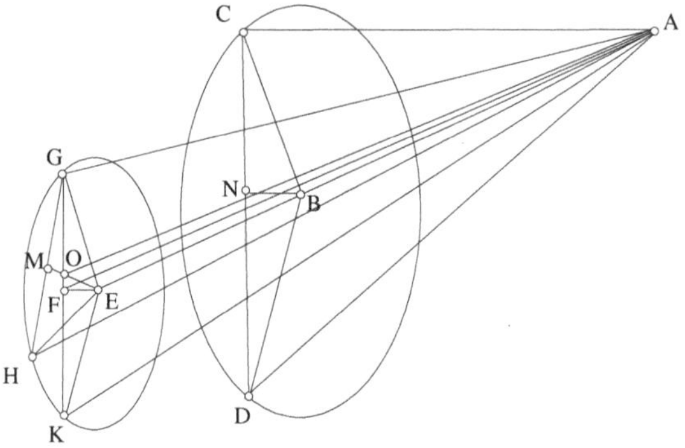

Fig. 3.33

But $G\hat{E}K > G\hat{E}H$, hence $E\hat{G}F < E\hat{G}M$, therefore GF cuts EM at a point O. However $G\hat{O}E$ is obtuse. AGK is isosceles and F is the midpoint of GK, therefore $AF \perp GK$, and it follows that $A\hat{O}G$ is obtuse; Ibn al-Haytham applies Lemma 6 for the ambiguous case: angle AOG becomes angle AEC of Lemma 6. But GM is parallel to the plane which is perpendicular to AEM, passing through AE, and therefore G and M are on the same side of the plane and the lemma is valid.

(2) $$\frac{\text{area } (EGM)}{\text{area } (EGO)} > \frac{\text{solid angle } (A, EGM)}{\text{solid angle } (A, EGO)}.$$

Angle AFG is a right angle, and from Lemma 7 we get

(3) $$\frac{\text{area }(EGF)}{\text{area }(EOF)} > \frac{\text{solid angle }(A, EGF)}{\text{solid angle }(A, EOF)};$$

from (3) we have

(4) $$\frac{\text{area }(EGO)}{\text{area }(EGF)} > \frac{\text{solid angle }(A, EGO)}{\text{solid angle }(A, EGF)},$$

and from (2) and (3), through multiplication member by member, we get

$$\frac{\text{area }(EGM)}{\text{area }(EGF)} > \frac{\text{solid angle }(A, EGM)}{\text{solid angle }(A, EGF)},$$

hence

(5) $$\frac{\text{solid angle }(A, EGF)}{\text{solid angle }(A, EGM)} > \frac{\text{area }(EGF)}{\text{area }(EGM)};$$

(1) can be written

(6) $$\frac{\text{solid angle }(A, BCD)}{\text{solid angle }(A, EGF)} > \frac{\text{area }(BCD)}{\text{area }(EGF)}, \text{ since } (EGK) = 2(EGF);$$

from (5) and (6), through multiplication member by member, we deduce

(7) $$\frac{\text{solid angle }(A, BCD)}{\text{solid angle }(A, EGM)} > \frac{\text{area }(BCD)}{\text{area }(EGM)}.$$

Solid angle A in P_2 and the base of P_2 are equimultiples of solid angle (A, BCD) and of the triangle BCD:

$$\text{Angle } A \text{ of } P_2 = n_2 \text{ angle } (A, BCD)$$
$$\text{Base of } P_2 = n_2 \text{ triangle } (BCD)$$

and similarly

$$\text{Angle } A \text{ of } P_1 = n_1 \text{ angle } (A, EGK) = 2n_1 \text{ angle } (A, EGM)$$
$$\text{Base of } P_2 = n_1 \text{ triangle } (EGK) = 2\, n_1 \cdot \text{triangle } (EGM).$$

From (7) we therefore deduce

$$\frac{\text{solid angle } A \text{ of } P_2}{\text{solid angle } A \text{ of } P_1} > \frac{\text{base of } P_2}{\text{base of } P_1}.$$

At the end of this long development and examination of previous lemmas, we may wonder whether Ibn al-Haytham might not have had any inkling about inequalities (as stated in Lemmas 7, 8, 10 and drawn from his research on solid angles) before attempting their proof using Lemma 6, then reducing it to a sub-lemma. His method does explain that Lemmas 7, 8, 10 are true independently of Lemma 6, as we have seen. And in support of this, it can be argued that in subsequent work, Ibn al-Haytham uses only Lemmas 8, 9 and 10, thus avoiding any error contained in Lemma 6.

Theorem of Proposition 5a. — *Of two regular polyhedra, with similar bases and equal surfaces, the one with the greater number of faces has the greater volume.*

Let A be the centre of the sphere circumscribed about the first polyhedron, AE the distance from A to the plane of one of the faces, S_A the total area of the polyhedron and V_A its volume, then $V_A = \frac{1}{3} S_A \cdot AE$.

Let B be the centre of the sphere circumscribed about the second polyhedron, BG the distance from B to the plane of one of the faces, then in the same way $V_B = \frac{1}{3} S_B \cdot BG$.

By hypothesis, we have $S_A = S_B$, and with n_A and n_B the number of faces of the polyhedra, then $n_B > n_A$.

The proof consists of comparing AE and BG.

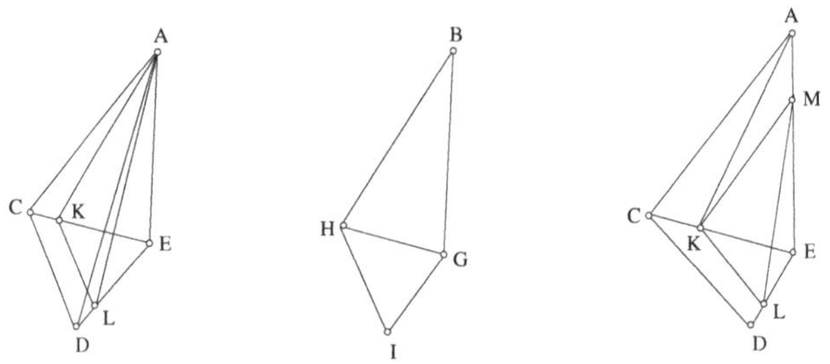

Fig. 3.34

Let us assume a base of pyramid A divided into triangles, and let CED be one of them. Let us proceed similarly for pyramid B and let GHI be the triangle obtained.

$$AE \perp \text{plane } (ECD) \text{ and } BG \perp \text{plane } (HGI).$$

Since the bases of pyramids are similar, then so are triangles ECD and HGI.

Hypotheses $S_A = S_B$ and $n_B > n_A$ mean that base A is larger than base B, therefore triangle ECD is greater than triangle HGI; they are isosceles, therefore $EC > GH$, $ED > GI$.

Let point K be on EC such that $GH = EK$ and point L on ED such that $EL = GI$. If it were that $AE = BG$, then pyramid $AEKL$ would be equal to pyramid $BGHI$ and then solid angle (A, EKL) = solid angle (B, GHI).

We know that (from Lemma 2 of Proposition 4)

$$\frac{\text{solid angle }(A, ECD)}{4\pi} = \frac{\text{pyr. }(AECD)}{V_A} = \frac{\text{area }(ECD)}{S_A}$$

and

$$\frac{\text{solid angle }(B, GHI)}{4\pi} = \frac{\text{pyr. }(BGHI)}{V_B} = \frac{\text{area }(GHI)}{S_B}.$$

Since $S_A = S_B$, we deduce

$$(1) \qquad \frac{\text{area }(ECD)}{\text{area }(GHI)} = \frac{\text{solid angle }(A, ECD)}{\text{solid angle }(B, GHI)},$$

then

$$\frac{\text{area }(ECD)}{\text{area }(GHI)} = \frac{\text{solid angle }(A, ECD)}{\text{solid angle }(A, EKL)},$$

which, according to Lemma 8, is absurd. Therefore $AE \neq BG$.

If $BG < AE$, M exists on AE such that $EM = BG$ and then

$$\text{solid angle }(M, EKL) = \text{solid angle }(B, GHI).$$

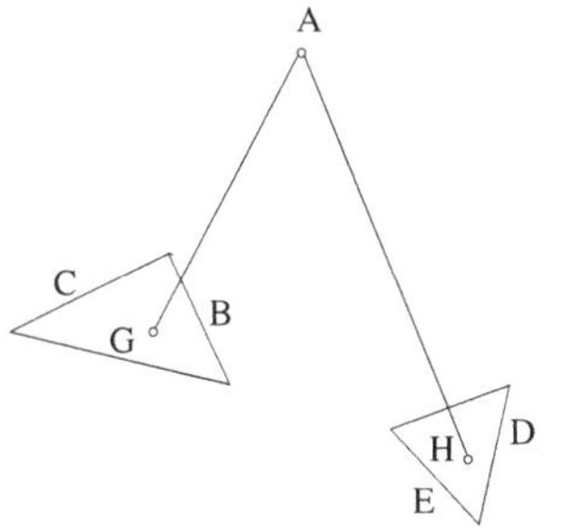

Fig. 3.35

But we would have $E\hat{M}K > E\hat{A}K$, $E\hat{M}L > E\hat{A}L$, $K\hat{M}L > K\hat{A}L$ (angles at the vertex of two isosceles triangles of same base KL); hence

$$E\hat{M}K + E\hat{M}L + K\hat{M}L > E\hat{A}K + E\hat{A}L + K\hat{A}L,$$

and it follows that solid angle $(M, EKL) >$ solid angle (A, EKL).

According to Lemma 8

$$\frac{\text{area }(ECD)}{\text{area }(EKL)} > \frac{\text{solid angle }(A, ECD)}{\text{solid angle }(A, EKL)};$$

we would therefore have here

$$\frac{\text{area }(ECD)}{\text{area }(EKL)} > \frac{\text{solid angle }(A, ECD)}{\text{solid angle }(M, EKL)},$$

that is

$$\frac{\text{area }(ECD)}{\text{area }(GHI)} > \frac{\text{solid angle }(A, ECD)}{\text{solid angle }(B, GHI)},$$

which is absurd according to (1).

We have necessarily $BG > AE$ and it follows that $V_B > V_A$.

Theorem of Proposition 5b. — *If the faces of two regular polyhedra are similar regular polygons and are inscribed in a same sphere, then the one with the greater number of faces has the greater surface and the greater volume.*

Let P_1 and P_2 be two polyhedra, S_1 and S_2 their surfaces, V_1 and V_2 their volumes, n_1 and n_2 the number of faces. Assume $n_1 > n_2$. If A is the centre of the sphere circumscribed about two polyhedra, then there would be n_1 regular equal pyramids with the same vertex A, associated with the faces of P_1, and n_2 regular equal pyramids associated with the faces of P_2.

Let α_1, s_1, h_1 be respectively the angle at the vertex, the area of the base and the height of regular pyramid P'_1 associated with P_1, and let α_2, s_2, h_2 be the same elements of regular pyramid P'_2 associated with P_2.

We have $n_1\alpha_1 = n_2\alpha_2 = 4\pi$, as $n_1 > n_2$, then $\alpha_1 < \alpha_2$.

We may assume that a pyramid P'_1 and a pyramid P'_2 have the same axis AH and since $\alpha_1 < \alpha_2$, the solid angle of P'_1 is inside the solid angle of P'_2, the edges of P'_1 cut the sphere beyond the plane of the base of P'_2. The planes of the two bases are parallel and cut the sphere following the circles circumscribed about these bases. From this it can be deduced that $s_1 < s_2$ and $h_1 > h_2$. Moreover,

$$\frac{\alpha_1}{4\pi} = \frac{s_1}{S_1} = \frac{1}{n_1} \quad \text{and} \quad \frac{\alpha_2}{4\pi} = \frac{s_2}{S_2} = \frac{1}{n_2},$$

hence

$$\frac{\alpha_2}{\alpha_1}=\frac{s_2}{S_2}\cdot\frac{S_1}{s_1}=\frac{s_2}{s_1}\cdot\frac{S_1}{S_2}.$$

But it was established (Proposition 9) that $\frac{\alpha_2}{\alpha_1}>\frac{s_2}{s_1}$, therefore $\frac{s_2}{s_1}\cdot\frac{S_1}{S_2}>\frac{s_2}{s_1}$, hence $S_1 > S_2$.

We know that $V_1 = \frac{1}{3}S_1h_1$ and $V_2 = \frac{1}{3}S_2h_2$; since $S_1 > S_2$ and $h_1 > h_2$, we have $V_1 > V_2$.

As explained in the introduction to this chapter, this theorem, although largely proved by Ibn al-Haytham's method, only applies in the case of tetrahedra, octahedra or icosahedra.

Corollary. — *Let two regular polyhedra be inscribed in a same sphere. These polyhedra* P_1 *and* P_2 *have respectively* n_1 *and* n_2 *faces, and the faces are regular polygons with* n'_1 *and* n'_2 *sides respectively*

If $n_1 > n_2$ and $n'_1 > n'_2$, then $S_1 > S_2$ and $V_1 > V_2$.

According to Lemma 10, the distance from the centre of the sphere to the faces of P_1 is greater than the distance to the faces of P_2.

The argument continues as before, using a result established in Lemma 10 – which refers to solid angles with the centre of the sphere as their vertex.

Commentary: Since the polyhedra in question are regular, this corollary signifies that if a regular tetrahedron, a cube and a dodecahedron are inscribed in a same sphere, their lateral surfaces and their volumes increase in that order.

From a remark by Hypsicles,[6] Apollonius compared the ratio of areas and the ratio of volumes of a dodecahedron and an isocahedron: these ratios are equal since the distances from the centre to the faces of these two solids are the same. Ibn al-Haytham also uses the comparison of distances from the centre to the faces, but in the other cases which he examined, he used a more general way.

[6] *The Thirteen Books of Euclid's Elements*, trans. and com. by T. L. Heath, 3 vols, 2nd ed., Cambridge, 1926, vol. III, p. 512.

3.3. *Translated text*

Al-Ḥasan ibn al-Haytham

On the Sphere which is the Largest of all the Solid Figures having Equal Perimeters and On the Circle which is the Largest of all the Plane Figures having Equal Perimeters

In the Name of God, the Forgiving, the Merciful

TREATISE BY AL-ḤASAN IBN AL-ḤASAN IBN AL-HAYTHAM

On the Sphere which is the Largest of all the Solid Figures having Equal Perimeters[1] and On the Circle which is the Largest of all the Plane Figures having Equal Perimeters

<*Introduction*>

One of the geometric notions on which is based the deduction that the sky is a sphere and that the entire Universe has a circular shape, is that the greatest of all the solid figures having equal perimeters, and the perimeter of each of which <consists> of equal parts, and which has the greatest volume,[2] is the figure of the sphere, and that the greatest of the plane figures having equal perimeters and the perimeter of each of which <consists> of equal parts, and which has the greatest area, is the figure of the circle, and that solid and plane figures having a shape close to being circular are greater than those whose shape is far from being circular. By 'that which has a similar perimeter', I mean that in which the parts of its perimeter are similar to each other. Among the solid figures, the sphere and those rectilinear figures whose bases have equal sides are similar, and among the plane figures, this includes the circle and the regular polygons.[3] Of these solid figures with bases, those whose shape is closest to being circular are those with the greatest number of bases, and their bases are similar to the bases of the other solid. The plane figures that are closest to being circular are those with the greatest number of sides.

[1] It is clear that in the case of 'volumes', we should read 'surface area'. However, as Ibn al-Haytham uses the same word for both solids and plane figures, a word which should be translated as 'that which encloses', we have chosen to use the same word, 'perimeter', in both cases in order to retain a single word translation.

[2] Lit.: measurement, which we translate in the case of a solid as 'volume', in the case of a surface by 'area', and in the general case by 'measurement'.

[3] Lit.: figures with straight lines whose sides and angles are equal.

Mathematicians have made mention of this notion and have made use of it. However, they have provided no formal proof of this notion, nor has any convincing argument of its truth come down to us from them. This state of affairs has prompted us to make a careful examination of this matter. A universal proof of this notion in all its forms has presented itself to us. We have therefore written this treatise.

<The circle: The greatest of all the plane figures>

<1> Any circle whose circumference is equal to the perimeter of a regular polygon has an area greater than that of the polygon.

Example: Let *AB* be the circumference of the circle and let this be equal to the perimeter *CDEGH* of the regular polygon.

I say that the area of the circle AB *is greater than that of the figure* CDEGH.

Proof: As the figure *CDEGH* is a regular polygon, there exists a circle inscribed within this polygon that is tangential to all its sides. If each of its angles is divided into two halves, and if the bisectors are drawn, then these will meet at a single point within the figure, and the triangles thus formed, whose vertices are all at this point, are all equal and similar. The perpendiculars drawn from this point to all the sides of the figure are equal. If this point is then taken as a centre and a circle is drawn around it with a radius equal[4] to one of these perpendiculars, then that circle will be tangential to all the sides of the figure.

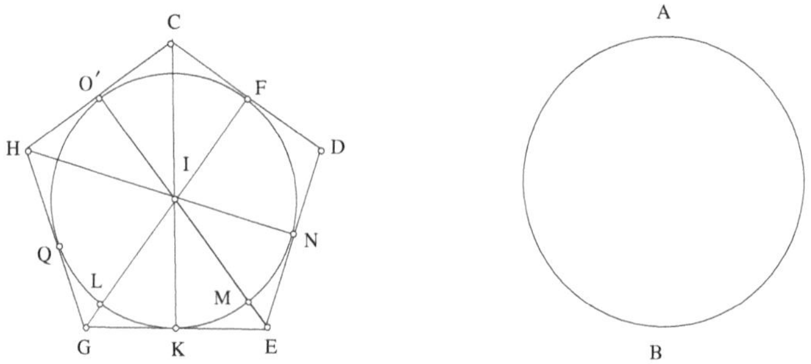

Fig. III.1

Let the circle that is tangential to the sides of the figure be the circle *KNFO′Q*, and let its centre be *I*. Join the two straight lines *IME* and *ILG* and draw the perpendicular *IK*. The product of the perpendicular *IK* and one half of *EG* is then equal to the area of the triangle *IEG*, and the product

[4] Lit.: at a distance. We shall translate this expression in this way from now on.

of *IK*, which is the half-diameter of the circle, and one half of the arc *ML* is equal to the area of the sector *IMKL*, and the triangle *IEG* is greater than the sector *IMKL*. The straight line *EG* is therefore greater than the arc *ML*. Similarly, we can show that each side of the figure is greater than the arc intercepted by the two straight lines drawn from the point *I* to the extremities of that side. The perimeter of the figure *CDEGH* is therefore greater than the circumference of the circle *KNFO′Q*, and the perimeter of the figure *CDEGH* is equal to the circumference of the circle *AB*. Therefore, the circumference of the circle *AB* is greater than the circumference of the circle *KNFO′Q*, and therefore the half-diameter of the circle *AB* is greater than the straight line *KI*. But the product of the half-diameter of the circle *AB* and one half of its circumference is equal to its area, and the product of the straight line *IK* and one half of the perimeter of the figure *CDEGH*, which is equal to the circumference of the circle *AB*, is equal to the area of the figure. The circle *AB* is therefore greater than the figure *CDEGH*. That is what we wanted to prove.

<2> If two regular polygons have the same perimeter and one has more sides than the other, then the area of the first is greater than the area of the second.

Example: Let there be two regular polygons *ABCDE* and *GHIKLM*, such that the sides of the figure *GHIKLM* are greater in number than the sides of the figure *ABCDE*, and let their perimeters be equal.

I say that the area of the figure GHIKLM *is greater than that of the figure* ABCDE.

Let the point *N* be the centre of the circle circumscribed around the figure *ABCDE*, and let the point *F* be the centre of the circle circumscribed around the figure *GHIKLM*. We draw the straight lines *NA*, *NB*, *NC*, *ND* and *NE*, and the straight lines *FG*, *FH*, *FI*, *FK*, *FL* and *FM*. We draw the perpendicular *NP* and the perpendicular *FU*.[5] If we draw perpendiculars from the point *N* to each of the sides of the figure *ABCDE*, this will generate a set of triangles with equal bases – each being equal to the triangle *END* – and whose angle at the summit[6] is equal to the angle *END*. Similarly, if perpendiculars are drawn from the point *F* to each of the sides of the figure *GHIKLM*, this will generate a set of triangles with equal bases – each being equal to the triangle *MFU* – and whose angle at the centre is equal to the angle *MFU*, such that the number of bases of the triangles – which are in each of the two figures – is equal to the number of angles at the centre of each of them. The ratio of the angle *ENP* to the sum of the

[5] *NP* perpendicular to *DE*, and *FU* perpendicular to *LM*.
[6] Lit.: centre.

angles at the centre *N* is therefore equal to the ratio of the straight line *EP* to the entire perimeter of the figure *ABCDE*, as the angles are equal, the bases of the triangles are equal, and the number of angles is equal to the number of bases, whose sum is equal to the perimeter of the figure.

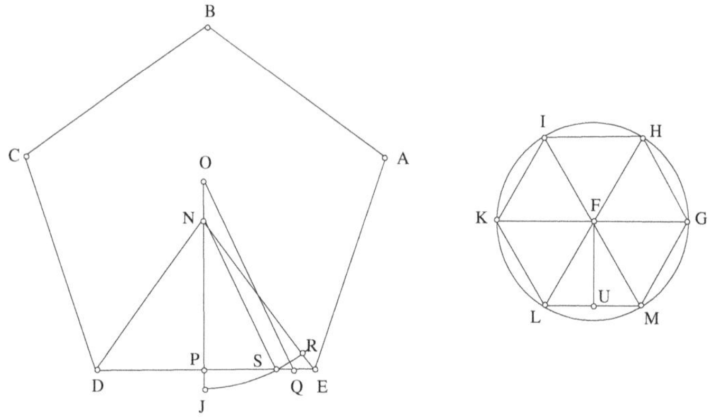

Fig. III.2

Similarly, the ratio of the angle *MFU* to the sum of the angles at the centre *F* is equal to the ratio of *MU* to the entire perimeter of the figure *GHIKLM*. Inverting, the ratio of the sum of the angles at the centre *F* to the angle *MFU* is equal to the ratio of the entire perimeter to *MU*. The sum of the angles at the point *N* is equal to the sum of the angles at the point *F*, as this sum is equal to four right angles. But the perimeter *ABCDE* is equal to the perimeter *GHIKLM*. Therefore the ratio of the angle *PNE* to the sum of the angles at the point *N* is equal to the ratio of the straight line *EP* to the entire perimeter of the figure *GHIKLM*, and the ratio of the sum of the angles at the point *F* to the angle *MFU* is equal to the ratio of the perimeter of the figure *GHIKLM* to the straight line *MU*. Using the equality ratio, the ratio of the angle *ENP* to the angle *MFU* is equal to the ratio of the straight line *EP* to the straight line *MU*. However, as the figure *GHIKLM* has more bases than the figure *ABCDE*, each of the angles at the point *F*, which is the vertex of the triangles, is less than each of the angles at the point *N*, as if each of these two figures were to be inscribed within a circle, each side of the figure *GHIKLM* would cut off from its circle a part whose ratio to the circle is less than the part <of the circle> cut off from its circle by a side of the figure *ABCDE*. Therefore, the angle *MFU* is less than the angle *ENP*. From the angle *ENP*, take away the angle *PNS* equal to the angle *MFU*. The ratio of the angle *ENP* to the angle *PNS* is then equal to the ratio of the straight line *EP* to the straight line *MU*. Setting *N* as the centre and the straight line *NS* as the radius, draw the arc <of a circle> *SJ*. The ratio of the angle *ENP* to the angle *SNP* is then equal to the ratio of the sector *RNJ* to the sector *SNJ*. But the ratio of the angle *ENP* to the angle *SNP* was equal to the ratio of the straight line *EP* to the straight line *MU*. Therefore, the ratio of the sector *RNJ* to the sector *SNJ* is equal to the ratio of the straight line *EP* to the straight line *MU*. But the ratio of the straight line *EP* to <the straight line> *PS* is greater than the ratio of the sector *RNJ* to the sector *SNJ*, as this ratio is the ratio of the triangle *ENP* to the triangle *SNP* which is greater than the ratio of the sector *RNJ* to the sector *SNJ*.[7] Therefore, the ratio of the straight line *EP* to the straight line *PS* is greater than the ratio of the straight line *EP* to the straight line *MU*, and therefore the straight line *SP* is less than the straight line *MU*. We mark off *PQ* equal to *MU* and we draw *QO* parallel to the straight line *NS*. The angle *POQ* is then equal to the angle *MFU*, as each of these two angles is equal to the angle *SNP*. But the two angles at the points *U* and *P* are right angles. Therefore the remaining angle in the triangle *OQP* is equal to the angle *FMU* and the side *QP* is equal to the side *MU*. Therefore, the triangle *OQP* is equal to the triangle *FMU*, and hence the straight line *OP* is equal to the perpendicular *FU*. But *OP* is greater than the perpendicular *NP*. Therefore, the perpendicular *FU* is greater than the perpendicular *NP*. The product of the perpendicular *FU* and one half of the perimeter of the figure *GHIKLM* is the area of the figure *GHIKLM*, and the product of the perpendicular *NP* and one half of the perimeter of the figure *ABCDE* is the area of the figure *ABCDE*. But the perimeters of the two figures are equal, and the perpendicular *FU* is greater than the perpendicular *NP*. Therefore, the area of the figure *GHIKLM* is greater than the area of *ABCDE*. That is what we wanted to prove.

This notion may also be proved in another way. If the proof leads to the conclusion that the straight line *SP* is less than the straight line *MU*, then it can be shown that the straight line *NP* is less than the straight line *FU*, as the angle *PNS* is equal to the angle *UFM* and each of the two angles at the points *P* and *U* are right <angles>. The two remaining angles are therefore equal, and hence the triangle *MFU* is similar to the triangle *SNP* and the straight line *MU* is greater than the straight line *SP*. The perpendicular *FU* is therefore greater than the perpendicular *NP*. But the perimeter of the figure *GHIKLM* is equal to the perimeter of the figure *ABCDE*. Therefore, the area of the figure *GHIKLM* is greater than the area of the figure *ABCDE*. That is what we wanted to prove.

From what we have shown, it is clear that, of all the figures with similar and <equal> perimeters,[8] the circle has the greatest <area> and, of all the

[7] This can be proved. See the mathematical commentary.

[8] i.e. regular polygons (see the mathematical commentary).

polygons, that which is closest to being circular in shape is greater than that which is less circular in shape.

<**3**> We also say that if two regular polygons are inscribed within the same circle such that one has more sides than the other, then the area of the figure with the greater number of sides is greater than the area of the other figure, and its perimeter is also greater than the perimeter of the other.

<**Lemma**> We first introduce the following lemma. Let there be two different arcs such that their sum is less than two thirds of a circle. Then the ratio of the larger arc to the smaller arc is greater than the ratio of the chord of the larger arc to the chord of the smaller arc.

Let there be two different arcs *AB* and *BC*, such that their sum which is the arc *ABC* is not greater than two thirds of a circle, and let *AB* and *BC* be their chords.

I say that the ratio of the arc AB *to the arc* BC *is greater than the ratio of the chord* AB *to the chord* BC.

Proof: We join *AC* and we make the angle *CBD* equal to the angle *BAC*, which is less than the angle *ABC* as the arc *BC* is less than the arc *CA* – the arc *BC* being less than one third of the circle and the arc *CA* not being less than one third of the circle. The angle *BDC* is equal to the angle *ABC*, and the two triangles *ABC* and *BDC* are therefore similar. Therefore, the ratio of *AC* to *CB* is equal to the ratio of *BC* to *CD*. But *AC* is greater than *CB*. Therefore, the straight line *BC* is greater than the straight line *CD*. Setting *C* as the centre, we draw an arc of a circle with *CB* as the radius, cutting the straight line *AC* between the two points *A* and *D* at the point *K*. We make the angle *DCE* equal to the angle *CBE*, which is equal to the angle *BAC*, which is less than the angle *BCA*. The point *E* therefore lies within the arc *BK*. We produce *CE* until it meets the arc at the point *H*. The ratio of *HC* to *CE* is therefore equal to the ratio of *BC* to *CE*, which is equal to the ratio of *BD* to *DC*, and equal to the ratio of *CD* to *DE*, as the two triangles *BCD* and *DCE* are similar. We draw *HG* parallel to the straight line *BD*. Then the ratio of *CG* to *CD* is equal to the ratio of *HG* to *ED* which is equal to the ratio of *HC* to *CE*, which is equal to the ratio of *BC* to *CE*, which is the ratio of *CD* to *CE*. The straight line *HG* is therefore equal to the straight line *DC*, and *BD* is greater than *DC* as its ratio to the latter is equal to the ratio of *AB* to *BC*. The straight line *BD* is therefore greater than the straight line *HG* and is parallel to it. We join *BH* and extend it until it meets the straight line *AC* at the point *I*. The ratio of *BI* to *IH* is therefore equal to the ratio of *BD* to *HG*, which is the ratio of *BD* to *DC*, which is the ratio of *AB* to *BC*. But the ratio of the sector *CBH* to the sector *CHK* is greater than the ratio of the triangle *CBH* to the triangle *CHI*.[9] Therefore, the ratio of the arc *BH* to the arc *HK* is greater than the ratio of the straight line *BH* to the straight line *HI*. Therefore, the ratio of the angle *BCH* to the angle *HCK* is greater than the ratio of the straight line *BH* to the straight line *HI*. Composing, the ratio of the angle *BCA* to the angle *ACH*, which is equal to the angle *BAC*, is greater than the ratio of the straight line *BI* to the straight line *IH*, which is the ratio of *BD* to *HB*, which is the ratio of *BD* to *DC*, which is the ratio of the straight line *AB* to the straight line *BC*. But the ratio of the angle *BCA* to the angle *BAC* is the ratio of the arc *AB* to the arc *BC*. The ratio of the arc *AB* to the arc *BC* is therefore greater than the ratio of the chord *AB* to the chord *BC*. That is what we wanted to prove.

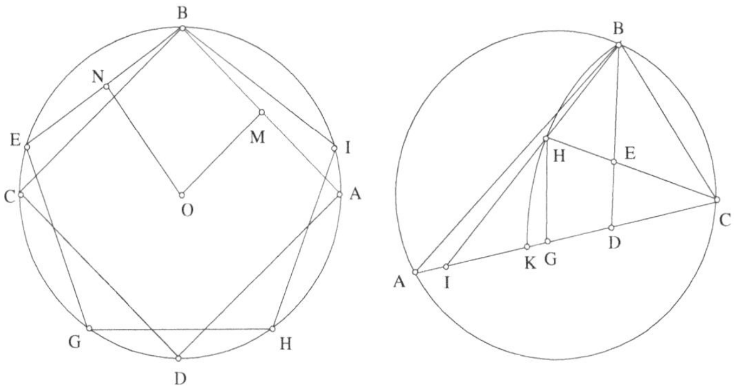

Fig. III.3

<Proof of Proposition 3> Having proved this, let there now be a circle *ABC* in which are inscribed two regular polygons, one of which has more sides than the other. Let these polygons be *ABCD* and *BEGHI*.

I say that the area of the figure BEGHI *is greater than the area of the figure* ABCD *and that its perimeter is greater than the perimeter of the other.*

Proof: There cannot exist in a circle any regular polygon with fewer sides than a triangle, each of whose sides subtends one third of the circle, and after the triangle, there cannot exist any regular polygon having fewer sides than a square, each of whose sides subtends one quarter of the circle. It is therefore impossible for there to exist in the same circle, two regular polygons whose respective sides subtend <arcs> on the circle that are greater than one third or one quarter of it. Let there be two regular polygons inscribed within a circle. Their respective sides will each subtend

[9] This is not obvious. See the mathematical commentary.

an arc[10] on the circle that is less than two thirds of the circle. <The sum of> the two arcs *AB* and *BE* is less than two thirds of the circle, and the arc *AB* is greater than the arc *BE*. Therefore, the ratio of the arc *AB* to the arc *BE* is greater than the ratio of the chord *AB* to the chord *BE*, and the ratio of the arc *AB* to the arc *BI* is greater than the ratio of the chord *AB* to the chord *BI*.[11] Ptolemy has proved this ratio in the first Book of the *Almagest* using a method different from that described here. We have taken up the proof of this ratio so that the notion, introduced here for the purposes of this lemma, should be clear without having first to read the *Almagest*.[12] The ratio of the arc *AB* to the arc *BI* is compounded of the ratio of the arc *AB* to the circumference of the circle and the ratio of the circumference of the circle to the arc *BI*. But the ratio of the arc *AB* to the circumference of the circle is equal to the ratio of the side *AB* to the perimeter of the figure *ABCD*, as the arcs and the sides are equal in number, and the ratio of the circumference of the circle* to the arc *BI* is equal to the ratio of the perimeter of the figure *BEGHI* to the side *BI*. The ratio of the arc *AB* to the arc *BI* is therefore compounded of the ratio of the side *AB* to the perimeter of the figure *ABCD* and the ratio of the perimeter of the figure *BEGHI* to the side *BI*. But the ratio of the arc *AB* to the arc *BI* is greater than the ratio of the side *AB* to the side *BI*. Therefore, the ratio compounded of the ratio of *AB* to the perimeter of the figure *ABCD* and the ratio of the perimeter of the figure *BEGHI* to the side *BI* is greater than the ratio of *AB* to *BI*. But the ratio of *AB* to *BI* is compounded of the ratio of *AB* to the perimeter of *ABCD* and the ratio of the perimeter of *ABCD* to *BI*. The ratio compounded of the ratio of *AB* to the perimeter of *ABCD* and the ratio of the perimeter of *BEGHI* to *BI* is therefore greater than the ratio compounded of the ratio of *AB* to the perimeter of *ABCD* and the ratio of the perimeter of *ABCD* to *BI*. Eliminating the ratio of *AB* to the perimeter of *ABCD*, which is common to both, there remains the ratio of the perimeter of *BEGHI* to *BI* greater than the ratio of the perimeter of *ABCD* to *BI*. Therefore, the perimeter *BEGHI* is greater than the perimeter *ABCD*.

But the perpendicular drawn from the centre of the circle* to the straight line *BI* is greater than the perpendicular drawn from the centre of the straight line *AB*, as *BI* is less than *AB*. But the product of the perpendicular drawn from the centre to *BI* and one half of the perimeter of the figure *BEGHI* is the area of this figure, and the product of the perpendicular drawn from the centre to the straight line *AB* and one half of

[10] That is, two arcs whose sum is less.
[11] From the lemma.
[12] Implying: without needing to look it up in the *Almagest*.
* The two asterisks indicate the repeated paragraphs in the Tehran manuscript.

the perimeter of the figure *ABCD* is the area of that figure. Therefore, the area of the figure *BEGHI* is greater than the area of the figure *ABCD* and the perimeter of the figure *BEGHI* is greater than the perimeter of the figure *ABCD*. That is what we wanted to prove.

<The sphere: The greatest of all the solid figures>

<4> *I also say that for any sphere whose lateral area is equal to the area of a regular polyhedron,*[13] *the volume of the sphere is greater than that of the regular polyhedron.*

<Lemmas> We first introduce the following lemmas. The eminent Archimedes has shown in his book *On the Sphere and the Cylinder*, that a sphere is two thirds of the cylinder whose base is the great circle inscribed within the sphere and whose height is the diameter of the sphere, and that the area of the sphere is four times that of the great circle inscribed within the sphere, and that the volume of the cylinder is the product of its height and its base. It necessarily follows from this that the product of the diameter of the sphere and two thirds of the great circle inscribed within the sphere is the volume of the sphere, and that the product of the half-diameter of the sphere and one and one third times the great circle inscribed within the sphere is its area, but one and one third times the great circle inscribed within the sphere is one third of the entire area of the sphere, as the area of a sphere is four times that of the great circle inscribed within the sphere. From all that, it is necessary that the volume of the sphere is the product of it half-diameter and one third of its area.

Any regular polyhedron inscribed within a sphere is such that, if planes are drawn from the centre of the sphere passing[14] through the sides of one of its bases, then these planes divide off a sector from the sphere whose ratio to the entire sphere is equal to the ratio of the spherical surface at the base of this sector to the entire surface of the sphere, and is also equal to the ratio of the solid angle, that is the angle at the centre of the sphere which is surrounded by the surfaces of a regular pyramid[15] whose lines are straight and whose base is one of the bases of the polyhedron, to the eight solid right angles which is the sum of all the solid angles at the centre of the sphere and

[13] Lit.: a solid figure whose bases are equal, whose sides are equal, and which are similar. From now on, we shall translate this expression as 'regular polyhedron'.

[14] Lit.: planes to the sides.

[15] Lit.: rectangular cone.

which are also at the centre of any regular polyhedron, as the sphere and the surface of the sphere are divided by these planes into equal parts.

As for the angles, if a great circle is drawn within the sphere, and if two diameters are drawn within this circle which cross each other at right angles, and if a perpendicular to this plane is drawn passing through its centre, and if this is extended on both sides until it meets the surface of the sphere, and if <perpendicular> straight lines are drawn from these two extremities onto the extremities of the two diameters, then they form eight equal pyramids within the sphere whose vertices are at the centre of the sphere and whose angles at the vertices are equal. Each of these angles is called a 'solid right angle', and the sum of these angles is equal to the sum of the angles[16] of any polyhedron inscribed within the sphere.

It necessarily follows that the product of the half-diameter of the sphere and one third of the area of the spherical surface forming the base of the spherical sector is the volume of the spherical sector.

Having proved this, now let there be a sphere A and a regular polyhedron B such that, regardless of the figure of the polyhedron, and regardless of the figure of its bases, the lateral area of the polyhedron is equal to the area of the sphere A.

I say that the volume of the sphere is greater than that of the polyhedron.

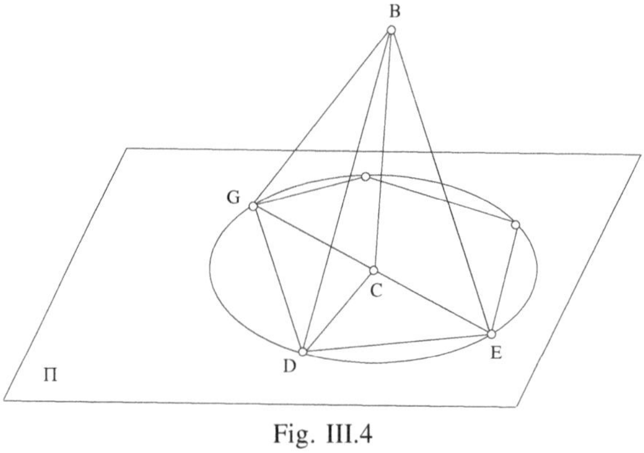

Fig. III.4

Proof <of Proposition 4>: Any regular polyhedron is inscribed within a sphere. Let the point B be the centre of the sphere in which is inscribed the polyhedron B, and let one of the bases of the polyhedron B be the plane DGE. We join the straight lines BG, BE and BD, which are equal. From the

[16] The sum of the solid angles whose vertices are at the centre of the polyhedron.

point *B*, we draw a perpendicular to the plane *DGE*, and let this be *BC*. The point *C* will be at the centre of the figure *DGE* as the point *C* is the centre of the circle circumscribing the figure *DGE*, and which is on the sphere circumscribing the polyhedron *B*. We imagine a sphere whose half-diameter is *BC*. This sphere is tangent to all the bases of the figure *B* as the perpendiculars drawn from the point *B* onto all the bases of the figure are equal as the figure *B* is inscribed within a sphere, and therefore its equal bases are inscribed within equal circles whose planes intersect the sphere. Therefore, the straight lines drawn from the centre of the sphere to the centres of these circles are equal. But they are perpendicular to their plane, and the faces of the pyramid *BCEG* cut off from the sphere[17] a section which is inside this pyramid. Therefore, the pyramid *BCEG* is greater than the spherical sector which is internal to it and which is a part of the sphere, as the pyramid is a part of the polyhedron. But the product of *BC* and one third of the area of *DEG* is equal to the volume of the pyramid, and the product of *BC* and one third of the spherical surface which is the base of the spherical sector, and which lies within the regular pyramid, is the volume of this sector as the ratio of the surface area of the sector to the entire surface area of the sphere is equal to the ratio of the sector to the entire sphere. The ratio of the product of the half-diameter of the sphere and one third of the base of the spherical sector to the product of the half-diameter of the sphere and one third of the entire surface area of the sphere is therefore equal to the ratio of the spherical sector to the entire sphere. But the product of the half-diameter of the sphere and one third of the entire surface area of the sphere is the volume of the sphere. Therefore, the product of the half-diameter of the sphere and one third of the base of the spherical sector is the volume of the spherical sector. The product of the straight line *BC* and one third of the surface *DEG* is therefore greater than the product of *BC* and one third of the base of the spherical sector. The surface *DEG* is therefore greater than the spherical surface which is the base of the <spherical> sector. Similarly, each base of the polyhedron is greater than the spherical surface cut off by the faces of the pyramid drawn from the centre of the sphere to the sides of this base.

We have therefore shown that the surface enclosing the polyhedron *B* is greater than the surface area of the sphere inscribed within this polyhedron whose half-diameter is the straight line *BC*. But the surface enclosing the polyhedron *B* is equal to the surface area of the sphere *A*. Therefore, the surface area of the sphere *A* is greater than the surface area of the sphere whose half-diameter is the straight line *BC*. Therefore, the half-diameter of the sphere *A* is greater than the straight line *BC*. The product of the half-

[17] This refers to this entire section of the inscribed sphere.

diameter of the sphere A and one third of the area of the sphere A is equal to the volume of the sphere A, and the product of the straight line BC and one third of the area of the surface enclosing the polyhedron B is the volume of the polyhedron B. But one third of the area of the sphere A is equal to one third of the area of the surface enclosing the polyhedron B, and the half-diameter of the sphere A is greater than the straight line BC. Therefore, the volume of the sphere A is greater than the volume of the polyhedron B. That is what we wanted to prove.

<**Proposition 5**> We also say that, given two regular polyhedra, the bases of one of which are similar to the bases of the other and the bases of one being more numerous than the bases of the other, this being true for polyhedra whose bases are equilateral triangles,[18] and if the surface enclosing one is equal to the surface enclosing the other, i.e. the sum of the bases of one is equal to the sum of the bases of the other, the volume of the polyhedron with the most bases is greater than the volume of the other.

We also say that, given two regular polyhedra, the bases of one of which are similar to the bases of the other and the bases of one being more numerous than the bases of the other, and which are inscribed within the same sphere, then the surface enclosing the polyhedron with the larger number of bases is greater than the surface enclosing the other, and the volume of the polyhedron with the larger number of bases is greater than the volume of the other.

We first introduce the following lemmas.

<**Lemma**> Let there be two pyramids located within the <same> sphere with their vertices at the centre of the sphere. Then the ratio of the angle[19] of one of the pyramids to the angle[20] of the other pyramid is equal to the ratio of the portion of the surface of the sphere intercepted[21] by the angle of one of the pyramids to the portion of the surface of the sphere intercepted by the angle of the other pyramid, and is equal to the ratio of the spherical sector whose base is one portion of the surface of the sphere to the spherical sector whose base is the other portion.

Proof: If we take equal multiples, regardless of what they are, for each of the two pyramids, then these multiples divide the sphere into equal sectors whose angles are equal and whose spherical surfaces are equal.

[18] i.e. the regular tetrahedron, octahedron and icosahedron.

[19] Solid angle.

[20] Solid angle.

[21] Lit.: the portion of the sphere that subtends the angle of the pyramid. We shall translate similar expressions in the same way from now on.

Therefore, if the portion of the sphere separated out by the whole multiple of one of the two pyramids was greater than the portion separated out by the whole multiple of the other pyramid, then the angle which is the multiple of the angle of one of the pyramids will be greater than the angle which is the multiple of the angle of the other pyramid, and the portion of the surface of the sphere intercepted by the larger angle will be greater than the portion of the surface of the sphere intercepted by the smaller angle. If the portion of the sphere separated out by the multiple of one of the pyramids was less than the portion of the sphere separated out by the multiple of the other pyramid, then the angle of one of the two portions will be less than the angle of the other portion, and the spherical surface intercepted by the smaller angle will be less than the spherical surface intercepted by the larger angle. And if one of the two portions is equal to the other, then one of the angles will be equal to the other, and one of the surfaces will be equal to the other. Therefore, if the angle which is the multiple of the angle of one of the pyramids were greater than the angle which is the multiple of the angle of the other pyramid, then the first spherical surface would be greater than the second spherical surface and the first sector, which is a portion of the sphere, would be greater than the other sector which is another portion of the sphere. And if the angle which is the multiple <of the angle of the first pyramid> were less than the angle which is the multiple <of the angle of the second pyramid>, then the first spherical surface would be less than the second spherical surface and the first sector would be less than the second sector. And if the angle which is the multiple of the first were equal to the angle which is the multiple of the angle of the second, then the first spherical surface would be equal to the second spherical surface and the first sector would be equal to the second sector. And if the multiple <of the pyramid> cannot be taken in a single sphere, then the multiple <of the pyramid> may be taken in two or more spheres.

The proof proceeds in the same way as in the case in which the spheres are equal. Therefore, the ratio of the angle of one of the two pyramids to the angle of the other pyramid is equal to the ratio of the spherical surface intercepted by the first angle to the spherical surface intercepted by the other angle, and is equal to the ratio of the first sector to the second sector. That is what we wanted to prove.

<Lemma 6> Let there be a pyramid with a triangular base such that one of its sides drawn from its vertex to one of the angles of its base encloses, with each of the two sides of the angle to which it was drawn, an angle which is not less than a right angle, and such that if a straight line is drawn from its vertex to one of the two sides of the base already mentioned and cuts this side, and if a straight line is drawn from the extremity of this straight line to the extremity of the other side separating from the base a triangle which is part of the base, and separating from the pyramid a pyramid which is part of the pyramid, and if one of the two angles of the triangle so formed – which are on the base – is not less than a right angle, then the ratio of the base of the larger pyramid to the base of the smaller pyramid is greater than the ratio of the angle of the larger pyramid to the angle of the smaller pyramid.

Example: Let there be a pyramid *ABCD* whose vertex is at the point *A* and whose base is the triangle *CBD*, and let the side *AB* enclose with each of the sides *BC* and *BD* an angle which is not less than a right angle. A straight line *AE* is drawn from the vertex and the straight line *EC* is joined, forming a pyramid *ABEC*, and one of the two angles *AEC* and *ACE* is not less than a right angle.

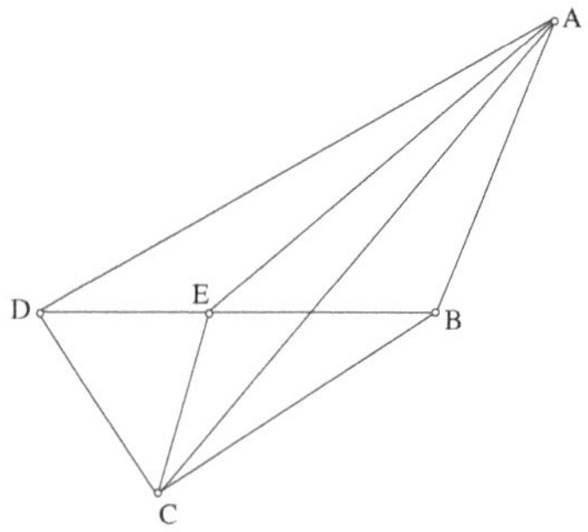

Fig. III.5

I say that the ratio of the triangle DBC *to the triangle* EBC *is greater than the ratio of the angle of the pyramid* ABCD, *which is at the point* A, *to the angle of the pyramid* ABEC *which is at the point* A.

Proof: We set *A* as the centre and *AB* as the radius, we draw a portion of a sphere. This will generate an arc of a circle in each of the planes of the pyramid. Let the arc formed in the plane *ABC* be the arc *BH*, the arc formed in the plane *ABD* be the arc *BLI*, the arc formed in the plane *ACD* be the arc *HI*, and the arc formed in the plane *ACE* be the arc *HGL*. We join the straight line *BL* and we extend it. It will meet the straight line *AD* as the angle *BAD* is acute, and the angle *ABL* is acute as *AB* is equal to *AL*. Let the point of meeting be the point *K*. The point *K* is therefore in the plane of the triangle *ACD* below the point *I*, and any straight line drawn from the point *B* to a point on the arc *HGL*, if extended, will meet the plane *ACD* as the plane in which this straight line and the straight line *AB* both lie cuts the plane *ACE* and forms with it a straight line which makes an acute angle with the straight line *AB*. Now, we imagine a cone whose vertex is at *B* and whose base is the sector *IHGL*, and we imagine it being extended in the direction of its base. It will therefore cut the plane *ACD*, forming with it a curved line whose extremities lie on the two points *H* and *K*. Let this line be *HFK*. This line will lie outside the spherical surface as any line drawn from the point *B* to a point on the arc *LGH*, if extended, will lie outside the spherical surface. It is for this reason that the spherical sector whose vertex is at the point *A* and whose base is the spherical surface bounded by the arcs *HI*, *HL* and *LI* lies within the surface of the circular pyramid[22] *AHIL*, and the circular pyramid *AHLB* lies within the spherical sector whose vertex is at the point *A* and whose base is the spherical surface bounded by the arcs *BH*, *BL* and *LH*. The ratio of the first pyramid to the second pyramid is greater than the ratio of the first spherical sector to the second spherical sector. Composing, the ratio of the circular pyramid *AHFKB* whose vertex is at the point *B* and whose base is *AHFK* to the circular pyramid *AHGLB* whose vertex is at the point *B* and whose base is the sector *AHGL* is greater than the ratio of the spherical sector whose vertex is at the point *A* and whose angle is the angle of the pyramid *ABCD* which is at the point *A* to the spherical sector whose vertex is at the point *A* and whose angle is the angle of the pyramid *ABCE* which is at the point *A*.

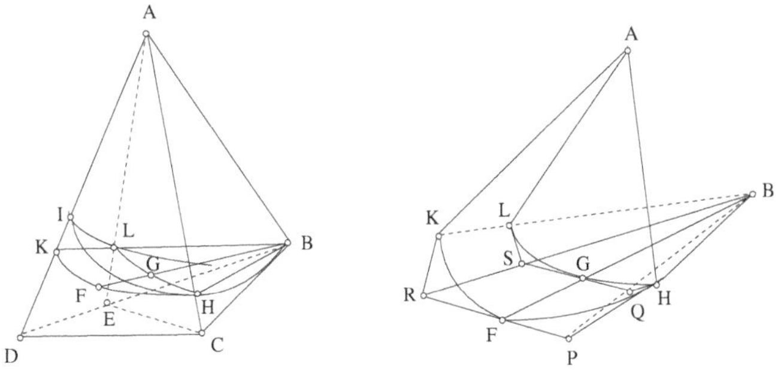

Fig. III.6

I say that the ratio of the triangle AEC *to the sector* ALGH *is not greater than the ratio of the triangle* ADC *to the sector* AKFH.

It could not be otherwise. If it were possible, then let the ratio of the triangle *AEC* to the sector *ALGH* be greater than the ratio of the triangle *ADC* to the sector *AKFH*. The ratio of the triangle *ADC* to the sector *AKFH* would then be equal to the ratio of the triangle *AEC* to a surface

[22] Although the word ‘circular’ does not appear in the text, we have included it in the translation whenever the outline of the base includes at least one arc.

greater than the sector *ALGH*. Let this surface be the surface L_a, and let the amount by which the surface L_a exceeds the sector *ALGH* be the surface *J*. It would then be possible to construct on the arc *LGH* a polygonal figure[23] whose sides are tangents to the arc *LGH*, such that the surface area by which it exceeds the sector *ALGH* is less than the surface *J*. Let this figure be that whose sides are the lines *LS*, *SQ* and *QH*, and let these lines be tangents to the arc *LGH* at the points *L*, *G* and *H*. The straight line drawn from the point *B* to the point *G* and extended reaches the curved line *HFK* at the point *F*. We join the straight lines *BH*, *BL*, *BS* and *BQ* to form a pyramid whose vertex is *B* and whose base is the polygonal figure *ALSQH*. If the faces of the pyramid *BALSQH* are extended, they meet the plane *ACD* forming straight lines tangential to the line *HFK* at the points *H*, *F* and *K*, as the faces of the polygonal pyramid only meet the faces of the circular pyramid[24] at the straight lines *BL*, *BG* and *BH*.[25] The bases[26] of this pyramid, which lie in the plane *ACD*, only meet the line *HFK* at a single point. Let these bases[27] be the straight lines *HP*, *PF*, *FR* and *RK*, and let the points of contact be the points *H*, *F* and *K*. The two points *P* and *R* lie outside the circular pyramid. We join the straight line *AS* and we extend it until it reaches the straight line *EC* at the point *N*. We join the straight line *AQ* and extend it until it reaches the straight line *EC* at the point *U*. We join *AR* and we extend it until it reaches the straight line *DC* at the point *V*. We join *AP* and we extend it until it cuts the straight line *DC* at the point *O*. As the straight line *BSR* cuts the straight lines *AR*, *AS* and *AB*, these straight lines must lie in the same plane. We then introduce *BNV*, which is in the plane of the triangle *CBD*, into the plane of these straight lines. The points *B*, *N* and *V* therefore lie on a straight line. We join this straight line and let it be *BNV*.

[23] The area of the polygonal figure defined in this way depends on the choice of the point *G* on the arc *LH*.

[24] This is the circular pyramid *BAHGL*.

[25] Remember that the straight line *BA* is an edge common to the two solids *BALSQH* and *BAHGL*.

[26] The sides of the base.

[27] The sides of the base.

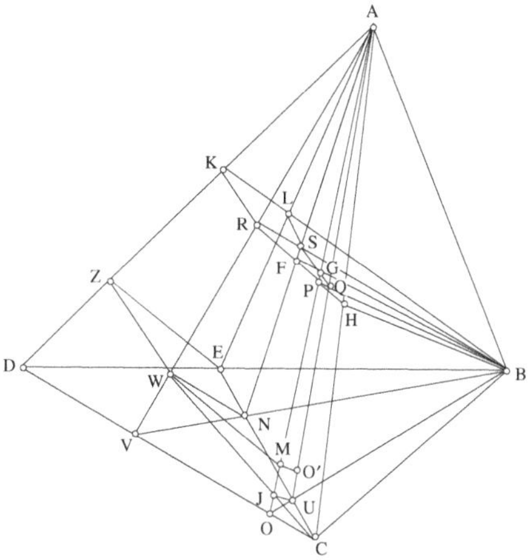

Fig. III.7

Similarly, we can show that the line *BUO* is also a straight line.

First, let the angle *AEC* be a right angle. The straight line *EC* will then be parallel to the straight line *LS*. From the point *N*, we draw a straight line parallel to the straight line *BSR*. This will cut the straight line *AV* between the two points *R* and *V* as it lies within the triangle *BRV*. Let this point be the point *W*. From the point *E*, draw a straight line parallel to the straight line *BLK*. This will cut the straight line *AD* between the points *K* and *D*. Let this point be the point *Z*. We join *WZ*. As *EZ* is parallel to the straight line *LK*, the ratio of *EA* to *AL* is equal to the ratio of *ZA* to *AK*. As *EN* is parallel to the straight line *LS*, the ratio of *EA* to *AL* is equal to the ratio of *NA* to *AS*. And, as *NW* is parallel to *SR*, the ratio of *NA* to *AS* is equal to the ratio of *WA* to *AR*. The ratio of *WA* to *AR* is therefore equal to the ratio of *ZA* to *AK*, and the straight line *WZ* is therefore parallel to the straight line *RK*. The ratio of the triangle *AWZ* to the triangle *ARK* is therefore equal to the ratio of the triangle *ANE* to the triangle *ASL*, as the square of the ratio of *ZA* to *AK* is equal to the square of the ratio of *EA* to *AL*. The ratio of the triangle *AVD* to the triangle *ARK* is therefore greater than the ratio of the triangle *ANE* to the triangle *ASL*.

From the point *N*, we draw the straight line *NO′* parallel to the straight line *SQ*. It will cut the straight line *AU* between the points *Q* and *U* as the angle *ANU* is obtuse and the angle *ASQ* is acute. Let this point be the point *O′*. From the point *O′*, draw the straight line *O′M* parallel to the straight line *BQP*. It will cut the straight line *AO*. Let this point be the point *M*. We join *MO′*. The ratio of *MA* to *AP* is equal to the ratio of *O′A* to *AQ*, the ratio of *O′A* to *AQ* is equal to the ratio of *NA* to *AS*, and the ratio of *NA* to *AS* is equal to the ratio of *WA* to *AR*. Therefore, the ratio of *WA* to *AR* is equal to the ratio of *MA* to *AP*, and therefore the straight line *MW* is parallel to the straight line *PR*, and the ratio of the triangle *AWM* to the triangle *APR* is equal to the ratio of the triangle *ANO′* to the triangle *ASQ*.

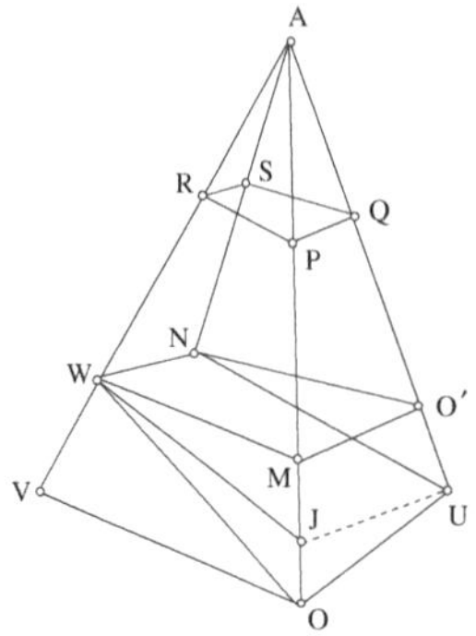

Fig. III.8

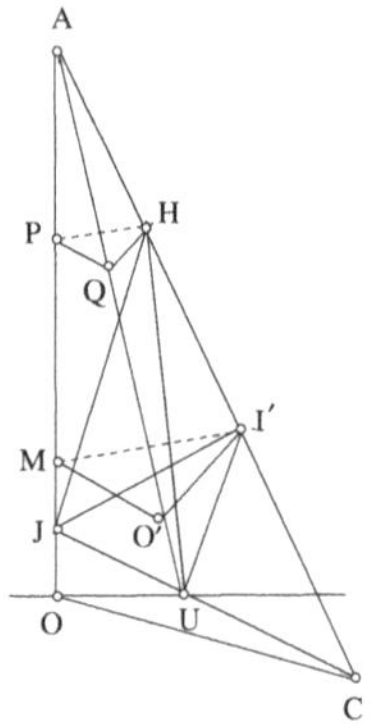

Fig. III.9

From the point *U*, we draw the straight line *UJ* parallel to the straight line *PQ*. It will cut the straight line *AO* between the points *P* and *O*. Let this point be the point *J*. Then the point *J* will lie between the points *M* and *O*, as *UJ* is parallel to *PQ*. We join *JW*. Then the ratio of the triangle *AWJ* to the triangle *AWM* is equal to the ratio of *JA* to *AM*, the ratio of *JA* to *AM* is equal to the ratio of *UA* to *AO′*, and the ratio of *UA* to *AO′* is equal to the ratio of the triangle *ANU* to the triangle *ANO′*. Therefore, the ratio of the triangle *AWJ* to the triangle *AWM* is equal to the ratio of the triangle *ANU* to the triangle *ANO′*, and the ratio of the triangle *ANO′* to the triangle *ASQ* is equal to the ratio of the triangle *AWM* to the triangle *APR*. Therefore, the ratio of the triangle *AWJ* to the triangle *APR* is equal to the ratio of the triangle *ANU* to the triangle *ASQ*. Therefore, the ratio of the triangle *AVO*[28] to the triangle *APR* is greater than the ratio of the triangle *ANU* to the triangle *ASQ*. From the point *O′*, we draw a perpendicular to the straight line *AC* and let this be *O′I′*. The point *I′* lies between the two points *A* and *C*[29] as the angle *AEC* is a right angle. The straight line *O′I′* is parallel to the straight line *QH* as the angle *AHQ* is a right angle as *HQ* is a tangent. We join *I′M*, *I′J* and *I′U*. Then the ratio of *MA* to *AP* is equal to the ratio of *O′A* to *AQ*. But the ratio of *O′A* to *AQ* is equal to the ratio of *I′A* to *AH*. Therefore, the ratio of *I′A* to *AH* is equal to the ratio of *MA* to *AP*.

[28] The triangle *AVO* is greater than *AWJ* as *J* lies between *M* and *O*, and *W* lies between *R* and *V*.

[29] *I′* also lies between *C* and *H*.

Therefore, the straight line *I′M* is parallel to the straight line *HP*. The ratio of the triangle *AMI′* to the triangle *APH* is equal to the ratio of the triangle *AO′I′* to the triangle *AQH*, the ratio of the triangle *AJI′* to the triangle *AMI′* is equal to the ratio of *JA* to *AM*, the ratio of *JA* to *AM* is equal to the ratio of *UA* to *AO′*, and the ratio of *UA* to *AO′* is equal to the ratio of the triangle *AUI′* to the triangle *AO′I′*. Therefore, the ratio of the triangle *AJI′* to the triangle *APH* is equal to the ratio of the triangle *AUI′* to the triangle *AQH*.

Return to the triangle *APH* and join *JC*. The ratio of the triangle *AJC* to the triangle *AJI′* is then equal to the ratio of *CA* to *AI′* and the ratio of *CA* to *AI′* is equal to the ratio of the triangle *ACU* to the triangle *AUI′*, and the ratio of the triangle *AUI′* to the triangle *AQH* is equal to the ratio of the triangle *AJI′* to the triangle *APH*. Therefore, the ratio of the triangle *AJC* to the triangle *APH* is equal to the ratio of the triangle *AUC* to the triangle *AQH*. The ratio of the triangle *AOC* to the triangle *APH* is therefore greater than the ratio of the triangle *AUC* to the triangle *AQH*. Therefore,[30] the ratio of the entire triangle *ADC* to the polygon *AKRPH* is greater than the ratio of the triangle *AEC* to the polygon *ALSQH*. But the ratio of the triangle *AEC* to the polygon *ALSQH* is greater than the ratio of the triangle *ADC* to the sector *AHFK*. Therefore, the ratio of the triangle *ADC* to the polygon *AKRPH* is greater than the ratio of the triangle *ADC* to the sector *AHGL*, and therefore the polygon *AKRPH* is less than the sector *AHFK*, which is impossible. Therefore, the ratio of the triangle *AEC* to the sector *AHGL* is not greater than the ratio of the triangle *ADC* to the sector *AHFK*. But the ratio of the triangle *AEC* to the sector *AHGL* is equal to the ratio of the pyramid *ABEC*, whose vertex is at the point *B*, to the pyramid *AHGLB* whose vertex is at the point *B*. The ratio of the pyramid *ABEC*, whose vertex is at the point *B*, to the pyramid *AHGLB*, whose vertex is at the point *B* and whose base is the sector *AHGL*, is not greater than the ratio of the pyramid *ABCD*, whose vertex is at the point *B*, to the pyramid *AHFKB* whose vertex is at the point *B* and whose base is the sector *AHFK*.

The ratio of the pyramid *ABCD* to the pyramid *AHFKB* is therefore either equal to the ratio of the pyramid *ABCE* to the pyramid *AHGLB*, or it is greater than it. By permutation, the ratio of the pyramid *ABCD* to the pyramid *ABCE* is therefore either equal to the ratio of the pyramid *ABHFK* to the pyramid *ABHGL*, or it is greater than it. But the ratio of the pyramid *ABHFK* to the pyramid *AHGLB* is greater than the ratio of the spherical sector, whose angle is the angle of the pyramid *ABCD* which is at the point *A*, to the spherical sector whose angle is the angle of the pyramid *ABCE* which is at the point *A*. The ratio of the pyramid *ABCD* to the pyramid *ABCE* is therefore greater than the ratio of the <first> spherical sector to

[30] See the mathematical commentary.

the <second> spherical sector. But the ratio of the pyramid *ABCD* to the pyramid *ABCE* is equal to the ratio of the triangle *CBD* to the triangle *CBE*, and the ratio of the larger spherical sector to the smaller spherical sector is equal to the ratio of the angle of the pyramid *ABCD* to the angle of the pyramid *ABCE*, both of which angles are at the point *A*. The ratio of the triangle *CBD* to the triangle *CBE* is greater than the ratio of the angle of the pyramid *ABCD* which is at the point *A* to the angle of the pyramid *ABCE* which is at the point *A*.

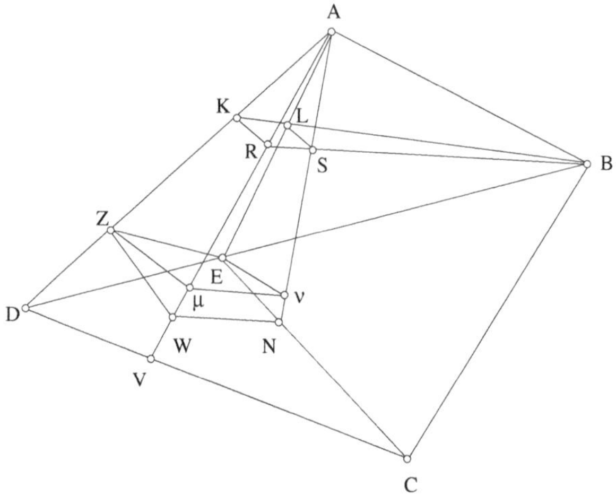

Fig. III.10

If the angle *AEC* is obtuse, we make the angle $AE\nu$ a right angle and, from the point ν, we draw the straight line $\nu\mu$ parallel to the straight line *BSR*. This will then be parallel to the straight line *NW*. The ratio of *NA* to $A\nu$ is equal to the ratio of *WA* to $A\mu$. We join $Z\mu$. The ratio of the triangle *AZW* to the triangle $AZ\mu$ is then equal to the ratio of the triangle *AEN* to the triangle $AE\nu$, and the ratio of μA to *AR* is equal to the ratio of νA to *AS*, and the ratio of νA to *AS* is equal to the ratio of *EA* to *AL*. The ratio of *EA* to *AL* is equal to the ratio of *ZA* to *AK*. Therefore, the ratio of *ZA* to *AK* is equal to the ratio of μA to *AR*, and hence the straight line $Z\mu$ is parallel to the straight line *KR*, and the ratio of the triangle *AZE* to the triangle *AKL* is equal to the ratio of the triangle $AZ\mu$ to the triangle *AKR*, and the ratio of the triangle $AZ\mu$ to the triangle *AKR* is equal to the ratio of the triangle $AE\nu$ to the triangle *ALS*. But the ratio of the triangle *AZW* to the triangle $AZ\mu$ was equal to the ratio of the triangle *AEN* to the triangle $AE\nu$. Therefore, the ratio of the triangle *AZW* to the triangle *AKR* is equal to the ratio of the triangle *AEN* to the triangle *ALS*. But the triangle *ADV* is greater than the triangle *AZW*. Therefore, the ratio of the triangle *ADV* to the triangle *AKR* is greater than the ratio of the triangle *AEN* to the triangle *ALS*.

From the point *v,* we draw a straight line parallel to the straight line *SQ* and we then follow the proof as described above. It can be shown that the ratio of the triangle *ADC* to the polygon lying within it is greater than the ratio of the triangle *AEC* to the polygon lying within it.

If the right angle or obtuse angle is the angle *ACE*, then we begin the construction from the point *C* and the proof proceeds as described above, as if the angle *ACE* is a right angle, then the straight line *CE* is parallel to the straight line *HQ*, and if it is obtuse, then the parallel to *HQ* cuts the angle *ACE*.

In all cases, if one of the two angles *AEC* or *ACE* is not less than a right angle, then the ratio of the triangle *CBD* to the triangle *CBE* is greater than the ratio of the angle of the pyramid *ABCD*, which is at the point *A*, to the angle of the pyramid *ABCE* which is at the point *A*. That is what we wanted to prove.

<Lemma 7> Let there be a pyramid whose base is a triangle, one of whose angles is not less than a right angle, and one of whose sides drawn from the vertex to one of the two acute angles of the triangle is perpendicular to the plane of the base. If a straight line is drawn from the vertex cutting the side of the base intercepted by the acute angle to which was drawn the perpendicular, and if the extremity of this line is joined to the foot of the perpendicular, dividing the base into two triangles and dividing the pyramid into two pyramids, then the ratio of the larger triangle to the smaller triangle, which is adjacent to the larger angle, is greater than the ratio of the angle of the larger pyramid to the angle of the smaller pyramid which is adjacent to the larger angle.

Example: Let the pyramid be *ABCD*, with its vertex at the point *A* and with its base being the triangle *BCD*, such that the angle *BCD* is not less than a right angle and the side *AB* is perpendicular to the plane of the base. From the point *A*, draw the straight line *AE* and join *EB*.

I say that the ratio of the triangle DBC *to the triangle* EBC *is greater than the ratio of the angle of the pyramid* ABCD *to the angle of the pyramid* ABCE.

Proof: We draw the perpendicular *AB* in the direction of *B* and we construct a right angle on the straight line *AE* and let this be *AEG*. The straight line *EG* will meet the straight line *AB* as the angle *EAB* is acute. Let this point be the point *G*. We join *CG* and *DG*. As the angle *BCD* is not less than a right angle, the straight line *EB* will be greater than the straight line *BC*, and as *AB* is perpendicular to the base, *AE* will be greater than *AC*. If

BC is drawn in the direction of *C*, cutting off <a straight line> equal to *BE*, and if its two extremities are joined to the two points *A* and *G*, it forms a right angle equal to the angle *AEG*. Therefore, the angle *ACG* must be obtuse as it lies within the right angle. But, as the angle *BCD* is not less than a right angle and the plane *ABC* is perpendicular to the plane *BCD*, the angle *ACD* cannot be less than a right angle. This is because, if the angle *BCD* were a right angle, then *DC* would be perpendicular to the plane *ABC* and the angle *ACD* would be a right angle, and if the angle *BCD* were obtuse, the straight line *CD* would be beyond the perpendicular drawn from the point *C* onto the plane *ABC* and the angle *ACD* would be obtuse.[31]

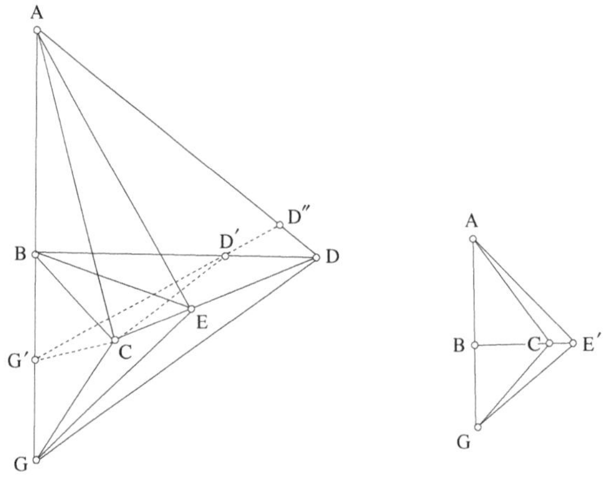

Fig. III.11

In all cases, the angle *ACD* cannot therefore be less than a right angle. But the angle *ACG* is obtuse, and therefore the pyramid *ACDG*, whose vertex is at the point *A* and whose base is the triangle *CGD*, is such that the side *AC* drawn from its vertex to the angle *GCD* encloses with each of the two straight lines *CB* and *CD* an angle that is not less than a right angle, and such that the straight line *AE* is drawn from its vertex to one of the two sides of the base which enclose the angle *GCD*, and a straight line *EG* is drawn from the extremity of this line that lies at the point *E*, the angle *AEG* being a right angle; therefore the ratio of the triangle *GCD* to the triangle *GCE* is greater than the ratio of the angle of the pyramid *ACGD*, which is at the point *A*, to the angle of the pyramid *AGCE* which is at the point *A*. But the ratio of the triangle *GCD* to the triangle *GCE* is equal to the ratio of *DC* to *CE*, and the ratio of *DC* to *CE* is equal to the ratio of the triangle *DBC* to the triangle *EBC*. But the angle of the pyramid *AGCD*, which is at the point

[31] See the mathematical commentary.

A, is the angle of the pyramid *ABCD*, and the angle of the pyramid *AGCE*, which is at the point *A*, is the angle of the pyramid *ABCE*. The ratio of the triangle *DBC* to the triangle *EBC* is therefore greater than the ratio of the angle of the pyramid *ABCD*, which is at the point *A*, to the angle of the pyramid *ABCE*, which is at the point *A*. That is what we wanted to prove.

<Lemma 8> Let there be a pyramid whose base is an isosceles triangle and whose side, drawn from its vertex to the vertex of the isosceles triangle, is perpendicular to the plane of the base. If a plane passing through its vertex cuts its base along a straight line parallel to the side of the base intercepted by the angle lying at the foot of the perpendicular, and if a pyramid is separated from the pyramid of which it forms a part, then the ratio of the base of the larger pyramid to the base of the smaller pyramid is greater than the ratio of the angle of the larger pyramid to the angle of the smaller pyramid, both of which angles are at the vertex of the pyramid.

Example: Let the pyramid be *ABCD*, with its vertex at the point *A* and with its base being the triangle *BCD*. The plane *AEG* passes through its vertex and cuts the triangle *BCD* along the straight line *EG* such that *EG* is parallel to the straight line *CD*.

I say that the ratio of the triangle CBD *to the triangle* EBG *is greater than the ratio of the angle of the pyramid* ABCD, *which is at the point* A, *to the angle of the pyramid* ABEG *which is at the point* A.

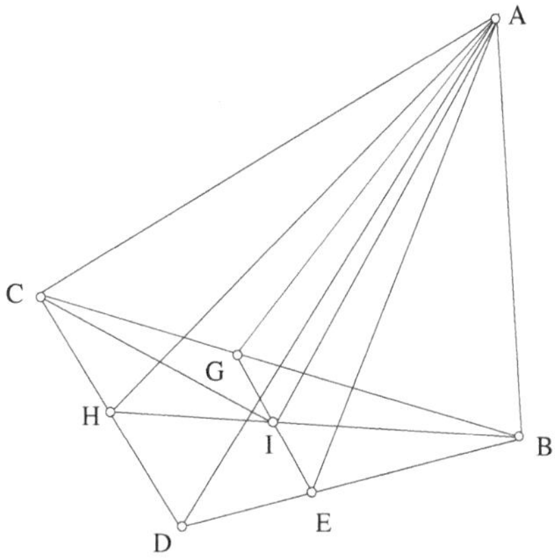

Fig. III.12

Proof: We divide *EG* into two halves at the point *I* and we join *BI*. This will be perpendicular to the straight line *EG*. Extend *BI* to *H*, dividing the straight line *CD* into two halves at the point *H*. Join the straight lines *AI*, *AH* and *CI*. As *AB* is perpendicular to the plane of the triangle *BCD*, the plane

ABC is perpendicular to the plane of the triangle *CBD*. As the plane *ABH* is perpendicular to the plane of the triangle *CBD* and the straight line *IG* is perpendicular to the straight line *BH* which is the intersection of the two planes, *IG* is perpendicular to the plane *ABH*. The angle *AIG* is therefore a right angle, and the angle *AIH* is obtuse. As *ABHC* is a pyramid and the side *AB* is perpendicular to the two straight lines *GC* and *BH* – the straight line *AI* has been drawn, *IC* has been joined, and the angle *AIC* is obtuse – the ratio of the triangle *BCH* to the triangle *BCI* is greater than the ratio of the angle of the pyramid *ABCH* to the angle of the pyramid *ABCI*. But as *ABCI* is a pyramid whose base is the triangle *BCI*, and the side *AB* is perpendicular to the two straight lines *BC* and *BI*, and the straight line *AI* encloses a right angle with *IG*, then the ratio of the triangle *CBI* to the triangle *IBG* is greater than the ratio of the angle of the pyramid *ABCI* to the angle of the pyramid *ABIG*, as has been proved in Lemma 6. Therefore, the ratio of the triangle *HBC* to the triangle *IBG* is greater than the ratio of the angle of the pyramid *ABCH* to the angle of the pyramid *ABIG*. But the triangle *DBC* is twice the triangle *HBC*, the angle of the pyramid *ABCD* is twice the angle of the pyramid *ABCH*, the angle of the pyramid *ABEG* is twice the angle of the pyramid *ABIG*, and the triangle *EBG* is twice the triangle *IBG*. Therefore, the ratio of the triangle *DBC* to the triangle *EBG* is greater than the ratio of the angle of the pyramid *ABCD* to the angle of the pyramid *ABEG*, both of which angles lie at the point *A*. That is what we wanted to prove.

<Lemma 9> Let there be two pyramids whose bases are two similar plane figures, one of which is greater than the other, and let them be enclosed by a sphere such that the vertices of the two pyramids are at the centre of the sphere. Then the ratio of the angle of the larger pyramid[32] to the angle of the smaller pyramid is greater than the ratio of the base of the larger pyramid to the base of the smaller pyramid.

Let the point *A* be the centre of the sphere. We imagine that the plane of the base of the larger pyramid cuts the sphere forming a circle on the sphere. Let the point *B* be the centre of this circle. We join <the straight line> *AB*. This will be perpendicular to the plane of the circle, which is the plane of the base of the pyramid as the sides of the pyramid are equal. We imagine straight lines drawn from the point *B* to the angles of the base of the pyramid. These will divide the base into equal triangles. Let one of these triangles be the triangle *BCD*. Now we also imagine that the plane of the base of the smaller pyramid cuts the sphere, also forming a circle on the sphere. And we imagine a straight line drawn from the point *A* to the centre

[32] The pyramid with the larger base.

of this circle. This will be perpendicular to the plane of the circle and to the plane of the base of the pyramid which lies within the circle. As the two bases of the two pyramids are similar and one is larger than the other, the circle on the sphere that encloses the larger base is greater than the circle which encloses the smaller base. The straight line drawn from the centre of the sphere to the centre of the smaller circle will therefore be greater than the straight line drawn from the centre of the sphere to the centre of the larger circle. But these two straight lines are the two perpendiculars to the planes of the two bases. The perpendicular drawn from the point *A* to the smaller base is greater than the straight line *AB*. Let the straight line *AE* be equal to this perpendicular. Imagine a set of straight lines drawn from the centre of this circle to the angles of the base that lies within this circle, dividing the base into equal triangles, each similar to the others and to the triangle *BCD*. We imagine a plane passing through the point *E* parallel to the plane of the triangle *BCD* forming a circle on the sphere that is equal to the smaller circle enclosing the base of the smaller pyramid. From the point *E*, we draw two straight lines, parallel to the straight lines *BC* and *BD* <respectively> and ending on the circumference of the circle whose centre lies at the point *E*. Let these two straight lines be *EG* and *EH*. We join *GH*. The triangle *EGH* will then be similar to the triangle *BCD* and equal to each of the triangles dividing the base of the smaller pyramid. If we draw a set of straight lines from the point *E* enclosing angles that are equal and equal to the angle *GEH*, and if we join the extremities of these straight lines, then they will form a figure equal and similar to the smaller pyramid within the circle whose centre is at the point *E*. And if we draw straight lines from the point *A* to the angles of the figure formed within the circle *E*, they form a pyramid equal and similar to the smaller pyramid whose angle at the point *A* is equal to the angle of this <smaller> pyramid. We join the straight lines *AC*, *AD*, *AH* and *AG*, and we draw a perpendicular from the point *B* onto the straight line *CD* dividing it into two halves. Let this be *BF*. We also draw a perpendicular from the point *E* onto the straight line *HG* dividing it into two halves. Let this be *EM*. Then, *BF* is greater than *EM* as the two triangles *BCD* and *EGH* are similar, and the triangle *BCD* is greater than the triangle *EGH*. We join *AF* and we extend it until it meets the surface of the sphere at the point *L*. We imagine that the plane of the triangle *ACD* cuts the sphere forming an arc of a circle on the surface of the sphere. Let this arc be *CLD*. We also join the straight line *AM* and we extend it until it meets the surface of the sphere at the point *N*. We produce the plane of the triangle *AGH*, such that it forms on the surface of the sphere the arc *GNH*. We join *FM* and we extend it. It will meet the straight line *AE* as the two straight lines *BF* and *EM* are parallel and *BF* is greater than *EM*. Let the point at which they meet be the point *K*. The ratio of *BK* to *KE* is then equal to the ratio of *BF* to *EM*, and the ratio of *BF* to *EM* is equal to the ratio of *BC* to *EG* as the two triangles are similar. The ratio of *BK* to *KE* is therefore equal to the ratio of *BC* to *EG*. *BC* and *EG* are parallel and they therefore lie in the same plane, as does the straight line *BEK*. If we join *C* and *G* by a straight line and we extend it, then it will reach the point *K*. Let us join them; let this straight line be *CGK*. Similarly, if we join *D* and *H* by a straight line, this can be extended to the point *K*. Let us join them; let it be *DHK*. It will therefore be in the same plane as the triangle *KCD*, but it cuts the two planes *EGH* and *BCD* which are parallels. The straight line *GH* is therefore parallel to the straight line *CD*. From the point *M*, we draw a straight line parallel to the straight line *FA*. This will meet the straight line *KA*. Let the point at which they meet be the point *O*. The point *O* will then lie between the two points *A* and *E*. We join *OG* and *OH*. They will lie in the same plane as the triangle *OGH*. As *GH* is parallel to the straight line *CD* and *MO* is parallel to the straight line *FA*, the plane of the triangle *HOG* is parallel to the plane of the triangle *ACD* and the two planes of the triangles *AKC* and *AKD* cut the two planes of the two triangles *ACD* and *OGH*. Therefore, the straight line *AC* is parallel to the straight line *OG*, the straight line *AD* is parallel to the straight line *OH*, and the angle *GOH* is equal to the angle *CAD*. We produce the plane of the triangle *OGH* until it cuts the sphere,[33] forming an arc on its surface that is similar to the arc *CLD*. Let this be the arc *GUH*. The sector *OGUH* is therefore similar to the sector *ACLD* and the ratio of the sector *ACLD* to the sector *OGUH* is equal to the square of the ratio of *AC* to *OG*, and the square of the ratio of *AC* to *OG* is equal to the ratio of the triangle *ACD* to the triangle *OGH*. The ratio of the sector *ACD* to the sector *OGH* is therefore equal to the ratio of the triangle *ACD* to the triangle *OGH* and is also equal to the ratio of the remaining portion *CLD* to the remaining portion *GUH*. By permutation, the ratio of the triangle *ACD* to the portion *CLD* is equal to the ratio of the triangle *OGH* to the portion *GUH*. The ratio of the pyramid *KACD* to the circular pyramid[34] *KCLD* is equal to the ratio of the pyramid *KOGH* to the circular pyramid *KGUH*. By permutation, the ratio of the pyramid *KACD* to the pyramid *KOGH* is equal to the ratio of the circular pyramid *KCLD* to the circular pyramid *KGUH*. But the ratio of the pyramid *KACD* to the pyramid *KOGH* is greater than the ratio of the pyramid *KACD* to the pyramid *KAGH*. Therefore, the ratio of the circular pyramid *KCLD* to the

[33] This is the sphere (*O*, *OG*).

[34] Although the word ‘circular’ does not appear in the text, we have included it in the translation whenever the outline of the base includes at least one arc.

circular pyramid *KGUH* is greater than the ratio of the pyramid *KACD* to the pyramid *KAGH*.

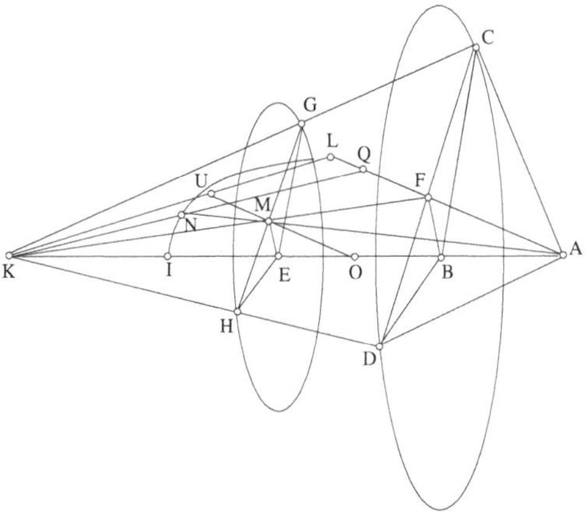

Fig. III.13

We imagine that the plane of the triangle *AKLF*[35] cuts the sphere forming an arc of a circle on its surface. This arc passes through the two points *U* and *N* as these two points are in the plane of the triangle *AKLF*. Let the arc be *LUNI*.[36] The point *N* will lie between the points *U* and *I*. As the triangle *ACD* is similar to the triangle *OGH* and the triangle *ACF* is similar to the triangle *OGM*, the ratio of *AC* to *OG* is therefore equal to the

[35] The point *F* lies on one of the sides of the triangle *AKL*. Here, *AKLF* designates the plane of this triangle.

[36] The point *I* has not been defined. It is assumed to lie on *AK*. We may note that Ibn al-Haytham seems to have assumed, mistakenly, that the point *U* lies on the arc *LNI* of the sphere with centre *A*; in fact, in general the straight line *KL* meets the arc in a point *U′* that is distinct from *U*. Ibn al-Haytham's reasoning is valid on condition that *N* lies between *U′* and *I*, that is if *M* lies between *O* and *V*, the point of intersection of *AU′* and *OU*. If this condition is not satisfied, the reasoning no longer holds, because *N* lies on the arc *LU* and *Q* lies beyond *L* on *AL* produced, and is thus outside the sphere. However, the lemma is true in all cases – see analytical commentary.

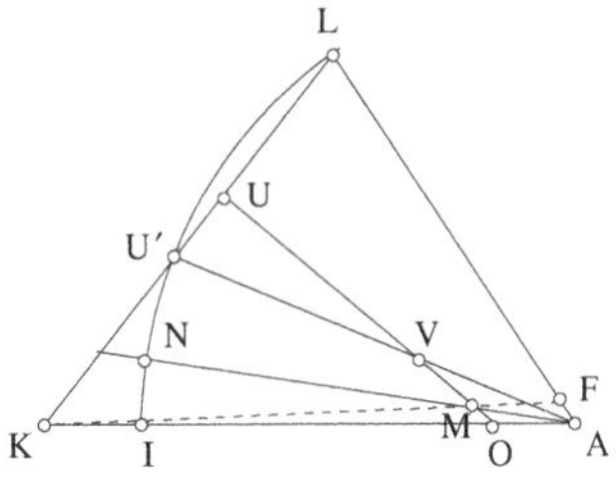

ratio of *AF* to *OM*. But *AC* is equal to *AL*, and *OG* is equal to *OU*. The ratio of *AL* to *OU* is therefore equal to the ratio of *AF* to *OM* and equal to the ratio of the remainder *FL* to the remainder *MU*, and the ratio of *FK* to *KM* is equal to the ratio of *AF* to *OM*. The ratio of *FK* to *KM* is therefore equal to the ratio of *FL* to *MU*.

If we join *L* and *U* by a straight line which we extend, it will reach the point *K*. Let us join them and let this straight line be *LUK*. But the two points *L* and *U* lie on the arc *LUI*. Therefore, the straight line *LUK* cuts the sphere and extends beyond it. The point *N*, which lies on the arc *UNI* between the two points *U* and *I*, is therefore below the straight line *UK*. From the point *K*, we draw a straight line to the point *N* and we extend it. It will cut the straight line *FL* at a point between the two points *F* and *L* as the points *M*, *N* and *U* are in the plane of the triangle *FKL* and the point *N* is in the plane of this triangle and between the straight lines *KF* and *KL*. Let the straight line *AG* cut the straight line *FL* at the point *Q*. This straight line cuts the plane of the portion *GUH* at a point inside the arc *GUH*, i.e. at a point in the direction of *E*. Let this point be the point *J*. Now, we imagine a circular pyramid whose vertex is at the point *K* and whose base is the sector *ANGH*. If extended, this circular pyramid will cut the plane of the sector *ACLD*, as it passes through the points *C*, *Q* and *D*. It will therefore generate a curved line on the plane of the arc *CLD*. Let this line be the line *CQD*. It also cuts the plane of the sector *OGUH*, generating a curved line on the portion *GUH*. Let this line be *GJH*. The ratio of the circular pyramid *KCLD* to the circular pyramid *KGUH* is equal to the ratio of the circular pyramid *KCQD* to the circular pyramid *KGJH*, as the planes of the two bases are parallel, and the ratio of the circular pyramid *KCLD* to the circular pyramid *KGUH* is greater than the ratio of the pyramid *KACD* to the pyramid *KAGH*. The ratio of the circular pyramid *KCQD* to the circular pyramid *KGJH* is therefore greater than the ratio of the pyramid *KACD* to the pyramid *KAGH*. The ratio of the entire circular pyramid *KACQD* to the circular pyramid *KAGNH* is greater than the ratio of the pyramid *KACD* to the pyramid *KAGH*. But the portion of the pyramid that lies between the two planes *ACQD* and *AGNH* is inside the spherical sector that lies between the two circular sectors *ACLD* and *AGNH*, and the spherical sector which lies between the point *I* and the sector *AGNH* is inside the circular pyramid *KAGNH*. The ratio of the sector *ACLDHGN* to the sector *AIHGN* is greater than the ratio of the portion of the pyramid *ACQDHNG* to the circular pyramid *KAHNG*. Composing, the ratio of the spherical sector whose angle is the angle of the pyramid *ABCD*, which is at the point *A*, to the spherical sector whose angle is the angle of the pyramid *AEGH*, which is at the point *A*, is greater than the ratio of the circular pyramid *KACQD* to the circular pyramid *KAGNH*, and the ratio of the circular pyramid *KAGQD* to the circular pyramid *KAGNH* is greater than the ratio of the pyramid *KACD* to the pyramid *KAGH*. The ratio of the spherical sector whose angle is the angle of the pyramid *ABCD*, which is at the point *A*, to the spherical sector whose angle is the angle of the pyramid *AEGH*, which is at the point *A*, is therefore greater than the ratio of the pyramid *AKCD* to the pyramid *AKGH*. But the ratio of the <first> spherical sector to the <second> spherical sector is equal to the ratio of its angle, which is at the point *A*, the centre of the sphere, to the angle of the other sector, which is at the point *A*. But the ratio of the pyramid *AKCD* to the pyramid *AKGH* is equal to the ratio of the triangle *KCD* to the triangle *KGH*, and the ratio of the triangle *KCD* to the triangle *KGH* is equal to the square of the ratio of *CK* to *KG*. But the square of the ratio of *CK* to *KG* is equal to the square of the ratio of *BC* to *EG*, and the square of the ratio of *BC* to *EG* is equal to the ratio of the triangle *BCD* to the triangle *EGH*. The ratio of the angle of the pyramid *ABCD*, which is at the point *A*, to the angle of the pyramid *AKGH*, which is at the point *A*, is therefore greater than the ratio of the triangle *BCD* to the triangle *EGH*. But the ratio of the angle of the pyramid of which the pyramid *ABCD* is a part, which angle is at the point *A*, to the angle of the pyramid *ABCD*, which angle is at the point *A*, is equal to the ratio of the entire base of the pyramid to the triangle *BCD*. But the ratio of the angle of the pyramid *AEGH* to the angle of the pyramid of which the pyramid *AEGH* is a part is equal to the ratio of the triangle *EGH* to the entire base of the pyramid. The ratio of the angle of the larger pyramid, which is at the point *A*, to the angle of the smaller pyramid, which is at the point *A*, is greater than the ratio of the base of the larger pyramid to the base of the smaller pyramid. That is what we wanted to prove.

<Lemma 10> Let there be two pyramids enclosed within a sphere such that the base of one is smaller than the base of the other and also has more sides than the other. Then the ratio of the angle of the pyramid which has the larger base to the angle of the pyramid with the smaller base is greater than the ratio of the base <of the first> to the base <of the second>.

This also applies to the cube and the dodecahedron.[37]

Let the point *A* be the centre de la sphere, and proceed for these two pyramids in the same way that we did for the two previous pyramids, i.e. divide the base of each into triangles. As one of the two bases is smaller than

[37] Lit.: that which has twelve bases. A cube may be inscribed within a sphere and can be divided into six pyramids with square bases and their vertices at the centre of the sphere. A regular dodecahedron may similarly be divided into twelve pyramids whose bases are regular pentagons.

the other, there exists a small base with the same number of sides as the larger, such that the two bases are similar. Then the circle which encloses the small base with the greater number of sides, and which is equal to the small base with fewer sides, is less than the circle which encloses the small base with fewer sides, and the circle which encloses it is very much smaller.[38] The perpendicular drawn <from the point *A*> onto the small base is therefore greater than the perpendicular drawn onto the large base.

Let the point *B* be the centre of the base of the pyramid with the larger base, and let the triangle *BCD* be one of its triangles. We join *AB*, *AC* and *AD*. Then *AB* will be perpendicular to the plane of the triangle *BCD*. We draw *AB*, and we make *AE* equal to the perpendicular in the other pyramid. We set the triangle *EGH* equal to one of the triangles in this other pyramid, and its plane parallel to the plane of the triangle *BCD*. We also set *EG* parallel to the straight line *BC*. *EH* will not be parallel to the straight line *BD* as the angle *GEH* is less than the angle *CBD*. We make the angle *GEK* equal to the angle *CBD*. Let the point *K* lie on the circumference of the circle that encloses the base. We join *AG*, *AH*, *AK*, *GH* and *GK*, we divide *CD* into two halves at the point *N*, and we join *BN*, which will therefore be perpendicular to the straight line *CD*. We divide *GH* into two halves at the point *M* and join *EM*. This will be perpendicular to the straight line *GH*. We join the straight lines *AN*, *AM* and *AF*, we divide *GK* into two halves at the point *F*, and we join *EF*, which will therefore be perpendicular to the straight line *GK*. We divide the angles of the three pyramids into two halves, i.e. the pyramid *ABCD*, the pyramid *AEGH*, and the pyramid *AEGK*. As the triangle *EGK* is similar to the triangle *BCD*, the ratio of the angle of the pyramid *ABCD* to the angle of the pyramid *AEGK* is greater than the ratio of the triangle *BCD* to the triangle *EGK*, as was proved in the previous proposition. The ratio of the angle of the pyramid *ABCD* to the angle of the pyramid *AEGF* is therefore greater than the ratio of the triangle *BCD* to the triangle *EGF*. But, as the angle *GEK* is greater than the angle *GEH*, the angle *GEF* is greater than the angle *GEM*. Therefore, the angle *EGF* is less than the angle *EGM*. The straight line *GF* therefore cuts the straight line *EM*. Let this point be the point *O*, and let us join *AO*.

[38] The circle that encloses the base of the second pyramid is less than that which encloses the base of the first pyramid.

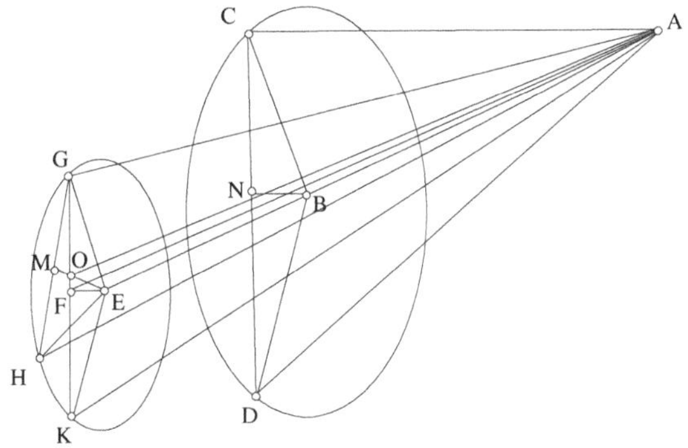

Fig. III.14

As the straight line AG is equal to the straight line AK – both are half-diameters of the sphere – and the straight line GF is equal to the straight line FK, the angle AFG is a right angle and therefore the angle AOG is obtuse. The ratio of the triangle EGM to the triangle EGO is therefore greater than the ratio of the angle of the pyramid $AEGM$ to the angle of the pyramid $AEGO$ by virtue of the proof given in the sixth lemma in this treatise. As the angle AFG is a right angle, the ratio of the triangle EGF to the triangle EOF is greater than the ratio of the angle of the pyramid $AEGF$ to the angle of the pyramid $AEOF$ by virtue of the proof given in the seventh lemma in this book. As the ratio of the triangle EGF to the triangle EOF is greater than the ratio of the angle of the pyramid $AEGF$ to the angle of the pyramid $AEOF$, the ratio of the triangle EGO to the triangle EGF is greater than the ratio of the angle of the pyramid $AEGO$ to the angle of the pyramid $AEGF$. The ratio of the triangle EGM to the triangle EGO is greater than the ratio of the angle of the pyramid $AEGM$ to the angle of the pyramid $AEGO$. The ratio of the triangle EGO to the triangle EGF is greater than the ratio of the angle of the pyramid $AEGO$ to the angle of the pyramid $AEGF$. Using the equality ratio, the ratio of the triangle EGM to the triangle EGF is greater than the ratio of the angle of the pyramid $AEGM$ to the angle of the pyramid $AEGF$. By inversion, the ratio of the angle of the pyramid $AEGF$ to the angle of the pyramid $AEGM$ is greater than the ratio of the triangle EGF to the triangle EGM. Therefore, the ratio of the angle of the pyramid $ABCD$ to the angle of the pyramid $AEGF$ is greater than the ratio of the triangle BCD to the triangle EGF. But the ratio of the angle of the pyramid $AEGF$ to the angle of the pyramid $AEGM$ is greater than the ratio of the triangle EGF to the triangle EGM. Using the equality ratio, the ratio of the angle of the pyramid $ABCD$ to the angle of the pyramid $AEGM$ is greater than the ratio of the triangle BCD to the triangle *EGM*. But we can show that the ratio of the angle of the entire pyramid with the larger base to the angle of the entire pyramid with the smaller base is greater than the ratio of the entire base of the <first> pyramid to the entire base of the <second> pyramid, as was proved in the previous proposition, as the entire angle of the first pyramid is a multiple of the angle of the pyramid *ABCD* and the base is a multiple of the triangle *BCD*, both multiples being the same. Similarly for the other pyramid, the ratio of the angle of the pyramid with the larger base to the angle of the pyramid with the smaller base and which has more sides is therefore greater than the ratio of the base <of the first> to the base <of the second>. That is what we wanted to prove.

<Proposition 5′> Now that we have introduced these lemmas, let us return to the proof for which these lemmas were required.[39]

We say that, given two regular polyhedra, such that the bases of one of which are similar to the bases of the other, and also such that the surface enclosing one is equal to the surface enclosing the other, then the volume of the polyhedron with the most bases is greater than the volume of the other.

Example: Let there be two regular polyhedra *A* and *B* such that the bases of one are similar to the bases of the other, and also such that the surface enclosing one is equal to the surface enclosing the other. Polyhedron *B* has more bases than polyhedron *A*.

I say that the volume of the polyhedron B *is greater than the volume of the polyhedron* A.

Proof: The perpendicular dropped from the centre of the sphere that circumscribes the polyhedron *B* onto one of the bases of the polyhedron *B* is greater than the perpendicular dropped from the centre of the sphere that circumscribes polyhedron *A*.

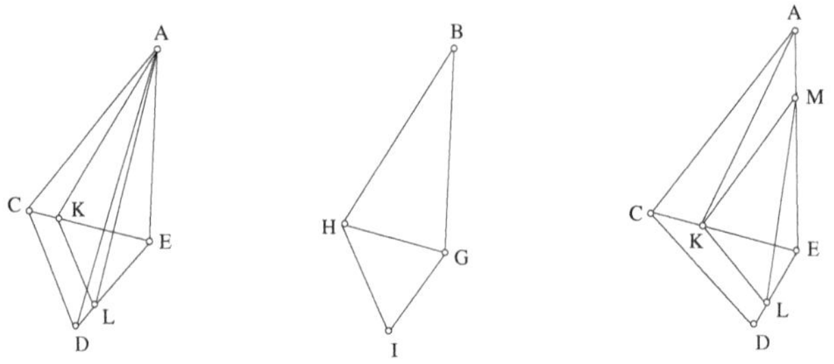

Fig. III.15

[39] i.e. Propositions 5′ and 5″.

Let the point A be the centre of the sphere circumscribed around the polyhedron A and let the point B be the centre of the sphere circumscribed around the polyhedron B, let the triangle ECD be one of the triangles into which one of the bases of the polyhedron A is divided, and let the triangle GHI be one of the triangles into which one of the bases of the polyhedron B is divided. We join AE and BG to form perpendiculars to the planes of the two triangles <respectively>, as has been shown previously. Let the perpendicular BG be equal to the perpendicular AE or less than it, if this is possible. But as the bases of the two polyhedra are similar, the triangle ECD will be similar to the triangle GHI and, as the bases of the polyhedron B are greater in number than the bases of the polyhedron A and the sum of the bases of the one is equal to the sum of the bases of the other, the triangle ECD will be greater than the triangle GHI. But both of them are isosceles. Therefore, each of the two straight lines EC and ED is <respectively> greater than each of the straight lines GH and GI. We separate the two equal straight lines EK and EL from the two straight lines GH and GI, and we join KL, AK and AL. If the perpendicular AE were equal to the perpendicular BG, then the pyramid $AEKL$ would be equal to the pyramid $BGHI$, and the angle of the pyramid $AEKL$, which is at the point A, would be equal to the angle of the pyramid $BGHI$, which is at the point B. But as the bases of the polyhedron A are all equal and similar, the ratio of the angle of the pyramid $AECD$, which is at the point A, to eight <solid> right angles is equal to the ratio of the pyramid $AECD$ to the entire polyhedron,[40] and is equal to the ratio of the triangle ECD to the sum of the bases of the polyhedron, which is the surface that encloses the polyhedron. Similarly, the ratio of the angle of the pyramid $BGHI$, which is at the point B, to eight <solid> right angles is equal to the ratio of the pyramid $BGHI$ to the entire polyhedron, and is also equal to the ratio of the triangle GHI to the entire surface that encloses the polyhedron. But the two surfaces that enclose the polyhedra are equal, the ratio of the triangle ECD to the entire surface enclosing the polyhedron A is equal to the ratio of the angle of the pyramid $AECD$, which is at the point A, to eight <solid> right angles, and the ratio of the surface enclosing the polyhedron B to the triangle GHI is equal to the ratio of eight <solid> right angles to the angle of the pyramid $BGHI$ which is at the point B. Using the equality ratio, the ratio of the triangle ECD to the triangle GHI is therefore equal to the ratio of the angle of the pyramid $AECD$, which is at the point A, to the angle of the pyramid $BGHI$ which is at the point B. But the pyramid $AEKL$ is equal and similar to the pyramid $BGHI$, and its angle, which is at the point A, is equal to the angle which is at the point B, and the triangle EKL is equal to the triangle GHI. The ratio of

[40] See the second lemma of Proposition 4.

the triangle *ECD* to the triangle *EKL* is therefore equal to the ratio of the angle of the pyramid *AECD*, which is at the point *A*, to the angle of the pyramid *AEKL*, which is at the point *A*. This is impossible, as it has been shown in the eighth lemma in this treatise that the ratio of the triangle *ECD* to the triangle *EKL* is greater than the ratio of the angle of the pyramid *AECD*, which is at the point *A*, to the angle of the pyramid *AEKL*, which is at the point *A*. Therefore, the perpendicular *AE* is not equal to the perpendicular *GB*.

If the perpendicular *BG* is less than the perpendicular *AE*, we separate out *EM* equal to *BG* and we join the two straight lines *MK* and *ML*. Then the pyramid *MEKL* will be equal to the pyramid *BGHI* and its angle, which is at the point *M*, will be equal to the angle which is at the point *B*. The ratio of the triangle *ECD* to the triangle *EKL* is then equal to the ratio of the angle of the pyramid *AECD*, which is at the point *A*, to the angle of the pyramid *MEKL* which is at the point *M*. But the plane angle *EMK* is greater than the angle *EAK*, the angle *EML* is greater than the angle *EAL*, and the angle *KML* is greater than the angle *KAL* as the two triangles are isosceles and the two sides *AK* and *AL* are greater <respectively> than the two sides *MK* and *ML*, and the base of the two triangles is the same. The angle *KML* is therefore greater than the angle *KAL*. Therefore, the <sum of the> plane angles enclosing the solid angle of the pyramid *MEKL*, which is at the point *M*, is greater than the <sum of the> plane angles enclosing the solid angle of the pyramid *AEKL*, which is at the point *A*. The solid angle of the pyramid *MEKL*, which is at the point *M*, is therefore greater than the angle of the pyramid *AEKL*, which is at the point *A*.[41] The ratio of the triangle *CED* to the triangle *KEL* is therefore greater than the ratio of the angle of the pyramid *AECD*, which is at the point *A*, to the angle of the pyramid *AEKL*, which is at the point *A*. Therefore, the ratio of the angle of the pyramid *ACED*, which is at the point *A*, to the angle of the pyramid *AEKL*, which is at the point *A*, is greater than the ratio of the angle of the pyramid *AECD*, which is at the point *A*, to the angle of the pyramid *MEKL*, which is at the point *M*. These two ratios were stated to be equal, which is impossible. Therefore, the perpendicular *BG* is not less than the perpendicular *AE*. But we have already shown that they are not equal. Therefore, the perpendicular *BG* is greater than the perpendicular *AE*. But the product of the perpendicular *BG* and one third of the sum of the bases of the polyhedron *B* is the volume of the polyhedron *B*, and the product of the perpendicular *AE* and one third of the sum of the bases of the polyhedron *A* is the volume of the polyhedron *A*. The hypothesis stated that the sum of the bases of the polyhedron *B* was equal to the sum of the bases of the polyhedron *A*, and

[41] See the mathematical commentary.

the perpendicular *BG* is greater than the perpendicular *AE*. Therefore, the volume of the polyhedron *B* is greater than the volume of the polyhedron *A*. That is what we wanted to prove.

<Proposition 5″> We also say that, given two regular polyhedra such that the bases of one are similar to the bases of the other, and the bases of one are greater in number than the bases of the other, and both of which are inscribed within the same sphere, then the surface that encloses the entire polyhedron whose bases are greater in number is greater than the surface that encloses the other polyhedron, and the volume of the polyhedron whose bases are greater in number is greater than the volume of the other polyhedron.

Let there be a sphere with its centre at the point *A*, and two polyhedra inscribed within it whose properties are those described above.

I say that the polyhedron with the greater number of bases has the greater surface and the greater volume.

Proof: The bases of one of the two polyhedra are similar to the bases of the other polyhedron. The solid angle which is at the centre of the sphere and which intercepts the base of the polyhedron with the greater number of bases is less than the solid angle which is at the centre of the sphere and which intercepts the base of the polyhedron with the lesser number of bases, as the ratio of each of these two solid angles to eight solid right angles is equal <for each polyhedron> to the ratio of the base intercepted by this angle to the sum of the bases[42] which are continuous with this base, and which are the surface that encloses the polyhedron. The ratio of the base of the polyhedron with the greater number of bases to the sum of its bases is less than the ratio of the base of the polyhedron with the lesser number of bases to the sum of its bases. Therefore, the solid angle which intercepts the base of the polyhedron with the greater number of bases is less than the solid angle which intercepts the base of the polyhedron with the lesser number of bases, and the number of plane angles that enclose one of the solid angles is equal to the number of plane angles which enclose the other solid angle. But the plane angles enclosing each of the two solid angles are equal. Therefore, each of the equal plane angles that enclose the smaller solid angle is less than each of the equal plane angles that enclose the larger solid angle as, if the plane angles that enclose the small solid angle were equal to the plane angles which enclose the large solid angle, then the two solid angles would be equal, and if the plane angles that enclose the small solid angle were greater than the plane angles that enclose the large solid angle, then the small solid angle would be greater than the large solid angle, which

[42] From Lemma 6.

is impossible. Therefore, each of the plane angles which enclose the small solid angle is less than each of the plane angles which enclose the large solid angle.[43] It is therefore possible for the entire small solid angle to fit within the large solid angle. If this is the case, then the circle circumscribed around the base intercepted by the small solid angle will be below the circle circumscribed around the base intercepted by the large solid angle. But if a circle lies below another circle, and if they are both in the same sphere, then the lower circle is less than the upper circle; and if it is smaller, then the straight line drawn from the centre of the sphere to the centre of the small circle is greater than the straight line drawn from the centre of the sphere to the centre of the large circle. But the straight line drawn from the centre of the sphere to the centre of any circle of the sphere – if the circumference of the circle lies on the surface of the sphere – is a perpendicular to the plane of the circle and is perpendicular to the plane of any figure in the circle. The perpendicular drawn from the centre of the sphere to the plane of the base of the polyhedron with the greater number of bases is greater than the perpendicular drawn from the centre of the sphere to the plane of the base of the polyhedron with the lesser number of bases. It is also clear from this proof that each of the bases of the polyhedron with the greater number of bases is less than each of the bases of the polyhedron with the lesser number of bases, as the bases of one of the two polyhedra are similar to the bases of the other polyhedron and the circle circumscribing the base of the polyhedron with the greater number of bases is smaller. The base that lies within the small circle is therefore less than the base which lies within the large circle.

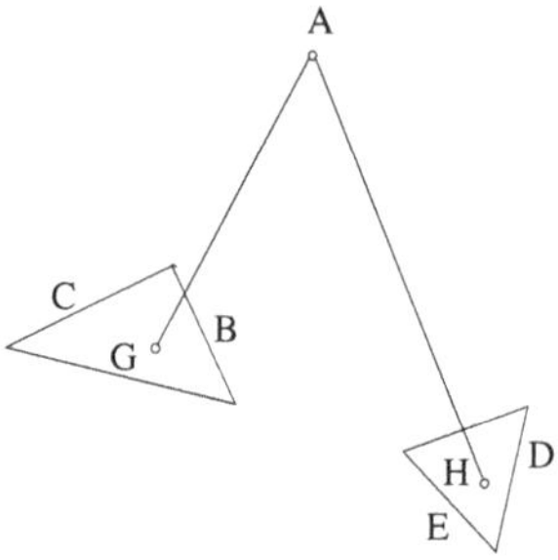

Fig. III.16

Let BC be the larger base and let DE be the smaller base. Let the perpendicular drawn from the centre of the sphere to the base DE be the

[43] This paragraph is not essential. Here, the author makes use of the reciprocal of the theorem used in the proof of Proposition 5″. See the mathematical commentary.

perpendicular *AH*. The ratio of the large solid angle to the small solid angle is therefore compounded of the ratio of the large <solid> angle to eight <solid> right angles and the ratio of eight <solid> right angles to the small <solid> angle. But the ratio of the large angle to eight right angles is equal to the ratio of the base *BC* to the sum of all the bases that are continuous with each other, and which are the entire surface of the solid. As the number of angles and the number of bases are equal, the angles are equal, the bases are equal, and the ratio of eight right angles to the small angle is equal to the ratio of the sum of all the bases of the solid that are continuous with the base *DE* to the base *DE*, therefore, the ratio of the large angle to the small angle is compounded of the ratio of the base *BC* to the sum of all the bases that are continuous with it and the ratio of the sum of all the bases that are continuous with the base *DE* to the base *DE*. But the ratio of the large angle to the small angle is greater than the ratio of the base *BC* to the base *DE*, as has been proved in the ninth lemma in this treatise. The ratio compounded of the ratio of the base *BC* to the entire surface of the polyhedron, of which *BC* is a base, and the ratio of the entire surface of the polyhedron, of which *DE* is a base, to the base *DE*, is therefore greater than the ratio of the base *BC*, to the base *DE*. But the ratio of the base *BC* to the base *DE* is compounded of the ratio of the base *BC* to the entire surface of the polyhedron, of which *BC* is a base, and the ratio of the entire surface of this polyhedron to the base *DE*. The ratio compounded of the ratio of the base *BC* to the entire surface of the polyhedron, of which *BC* is a base, and the ratio of the entire surface of the polyhedron, of which *DE* is a base, to the base *DE*, is greater than the ratio compounded of the ratio of the base *BC* to the entire surface of the polyhedron, of which *BC* is a base, and the ratio of the entire surface of this polyhedron to the base *DE*. We take off the common ratio, and the ratio of the entire surface of the polyhedron, of which *DE* is a base, to the base *DE* remains therefore greater than the ratio of the entire surface of the solid, of which *BC* is a base, to the base *DE*. The entire surface of the polyhedron, of which *DE* is a base, is greater than the entire surface of the polyhedron, of which *BC* is a base. But the perpendicular *AH* is greater than the perpendicular *AG*. Therefore, the product of the perpendicular *AH* and one third of the entire surface of the polyhedron, of which *DE* is a base, is greater than the product of *AG* and one third of the entire surface of the polyhedron, of which *BC* is a base. But the product of the perpendicular and one third of the sum of all the bases of the regular polyhedron inscribed within the sphere is the volume of this polyhedron. Therefore, the volume of the polyhedron whose bases are greater in number is greater than the volume of the polyhedron whose bases are fewer in number.

Therefore, for two regular polyhedra inscribed within a sphere, the bases of one being similar to the bases of the other, and the bases of one being greater in number than the bases of the other, then the surface that encloses the polyhedron with the greater number of bases is greater than the surface of the polyhedron whose bases are fewer in number, and the volume of the polyhedron with the greater number of bases is greater than the volume of the polyhedron with the lesser number of bases. That is what we wanted to prove.

<*Corollary*> If the bases of the polyhedron with the greater number of bases have more sides that the bases of the other polyhedron, and if the perpendicular drawn to the base of the polyhedron with the greater number of bases is greater than the perpendicular drawn to the base of the other polyhedron, then the surface of the polyhedron with the greater number of bases is greater than the surface of the other polyhedron, and the volume of the first is greater than the volume of the second.

We can show, as we have shown in the previous proposition, that the ratio of the solid angle to the other solid angle is compounded of the ratio of the base to the sum of the bases that are continuous with it and the ratio of the sum of the bases that are continuous with the other base to this base. But the ratio of the solid angle to the other solid angle is greater than the ratio of the base to the other base, as has been proved in the tenth lemma in this treatise.

We can show, as we have shown previously, that the surface that encloses the polyhedron with the greater number of bases is greater than the surface that encloses the polyhedron with the lesser number of bases, and that the volume of the polyhedron with the greater number of bases is greater than the volume of the other polyhedron.

From all that we have proved earlier in this treatise, it is clear that the sphere is greater than all the solid figures with equal perimeters, that the circle is the greatest of all the plane figures having equal perimeters, and that of these figures, that which more closely approaches the circular in shape is greater than that which is less circular in shape. That is what we intended to prove in this treatise.

The treatise is complete. Thanks be unto God, the Lord of worlds, and blessings be upon His envoy Muḥammad, the elect, and his Household.

APPENDIX

THE APPROXIMATION OF ROOTS

1. MATHEMATICAL COMMENTARY

Like many of his contemporaries, Ibn al-Haytham was concerned with the extraction of square and cube roots. It is clear that he encountered the familiar problem of approximation; but this time, in contrast to other areas where the author deals with infinitesimal determinations, he refers neither to what is known as Archimedes' axiom nor to the exhaustion method: the two concepts which go some way to unify the diverse works of research on the infinitesimal are absent here. The approximation of roots would therefore seem to constitute a separate area, which would only later be included in this research. It will be shown in my third volume that algebra is instrumental in introducing this work, firstly in the twelfth century with al-Samaw'al and Sharaf al-Dīn al-Ṭūsī, then five centuries later, in quite a distinct branch of study, where it developed in a completely different direction. This is what led us to take up Ibn al-Haytham's texts herein, but in an appendix in order to highlight the differences in their status: the two texts which have come down to us and which have only recently been found, are devoted to the square and the cube root respectively. If the evidence of ancient biobibliographers is to be accepted, then these are the only texts which Ibn al-Haytham wrote on this theme.

Let us start by defining Ibn al-Haytham's method in a quite deliberately different language in order to understand the ideas which underpin it.[1] This will demonstrate an algorithm which will lead eventually to Ruffini-Horner's. Even if Ibn al-Haytham seems to share this algorithm with his contemporaries, he is distinguished from them by his efforts to mathematically justify it in the case of the square root.

Let the polynomial with integer coefficients be $f(x)$ and the equation be

(1) $\quad f(x) = N.$

[1] Sharaf al-Dīn al-Ṭūsī, *Œuvres Mathématiques. Algèbre et géométrie au* XII*e* *siècle*, Text edited and translated by R. Rashed, 2 vols, Paris, 1986, vol. I, pp. LXXX–LXXXIX.

Let s be a positive root of this equation and let us assume $(s_i)_{i \geq 0}$ a sequence of positive integers such that the partial sums are

$$\sum_{i=0}^{k} s_i \leq s.$$

We say that the s_i are parts of s.

It is evident that the equation

(2) $f_0(x) = f(x + s_0) - f(s_0) = N - f(s_0) = N_0$

has for roots those of equation (1) diminished by s_0.

For $i > 0$, let us form by recurrence the equation

(3) $f_i(x) = f(x + s_0 + \dots + s_i) - f(s_0 + \dots + s_i)$

$$= [N - f(s_0 + \dots + s_{i-1})] - [f(s_0 + \dots + s_i) - f(s_0 + \dots + s_{i-1})]$$
$$= N_i;$$

thus, by example for $i = 1$, we have

$$f_1(x) = f(x + s_0 + s_1) - f(s_0 + s_1) = [N - f(s_0)] - [f(s_0 + s_1) - f(s_0)]$$
$$= N_0 - [f(s_0 + s_1) - f(s_0)] = N_1.$$

The method applied by Ibn al-Haytham, used by Kūshyār ibn al-Labbān[2] and said to be by Ruffini-Horner, provides an algorithm which gives the coefficients of the ith equation starting with coefficients of the $(i - 1)$th equation. This is the principal idea of this method.

Let us begin with the extraction of the nth root (which can already be found in the twelfth century, if not before). Then

$$f(x) = x^n;$$

[2] Kūshyār ibn al-Labbān, *Fī uṣūl ḥisāb al-hind* (*Principles of Hindu Reckoning*) has been translated into English and analysed by Martin Levey and Marvin Petruck, Madison, Milwaukee, 1965. A. S. Saidan edited the Arabic text in *Majallat Ma'had al-Makhṭūṭāt al-'Arabiyya* 13, 1976, pp. 55–83.

if we know the binomial formula given in the tenth century by al-Karajī, there is no need to know the Horner table. The coefficients of the ith equation will in this case therefore be

$$\binom{n}{k}(s_0 + \dots + s_{i-1})^{n-k}, \quad \text{for } k = 1, \dots, n,$$

(4) and

$$N_i = N_{i-1} - \sum_{k=1}^{n}\binom{n}{k}(s_0 + \dots + s_{i-1})^{n-k} s_i^k.$$

After this preliminary investigation, let us return to the text attributed to Ibn al-Haytham for the square and cube root. Let

$$f(x) = x^2 = N;$$

we therefore have two cases:

First case: N is the square of an integer. Let us assume the root as

$$s = s_0 + \dots + s_h, \qquad \text{with } s_i = \sigma_i\, 10^{h-i} \qquad (0 \le i \le h).$$

The task of the eleventh-century mathematicians was partly to determine h and the figures σ_i. Formula (4) can be rewritten as

$$2\,(s_0 + \dots + s_{i-1}), \quad 1, \quad N_i = N_{i-1} - [\,2(s_0 + \dots + s_{i-1})\, s_i + s_i^2\,].$$

Thus σ_0 can be determined by inequalities

$$\sigma_0^2 10^{2h} \le N < (\sigma_0 + 1)^2 10^{2h}$$

and $\sigma_1, \dots, \sigma_h$ by

$$\sigma_i = \frac{N_i}{2(s_0 + \dots + s_{i-1}) \cdot 10^{h-i}}.$$

In these expressions the N_i, for $(1 \le i \le h)$, are calculated starting from N_{i-1}, subtracting $[2(s_0 + \dots + s_{i-1})\, s_i + s_i^2\,]$. For $i = h$, we find $N_h = 0$.

The author describes and proves an algorithm of the extraction of the square root. Let us summarize this method by separating these two tasks. Therefore in order to extract the square root or its whole part, we take the following steps:

1. Place number N, whose square root is to be extracted, in a table.
2. Place σ_0 in decimal position $2h$.
3. Subtract σ_0^2 in this position of N, which leads to $N - s_0^2$.
4. Multiply σ_0 by 2 and move the product one decimal place to the right.
5. Find σ_1 and place under decimal place $2h - 2$.
6. Multiply σ_1 by $2\sigma_0$ in $2\sigma_0$ place.
7. Subtract the product of N_0.
8. Subtract σ_1^2 – in position σ_1 from N_0 –, which finally leads to

$$N_1 = N_0 - (2s_0 + s_1)\, s_1$$

9. Multiply σ_1 by 2 and move $2\sigma_0$ and $2\sigma_1$, placed under decimal places $2h - 1$ et $2h - 2$ respectively, by one decimal place.
10. Find σ_2 and place it under decimal place $2h - 4$.
11. Multiply σ_2 by $2\sigma_1$ and $2\sigma_0$ in their respective positions.
12. Subtract the products of N_1.
13. Subtract σ_2^2 in position σ_2 of N_1 – which leads to

$$N_2 = N_1 - [2\,(s_0 + s_1)\, s_2 + s_1^2].$$

14. Repeat until $N_h = 0$, then divide each of the numbers $2\sigma_0$, $2\sigma_1$, …, $2\sigma_h$ in their positions by 2 to obtain the square root desired.

The author's proof of his algorithm is formulated in general terms from the Euclidean number theory. It relies heavily on the properties of the geometric progression given in Book IX of the *Elements*, particularly Proposition 8, which he specifically refers to.

Second case: N is not the square of an integer. Ibn al-Haytham uses the same method to determine the whole part of the root. He then gives as an approximation formula the one used by al-Khwārizmī as well as a 'conventional approximation' which can be rewritten respectively as

$$(s_0 + \ldots + s_h) + \frac{N_h}{2(s_0 + \ldots + s_h)}$$

and

$$(s_0 + \ldots + s_h) + \frac{N_h}{2(s_0 + \ldots + s_h) + 1}.$$

This is not merely a description of algorithms as with Kūshyār, but an attempt to give proper mathematical reasons, and to justify how these two approximations surround the root.

To extract a cube root from an integer, the method is the same. Let

$$f(x) = x^3 = N\,;$$

here too, there are two cases.

First case: N is the cube of an integer. In this case s_0 is determined in such a way that $s_0^3 < N$. Ibn al-Haytham, in the same way as his contemporaries, therefore takes $s_1 = s_2 = \ldots = s_h = 1$.

The coefficients of the ith equation can be rewritten as

$$3(s_0 + i)^2,\ 3(s_0 + i),\ 1,\ N_i = N_{i-1} - [3\,(s_0 + (i-1))^2 + 3(s_0 + (i-1)) + 1].$$

If N_i is the cube of an integer, then there exists a value k of i such that $N_k = 0$; that is, such that $(s_0 + k)$ is the required root. Ibn al-Haytham puts forward this algorithm:

1. Choose s_0 such that $s_0^3 \leq N$.
2. If $s_0^3 = N$, then the problem is solved; if not, we must carry on.
3. Formulate $N_1 = N - s_0^3$.
4. Take $s_1 = s_0 + 1$.
5. Formulate $N_2 = N_1 - (3s_0^2 + 3s_0 + 1) = N - s_1^3$.
6. If $N_2 = 0$, then s_1 is the required root; if not, we must carry on.
7. Take $s_2 = s_1 + 1$.
8. Formulate $N_3 - N_2 - (3s_1^2 + 3s_1 + 1) = N - s_2^3$.
9. If $N_3 = 0$, then s_2 is the required root; if not, we must start again from the beginning.

It is clear that this algorithm is based on the idea previously stated: if N is the cube of an integer and s_0 is such that $s_0^3 < N$, then the cube root s can be written as $s_0 + k$, with $k \in \mathbf{N}^*$, an unknown number. If $s_1 = s_0 + 1$, then

$$(x + s_1)^3 = N,$$

whose root is $k - 2$. This last equation can be rewritten

$$(x + s_2)^3 - s_1^3 = N - s_1^3 = N_1 - (3s_0^2 + 3s_0 + 1) = N_2.$$

If we carry on, we come to the equation

$$(x + s_k)^3 = N$$

with

$$s_k = s_{k-1} + 1 = s_0 + k,$$

whose root is 0. Therefore s_k is the required root.

Second case: N is not the cube of an integer. Ibn al-Haytham uses the same method to determine the whole part of the root. He takes in this case the approximate value

$$(s_0 + \ldots + s_h) + \frac{N_h}{3(s_0 + \ldots + s_h)^2}.$$

But the manuscript of this text which comes down to us has sadly been mutilated, and we wonder if Ibn al-Haytham gave the second approximation in the lost part of the text, since this was known to his contemporaries – *viz.*

$$(s_0 + \ldots + s_h) + \frac{N_h}{3(s_0 + \ldots + s_h)^2 + 3(s_0 + \ldots + s_h) + 1}.$$

This is highly likely: firstly, for reasons of symmetry, since with the square root, he gave the two approximations. Secondly, this last approximation was well known to his contemporaries. (It is also likely that the lost part contained a piece on the method of Hindu reckoning.)

Whatever the case may be, we can see that by the turn of the tenth century, the so-called 'Ruffini-Horner method' was known, and there had even been attempts to justify its usage mathematically. It went on to be more widely used by algebrists, when, again at the turn of the century, al-Karajī's work on binomial formulae and coefficient tables became known. And so there were several possible candidates for this wider application, such as al-Bīrūnī and al-Khayyām.

4.2. *Translated texts*

Al-Ḥasan ibn al-Haytham

4.3.1. *On the Cause of the Square Root, its Doubling and its Displacement*

4.3.2. *On the Extraction of the Side of a Cube*

In the Name of God, the Forgiving, the Merciful

TREATISE BY ABŪ ʿALI AL-ḤASAN IBN AL-ḤASAN IBN AL-HAYTHAM

On the Cause of the Square Root, its Doubling and its Displacement

The cause of this: Why is it that the first rank has a square root, the second has no square root, the next has a square root, and the next has no square root? The cause is that each rank in the Indian calculus is ten times greater than the preceding rank, and the first rank is unity. Hence, all the ranks are in the same ratio to their neighbours, and each is a proportional number beginning with unity. The third rank, counting from the first or unity, is a square. The next rank is not a square, and the one after that is a square.[1] This is clear from Proposition eight in Book nine of the works of Euclid.

But why should we double the fixed number? We take the double so that, if we take a number that precedes it and then multiply it by the doubled number, then we obtain twice the product of the second number by the first. But why do we reduce the rank of the doubled number by one rank? This is because the rank of the first fixed number is the rank of the side of the square above it. Therefore, the rank of the side of the square

[1] For $n \in N$,

$$\frac{a^{n-1}}{a^n} = \frac{a^n}{a^{n+1}};$$

hence

$$a^{n+1} = \frac{\left(a^n\right)^2}{a^{n-1}}.$$

Therefore if a^{n-1} is a square, then a^{n+1} is too. But since $a^0 = 1$ is a square, all the odd number expressions are squares. But since a^1 is not a square, the odd number expressions are not squares. Note that the word 'rank' (*martaba*) describes the term 10^n or $a_n 10^n$ as well as the number itself, which is $(n + 1)$.

which is above it[2] is the intermediate rank between the other square and the first square, which is unity. The ratio of unity to the intermediate number is equal to the ratio of the intermediate number to the other square. Hence, the product of the intermediate number and itself is the other square, and the intermediate number is the side of the other square. Similarly, the number that precedes the intermediate <number> is the side of the second square which precedes the other square, as it is intermediate between unity and the other squared number, as the number of ranks, the first of which is unity and the last of which is the second square, is an odd number and is two less than the first odd number, the <place> of whose extremity is the other square. Its intermediate is therefore the number immediately following the first intermediate, and therefore the ratio of unity to it is equal to the ratio of it to the second square. It is therefore the side of the second square. In this way, we have shown that the successive powers of ten following unity are the sides of the successive squares following unity. If we multiply the first fixed number below the other square by itself, and then subtract <the product of> the other square which is above it, then its rank is the rank of the intermediate number between the other square and unity, which is the last of the sides of the successive squares. The same applies to the second given number below the second square which precedes the other square. Its rank is the rank of the side of the second square, and its rank is that of the power of ten which comes immediately after the power of ten which is the side of the other square. The same applies to any given number below the successive squares. The rank of each of them is that of the power of ten – among the first powers of ten – which come after <that of> the side of this square.

It necessarily follows that the product of the side of the first square and the side of the second square is the power of ten lying between the two successive squares, as each of the successive powers of ten is ten times the power of ten that precedes it. If the power of ten is multiplied by the power of ten that succeeds it, then the product is ten times greater than the square of the first power of ten. Hence, for each of the successive squares, the power of ten that follows the square is ten times it, and the power of ten that follows that is the intermediate power of ten between it and the next

[2] For every $n \in \boldsymbol{N}$, 10^n is the root of the square 10^{2n}, and 10^n is equidistant between the first square $10^0 = 1$ and 'the other' square which is 'above it' since

$$\frac{1}{10^n} = \frac{10^n}{10^{2n}}.$$

The expression 10^{n-1} which precedes 10^n is the root of the square 10^{2n-2} which precedes 10^{2n}, since $\frac{1}{10^{n-1}} = \frac{10^{n-1}}{10^{2n-2}}$.

square. It is for this reason that the rank between the two squares becomes the rank of the power of ten obtained by multiplying the side of the first square by the side of the square that follows it. Hence, the first number fixed for the square root is always one rank lower so that, if we take a number below the second square which precedes the last square, and if we multiply this by the number reduced in rank, then the product will belong to the type of rank which lies between the second square and the last square. Therefore, if we subtract the product from that which is above the number reduced in rank, we are subtracting it from the power of ten which lies in its rank. It is for this reason that we reduce the rank of the doubled number by one rank.

But why do we raise the second fixed number <to a square>, and then subtract it from that which lies above it? As for this subtraction, it is so that the sum of the <terms> subtracted from the two squares is the square of the sum of the two fixed numbers. For any two numbers, <the sum of> their squares and twice the product of one and the other is the square of their sum. As for the subtraction of the square of the second number, this lies in the rank of the number which is above it, as two successive numbers are the two square roots of two successive squares. Therefore, if the first number is raised to the square, and if <this square> is subtracted from that which is above it, if it is doubled, if it is reduced in rank by one rank, if the second number is multiplied by the doubled number, if the <product> of that which is above it is subtracted, if the second number is raised to the square, and if <the square> is subtracted from that which is above it, then the sum of the numbers subtracted is the square of the number that is the sum of two numbers that are successive powers of ten, and such that each of the subtracted numbers is subtracted from its own rank. If the second number is then doubled, and if all are reduced in rank by one rank, and if one takes a given number under the third square, and if this number is multiplied by the two numbers that were doubled and reduced in rank and by itself, and if each of the terms of this product is subtracted from the number which is above it, then each of the subtracted numbers is subtracted from its own rank. The third given number is the side of the third square. Therefore, if it is multiplied by the second number – which is the side of the second square – the product will belong to the type of rank that lies between the third square and the second square, as has been shown previously. Therefore, if it is multiplied by the first number – which is the side of the other square – the product will belong to the type of rank of the second square, as the rank of the first number is ten times the rank of the second number, and the rank of the second square is ten times the rank which lies between the second square and the third square. If the third number is then doubled, and if all are reduced in rank by a single rank, and if the given number preceding it is multiplied by the three numbers and by itself, then each of the two products will belong to the rank which is above it, as has been shown for the three numbers, and so on until the last number arrives in the first rank, which is unity. If one arrives at the first rank, the doubled numbers are divided into two halves. The remainders of these numbers after having been divided in two, with the number which is above the first rank, is the square root of the given number, if this has disappeared in the successive subtractions, and their ranks are the ranks of the powers of ten obtained after the division by two, the first of which is unity.

Let the number obtained be the square root of the given number that disappeared after the subtractions. This is so because if any given number is multiplied by itself, and another given number preceding it is multiplied by itself and by twice the first number, then the sum of the numbers produced and the square of the first number is the square of the sum of the two numbers. Similarly, if another given number prior to the second is multiplied by itself and by twice <the sum of> the two first numbers, then <the sum of> the products plus the first subtracted numbers is the square of the sum of the three numbers. The same applies to all given numbers. They all have this property. The numbers obtained after dividing by two, plus the number which is above the first rank, are the given numbers which have been multiplied with each other in accordance with the property that we have just mentioned. The square <of their sum> is therefore the sum of the subtracted numbers. If the sum of the subtracted numbers is the given number whose square root we are trying to find, then <the sum of> the numbers obtained, the first of which is under the first rank, is the square root of the given number whose square root we are trying to find if this number has entirely disappeared during the subtractions. Let their ranks be the ranks of the powers of ten obtained after the division by two, and let the first be the rank of unity. This is because the powers of ten obtained after the division by two are the sides of the successive squares. But the sides of the successive squares are the first successive powers of ten, the first of which is unity, as has already been shown. The ranks of the powers of ten of the numbers obtained after the division by two and the number which is below the first rank are the ranks of the successive powers of ten beginning with unity.

The number whose square root we are trying to find is either a square or not a square. If it is a square, then there will be no remainder after we have taken the square root. If it is not a square, then there will be a remainder after we have taken the square root, and that remainder cannot have a square root. The whole number <that remains> is always less than twice the square root obtained plus one, as, if twice the square root obtained plus one is added to the subtracted numbers, the sum will be a square number, as has been shown in the preliminaries. If the given number is not a square, then if the square root is taken from that part of it that has a whole square root, and if the number whose square root has just been taken is subtracted from the given number, the remainder will be less than twice the square root thus obtained plus one. This is because if twice the square root plus one is added to the subtracted numbers, then the sum will be a square number. That which has this property in no way has an exact square root, as any number which does not have an exact square root is a number that is not a square. If this approximation is continued in order to take the square root of the remainder, the number remaining after the root has been taken is divided, as a whole number, by twice the square root plus one or by twice the square root alone. The result will be a part of unity which is added to the obtained square root, giving a whole number plus some parts of unity. If this sum is multiplied by itself, the product will be the given number whose square root we are trying to find, less a small amount or plus a small amount. If you divide the remaining number by twice the square root plus one, the quotient will be a part of twice the square root and a part of unity,[3] such that the ratio of this part to unity is the ratio of the divided number to twice the square root plus one. Hence, when this number is multiplied by twice the square root plus one, we get back to the divided number. But the product of the part and twice the square root is a part of twice the square root and a part of unity, and it is greater than the square of this part. But if the part of twice the square root plus the square of this part is added to the square of the square root, then the result will be a square. Therefore, if the product of the part and twice the square root plus one is added to the square of the square root, the sum will be greater than the square of the square root plus this part. But the amount by which it exceeds it is the number obtained by multiplying the amount by which unity exceeds the part by the part. It is therefore a very small amount.

If the number remaining is divided by twice the square root alone, the quotient is a part of twice the square root alone. Hence, when this number is multiplied by twice the square root, the result is the divided number. But if the product of the part and twice the square root is added to the square of the square root, then the sum will be a non-square number, and will be less than a square number by the square of the part. If the product of the part and twice the square root plus the square of the part is added to the square of the square root, then the result will be a square number. Hence, if <the sum of> the square root and the part is multiplied by itself, the product will be a square number which exceeds the given number by the square of the part, which is a small quantity. If this is continued for the remainder

[3] That is, a part of the parts whose number is twice the square root plus one.

<dividing> by twice the square root plus one or by twice the square root alone, and if the whole number obtained when taking the square root is added, then the sum is by approximation the square root of the given number whose square root we were looking for. Which of the two methods mentioned in the approximation is preferable, and which should be chosen? In order to decide, we have to experiment with each of the two methods and choose that with the lesser difference. This is what we wished to explain in relation to the causes of the displacement of square roots and their doubling in the Indian calculus, with the Grace of God.

We completed this copy on the 11th of the Second Jumādā in the year 721 in al-Sulṭāniyya.

In the Name of God, the Forgiving, the Merciful

TREATISE BY AL-ḤASAN IBN AL-ḤASAN IBN AL-HAYTHAM

On the Extraction of the Side of a Cube

A cubic number is obtained by multiplying a number by the product of that number and itself. The extraction of the side of the cube is carried out when there is a given cubic number whose side is not known, and when there is a need to know its side, that is, the number which when multiplied by itself and then by itself again gives the given cubic number.

Arithmeticians usually extract the side of a cube using the Indian calculus. We are not aware that any of them has ever described a method for extracting the side of a cube, other than the Indian calculus. But when we examined the properties of this number, it became apparent to us that it is possible to extract the side of this number by the transactions method <of calculation>, without requiring the Indian <calculus>. We have therefore written this treatise. In this treatise, we show how to extract the side of a cube by the transactions method <of calculation>, and also how to extract it using the Indian method, as those reading this treatise may not be familiar with the Indian way and may wish to know of it when it is mentioned. We therefore include it with the transactions method so that those who wish to know this number may learn how to do so by two methods at the same time.

The method for extracting the side of a cubic number, if the cubic number is given, is to take a number – any number – multiply it by itself, and then multiply this product by the first number. The product is then equal to the given number, or is not equal to it. If it is equal to it, then the first number taken is the side of the given cube. If the product is not equal to the given number, then it is either less than or greater than it. If it is greater, the number taken is rejected and another smaller number is taken and this is multiplied by itself. Then, it is multiplied by its square – its square being the product of its multiplication by itself – until the product is less than the given number or is equal to it. If it is equal to it, then the number taken is the side of the given number. If it is less than the given number, the number taken is multiplied by three and its square is multiplied by three, and these are added, and one is added, and this sum is added to the product of the number taken and its square. If this sum is equal to the given number, then one is added to the first number taken, and the sum of the first number and one is the side of the given cube. If the number obtained is not equal to the given number, then it is less than it and it is not greater than it if the given number is a cube. If it is less than it, the number obtained is multiplied – adding one to the first number – by three, and its square is also multiplied by three. The results are all added together, increased by one, and this result is added to the first number obtained which is less than the given number. If that which is obtained is equal to the given number, another one is added to the number obtained from the first number plus one. This will then be the side of the given cube. If it is not equal to it, then it is less than it. We then apply to the second acquired number the same process as was applied to the first acquired number, and we continue to proceed in this way, multiplying the acquired number by three and its square by three, adding one to the sum and adding the result to the first number. Each time, the acquired number is increased by one, until the sum is equal to the given cubic number. If it is equal to it, then the acquired number is the side of the given number, and the number that we have called the 'acquired number' is the number obtained from the first number taken plus the unities successively added to it. The process may also be reduced to the following: The first number taken is multiplied by itself, and then multiplied by its square. This product is then subtracted from the given cubic number. If there is a remainder from the cube, the number taken is multiplied by three, its square is multiplied by three, and all are added together. One is added to the sum, and the result is subtracted from the remainder obtained from the cube, then one is added to the number taken. If there is a second remainder from the cube, the acquired number is multiplied by three, its square is multiplied by three, and one is added to the sum before subtracting the result from the second remainder obtained and adding one to the acquired number. We continue to proceed in this way until the given cubic number disappears and nothing remains. If the cubic number disappears, then the acquired number is the side of the cubic number. If the given number whose side is sought is a cube – and if you follow the procedure that we have described – then the cubic number will eventually disappear leaving no remainder.

Here is an example of the process that we have just described: Let the given cubic number be one thousand, seven hundred and twenty-eight, and we wish to extract its side. We take the number ten which we multiply by itself to give one hundred. Then, we multiply ten by the square and the result is one thousand. Comparing this with the given number – which is one thousand, seven hundred and twenty-eight – we see that it is less than it. We then multiply ten by three, giving thirty, and we multiply one hundred by three, giving three hundred. Adding them gives three hundred and thirty, and adding one gives three hundred and thirty-one. Adding this to one thousand gives one thousand, three hundred and thirty-one, which is still less than the given number. We therefore add one to ten giving eleven, which is the side of the cube one thousand, three hundred and thirty-one as, if eleven is multiplied by itself and the product is then multiplied by eleven, then the result is one thousand, three hundred and thirty-one. We then multiply eleven by itself, giving one hundred and twenty-one. We multiply eleven by three, giving thirty-three, and we multiply one hundred and twenty-one by three, giving three hundred and sixty-three. Adding these two products and adding one gives three hundred and ninety-seven, which we add to the first number obtained – which was one thousand, three hundred and thirty-one – to give one thousand, seven hundred and twenty-eight, which is equal to the given number. Adding one to eleven gives twelve, which is the side of the given cube – one thousand, seven hundred and twenty-eight. If we subtract one thousand from one thousand, seven hundred and twenty-eight, and then subtract three hundred and thirty-one from the remainder, and then subtract three hundred and ninety-seven from the remainder, until the given number disappears, and if we add one to the first number each time, then the process gives the same single result.

In order to verify the validity of this procedure, we multiply the final acquired number – which is twelve in this example – by itself to give one hundred and forty-four, and then multiply twelve by one hundred and forty-four, the result is one thousand, seven hundred and twenty-eight.

But not all numbers are cubic numbers, and the given numbers whose side we are looking for are not all cubic numbers. Any non-cubic number will not have an exact cubic side. It is, however, possible to extract the cubic side of a number that is not a cube by approximation, in the same way that it is possible to extract the square root of a number that is not a square by approximation. Therefore, if we have a given number and we wish to extract its cubic side, we begin by applying the procedure that we have just described. If the number is a cube, it is necessary that the procedure that we have laid out results in a number equal to the given number, such that if the given number is subtracted from it, the number disappears without leaving a remainder. If the number is not a cube, it is necessary that there is a remainder, such that if the acquired number is multiplied by three, and if its square is multiplied by three, and if the two <products> are added together and added to one, then the sum is greater than the remainder. If the process ends in this way, we multiply the acquired number by three, then multiply its square by three, and then divide the remainder of the given number by the square multiplied by three. The results are parts of unity. If these parts are added to the acquired number, the sum will be the cubic side of the given number by approximation.

Example: Let the given number be one thousand eight hundred and we wish to find its cubic side. We apply the procedure described until we obtain twelve, whose cube is one thousand, seven hundred and twenty-eight. If we subtract this number from one thousand eight hundred, either in one operation in the first method, or in several operations in the subtraction method, there remains seventy-two from the one thousand eight hundred, such that, if we multiply twelve by three, and if we multiply its square – which is one hundred and forty-four – by three, and if we add the two <products>, and if we add one to them, then the sum will be four hundred and sixty-nine, which is greater than the remainder which is seventy-two. We multiply the square of twelve – which is one hundred and forty-four – by three giving four hundred and thirty-two, and we divide seventy-two by four hundred and thirty-two giving seventy-two parts out of four hundred and thirty-two, which is equal to one sixth. We then add one sixth to twelve giving twelve and one sixth, which is approximately equal to the cubic side of one thousand eight hundred. In order to verify this, we multiply twelve plus one sixth by twelve plus one sixth, giving one hundred and forty-eight and one part of thirty-six parts. We then multiply twelve plus one sixth by one hundred and forty-eight and one part of thirty-six parts.

SUPPLEMENTARY NOTES

[1] *On the Arithmetic of Transactions*

The case of the treatise *On the Arithmetic of Transactions* is a rather delicate one for two reasons: firstly, the generality of the title, which indicates an area rather than a particular work; and secondly the number of works attributed to Ibn al-Haytham on this subject. The aim of this note is simply to clarify the problem without claiming to solve it.

There are two books attributed to Ibn al-Haytham which deal with this kind of calculation. The first is entitled *al-Mu'āmalāt fī al-ḥisāb (Arithmetic Transactions)*. Two manuscripts of this book are known to have been in existence at this time in Istanbul: Feyzullah 1365/22 fols 73^{v}–164^{r}, and Nur Osmaniye 2978 fols 39–125. Here authorship is spelled out on the first page, with a declaration: 'written by al-Shaykh al-Imām and great scholar al-Ḥasan ibn al-Haytham of Baghdad – may God have mercy on him'. His name also appears a few folios later. 'Shaykh Abū al-Ḥasan ibn al-Haytham said....', followed by some fifteen lines of quotation from the second book. However, it should be understood that the mere mention of his name does not necessarily imply authorship. Although certain modern bibliographers have been tempted to do so, attribution of works to Ibn al-Haytham on this basis, without any other proof, will not bear even cursory examination. In the first place there is no record in any ancient bibliography of Ibn al-Haytham's works under this title, nor any reference to it in any of his other works. Secondly, this is a compilation, not a monograph on the subject. In fact, it begins with a short tract on the use of 'Coptic' figures in calculation, then quotes the fifteen lines already referred to and continues with elementary commercial arithmetic – conversion of weights and money, grain measures, etc. – various purchase problems and practical geometry, etc. Neither the style of composition nor the level of mathematics is anything like that of Ibn al-Haytham. Here is one (not the worst) example: how is stated the area of a circle: 'the area of a circle with diameter seven and circumference twenty-two is calculated by multiplying the diameter by half the circumference, that is three and a half by eleven, which is thirty-eight and a half; that is, the area' (ms. Feyzullah, fol. 129^{v}). Although the result is correct, such a style was not used by Ibn al-Haytham to expound mathematics, not even in cases of practical geometry, such as those in his treatise *On the Principles of Measurement.*

The second treatise is entitled *Discourse of al-Shaykh Abū 'Alī al-Ḥasan ibn al-Ḥasan ibn al-Haytham, renowned for his 'astounding' work on the Arithmetic of Transactions*, and is very different. It comes down to us in two manuscripts; Istanbul, 'Atif 1714/13, fols 116^r–125^r and Berlin Oct. 2970/17 fols 117^r–186^r. There is doubt over the term '*al-gharīb*' – meaning 'astounding'. Al-Ḥasan ibn al-Haytham was not in the habit of using such expressions to describe his works (and this can be proved). On the other hand, this word can be read as '*al-qarīb*' which means 'near' or 'easy'. Moreover, the colophon reads: *Tamma al-kitāb fī ḥisāb al-mu'āmalāt* (The Book on the Arithmetic of Transactions is completed...') without any mention of 'astounding'. And it is this exact title which appears on the two lists of al-Ḥasan and of Muḥammad. The book itself deals with 'the fundamentals of this art', that is the non-demonstrative study of the operations of this arithmetic: proportion, multiplication, division, addition of fractions. The way the text is presented, it could well be the work of al-Ḥasan ibn al-Haytham, but without supporting evidence, it is impossible to be certain.

[2] *The Configuration of the Universe*: a Book by al-Ḥasan ibn al-Haytham?[1]

> 'Our statements on all the motions are only according to the viewpoints of Ptolemy, and according to his opinion' (*Configuration of the World* ed. Langermann, ar. p. 6).
>
> 'From what emerges from the remarks by His Excellency the Shaykh, it is clear that he believes in the word of Ptolemy in all that he says, without relying on a demonstration and without invoking any proof, but by pure imitation; it is thus that the specialists of the prophetic tradition have faith in the Prophets, may God bless them. Yet it is not in this way that mathematicians have faith in the specialists of demonstrative sciences. And I have observed that it is painful to him that I have contradicted Ptolemy, and that he feels resentment because of this; his remarks make it clear that error is foreign to Ptolemy. But there are many errors in Ptolemy, in many passages of his books. Among others, there is what he says in the *Almagest*:

[1] This Supplementary Note has been written after the publication of *Les Mathématiques infinitésimales*, vol. V. It is the development of the initial note entitled '*Treatise on the Configuration of the Universe*, attributed to Ibn al-Haytham'. It was published in the *Revue d'histoire des sciences,* 60, no. 1, janvier–juin 2007, pp. 47–63.

if one examines it attentively, one discovers many contradictions in it. Indeed, he has affirmed principles for the configurations he mentions, then for the motions he proposed configurations in contradiction with the principles he affirmed' (al-Ḥasan ibn al-Haytham, *On the Resolution of Doubts relating to the Winding Movement – Fī ḥall shukūk ḥarakat al-iltifāf*, ms. Leningrad, B1030, fol. 19^v).

Translated into Hebrew, and then into Latin from the Hebrew translation, *The Configuration of the Universe* (*Fī hay'at al-'ālam*) was one of the references for medieval astronomy, as has been shown by P. Duhem, C. Nallino, F. J. Carmody, and many others.[2] The impact of this work on Arabic astronomy was quite weak. It was used only by a few low-ranking astronomers, such as al-Khiraqī, as we shall see below.

Since Duhem, however, a common opinion has become entrenched, according to which the *Configuration of the Universe* represents an essential dimension of Ibn al-Haytham's contribution. Among other reasons for the success of this opinion, commonly widespread among historians, as well as of the book itself during the Middle Ages, are quite certainly the simplicity of its contents, the absence of any mathematical technicality, and above all the combination of the planetary theory of the *Almagest* with a specific cosmology. This success was all the more resounding in that the book bears the name of the prestigious mathematician and physicist Ibn al-Haytham. It is not rare, however, that a great success is the effect of misunderstanding, if not of a mistake. This is precisely what I shall establish here.

Three Arabic manuscripts[3] have come down to us, which give al-Ḥasan ibn al-Haytham as the author; thus, the book as it has been handed down to

[2] P. Duhem, *Le système du monde*, vol. 2: *Histoire des doctrines cosmologiques de Platon à Copernic*, Paris, 1965, pp. 119–26. Y. T. Langermann has given an edition of the Arabic text on the basis of two manuscripts, that of London and that of Turkey (Kastamonu), as well as an English translation. The ensemble, preceded by a 50-page introduction, and an Arabic, Latin and Hebrew glossary, is entitled *Ibn al-Haytham's On the Configuration of the World*, New York/London, 1990.

[3] This book has come down to us in three manuscripts: 1) London, India Office, Loth 734, fols 101–116. It is thus part of collection that was written late, approximately in the seventeenth century. 2) Kastamonu (Turkey) 2298, fols 1–43. We do not know the date of the transcription of this manuscript. As Y. T. Langermann has observed, it is marred by serious omissions. 3) Rabat, Bibliothèque Royale, no. 8691, fols 29^r–48^r.

It can be verified that these three manuscripts differ in pairs, and that some of these differences are irreducible; all these elements show that the transmission of the text raises serious problems, which remain to be studied. What is even more serious is that the London manuscript includes a gloss, added at the end, which reads as follows: 'gloss (*ta'līq*) we found in the hand of the Shaykh, may God prolong his life; we have therefore

us would be the work of the eminent eleventh-century mathematician. However, the latter's name has undergone a transformation in two of these three manuscripts. Thus, instead of al-Ḥasan ibn al-Ḥasan, we find Abī al-Ḥasan, which already shows that we are dealing with the act of a scribe. As far as the third manuscript is concerned, which is relatively late,[4] it contains several treatises by al-Ḥasan ibn al-Haytham, which may have induced the scribe to normalize the name. As we shall see, however, this attribution raises numerous epistemic and historical problems, problems which are of major importance, but are never tackled head-on by historians.

Some fifteen years ago, on the occasion of the first critical study of the biobibliographical and historical sources concerning the activities and titles of al-Ḥasan ibn al-Haytham, a study which was preliminary to the critical edition of his mathematical works, we had shown that there were two personages who had eventually been confused by the vicissitudes of history: the mathematician al-Ḥasan ibn al-Haytham and the philosopher and physician Muḥammad ibn al-Haytham. We had also raised doubts concerning the validity of the attribution to al-Ḥasan ibn al-Haytham of the *Configuration of the Universe.*[5] A few years later, however, the appropriateness of our method was questioned, and it has been thought possible to affirm that the two personages were in fact one and the same, and the authenticity of the attribution of *The Configuration of the Universe* to al-Ḥasan ibn al-Haytham has been clearly proclaimed.[6] Several arguments were advanced, only one of which deserves to be considered, since all the others are either purely rhetorical, or else the result of

---

copied it as we have found it' (fol. 116^{r}; ed. Langermann, ar. p. 66). On the identity of this Shaykh, who was still alive since the scribe wishes that his life should be prolonged, we know nothing whatsoever. This attributed gloss, which is uncertain at best, sets forth common ideas on celestial motions, as well as a few vague elements of an Aristotelian cosmology. Without any precaution, people have hastened to attribute it to al-Ḥasan ibn al-Haytham, at the conclusion of a comparison, as arbitrary as it is vague, with a few general phrases by Ibn al-Haytham in *The Light of the Moon*. It was this gloss that misled a scholar as great as the late M. Schramm (*Ibn al-Hayt͟hams Weg zur Physik*, Wiesbaden, 1963, pp. 63ff.).

[4] The London manuscript, India Office, Loth 734.

[5] *Vide supra*, pp. 1–17. See also vol. III: *Ibn al-Haytham. Théorie des coniques, constructions géométriques et géométrie pratique*, London, 2000; and particularly the *addenda*, pp. 937–41, entitled 'Al-Ḥasan ibn al-Haytham et Muḥammad ibn al-Haytham: le mathématicien et le philosophe'; vol. IV: *Méthodes géométriques, transformations ponctuelles et philosophie des mathématiques*, London, 2002, pp. 957–9, on al-Ḥasan ibn al-Haytham and Muḥammad ibn al-Haytham: the mathematician and the philosopher: *On Place*.

[6] Cf. A. I. Sabra, 'One Ibn al-Haytham or two?', *Zeitschrift für Geschichte der arabischen - islamischen Wissenschaften*, 12, 1998, pp. 1–51, particularly pp. 19–21.

misunderstanding the mathematical contents of Ibn al-Haytham's work: we have, moreover, already refuted the main ones.[7] This argument relies primarily on the reproduction of the colophon of one of the three manuscripts of the *Configuration of the Universe*, where the scribe attributes it explicitly to al-Ḥasan ibn al-Haytham – note that the colophons of the other two manuscripts do not provide any information.[8] We will show in what follows that this argument is also erroneous and that the objections raised are based on a very weak argument.

If we insist on replying here to this question of authenticity, it is not only in order to rectify a historical error, but because it concerns our conception of the astronomy of al-Ḥasan ibn al-Haytham. Indeed, if it is demonstrated that this book is not, and cannot belong, among the latter's works, then we must conclude that the interpretations given of his astronomy by historians on the basis of this book are to be rejected. According to these interpretations, Ibn al-Haytham's astronomy is descriptive rather than demonstrative, like the *Configuration of the Universe*, where the author combines the planetary theory of the *Almagest*, just as it is, with a cosmology of Aristotelian origin. In fact, this is an inaccurate image of Ibn al-Haytham's astronomy.[9] Let us begin by recalling a few well-established facts, taking care to distinguish them from conjectures.

The biography as well as the bibliograpecounted by several ancient authors: the principal ones among them are al-Qifṭī (568/1172–646/1248), Ibn Abī Uṣaybi'a (596/1200–668/1270), and an Anonymous, whose text is found in a manuscript of Lahore.[10] The latter is the oldest one, since it was copied in 1161 at Niẓāmiyya in Baghdad.

A second fact, which has not been noticed, has given rise to serious confusion: al-Qifṭī gives only the list of writings of al-Ḥasan ibn al-Haytham, and never mentions the name of 'Muḥammad' ibn al-Haytham. The list given by al-Qifṭī is almost exclusively of his writings in the mathematical sciences, and once a verification is made, all the works of al-Ḥasan that have

[7] See note 4, particularly vols III and IV.

[8] Indeed, the London manuscript does not contain a colophon. As far as the Moroccan manuscript is concerned, it informs us that the copy was completed on 'Sunday the third of Rajab one thousand two hundred ninety-one' of the Hejira, that is in 1874. In this manuscript as in the one from Kastamonu, Abū al-Ḥasan is written instead of Ibn al-Ḥasan.

[9] This debate is all the more urgent in that we have just completed the *editio princeps* as well as the first translation of the main work of Ibn al-Haytham on astronomy, spherical geometry and trigonometry (*Les Mathématiques infinitésimales*, vol. V).

[10] We shall call him the Anonymous of Lahore.

come down to us – with two exceptions – also appear on this list by al-Qifṭī.

The Anonymous of Lahore and Ibn Abī Uṣaybiʿa had the same source: an autograph manuscript of Muḥammad ibn al-Ḥasan in which the latter, after having narrated a few elements of his biography, or rather of his philosophical biography, gives a list of his writings down to the year 417/1026. Immediately after, he gives a complementary list of his writings down to the year 419/1028. However, there are two important differences, which have gone unnoticed, between the Anonymous of Lahore and Ibn Abī Uṣaybiʿa. Immediately after this list, the Anonymous of Lahore transcribes a list of the writings of al-Fārābī, after a copy by the Baghdad judge Ibn al-Murakhkhim. It is after this list that he transcribes a list of the writings of al-Ḥasan ibn al-Haytham. In other words, the Anonymous of Lahore has not confused either al-Ḥasan and Muḥammad, or their lists.

The exposition by Ibn Abī Uṣaybiʿa is quite different. He is not a scribe, but a biobibliographer, like al-Qifṭī, and he therefore intends to write an article on Ibn al-Haytham in his dictionary. He begins his article with an introduction in which he borrows, particularly from al-Qifṭī, some facts which the latter attributes to al-Ḥasan ibn al-Haytham. However, since he will soon follow this introduction by the philosophical autobiography as well as the lists of Muḥammad ibn al-Ḥasan, he combines the two personages and composes a new portrait, that of an alleged 'Abū ʿAlī Muḥammad ibn al-Ḥasan ibn al-Haytham'. No author and no source before Ibn Abī Uṣaybiʿa mention such a personage, and no work by al-Ḥasan ibn al-Haytham, any more than a commentator on his works, as we have shown, mentions such a name. It is this confusion that Ibn Abī Uṣaybiʿa fell into that has misled the biobibliographers and historians.[11]

Whatever may be the case for the examination of the list of works by Muḥammad ibn al-Ḥasan given by Ibn Abī Uṣaybiʿa, it clearly emerges that he had at his disposition the same autograph manuscript that was available to the Anonymous of Lahore. It remains true that once he has copied these lists, Ibn Abī Uṣaybiʿa follows them with the list of the works of al-Ḥasan ibn al-Haytham. He introduces this list with the words: 'this is also a list

[11] Anton Heinen edited the text of the manuscript of Lahore while eliminating the list of al-Fārābī that separates the work by Muḥammad ibn al-Ḥasan and the list of al-Ḥasan. He simply remarked in a note: 'An dieser Stelle, auf derselben Seite und in derselben Hand, folgt das Verzeichnis der Werke al-Fārābīs' ('Ibn al-Haiṯams Autobiographie in einer Handschrift aus dem Jahr 556 H./1161 A.D.', in Ulrich Haarmann und Peter Bachmann (eds), *Die islamische Welt zwischen Mittelalter und Neuzeit, Festschrift für Hans Robert Roemer zum 65. Geburtstag*, Beirute Texte und Studien, vol. 22, Beirut, 1979, pp. 254–77, at p. 272, n. 27). Following Ibn Abī Uṣaybiʿa, he has confused the two authors.

(*fihrist*) that I found of the books of Ibn al-Haytham, down to the end of the year four hundred twenty-nine'.[12] Except for some differences in the order,[13] this list resembles the one given by the Anonymous of Lahore and by al-Qifṭī. The fact that Ibn Abī Uṣaybiʿa adds this list at the end, and introduces it in these terms, shows that it existed independently of the autograph copy of Muḥammad ibn al-Haytham, since it had been found in the manuscript of Lahore.

A third fact is equally indisputable: the examination of these two lists, that of Muḥammad and that of al-Ḥasan, shows that the writings of Muḥammad deal with philosophy, and particularly medicine, or are commentaries, didactic in intention, on ancient scientific texts, such as the *Elements*, *Almagest*, as well as a work by Menelaus, whereas the works of al-Ḥasan deal with research problems, often cutting-edge, in the mathematical sciences. To take only one example, that of analysis and synthesis, which is significant because both men deal with it: whereas Muḥammad wrote his treatise 'by the method of examples for students',[14] al-Ḥasan[15] tackled research problems that were still alive in the eighteenth century, such as the reciprocal of Euclid's theorem on perfect numbers, or the problem of three tangent circles: that is, problems of advanced research, far removed from any didactic intention. We can see the distance that separates the two projects.

A fourth fact is particularly important: the commentaries on the ancients by Muḥammad that have come down to us under his name – those on the *Almagest* and on the work by Menelaus – are repetitive and didactic paraphrases. Now, of the works by al-Ḥasan ibn al-Haytham that appear on the list, the only ones – and they are rare – that could belong to the genre of the commentary are *rectificative*: that is, they deal with the solution of aporias by Euclid or Ptolemy, and *foundational*, in the sense that they go back to the very foundations, like his commentary on the postulates of the *Elements*.

A fifth fact, which has also gone unnoticed, is the testimony of ancient authors who had access to the works both of Muḥammad and of al-Ḥasan: thus, the philosopher Fakhr al-Dīn al-Rāzī does distinguish the two authors.[16]

[12] Ibn Abī Uṣaybiʿa, *ʿUyūn al-anbāʾ fī ṭabaqāt al-aṭibbāʾ*, ed. N. Riḍā, Beirut, 1965, p. 559.

[13] *Vide supra*, pp. 8–12.

[14] Ibn Abī Uṣaybiʿa, *ʿUyūn al-anbāʾ fī ṭabaqāt al-aṭibbāʾ*, p. 555.

[15] *Les Mathématiques infinitésimales*, vol. IV.

[16] *Les Mathématiques infinitésimales*, vol. III.

These facts, which are far from being the only ones, all lead to the same conclusion: there were two homonymous personages, al-Ḥasan ibn al-Ḥasan ibn al-Haytham and Muḥammad ibn al-Ḥasan ibn al-Haytham. The first was the famous mathematician; as far as the second one is concerned, he was a philosopher-doctor, familiar with the sciences, like many philosophers in the tradition of al-Kindī, but without being an inventive scientist himself. Homonyms, these two personages are also contemporaries, have the same origin (southern Iraq), and were perhaps related. Yet whereas the mathematician emigrated to Cairo, the philosopher remained in Iraq.

Since confusions tend to be long-lived, a meticulous examination of the authenticity of the *Configuration of the Universe* is essential. Once again, let us recall a few facts.

Whereas the three ancient biobibliographers inscribe the title of the *Configuration of the Universe* on the list of al-Ḥasan, Ibn Abī Uṣaybi'a and the Anonymous of Lahore inscribe it twice, once on the list of Muḥammad, and once on that of al-Ḥasan – which might have alerted historians and induced them to discuss the authenticity of the attribution of the *Configuration of the Universe*. Yet nothing of the sort occurred. However we interpret this double attribution, if we persist in considering that Muḥammad and al-Ḥasan are one and the same person, we can only end up with an absurdity. Indeed, in that case we would have to accept that the same author wrote two different books, at two different times, with the same title, and without pointing it out: a conclusion which is all the less plausible in that it is not supported by any argument. To be sure, we could impute the responsibility for this to the author of the source (or the sources) of the works of al-Ḥasan ibn al-Haytham consulted by Ibn Abī Uṣaybi'a and the Anonymous of Lahore, or else to these authors themselves. Yet since this title appears on the list of the works of al-Ḥasan established by al-Qifṭī, without any link to the source on which Ibn Abī Uṣaybi'a and the Anonymous of Lahore depend, there is nothing to induce us to conclude that the last two bibliographers or their source could have committed such an error. For the moment, we must restrict ourselves to the bibliographical fact, prepared to discuss it later on: there are two books entitled *Configuration of the Universe*, one attributed to Muḥammad and the other to al-Ḥasan.

The arguments advanced in favour of the authenticity of the *Configuration of the Universe* such as it has come down to us boil down to two: 1) the title cited by the ancient biobibliographers; and 2) the colophon of one of the manuscripts of the *Configuration of the Universe*.

1) The three ancient biobibliographers indeed cite the title of the *Configuration of the Universe* among the works of al-Ḥasan. Let us recall a

well-known fact, however: before the beginnings of printing, titles were scarcely stable, and the variations they underwent were sometimes considerable. Among many other examples, let us take the list of the writings of al-Ḥasan established by al-Qifṭī, the great majority of whose titles are authenticated. Al-Qifṭī cites a book by al-Ḥasan which he entitles *The Sphere is the Greatest of the Solid Figures* (*al-kura awsaʿ al-ashkāl al-mujassama*),[17] instead of giving it its real title: *On the Sphere, which is the Greatest of all the Solid Figures having Equal Perimeters, and on the Circle which is the Largest of all the Plane Figures having Equal Perimeters.*

It is easy to see that the title given by al-Qifṭī is incomplete and less detailed. The argument from the title, especially when invoked by an ancient biobibliographer, should be handled with infinite precaution. In the case of the *Configuration of the Universe*, attributed to al-Ḥasan because he had written another book whose title begins with the same expression, 'On the configuration of...', viz. *On the Configuration of the Movements of Each of the Seven Planets* (*Fī hay'at ḥarakāt kull wāḥid min al-kawākib al-sabʿa*), a title that is not cited by any ancient bibliographer, any more than his capital work on *The Completion of Conics*, increased precautions are called for.

We now come to the colophon of one of the three manuscripts that have come down to us.

Here is the colophon:

وكتب هذا الكتاب من النسخة التي نُسخ [كذا] من نسخة الشيخ أبي القسم السميساطي بخطه
ذكر أنه نقلها من نسخة بخط مصنف الكتاب الشيخ أبى على الحسن بن الحسن بن الهيثم وقابل
عليها من أولها إلى آخرها في رجب من سنة ست وسبعين وأربعمائة.

This book was transcribed from the copy that was transcribed from the copy of the Shaykh Abū al-Qasam al-Sumaysāṭī in his hand. He (al-Sumaysāṭī) mentioned that he had transcribed it from a copy in the hand of the author of the book, Abū ʿAlī al-Ḥasan ibn al-Ḥasan ibn al-Haytham; and that he (al-Sumaysāṭī) compared it with the latter from the beginning to the end, in the <month> of Rajab of the year four hundred seventy-six <1083>.[18]

[17] Al-Qifṭī, *Ta'rīkh al-ḥukamā'*, ed. J. Lippert, Leipzig, 1903, p. 168.

[18] Ms. Kastamonu 2298, fol. 43^r. The meaning of the Arabic is clear. *qābala* cannot be translated by 'was checked'. There is no need to be a great philologist in order to grasp that the subject of this active verb is the same as that of *dhakara* and of *naqala*, that is, al-Sumaysāṭī. However, it is this strange error in translation (which is not made by Langermann, cf. p. 43), that compromises Sabra's entire argumentation, pp. 19–20 and n. 34.

According to this colophon, the grandfather of the Kastamonu manuscript would be the one belonging to al-Sumaysāṭī, which is supposed to be have been copied by him in 476/1083, after the autograph of al-Ḥasan ibn al-Haytham. If this information could be verified, one would have a strong argument in favour of the authenticity of the attribution of the *Configuration of the Universe* to al-Ḥasan ibn al-Haytham; all the more so in that al-Sumaysāṭī is a younger contemporary of the latter. But nothing could be further from the truth: the colophon is more than dubious. Abū al-Qasam al-Sumaysāṭī is not an unknown figure. He has left us a short treatise on isoperimetric figures (mss Istanbul, Carullah 1502 and Beshir Aga 440).[19] In addition, the ancient biographers and historians inform us about his dates and some of his activities. Thus, Ibn al-'Imād, in his *Shadharāt al-dhahab*, lists at the same time among personalities who died in the course of the year 453/1061, the doctor Ibn Riḍwān, a contemporary of al-Ḥasan ibn al-Haytham:

وفيها (توفي) أبو القسم السميساطي واقف الخانكاه قرب جامع بني أمية بدمشق. [...] على بن محمد بن يحيى السلمى الدمشقى، روى عن عبد الوهاب الكلابي وغيره، وكان بارعا فى الهندسة والهيئة، صاحب حشمة وثروة واسعة عاش ثمانيين سنة.

Abū al-Qasam al-Sumaysāṭī, who has bequeathed a hospice, next to the mosque of the Omeyyads at Damascus, as a religious bequest [...], 'Alī ibn Muḥammad ibn Yaḥyā al-Sulamī of Damascus, who recited <the prophetic word> according to 'Abd al-Wahāb al-Kilābī, and others; he excelled in

---

This colophon is followed by a gloss, clearly separated from it in the manuscript, of which the following is a translation, as literal as possible: 'the copy from which this copy has been transcribed has been compared with the copy of above-mentioned origin, which is in the hand of the Shaykh Abū 'Alī ibn al-Haytham. It was rectified, thanks be to God, the Lord of worlds, and transcribed in the month of Rajab of the aforementioned year' (ms. Kastamonu 2298, fol. 43^r).

والنسخة المكتوبة منه [كذا] هذه النسخة عورض بها النسخة الأصل المذكور وهو بخط الشيخ أبي علي [كذا] بن الهيثم وصحح والحمد لله رب العالمين وكتب في رجب من السنة المذكورة.

Despite the mistakes in Arabic, as numerous as they are gross, which disfigure these few lines (and which can only confirm our doubts with regard to the quality of the information transmitted), it seems that the scribe confines himself to extracting from the colophon the information necessary for specifying the relationship between the manuscript he has just copied and the autograph of the author. He therefore suppresses the mention of the name of al-Sumaysāṭī, which is useless for his purposes, and contents himself with saying that the copy [i.e. that by al-Sumaysāṭī] on the basis of which 'this copy' [i.e. what is now the ms. Kastamonu] was transcribed, that is, from which it *derives*, is a copy made after the autograph of the text, then compared with it.

[19] See *Les Mathématiques infinitésimales*, vol. I, pp. 830–2.

geometry and in astronomy. He had a retinue of servants and a great fortune, and he lived for eighty years.[20]

This information is confirmed by other classical historical and biobibliographical sources, for instance Ibn ‘Asākir, Yāqūt, al-Dhahabī, al-Nu‘aymī. Thus, in his *History of Damascus*, Ibn ‘Asākir evokes Dār al-Sūfiyya, that is, the house of the Sufis, the one bequeathed by al-Sumaysāṭī.[21] More important is the article devoted to the town of Sumaysāṭ by Yāqūt in his *Dictionary of the Lands* (*Mu‘jam al-buldān*). Thus, after having given its geographical coordinates, he writes:

> وإليها ينسب أبو القاسم علي بن محمد السميساطي السلمي المعروف بالجميش، مات بدمشق في شهر ربيع الآخر سنة ٤٥٣ ودفن في داره بباب الناطفانيين، وكان قد وقفها على فقراء المسلمين والصوفية.

> Related to the latter is Abū al-Qāsim ‘Alī ibn Muḥammad al-Sumaysāṭī al-Sulamī, known as al-Jamīsh, who died at Damascus in the month of Rabī‘ al-ākhar of the year 453. He was buried in his house at Bāb al-Nāṭafā’iyyin, which he had bequeathed as a religious bequest to Muslims and to poor Sufis.[22]
> And he continues :

> وكان يذكر أن مولده في رمضان سنة ٣٧٧.

> He (al-Sumaysāṭī) mentioned that his birth was <in the month> of Ramaḍān of the year 377.[23]

Al-Dhahabī provides the same information, but with the difference that he gives as a birth date the month of Ramaḍān, in the year three hundred seventy-four.[24] As far as al-Nu‘aymī[25] and Ibn Taghrībardī[26] are concerned, they repeat the same information.

[20] *Shadharāt al-dhahab fī akhbār man dhahab*, Beirut, n.d., vol. II, p. 291.

[21] Ibn ‘Asākir, *Tārīkh Madīnat Dimashq*, vol. 43, ed. Sakīna al-Shīrabī, Damascus, 1993, p. 13.

[22] *Mu‘jam al-buldān*, Beirut, n.d., vol. III, p. 258.

[23] Yāqūt writes ‘Abū al-Qāsim’ instead of ‘Abū al-Qasam’. This confusion may be due to the scribes, or else to Yāqūt himself; we find it in other biobibliographers, but it is not consequential, since the first names are the same, as are the last names and the dates.

[24] Al-Dhahabī, *Siyar a‘lām al-nubalā’*, ed. S. al-Aranu’ūṭ *et al.*, Beirut, 1984, vol. XVIII, pp. 71–2.

[25] Al-Nu‘aymī, *al-Dāris fī ta’rīkh al-madāris*, ed. Ja‘far al-Ḥasanī, Damascus, 1951, vol. II, pp. 151–2. He gives al-Sumaysāṭī’s birth date as 373/983–984, which confirms the date given by Yāqūt.

[26] Ibn al-Taghrībardī, *al-Nujūm al-zāhira fī muluk Miṣr wa-al-Qāhira*, 12 vols, Beirut, 1992, vol. V, pp. 70–2.

Other historians inform us about certain facts concerning al-Sumaysāṭī, such as al-Dhahabī, and all of them agree on the date of his death: the year 453.

Thus, born in 374/984 or 377/987, al-Sumaysāṭī lived for 79 or 76 lunar years before dying in 453/1061, at Damascus. This confirms the global estimate of Ibn al-ʻImād: 80 lunar years of life.

Now these dates are in flagrant contradiction with the colophon. Indeed, if we accept, according to the colophon, the date of 476/1083 for the transcription of the manuscript, al-Sumaysāṭī would have carried it out at the age of 102 or 99 lunar years. Yet this fact, in its singularity, would not have struck historians. This is more than improbable, even impossible.

If, on the other hand, we are to believe the unanimous biographers, we should also accept that al-Sumaysāṭī would have transcribed the *Configuration of the Universe* twenty-two years after his death. This is unlikely.

However we envisage the date given by the colophon, we observe that it is severely erroneous. Is this the scribe's mistake, or the intervention of a forger, as sometimes happens? It is impossible to decide, since we know nothing of the scribes, their dates, and their places of activity. In any case, nothing can be established on the basis of such a fanciful colophon.

It is thus obvious that neither the title, nor the colophon, allow us to discuss the validity of the attribution of the *Configuration of the Universe* to al-Ḥasan. We must therefore turn towards the book itself and its contents, in order to compare it with al-Ḥasan's other works on astronomy.

Now, as soon as we study the *Configuration of the Universe* as the book has come down to us, as well as its contents, its attribution to al-Ḥasan seems indefensible.

According to the dates given by Ibn Abī Uṣaybiʻa and the Anonymous of Lahore, the *Configuration of the Universe* attributed to Muḥammad was written before 417/1027, when the philosopher was 63 years old. Still according to the same sources, the *Configuration of the Universe* attributed to al-Ḥasan was written before 429/1038. If, therefore, we suppose that Muḥammad and al-Ḥasan are one and the same person, we would necessarily have to admit that this last writing – that is, that of the *Configuration of the Universe* mentioned in the list of al-Ḥasan – was carried out between 1027 and 1038, that is, between the author's sixty-third and his seventy-fourth year. If, moreover, we recall that al-Ḥasan died soon after 1040, these would be the last years of his life. Yet this hypothesis is not only adventurous, but leads to irreducible contradictions.

Indeed, other testimonies[27] inform us that al-Ḥasan had written his book entitled *The Resolution of Doubts Concerning the Almagest* in this same period of time (between 1027 and 1038), that is, *after* 1028. In this work, he declares without the slightest ambiguity that 'the aporias (in the *Almagest*) are much too numerous to be enumerated'. We should also note that in this work, al-Ḥasan ibn al-Haytham cites his *Book of Optics*, which contains the reform with which we are familiar, as well as a radical critique of the visual ray doctrine. Now, in the *Configuration of the Universe*, which, according to Ibn Abī Uṣaybiʿa and the Anonymous of Lahore, should have been written in the period 1027–1038, the author adheres to this refuted doctrine without any nuances. Here is what we read: 'The ray emanates from our eyes in the form of a cone whose summit is the point of the eye, and whose base is the surface of the visible object.'[28] We also know that is his *Treatise on the Light of the Moon*, written relatively early, since the doctor Ibn Riḍwān had copied it at Cairo on Friday August 7, 1031, Ibn al-Haytham criticized the doctrine according to which the moon is a polished body that reflects the light of the sun. Yet this is precisely the doctrine adopted by the author of the *Configuration of the Universe*. Indeed, he writes that the moon 'is a polished body which, if the sun is facing it, receives its light, and this light is reflected on its surface towards the earth'.[29] Thus, in Proposition 7 and those that follow of his treatise *On the Light of the Moon*, al-Ḥasan ibn al-Haytham demonstrates that 'the light emanating from the moon to the earth is not by reflection'. In these conditions, we should conclude that he is in full contradiction, which is absurd.

In addition, according to Ibn Abī Uṣaybiʿa and the Anonymous of Lahore – if we stick to the hypothesis identifying him with Muḥammad – he had written several books, *all of which* were critical of Ptolemy (*Doubts on Ptolemy, On the Winding Motion, Resolution of Doubts relating to the Winding Motion*),[30] between 1027 and 1038. In the *Configuration of the Universe*, however, the author's starting point is perfectly transparent. He writes: 'Our statements on all the motions are only according to the

[27] See above.

[28] Ed. Langermann, ar. p. 42:

والشعاع يخرج من أبصارنا على شكل مخروط رأسَه نقطة البصر وقاعدته سطح جرم المبصر.

[29] Ed. Langermann, ar. p. 44:

وذلك أن القمر لا نور له وإنما يكتسب النور من ضوء الشمس وهو جسم صقيل إذا قابلته
الشمس قبل نورها واستنار بضوءها وانعكس ذلك النور من سطحه إلى الأرض فأنارت به.

[30] See *Les Mathématiques infinitésimales*, vol. V.

viewpoints of Ptolemy, and according to his opinion',[31] that is, without any possible dispute, according to the planetary theory set forth in the *Almagest.* And in fact, the author follows Ptolemy's work step by step: he speaks of the *prosneusis* (*al-muḥādhāt*[32]) of the moon, whereas al-Ḥasan gets rid of it in his writings; and of the equant, whereas the latter rejects it, etc. In other words, if we stick with this position, the mathematician Ibn al-Haytham, in the course of the same years, would have written one thing and its contrary.

Yet this is not the only absurd conclusion. Indeed, the explicit goal of the *Configuration of the Universe* is to present, on the basis of Ptolemy, the orbs of the planets in terms of simple and continuous motions of the solid spheres. The goal is thus to wed the planetary theory of the *Almagest* to a cosmology inspired by Aristotelian philosophy, yet without raising any of the technical problems raised by such a project, and without solving any of the difficulties in mathematical astronomy that follow from it. Yet it suffices to skim through the works of al-Ḥasan, in astronomy as well as in optics or in statics, to observe that these technical questions are always important to him, and that, at any rate, these works are at a theoretical and technical level that is incomparably higher than that of the *Configuration of the Universe.* In all his astronomical works, without exception, al-Ḥasan deals with all the required technicality, with the problem of the combination of geometrical models with the terms of a description of celestial motions. To be sure, it sometimes happens, as in his work on the winding motion,[33] that he studied the combination of the geometrical model with a physical description of the motion, but always with the technicality the subject demands. He always behaves as a mathematical astronomer, whereas the *Configuration of the Universe* is rather the work of a philosopher.

To these numerous and irreducible differences between the project, the method and the style of al-Ḥasan and those of the author of the *Configuration of the Universe*, we can add a few other arguments based on reliable data, which are just as flagrant. The parameters mentioned by the

[31] Ed. Langermann, ar. p. 6.

وقولنا في كل الحركات إنما هو بحسب رأي بطلميوس فيها واعتقاده.

[32] Ed. Langermann, ar. p. 42.

[33] Indeed, in *The Winding Motion*, he engages in a technical discussion to show the error committed by Ptolemy when he supposes that portions of the sphere cause the epicycle to move. He demonstrates that such an assumption leads either to one of the spherical portions straying from its location, or else to the epicycle's being subject to a tipping motion; in other words, to two impossibilities. It is always problems of mathematical astronomy that he raises, as for instance that of the diameters that remain in the vicinity of the centre of the ecliptic. There is nothing in common here with the style of the *Configuration of the Universe*.

author of this treatise are those of Ptolemy, with no reference to the works of astronomers of the ninth and tenth centuries:

1) In §144, the obliquity of the ecliptic is given as 'close to 24 degrees'. Ptolemy had in fact given a value of 23°51′ whereas Arab astronomers, from the very beginning, had found this value to be equal to 23°33′ (or 23°35′, depending on the author) and the value of 23°51′ had been discarded.[34] However, this argument in isolation is not sufficient, given that his book *On the Lines of Hours*, al-Ḥasan ibn al-Haytham uses the value of 24 degrees.

2) In §195, the apogee of the sun on the ecliptic is given as described by Ptolemy (after Hipparchus) as 24°30′ from the summer solstice, in the opposite direction to the signs. This value was recalculated at the beginning of the ninth century and was found to equal 9°15′ from the same point. Then al-Battānī, at the beginning of the tenth century, improved the calculation and found it to be 7°43′. Furthermore, the author of the treatise in question also recalled that Ptolemy had confirmed that this apogee was fixed on the ecliptic, whereas 'more recent astronomers' had found that this apogee was moveable in the direction of the signs, but without giving further details. It was accepted from the beginning of the ninth century that the apogee of the sun was inextricably linked to the movement of precession.

3) The rate of precession, for Ptolemy, was one degree per century; this value is referred to a further three times (§286, 350, 361), whereas it had been accepted, at the beginning of the ninth century, after works had been regrouped in the 'Verified Table' (*al-zīj al-mumtaḥan*), that the value of precession was about one and a half degrees per century.

It is very clear that these precise points were not known to the author of this treatise, despite the fact that the corresponding results had been made known in all scientific circles of eleventh century, including one would have thought, Ibn al-Haytham's.

Moreover, in §381, the author of the *Configuration of the Universe*[35] calculates the celestial motions that figure in the *Almagest*. He counts forty-seven: one for diurnal motion, one for precession, eighteen for the three upper planets, two for the sun, eight for Venus, nine for Mercury, six for the moon, and two for the sublunary world (the heavy and the light).[36] Here

[34] Parameters defined in the ninth century, cf. Thabit ibn Qurra, *Œuvres d'astronomie,* ed., trans. and com. by Régis Morelon, Paris, 1987, p. 8 for the obliquity of the ecliptic and pp. 240–67 for the position of the apogee of the sun and the value of the constant of precession, with an introduction and corresponding supplementary notes.

[35] Ed. Langermann, §138, p. 25.

[36] Ed. Langermann, ar. p. 65.

again, the author recalls that he is relying on 'his [Ptolemy's] research and his observations for all the celestial motions'. Now, in the *Doubts on Ptolemy* (*Fī al-shukūk 'alā Baṭlamiyūs*),[37] al-Ḥasan makes the same calculation, but only for the motions of the seven stars; he finds thirty-six motions. Indeed, he does not count the first two, and obviously leaves aside the last two; and he counts one less motion for each of the stars, since he naturally leaves out the diurnal motion for each of them, insofar as this motion is inclusive. This difference in a simple reckoning of motions is enough to distinguish, on the one hand, al-Ḥasan ibn al-Haytham, who understood what was at issue; and on the other the author of the *Configuration of the Universe*, a commentator in the tradition of Ptolemy, like Muḥammad ibn al-Haytham. Still other facts of the same nature could be added, some of which have not ceased to intrigue the editor of the *Configuration of the Universe.*

Projects, methods, styles and scientific facts, whether in astronomy or in optics: everything opposes the writings of al-Ḥasan and the *Configuration of the Universe.* The historical argument – the only one set forth – that of the colophon, is inconsistent and fallacious. To attribute the *Configuration of the Universe* to al-Ḥasan is to maintain the confusion between the authors and the writings; but it is also to falsify the interpretation of the latter's astronomy, and it is finally to accuse him of a serious scientific schizophrenia, which has never manifested itself in any of the numerous other domains he engaged in. To maintain, as has recently been done, arbitrarily and without the shadow of a proof, that we are in the presence of a work of the author's youth, does not resist the argument of the dates given by the ancient biobibliographers. We have recalled that after 1028, al-Ḥasan was struggling with the aporias of the *Almagest*, and therefore far from following Ptolemy slavishly, as is done by the author of the *Configuration of the Universe* both in astronomy and in optics. The contradiction is even more striking if, like Ibn Abī Uṣaybi'a and the Anonymous of Lahore, we place its writing between 1027 and 1038. In addition, to show that this is a work of the author's youth, one would have to explain rigorously by what paths it leads to the works of maturity. But none of those who hazard such an affirmation has ever set out in search in of these paths, none of which, moreover, in our opinion, links the *Configuration of the Universe* to the other works of al-Ḥasan ibn al-Haytham.

When did this confusion of attribution take place? Everything indicates that it is already present in the biobibliographers – in Ibn Abī Uṣaybi'a – as

[37] Text edited by A. I. Sabra and N. Shehaby, Cairo, 1971, pp. 39–41.

well as in such third-class astronomy professors as al-Khiraqī.[38] It is to be noted, however, that no great astronomer has ever, to my knowledge, fallen into this confusion. Thus, al-ʻUrḍī cites the *Doubts*, al-Ṭūsī evokes the *Winding Motion*, but neither one nor the other, nor any other astronomer of their stature, associates the name of al-Ḥasan with the *Configuration of the Universe*.

We shall thus affirm clearly, and with no risk of being proven wrong, that the *Configuration of the Universe* as we have it is not a work of al-Ḥasan ibn al-Haytham, but is quite probably a book by Muḥammad ibn al-Haytham. As far as the title *Configuration of the Universe*, attributed to al-Ḥasan, is concerned, it would be that of a book that has never come down to us or the result – and here we enter the realm of conjecture – of a modification of the title of his book *On the Configuration of the Motions of Each (kull,* all*) of the Seven Stars*, which could have been written *On the Configuration of the Motions of the all* ... Yet while it is true that this conjecture must await the confirmation of future research, the attribution of the *Configuration of the Universe* to al-Ḥasan is henceforth indefensible.

One may well be astonished that the *Configuration of the Universe* could have been attributed to al-Ḥasan ibn al-Haytham, all the more in that the case is far from being unique. Recently, in fact, and without the slightest hesitation, the eminent mathematician and astronomer has been seen as the author of a *Commentary on the Almagest*, which is of pure Ptolemaic obedience, and what is more, explicitly attributed to Muḥammad ibn al-Haytham.[39] In conclusion, let us note that here as elsewhere, we have been forced, in order to undertake a critical philological examination and carry out a rigorous history of the textual tradition, to have recourse to the conceptual tradition, that is, to the examination of the scientific contents of the text. Is there any other way?

[3] Ibn Sinān and Ibn al-Haytham on the subject of 'shadow lines'

Other works of al-Ḥasan ibn al-Haytham offer additional arguments to confirm – if confirmation were needed – that this abridged version *(ikhtiṣār)* or summary (*talkhīṣ*) of Ibn Sinān's book on the *Instruments for Shadows* could not, either in letter or in spirit, be by al-Ḥasan. These arguments confirm the conclusions on the *Commentary of the Almagest*, as well as the

[38] *Muntahā al-idrāk fī taqāsīm al-aflāk*, ms. Paris, BNF Ar. 2499, fol. 2^v.

[39] This error is the consequence of another one, which has led A. Sabra (*Dictionary of Scientific Biography*, vol. VI, pp. 206–8) to attribute to al-Ḥasan ibn al-Haytham a 'summary' by Muḥammad ibn al-Haytham of the book by Ibn Sinān on *The Lines of the Shadows* (see above and Supplementary Note [3]).

distinction between its author Muḥammad and the eminent mathematician al-Ḥasan.

Firstly, according to the author himself, Ibn Sinān's book was about sundials, and hence hour lines. This is what he writes in the introduction to his book:

> I have seen that the mathematicians who precede us were interested in these instruments, but only in a particular way. In order to construct astrolabes, a group of mathematicians wrote books, containing all their knowledge; but where sundials are concerned, I have found no one who was satisfied with his work on the subject. On water instruments and observation instruments, the Ancients did write satisfactory works. I have undertaken to write this book specifically on sundials and I have called it *The Book on Instruments for Shadows.*[40]

It goes without saying that there is absolutely no link between this book and the one that al-Ḥasan devotes to the *Formation of Shadows* (*Maqāla fī kayfiyyat al-aẓlāl*) that is, shadows as optical phenomena.

There are two pieces of al-Ḥasan's writing whose authenticity is unquestionable, which should also be mentioned here if we are to tell whether al-Ḥasan could have written an abridged version or a summary of Ibn Sinān's book. It really centres on his book on *Sundials* (*Fī al-rukhāmāt*)[41] and a treatise of particular importance *On the Lines of Hours* (*Fī khuṭūṭ al-sā'āt*).[42] This is his own work as opposed to Ibn Sinān's, and really answers the question which is raised here. At the end of his book on sundials, al-Ḥasan does promise to write a treatise on instruments of shadows. However, the aim of this work would definitely not be to write the 'shortened version-summary': on the contrary, it is 'to explore all the notions, properties and constructions necessitated by this art'. Furthermore, nothing tells us that the promised book was in fact ever written. Al-Ḥasan does not refer to it in his other works, and bibliographical sources seem unaware of it. In any case, his declared intention was surely not that of an author about to repeat a resume of Ibn Sinān's book.

More importantly for us is his book *On the Lines of Hours*, where al-Ḥasan ibn al-Haytham makes his own contribution as opposed to Ibn Sinān's, and sheds some light on his project: he intended to pursue research on sundials, using Ibn Sinān as a starting point, but then taking a different

[40] See our edition in R. Rashed and H. Bellosta, *Ibrāhīm ibn Sinān. Logique et géométrie au Xᵉ siècle*, Leiden, 2000, p. 339.

[41] See List of Ibn al-Haytham's Works.

[42] See our edition, French translation and commentary in *Les Mathématiques infinitésimales du IXᵉ au XIᵉ siècle.* Vol. V: *Ibn al-Haytham: Astronomie, géométrie sphérique et trigonométrie*, London, 2006, Second part, Chapter I.

direction; he did not intend to produce of some kind of shortened version of an existing work. Here, at some length, are al-Ḥasan's own words:

> When we examined the book by the geometer Ibrāhīm ibn Sinān *On Instruments for Shadows*, we noticed that he criticizes the opinion of earlier writers who suppose that the lines that define the edges of seasonal hours on the planes of sundials are straight lines, and who believes that on each day of the year the tip of the shadow of the gnomon, at the end of the same seasonal hour and at the beginning of the hour that follows it, lies close to a straight line. He stated that one straight line in the plane of a horizontal sundial does not define the edge of the same seasonal hour except for three of the seasonal circles – one of which is the equator, while the two others lie on either side of the equator and at equal distances from it; and that the straight line that lies in the plane of a horizontal sundial and defines the edge of the same seasonal hour in the three circles we have just mentioned is the intersection of the plane of the dial and the plane of a great circle that passes through the tip of the gnomon and through the points that indicate the edges of the same seasonal hour on the three circles. This statement is true and cannot be doubted. He went on to state that this great circle does not cut any of the remaining hour circles in a point that marks the edge of the seasonal hour associated with the circle in question. This statement is also a true one; however he was not able to prove it, for when he came to give a proof of his statement, he showed correctly that one great circle cuts the circumferences of the three circles in three points that mark the edges of the same seasonal hour. He next wanted to prove that the great circle that cuts off a seasonal hour on the three circles, does not cut off this same seasonal hour on any other remaining hour circle. He then presents a proof that does not show this idea is true. He has in fact imagined two great circles that cut off two seasonal hours from the three circles; he went on to draw a fourth hour circle and he showed that these two great circles cut off two different arcs on the fourth circle, but he did not show that, of these two different arcs, neither is a seasonal hour; thus the result <established by> his proof is different from what is set out clearly in his statement; moreover, the result established by the proof does not make it impossible for one of the two different arcs to be a seasonal hour. It is as if he had stated that none of the hour lines is straight, and proved that not all the hour lines are straight. So what he said about this idea falls short of what he intended, and furthermore does not show the idea is a true one.
>
> Similarly, he has not shown what is the magnitude of the distance by which the tips of the shadows at the seasonal hours depart from the <straight> line given for that hour. It is possible that the tips of the shadows depart very little from the straight line given for that hour, so that this deviation is insensibly small. And the proof indeed depends on the fact that a mathematical straight line is a length that has no width, whereas the line drawn on the plane surface of the sundial is one that has a noticeable

thickness, which could take in the deviation of the shadows, if this deviation is insensibly small, or is less than it <sc. the thickness of the line> by some negligibly small amount.

In the same way, all instruments constructed for <observing> the Sun and the planets are constructed in a manner that is approximate and not absolutely exact. The astrolabe divides its circles into three hundred and sixty parts. If we take a height with this instrument, we obtain it only in whole degrees; now a height is never a whole number of degrees, instead, on most occasions one can have minutes along with the whole degrees; now these minutes do not appear on the astrolabe; it is even possible that the minutes are numerous, but, despite their number, they do not appear. In the same way the lines that serve to divide the circles of the astrolabe each have a perceptible width; this width is a part of the degree cut off by each line, and it is a part that has a magnitude, for the parts of a circle on an astrolabe are small, and especially so if the astrolabe is small. However we do not take into account the width of the lines that mark the divisions on an astrolabe.

These notions apply equally to an armillary sphere, a quadrant used to observe the Sun and all the instruments used to observe the Sun and the planets. It is possible that our predecessors supposed that the hour lines <on a sundial> are straight lines, while at the same time knowing how far they deviated <from straight lines>, given that what they are aiming to achieve by their assumption is an approximation, and not the ultimate exactitude, that they aimed for in the construction of the astrolabe and of observing instruments. Since we found this idea unclear, because Ibrāhīm ibn Sinān had not succeeded in showing it was true; and since it can be accepted by way of approximation, we decided to go deeper into investigating the truth of this idea, and to allow ourselves to discuss it, as well as to find out about the boundaries between seasonal hours on the surfaces of horizontal sundials. So we reflected on these matters and pursued our researches until the truth was clear. It thus became apparent that our predecessors had been right to suppose that the hour lines are straight lines, that this is by way of an approximation, and the best approximation, and that there is no other way of drawing the boundaries between hours on the surfaces of sundials.

From what we have proved it is clear that Ibrāhīm ibn Sinān had been right in one respect, and mistaken in another respect, and this in fact happened because he employed mathematical procedures without thinking about physical ones; so he was right from the point of view of imagination, but wrong from the point of view of sensory perception, because he chose to prove the result he had stated as if the lines drawn on sundials were imagined lines, that is to say having length without breadth; but the lines drawn on sundials have breadth; thus he did not distinguish an imagined line from one perceived by the senses: so he was completely mistaken.

> Once we had come to this idea that we have described, we composed this treatise to provide a justification for our predecessors' opinions on the subject, to give an argument in support of what they had supposed to be true, and to indicate where Ibrāhīm ibn Sinān went astray.
>
> Before the treatise we have given lemmas that are themselves new results, results that none of those who preceded us has mentioned – as it seems to us – and thanks to these lemmas we can go on to derive all the ideas that we have expressed in this treatise. So let us now begin to speak of them, with God's help in everything.[43]

Reading this introduction – as with the rest of the book – there cannot be the least confusion between al-Ḥasan's book and the summary or abridged version by Muḥammad.

[4] Commentary in the *Resolution of Doubts* ... by Ibn al-Haytham on Proposition X.I of the ***Elements***

Having written a short treatise (translated here) on Proposition X.I of the *Elements*, it seems that Ibn al-Haytham returns to the same question, in his *Resolution of Doubts*...; in fact, he goes right back to the original text of the treatise, including several variants. His commentary in the *Resolution of Doubts*... does include an introduction. He uses the introduction from the short treatise, with some variations, before going on to quote the proof already given in it. We have therefore translated the introduction, where the reader will recognize whole phrases from the treatise. As for the proof, we have only noted the variants as it is quoted directly from the text of the short treatise. For the edition of the text as well as for the variants, we have used the following manuscripts from the *Resolution of Doubts*...[44]

1. Istanbul, University 800, noted here as A; date of the copy is not mentioned; it could be between 6th and 7th century Hijra, which is 13th to 14th century.
2. Bursa Haracci 1172, noted here as B; copied in 477/1084.
3. Tehran, Malik 3433, noted here as T; copied same year as above.

> The meaning of this proposition might seem equivocal for people. The majority of mathematicians think that it is particular, as had been commented by Euclid, and that it is not true except in the way Euclid mentioned. The reality, however, is different from what people think. In fact, Euclid had limited himself to this one particular meaning – (that the <magnitude> taken away is greater than the half) – because this is the meaning he used in his own work; he restricted himself to this meaning because he needed it.

[43] *Les Mathématiques infinitésimales,* vol. V, pp. 733–7.

[44] See the Arabic text in *Les Mathématiques infinitésimales,* vol. II.

When we examined this notion carefully and tested its validity, we found it to be a universal idea; one of the properties of proportions, that is if we let the ratio of the subtracted <magnitude> to the greater magnitude be any ratio, and if we let all the subtracted <magnitudes> follow this ratio, the division necessarily ends up with a magnitude which is less than the smallest magnitude. When the notion became clear, we decided to explain it, so that the previously mentioned opinion (that this notion is particular) would become obsolete. A discussion was organized to demonstrate that, although it is an extremely concise and brief argument, it is a universal argument. It was written before *Resolution of Doubts* was conceived. When we began the *Resolution of Doubts* and the explanation of the equivocal notions in this book and when we obtained this proposition, we have had to explain here this notion – because it formed a part of the collection of ideas in the book which ought to be explained – and to summarize the proof so that it accompanies this proposition. We mention now the proof of this notion.

Ibn al-Haytham and the criticism of Ibn al-Sarī: Proposition X.1 of *The Elements*

Ibn al-Sarī – alias Ibn al-Ṣalāḥ – (Najm al-Dīn Abū al-Futūḥ Aḥmad ibn Muḥammad) comes from Hamadān, according to Ibn Abī Uṣaybiʿa[45] or from Sumaysāṭ, according to al-Qifṭī.[46] The two biobibliographers agree that he lived in Baghdad before leaving for Damascus, where he died at the end of the year 584, that is 1153/1154.[47]

The writings of Ibn al-Sarī which we have seen allow us to unveil the portrait of a scholar and philosopher who knew logic, part of the line of scholar-philosophers going back to al-Kindī. This double interest, in both mathematics and logic, might explain, if only partially, the critical style of the author, as well as the subject of his writings. His style has identified him as the author of the short tract *On the Fourth Figure of Categorical Syllogism which is Attributed to Galen*,[48] and other works in logic[49] as well as in

[45] Ibn Abī Uṣaybiʿa, *ʿUyūn al-anbāʾ*, Beirut, 1965, pp. 638–41.

[46] Al-Qifṭī, *Taʾrīkh al-ḥukamāʾ*, p. 428.

[47] *Ibid.*

[48] N. Rescher, *Galen and the Syllogisms*, Pittsburgh, 1966, includes editing and translation of the Arabic text. A. Sabra, 'A twelfth-century defence of the figure of the syllogism', *Journal of the Warburg and Courtauld Institutes* XXVIII, 1965, pp. 14–28.

[49] Mubahat Türker Küyel, 'Ibn uṣ-Salâḥ comme exemple à la rencontre des cultures', *Araştirma,* VIII, 1972 (appeared in 1973); as well as his edition and translation into Turkish 'Aristoteles' in 'Burhân Kitabi'nin ikinci makalesi'nin sonundaki kismin

physics[50] and astronomy.[51] His mathematical writings are in a critical tone, which is common to his work. He corrects an error of al-Qūhī regarding the determination of the ratio of the diameter of a circle to its circumference.[52] He also wrote three short tracts refuting Ibn al-Haytham's criticism of Euclid's *Elements*.[53] The text translated here is a part of this group of writings and concerns Ibn al-Haytham's work on Proposition X.1 of the *Elements*. Ibn al-Sarī knew this work. He also knew the book entitled: *On the Resolution of Doubts in Euclid's Elements*,[54] where Ibn al-Haytham takes up the essential points of his first redaction. Before looking at the content of the text of Ibn al-Sarī, we should understand that this is extremely valuable evidence of the knowledge available from Arabic versions of Archimedes' works in the first half of the twelfth century.

In this short treatise, Ibn al-Sarī levels two major criticisms at Ibn al-Haytham:

1. According to Ibn al-Sarī, Ibn al-Haytham should have considered that his own proposition was Universal, whereas that established by Euclid was Particular. He should therefore have concluded that he should replace Euclid's proposition with his own. According to Ibn al-Sarī, Ibn al-Haytham made two mistakes: over the Universality of his own proposition and the Particularity of Euclid's and over the exact meaning of a Universal proposition.

2. In the proposition, Ibn al-Haytham assumes a constant ratio α, such that $0 < \alpha < 1$, instead of a sequence $(\alpha_i)_{i\geq 1}$ of variable ratios such as $\frac{1}{2} < \alpha_i < 1$, for $i = 1, 2, \ldots$ Ibn al-Haytham's assumption would certainly have stopped him from proving certain propositions in Book XII of the *Elements*

şerhi ve oradaki yanlişin düzeltilmesi hakkinda', *Araştirma,* VIII, 1972 (appeared in 1973).

[50] M. Türker, 'Les critiques d'Ibn al-Ṣalāḥ sur le *De Caelo* d'Aristote et sur ses commentaires', *Araştirma* II, 1964, pp. 19–30 and 52–79.

[51] P. Kunitzsch, *Ibn aṣ-Ṣalāḥ. Zur Kritik der Koordinatenüberlieferung im Sternkatalog des* Almagest, Abhandlungen der Akademie der Wissenschaften in Göttingen, Philologisch-Historische Klasse, series 3, no. 94, Göttingen, 1975.

[52] Ms. Aya Sofya 4845, fols 36^v–40^r.

[53] *Jawāb li-Aḥmad ibn Muḥammad ibn Sarī 'an burhān mas'ala muḍāfa ilā al-maqāla al-sābi'a min kitāb Uqlīdis fī al-Uṣūl*, ms. Aya Sofya 4830, fols 139^r–145^v; *Fī bayān mā wahama fīhi Abū 'Alī ibn al-Haytham fī kitābihi fī al-shukūk 'alā Uqlīdis*, ms. Aya Sofya 4830, fols 146^v–149^v; *Fī kashf li-l-shubha allatī 'araḍat li-jamā'a mimman yansib nafsahu ilā 'ulūm al-ta'ālīm 'alā Uqlīdis fī al-shakl al-rābi' 'ashar min maqāla al-thāniya 'ashar min kitāb Uqlīdis*, ms. Aya Sofya 4830, fols 151^v–154^v.

[54] Ms. Istanbul University, 800, fols 143^v–145^r. Cf. Supplementary Note [4].

(actually established using Proposition X.1), and would have ruined the Universality of his proposition.

In order to understand Ibn al-Sarī's criticism, and its real impact, let us recall, at the risk of repeating ourselves, Euclid's and Ibn al-Haytham's propositions. Let us start with Euclid's, which can be written as follows.

Let A and a be two magnitudes of the same kind such that $A > a$ and $(\alpha_i)_{i \geq 1}$ and a sequence of ratios equal or unequal such that $\frac{1}{2} < \alpha_i < 1$ for $i = 1, 2, \ldots$

Let us consider the magnitudes

$$A_1 = (1-\alpha_1)A,\ A_2 = (1-\alpha_2)A_1 = (1-\alpha_1)(1-\alpha_2)A,\ \ldots,\ A_k = \prod_{i=1}^{k}(1-\alpha_i)A,$$

then $N \in \mathbf{N}$ exists such that for every $n > N$, we have $A_n < a$.

Let us note that the Greek text of Proposition X.1 is followed by a porism ὁμοίως δὲ δειχθήσεται, κἂν ἡμίση ᾖ τὰ ἀφαιρούμενα which means 'this proposition can be proved in the same way, if the subtracted parts are halves'. However, this porism does not seem to have been known to the mathematicians who worked any of the many Arabic translations of Euclid.

Ibn al-Haytham's proposition deals with the same problem, in the hypothesis: $\alpha_i = \alpha$ for $i = 1, 2, \ldots,$ constant ratio, with $0 < \alpha < 1$; thus $A_n = (1-\alpha)^n A$ and for $\alpha = \left(\frac{1}{2}\right)$ $A_n = \left(\frac{1}{2}\right)^n A$, $N \in \mathbf{N}$ exists such that for every $n > N$, we have $A_n < a$.

In other words, Ibn al-Sarī's criticism amounts to the observation that it was insufficient to show, as Ibn al-Haytham had, that

$$\lim_{n \longrightarrow \infty} (1-\alpha)^n = 0,$$

but that it was necessary to prove that

$$\lim_{n \longrightarrow \infty} \prod_{i=1}^{n}(1-\alpha_i) = 0$$

since the ratios could be equal or unequal.

The formulation is equivalent to the one done by Ibn Qurra[55] and Ibn al-Sarī adds nothing to it. He did not notice that the condition obtained here

[55] Ibn Qurra, *On the Measurement of Paraboloids,* Proposition 30. See vol. I.

is slightly more general than in al-Haytham's, if we say that, for ($0 < \alpha < \alpha_i < 1$)

$$(1-\alpha_i) < (1-\alpha) \Rightarrow \prod_{i=1}^{n}(1-\alpha_i) < (1-\alpha)^n.$$

If therefore only α_i are considered as satisfying $\frac{1}{2} \leq \alpha_i < 1$, the result established by Euclid – and *a fortiori* the result established by Ibn Qurra – shows Ibn al-Haytham's result is a particular case; but, for $0 < \alpha_i < \frac{1}{2}$, the result established by Ibn al-Haytham allows the generalization of Euclid's results, which can then be applied for $0 < \alpha_i < 1$.

Let us now turn to Ibn al-Sarī's second criticism; Ibn al-Haytham's proposition does not apply to Propositions 2, 5, 10 or 11 of Book XII of *Elements*. Let us consider Proposition XII.2 in order to understand Ibn al-Sarī's reasoning:

If C and C_1 are areas of two circles with respective diameters D and D_1, then $\frac{C}{C_1} = \frac{D^2}{D_1^2}$. Let us assume that $\frac{D^2}{D_1^2} > \frac{C}{C_1}$, therefore an area S, $S < C_1$ exists such that $\frac{D^2}{D_1^2} = \frac{C}{S}$. Say $S + \varepsilon = C_1$.

Let S_i ($i = 2, 3, \ldots, n$) be the areas of polygons inscribed in circle of area C_1, having 2^i sides: then successively:

$$S_2 \quad > \frac{1}{2}C_1, \text{ hence } A_1 = C_1 - S_2 < \frac{1}{2}C_1$$

$$S_3 - S_2 \quad > \frac{1}{2}A_1, \text{ hence } A_2 = C_1 - S_3 = A_1 - (S_3 - S_2) < \frac{1}{2}A_1 < \left(\frac{1}{2}\right)^2 C_1,$$

$$\ldots$$

$$S_m - S_{m-1} > \frac{1}{2}A_{m-2},$$

hence

$$A_{m-1} = C_1 - S_m = A_{m-2} - (S_m - S_{m-1}) < \frac{1}{2}A_{m-2} < \left(\frac{1}{2}\right)^{m-1} C_1.$$

By applying X.1, starting with these inequalities, it is possible to find n such that $A_n < \varepsilon$. But $A_n + S_n = C_1$, therefore $S_n > S$.

However the ratios which are used in Euclid's method are not equal and are always higher than $\frac{1}{2}$:

$$\frac{S_2}{C_1}=\frac{2}{\pi};\quad \frac{S_3-S_2}{C_1-S_2}=\frac{2(1-\sqrt{2})}{\pi-2};\quad \frac{S_4-S_3}{C_1-S_3}=\frac{2\left(2\sqrt{2-\sqrt{2}}-\sqrt{2}\right)}{\pi-2\sqrt{2}}.$$

Ibn al-Sarī then states that Euclid's proposition allows for a conclusion whereas Ibn al-Haytham's does not. This criticism is unfounded, since from the successive inequalities used by Euclid

$$S_2>\frac{1}{2}C_1;\quad S_3-S_2>\frac{1}{2}A_1;\ \ldots\ ;\ S_{m+1}-S_m>\frac{1}{2}A_{m-1};$$

from this can be deduced

$$A_1<\frac{1}{2}C_1;\quad A_2<\frac{1}{2}A_1;\ \ldots;\ A_m<\frac{1}{2}A_{m-1},$$

hence the recurrence relation

$$\forall n\in \boldsymbol{N},\quad A_n<\left(\frac{1}{2}\right)^n C_1.$$

Ibn al-Haytham has established in another formal language that $\lim\limits_{n\longrightarrow\infty}\left(\frac{1}{2}\right)^n C_1=0$; we can immediately deduce that $\lim\limits_{n\longrightarrow\infty} A_n=0$.

Although Ibn al-Sarī's criticism is inexact, it does show a certain perspicacity and attests to the interest shown by mathematicians and mathematician-philosophers in Proposition X.1, which formed the basis of the approximation method. Not only did Ibn al-Sarī not understand the point of Ibn al-Haytham's proposals, he also failed to understand the depth of his methods.

In the name of God the Clement and the Merciful

TREATISE OF AL-SHAYKH ABŪ AL-FUTŪḤ AḤMAD IBN MUḤAMMAD IBN AL-SARĪ
– May God have Mercy on him –

Explanation of Abū ʿAlī ibn al-Haytham's error regarding the first proposition of the 10th book of Euclid's *Elements*

He said: I have seen the treatise of Abū ʿAlī ibn al-Haytham, entitled the Division of two different[56] magnitudes mentioned in the 10th book of Euclid's *Elements* and I found that he evoked in his introduction the opinion of numerous mathematicians who say that the meaning of Proposition X.1 of Euclid's *Elements* is particular[57] and is only true in the way mentioned by Euclid – that is to say that if the two magnitudes are different, then one subtracts from the larger one a magnitude greater than its half and from the remainder a magnitude greater than its half, and so on until a magnitude less than the smallest given magnitude remains. The reality however is different from that which may be imagined by this group; if Euclid had limited himself to such a particular notion (that the <magnitude> subtracted is greater than half), it is because he used this notion in his book; he restricted himself to this, because he had need of it. He (Ibn al-Haytham) then recalled his own need in certain of his geometric deductions to take away half from the greater of the two different magnitudes, then half again from the remaining magnitude and so on continuously until the division necessarily culminated in a magnitude less than the smallest magnitude; and how in this way he had defined the notion he needed. He then claimed that, when he had carefully examined this notion, he found it to be universal: that one of the properties of proportions meant that if we let the ratio of the subtracted <magnitude> to the greater magnitude be any ratio, and if we say let all the subtracted <magnitudes> follow this ratio, the division necessarily culminates in a magnitude less than the smallest magnitude. He decided to elucidate this notion and make it known, so that it could be used and the opinion that this notion is particular[58] should become obsolete. He then proceeded to prove that this was a universal notion.

Abū ʿAlī goes on to refer both to this proposal and to the proof in his book *On the Solution of the Difficulties in the 10th Book of Euclid's*

[56] Two homogeneous and unequal magnitudes.
[57] 'universal' in the manuscript.
[58] 'universal' in the manuscript.

Elements. He recalls here how he had written his own treatise and he refers to this treatise.

When I meditated upon the proposal of this man, as soon as I began to examine it, I found that he had committed several sorts of errors. In the first place, misunderstanding the meaning of 'universal' as opposed to 'particular'. Secondly, misunderstanding the meaning of Euclid's proposal and propositions where he applied Proposition X.1 and his opinion that his proposition takes the place of Euclid's proposition. Thirdly, limiting himself to the proposition revealed by Euclid, to Euclid's book alone, and thinking only of his need for this notion to the exclusion of all others.

When I saw that, I indicated the fault apparent in his proposals, so that this would not be the cause of doubt for those who would study him, and in order to preserve Euclid's proposition which is uniquely particular in nature and which proves all geometric ideas regarding surfaces of non-homogeneous bodies. By non-homogeneous, I mean rectilinear figures, circles and solid figures surrounded by plane surfaces and the sphere.

This is what we have to say on the subject. As for his error in understanding the meaning of 'universal' as opposed to 'particular', this is evident, since universal and particular are among the things which have a correspondence with each other, where the first says of the second, that, in certain cases, the attributes and conditions of the general may be found in the particular. However, the converse does not necessarily follow; that is, that all attributes and conditions of the particular are to be found in the general. For example: take the rectilinear figure in general for the triangle and the square; and the number in general for the odd and the even. It is true that every rectilinear <figure> is a figure, but the proposition does not have a reciprocal one, whereby every figure is rectilinear; similarly, all triangles are figures, but all figures are not necessarily triangles; and all even <numbers> are numbers, but not all numbers are even. If we look for a definition of the universal and the particular, we will not find it in his proposition; in fact he used his proposition, which he claimed was universal, by adding a condition; that subtracted <magnitudes> all follow the same ratio. But Euclid himself had explained, simply and without any condition, that magnitudes might be proportional or non-proportional. By this I mean that, Euclid's proposition, whether the subtracted <magnitudes> are proportional or non-proportional, will culminate in a magnitude less than the smallest magnitude. Ibn al-Haytham's proposals, if we add a condition, and also his assertion, are very obvious to anyone with even the slightest idea about geometry. Abū ʿAlī's proposition can only be understood starting from Euclid's proposition, if the <magnitudes> subtracted follow the same ratio, which is much easier, but are not proportional, which is harder, then Ibn al-Haytham's proposition is not universal, when compared with this proposition, nor is there is any reciprocity between the two – nor is the one contained in the other. In fact in Euclid's proposition, the subtracted <magnitudes> are greater than half and they are absolute according to the ratio, which means that they might be proportional or not, whereas in Abū ʿAlī's proposition, the subtracted <magnitudes> might be greater than half, smaller than half or equal to half, but they are limited by the condition that they should be proportional.

As for his error in the comprehension of Euclid's proposition and other propositions where this proposition has been applied, this is because he wished to substitute his proposition for Euclid's – this is why he mentioned it separately in his book *On the Resolution of Doubts*. He considered Euclid's proposition not useful and he substituted this proposition. But when we want to substitute his proposition for Euclid's, we cannot show any of the propositions where Euclid applied this proposition. In fact the propositions where Euclid applied this proposition number only four among the propositions in Book XII of *Elements*; these are the 2nd, 5th, 10th and 11th, and it is not correct to apply Abū ʿAlī's proposition in any of these propositions.

Proof: The first time Euclid used this proposition as a lemma is in the second proposition of Book XII, where he declared that for two circles the ratio of one to the other is equal to the ratio of the squares of their diameters: we prove this by stating that if this was not the case, then the ratio of the square to the square is greater or smaller than the ratio of the surface of the circle to the surface of the circle. First, we assume the ratio of square to square[59] is greater than the ratio of circle to circle. Then we assume a smaller magnitude whose ratio to the circle is equal to the ratio of the square to the square; then we assume another magnitude such that this, together with the smaller magnitude to which it is related, have a sum equal to the next circle in the ratio and by comparison to which the small magnitude is smaller. If we then draw a square in this circle, which is the last in the ratio, this square equals more than half the circle. If we also draw an octagon in this circle, we know that the excess of this octagon beyond the square is greater than half the excess of the circle over the square; similarly, if we construct a figure with sixteen sides, then we show that the excess of this figure over the octagon is greater than half the excess of the circle over the octagon. If we continue to construct figures whose number of sides successively increase by two[60] and we establish the same proof for excesses, it follows therefore from the last <stage> that excesses end in an excess which

[59] Ms. smaller. See also Euclid, Book 12, Proposition 2, the first part of which is summarized here.

[60] The number of sides is 2^n.

is less than the smallest given magnitude; that is, the magnitude which we assumed equal to the excess of the next circle in the ratio, over the magnitude for which the ratio of the previous circle in the ratio, is equal to the ratio of the square to the square.

Even if we want to prove this assertion using the proposition which Ibn al-Haytham considered to replace Euclid's proposition and which he considered should be put forward to it, we cannot, since it is necessary to prove not only that the ratio of the square to the circle is equal to the ratio of the excess of the octagon over the square to the excess of the circle over the square but that it is also equal to the ratio of the excess of the16-sided figure over the octagon to the excess of the circle over the octagon. This continues in the same was as with the other subtracted <magnitudes>. However, it is not necessary here for these subtracted <magnitudes> to be proportional. Therefore Abū ʿAlī's proposition cannot be correctly applied in this proposition because there is no necessity for proportionality of the excesses; it is of no use here. Leading on from this point, it is clear that the application of his proposition would not be correct and not be useful for other propositions in the aforementioned book either.

His error becomes apparent <when he affirms> that this proposition had been introduced by Euclid as a lemma, because of the need he had for it in his book and not as a fundamental proposition, and when he ignores other books. We have in fact pointed out before the proof that this proposition is fundamental; in fact this proposition is so fundamental as to be necessary for the understanding of what is contained in Euclid's and in other books both Ancient and Modern. As for the Ancients, such as Archimedes' book *On Measurement of a Circle*, he applies it to prove this proposition and to show that it is true and that the proof is in order. As for the Moderns, it is in fact Ibrāhīm ibn Sinān ibn Thābit ibn Qurra's book that is to <establish> that the measurement of a section of a parabola is one and a half times the triangle whose base is the base of the section and whose height is its height. This proposition was actually mentioned by Archimedes in the introduction of his book *On the Sphere and Cylinder* and although he indicated that he wrote a book on the subject, the book has not come down to us: this is why we have attributed it to the Moderns. If I listed all the books, apart from Euclid's, where this proposition has been used, and all the works which would not be valid without this proposition, it would be more than required here: a short tract such as this one would not be long enough. We merely mention this proposal in order to draw attention to its omission. Peace be upon you.

The fifth treatise has been finished with the help of God.

[5] List of Ibn al-Haytham's works

The following table contains the titles of al-Ḥasan ibn al-Haytham's works taken from three ancient biobibliographies – al-Qifṭī (I), Ibn Abī Uṣaybiʿa, (II) the Lahore list (III) – as well as manuscripts which have come down to us, some of which are indicated here for the first time. Having examined these manuscripts, we have picked out the name of the author, his own references to other works, references to him by his successors and the name by which they have referred to him.

No.	**Treatise of al-Ḥasan ibn al-Ḥasan ibn al-Haytham**	**Variants of his name**
1	***Fī ādāb al-kuttāb*** *On the Culture of Government Officers*	
2	***Fī a'dād al-wifq*** *On Magical Squares*	
3	***Fī aḍwā' al-kawākib*** *On the Light of the Stars* Berlin, Oct. 2970/16, fols 173^{v}–176^{v} Berlin 5668, fols 11^{r}–14^{r} Istanbul, 'Āṭif 1714/12, fols 112^{r}–115^{v} Istanbul, Fātiḥ 3439, fols 131^{v}–136^{v} St. Petersburg 600 (= Kuibychev), fols 295^{v}–298^{v} London, India Office 1270, fols 10^{v}–12^{r} Oxford, Seld. A. 32, fols 128–133 Tehran, Majlis Shūrā 2998, fols 158^{v}–163^{r}	 Ibn al-Haytham *Abū 'Alī* al-Ḥasan ibn *al-Ḥusayn* ibn al-Haytham al-Ḥasan ibn al-Ḥasan ibn al-Haytham Ibn al-Haytham *al-Ḥusayn* ibn al-Ḥasan ibn al-Haytham al-Ḥasan ibn al-Ḥasan ibn al-Haytham al-Ḥasan ibn al-Ḥasan ibn al-Haytham *Abū 'Alī* al-Ḥasan ibn *al-Ḥusayn* ibn al-Haytham
4	***Fī aḥkām al-nujūm*** *(Kitāb mā yarāhu al-falakiyyūn —)* *On the Opinions of Astronomers on Astrology*	
5	***Fī al-akhlāq*** — *On Ethics* Tehran, Majlis Shūrā 1397, fols 33–86	 *Abū 'Alī* al-Ḥasan ibn al-Ḥasan ibn al-Haytham
6	***Fī 'amal al-binkām*** *On the Construction of the Water-Clock* Istanbul, Askari Müze 3025, 6 fols Istanbul, 'Āṭif 1714/ 8, fols 77^{r}–82^{v} Istanbul, Fātiḥ 3439, fols 138^{r}–140^{r}	 al-Ḥasan ibn al-Ḥasan ibn al-Haytham al-Ḥasan ibn al-Ḥasan ibn al-Haytham al-Ḥasan ibn al-Ḥasan ibn al-Haytham
7	***Fī 'amal mukhammas fī murabba'*** *On the Construction of a Pentagon in a Square*	

I	II	III	Own references	References by other scholars
	89			
17	51	46		
43	48	47	**quoted in:** *Fī māhiyyat al-athar alladhī fī wajh al-qamar* Cairo, Taymūr 78, fol. 18	
69				
59	88			**quoted by** al-Bayhaqī, *Tārīkh ḥukamā' al-Islām*, p. 85
66	76			
	45	35		

No.	**Treatise of al-Ḥasan ibn al-Ḥasan ibn al-Haytham**	**Variants of his name**
8	***Fī 'amal al-musabba' fī al-dā'ira*** *On the Construction of the Heptagon in the Circle* Istanbul, Askari Müze 3025, 10 fols Istanbul, 'Āṭif 1714/19, fol. 200^{r}–210^{r}	al-Ḥasan ibn al-Ḥasan ibn al-Haytham al-Ḥasan ibn al-Ḥasan ibn al-Haytham
9	***Fī 'amal al-quṭū'*** *On the Construction of <Conic> Sections*	
10	***Fī anna al-kura awsa' al-ashkāl al-mujassama allatī iḥāṭātuhā mutasāwiya wa-anna al-dā'ira awsā' al-ashkāl al-musaṭṭaḥa allatī iḥāṭātuhā mutasāwiya*** — *On the Sphere which is the Largest of all the Solid Figures having Equal Perimeters, and on the Circle which is the Largest of all the Plane Figures having Equal Perimeters* Berlin, Oct. 2970/9, fols 84^{r}–105^{r} Istanbul, 'Āṭif 1714/18, fols 178^{r}–199^{v} Tehran, Majlis, Tugābunī 110, fols 462–502	Ibn al-Haytham al-Ḥasan ibn al-Ḥasan ibn al-Haytham al-Ḥasan ibn al-Ḥasan ibn al-Haytham
11	***Fī anna mā yurā min al-samā' huwa akthar min niṣfihā*** — *What is Seen from the Sky is Greater than its Half* Alexandria, Baladiyya 2099, fol. 12^{r}–13^{r} Oxford, Bodl., Thurst 3, fol. 104^{r}, 116^{r} Oxford, Marsh 713, fol. 232^{r-v}	— Ibn al-Haytham Ibn al-Haytham
12	***Fī al-ashkāl al-hilāliyya*** ***(Maqāla mustaqṣāt*** *—)* *Exhaustive Treatise on the Figures of Lunes* Berlin, Oct. 2970/3, fols 24^{r}–43^{v} Istanbul, 'Āṭif 1714/17, fols 158^{r}–177^{v} Istanbul, Fātiḥ 3439, fols 115^{r}–117^{r} St. Petersburg, B 1030, fols 50^{r}–72^{v}, 133^{v}–144^{r} London, India Office 1270, fols 70^{r}–78^{v}	al-Ḥasan ibn *al-Ḥusayn* ibn al-Haytham al-Ḥasan ibn *al-Ḥusayn* ibn al-Haytham ms. truncated al-Ḥasan ibn al-Ḥasan ibn al-Haytham al-Ḥasan ibn al-Ḥasan ibn al-Haytham

I	II	III	Own references	References by other scholars
20	74		**quotes**: *Fī muqaddimat ḍil' al-musabba'* 'Āṭif 1714/19, fol. 200ᵛ	
			quoted in: *Fī al-marāyā al-muḥriqa bi-al-quṭū'* India Office 1270, fol. 21ʳ	
28	26	25	**quoted in**: *Fī ḥall shukūk fī kitāb al-Majisṭī* Aligarh 678, fol. 23ᵛ *Fī al-makān* India Office 1270, fol. 26ʳ	
26	37	40	**quoted in**: *Fī taṣḥīḥ al-'amal al-nujūmiyya* Oxford, Seld A. 32, fol. 163ᵛ	
	21	21	**quoted in:** *Fī ḥall shukūk Uqlīdis fī al-Uṣūl wa-sharḥ ma'ānīhi* Istanbul, University 800, fol. 3ᵛ, 13ʳ, 167ʳ **quotes**: *Fī al-hilāliyyāt* Berlin, Oct. 2970, fol. 24ᵛ	**quoted by** : al-Shīrāzī, *Fī ḥarakat al-daḥraja*, Gotha 158, fol. 79ᵛ

No.	Treatise of al-Ḥasan ibn al-Ḥasan ibn al-Haytham	Variants of his name
13	***Fī a'ẓam al-khuṭūṭ allatī taqa' fī qiṭ'at al-dā'ira*** — *On the Greatest Line Lying in a Segment of Circle*	
14	***Fī birkār al-dawā'ir al-'iẓām*** *On the Compasses of the Great Circles* ***Maqāla mukhtaṣara*** (short) — ***Maqāla mashrūḥa*** (expanded) — Aligarh 678, fols 29^{r}, 8^{r}–10^{r} Leiden Or. 133/6, fols 106–111 St. Petersburg, B 1030, fols 125^{v}–131^{r} London, India Office 1270, fols 116^{v}–118^{r} Rampur 3666, fols 436–442	 *Abū 'Alī* al-Ḥasan ibn al-Haytham *Abū 'Alī* ibn al-Haytham (*incipit*) and Abū *'Alī* al-Ḥasan ibn al-Ḥasan ibn al-Haytham (*explicit*) al-Ḥasan ibn al-Ḥasan ibn al-Haytham Ibn al-Haytham (*explicit*) Ibn al-Haytham (*explicit*)
15	***Fī birkār al-quṭū'*** *(2 maqāla)* *On the Compasses for Conic Sections*	
16	***Fī al-ḍaw'*** — *On the Light* Berlin 5668, fols 1^{r}–10^{r} Berlin, Oct. 2970/15, fols 163^{r}–173^{r} Istanbul, 'Āṭif 1714/11, fols 102^{r}–111^{v} London, India Office 1270, fols 12^{v}–17^{v} Tehran, Majlis Shūrā 2998, fols 134–156	 al-Ḥasan ibn *al-Ḥusayn* ibn al-Haytham Ibn al-Haytham al-Ḥasan ibn al-Ḥasan ibn al-Haytham al-Ḥasan ibn al-Ḥasan ibn al-Haytham al-Ḥasan ibn *al-Ḥusayn* ibn al-Haytham
17	***Fī ḍaw' al-qamar*** — *On the Light of the Moon* London, India Office 1270, fols 32^{v}–47^{v}	 *Abū 'Alī* al-Ḥasan ibn al-Ḥasan ibn al-Haytham
18	***Fī al-hāla wa-qaws quzaḥ*** *On the Halo and the Rainbow* Berlin 2970/10, fols 106^{r}–117^{v} Istanbul, 'Āṭif 1714/14, fols 126^{r}–138^{r}	 al-Ḥasan ibn al-Haytham al-Ḥasan ibn al-Haytham

I	II	III	Own references	References by other scholars
	81			
44	22 23	22 15	**quoted in**: *Fī al-marāyā al-muḥriqa bi-al-dawā'ir* India Office 1270, fol. 24ᵛ	
	13	11		
	60	53	**quotes:** *Fī al-manāẓir* India Office 1270, fol. 13ʳ	**quoted by**: - Fatḥ Allāh al-Shirwānī, Tehran, Millī 799, fol. 4ᵛ - al-Fārisī, *Kitāb tanqīḥ al-manāẓir*, vol. I, p. 401
57	6	5	**quoted in**: *Fī māhiyyat al-athar* Cairo, Taymūr 78, fol. 9, 10	
36	8	7	**quoted in**: *Fī ḥall shukūk al-Majisṭī* Istanbul, Beyazit 2304, fol. 8ᵛ	**quoted by**: - Ibn Rushd, *Talkhīṣ al-āthār al-'ulwiya,* Paris 1800 Heb., fol. 82ᵛ - al-Fārisī, *Kitāb tanqīḥ al-manāẓir*, vol. II, pp. 258, 279…

No.	**Treatise of al-Ḥasan ibn al-Ḥasan ibn al-Haytham**	**Variants of his name**
19	***Fī ḥall shukūk ḥarakat al-iltifāf*** *On the Resolution of Doubts Relating to the Winding Movement*	
	Berlin, Oct. 2970/11, fols 118^{r}–127^{r}	al-Ḥasan ibn al-Ḥasan ibn al-Haytham
	Istanbul, ʿĀṭif 1714/15, fols 139^{r}–148^{v}	al-Ḥasan ibn al-Ḥasan ibn al-Haytham
	St. Petersburg, B1030/1, fols 1^{v}–20^{v}	al-Ḥasan ibn al-Ḥasan ibn al-Haytham
20	***Fī ḥall shukūk fī kitāb al-Majisṭī yushakkiku fīhā baʿḍ ahl al-ʿilm*** *On the Resolution of Doubts on the Book of the* Almagest, *Raised by a Scholar*	
	Aligarh, ʿAbd Ḥayy 21, fols 19^{v} (incomplete)	*Abū ʿAlī* al-Ḥasan ibn al-Ḥasan ibn al-Haytham
	Istanbul, Beyazit 2304, fols 1^{v}–20^{v}	Ibn al-Haytham
	Istanbul, Fātiḥ 3439, fols 142^{r}–154^{v}	—
21 *a*	***Fī ḥall shukūk kitāb Uqlīdis fī al-Uṣūl wa-sharḥ maʿānīhi*** — *On the Resolution of Doubts on Euclid's Elements and the Explanation of its Concepts*	
	Cairo, Dār al-Kutub, Khalīl Aghā 1 (incomplete)	
	Istanbul, Univ. 800, 181 fols	al-Ḥasan ibn al-Ḥasan ibn al-Haytham
	Bursa, Haraççi 1172/2, fols 83^{r}–226^{v}	al-Ḥasan ibn al-Ḥasan ibn al-Haytham
	Istanbul, Fātiḥ 3439, fols 66^{r}–117^{r}	al-Ḥasan ibn al-Ḥasan ibn al-Haytham
	Kasan, KGU, Arab 104, fols 1–150	al-Ḥasan ibn al-Ḥasan ibn al-Haytham
	Leiden, Or 516, fols 184^{v}–208^{r}	al-Ḥasan ibn al-Ḥasan ibn al-Haytham
	Peshāvar 323, 112 fols	al-Ḥasan ibn al-Ḥasan ibn al-Haytham
	Tehran, Milli Malek 3433, 157 fols	al-Ḥasan ibn al-Ḥasan ibn al-Haytham
b	*Fī ḥall shukūk al-maqāla al-ūlā min kitāb Uqlīdis*	
c	*Fī ḥall shakk fī al-maqāla al-thāniya ʿashar min kitāb Uqlīdis*	
d	*Fī ḥall shakk fī mujassamāt kitāb Uqlīdis*	
e	*Fī ḥall shakk min al-mujassam*	
f	*Fī ḥall shakk min Uqlīdis*	
22	***Fī ḥarakat al-iltifāf*** *On the Winding Movement*	

I	II	III	Own references	References by other scholars
53	63	60	**quotes**: *Fī al-shukūk 'alā Baṭlamiyūs* Istanbul, 'Āṭif 1714, fol. 139^{v} *Fī ḥarakat al-iltifāf* Istanbul, 'Āṭif 1714, fol. 140^{r}, 143^{v}	
55	38	33	**quotes**: *Fī al-manāẓir* Aligarh, fol. 21^{r} *Fī qaws quzaḥ* Beyazit, fol. 8^{v} *Fī anna al-kura awsa'* Aligarh, fol. 23^{v}	
4			**quotes**: *Fī sharḥ muṣādarāt kitāb Uqlīdis* Istanbul, University 800, fol. 3^{v}, 13^{r}... *Fī al-ashkāl al-hilāliyya* Istanbul, University 800, fol. 3^{v}, 13^{r} *Fī (uṣūl) al-misāḥa* Istanbul, University 800, fols 87$^{r\text{-}v}$ *Fī qismat al-miqdārayn al-mukhtalifayn*	**quoted by**: - al-Fārisī, *al-Zāwiya* - Ibn al-Sarī, Aya Sofya 4830, fol. 139^{v}, 146^{r}, 150^{r}, 150^{v}, 152^{r} (see Supplementary Note [4]) - Naṣīr al-Dīn al-Ṭūsī, *al-Risāla al-shāfiyya,* Ahmet III 3342, fol. 248^{v} - al-Qalqashandī, *Ṣubḥ al-a'shā*, vol. I, p. 480; vol. XIV, p. 227. - Fakhr al-Dīn al-Rāzī, *al-Mulakhkhaṣ*, Majlis Shūrā 827 -al-Shīrāzī, *Fī ḥarakat al-daḥraja*, Gotha 158, fol. 80^{v}
	56	55		**quoted by** al-Khayyām, *Sharḥ mā ushkila min muṣādarāt Uqlīdis*, Paris 4946/4, fol. 40^{r}
	55	51		
	39	37		
21				
22				
52	61	57	**quoted in**: *Fī ḥall shukūk ḥarakat al-iltifāf,* 'Āṭif 1714, fol. 140^{r} and 143^{v}	**quoted by** : Naṣīr al-Dīn al-Ṭūsī, *al-Tadhkira*, Leiden, Or. 905, fol. 49^{r}, 50^{r} Ibn al-Shāṭir, *Nihāyat al-sūl,* Marsh 139, fol. 31^{v}

No.	Treatise of al-Ḥasan ibn al-Ḥasan ibn al-Haytham	Variants of his name
23	***Fī ḥarakat al-qamar*** *On the Motion of the Moon* Istanbul, Fātiḥ 3439, fols 158^{r}–159^{v} Oxford, Bodl., Seld. A. 32, fols 100^{v}–107^{r} St. Petersburg, B 1030, fols 81^{v}–89^{v}	 al-Ḥasan ibn al-Ḥasan ibn al-Haytham al-Ḥasan ibn al-Ḥasan ibn al-Haytham al-Ḥasan ibn al-Ḥasan ibn al-Haytham
24	***Fī hay'at al-'ālam*** *On the Configuration of the Universe* (Apocryphal, see Introduction and Supplementary Note [2]) London, India Office 1270, fols 101^{r}–116^{r} Kastamonu, Genel 2298, fols 1–43 Rabat, Ḥasaniyya, Malik 8691, fols 190–228	 al-Ḥasan ibn al-Ḥasan ibn al-Haytham *Abū 'Alī* al-Ḥasan ibn al-Ḥasan ibn al-Haytham *Abū* al-Ḥasan *'Alī* ibn al-Ḥasan ibn al-Haytham
25	***Fī hay'at ḥarakāt kull wāḥid min al-kawākib al-sab'a*** *On the Configuration of the Movements of Each of the Seven Planets* St. Petersburg 600 (= Kuibychev), fols 368^{v}, 397^{v}, 397^{r}–401^{v}, 402^{v}, 402^{r}, 403^{r}–408^{v}, 369^{r}–396^{v}, 409^{r}–420^{v}	 Abū 'Alī ibn al-Haytham
26	***Fī al-hilāliyyāt*** *On Lunes* Aligarh, 'Abd al-Ḥayy 678/55, fols 14^{v}–16^{v}	 al-Ḥasan ibn al-Ḥasan ibn al-Haytham
27	***Fī ḥisāb al-khaṭa'ayn*** *On the Rule of Two False Positions*	
28	***Fī ḥisāb al-mu'āmalāt*** *(al-qawl al-ma'rūf bi-al-gharīb —)* *On the Arithmetic of Transactions* Istanbul, 'Āṭif 1714/13, fols 116^{r}–125^{r} Berlin, Oct. 2970/17, fols 177^{r}–186^{r}	 *Abū 'Alī* al-Ḥasan ibn al-Ḥasan ibn al-Haytham *Abū 'Alī* al-Ḥasan ibn al-Ḥasan ibn al-Haytham
29	***al-ikhtilāf fī irtifā'āt al-kawākib*** *(Fī mā ya'riḍ min —)* *On the Differences in the Heights of the Stars* Istanbul, Fātiḥ 3439, fols 151^{r}–155^{r}	 al-Ḥasan ibn al-Ḥasan ibn al-Haytham
30	***Fī ikhtilāf al-manāẓir*** *On the Parallaxes*	

I	II	III	Own references	References by other scholars
23	82			
31	1	1		
6	20	18	**quoted in**: *Fī tarbī' al-dā'ira,* Aligarh 678, fol. 10v *Fī al-ashkāl al-hilāliyya,* Berlin 2970, fol. 24v	
48	57	58		
35	10			**quoted in**: *Fī al-mu'āmalāt fī al-ḥisāb* Feyzullah 1365, fol. 76v See Supplementary Note [1]
63	9	8		
56	41	34		

No.	Treatise of al-Ḥasan ibn al-Ḥasan ibn al-Haytham	Variants of his name
31	***Fī ikhtilāf manẓar al-qamar*** *On the Parallax of the Moon* London, India Office 1270, fol. 120^{r-v} St. Petersburg, B 1030, fols 122^{r}-125^{r} Tehran, Malik 3086/3, fols 56^{v}–59^{v}	*Abū ʿAlī* al-Ḥasan ibn al-Ḥasan ibn al-Haytham al-Ḥasan ibn al-Ḥasan ibn al-Haytham
32	***Fī ʿillat al-jadhr wa-iḍʿāfihi wa-naqlihi*** — *On the Cause of the Square Root, its Doubling and its Displacement* Aligarh 678, fols 17^{r}-19^{r}, 13^{v}–14^{v}	*Abū ʿAlī* al-Ḥasan ibn al-Ḥasan ibn al-Haytham
33	***Fī istikhrāj aʿmidat al-jibāl*** — *On the Determination of the Altitude of Mountains* Oxford, Seld. A.32, fol. 187^{r}–188^{r}	al-Ḥasan ibn al-Ḥasan ibn al-Haytham
34	***Fī istikhrāj arbaʿa khuṭūṭ*** *On the Determination of Four Straight Lines*	
35	***Fī istikhrāj ḍilʿ al-mukaʿʿab*** *On the Extraction of the Side of a Cube* St. Petersburg 600 (= Kuibychev), fols 400^{v}–401^{r}	*al-Ḥusayn* ibn al-Ḥasan <ibn> al-Haytham
36	***Fī istikhrāj irtifāʿ al-quṭb ʿalā ghāyat al-taḥqīq*** — *On the Determination of the Height of the Pole with the Greatest Precision* Berlin, Oct. 2970/6, fols 60^{r}–65^{r} Istanbul, ʿĀṭif 1714/4, fols 26^{v}–30^{v} Istanbul, Fātiḥ 3439, fols 140^{r}–142^{v} Leiden Or. 14/11, fols 246–254 London, Br. Mus., Add. 3034, fols 3–13 Oxford, Bodl. Seld. A. 32, fols 121–128 New York, Smith Or. 45/3, fols 35–46	Ibn al-Haytham al-Ḥasan ibn al-Ḥasan ibn al-Haytham al-Ḥasan ibn al-Ḥasan ibn al-Haytham al-Ḥasan ibn *al-Ḥusayn* ibn al-Haytham al-Ḥasan ibn *al-Ḥusayn* ibn al-Haytham al-Ḥasan ibn *al-Ḥusayn* ibn al-Haytham al-Ḥasan ibn *al-Ḥusayn* ibn al-Haytham
37	***Fī istikhrāj jamīʿ al-quṭūʿ bi-ṭarīq al-āla*** — *On the Determination of all the Conic Sections by means of an Instrument*	

I	II	III	Own references	References by other scholars
10				
25	70			
	69			
51	29	32		al-Khayyām, *al-Jabr*, India Office 1270, fol. 55^{r-v}
24	47	43		
62	75			
			quoted in: *al-marāyā al-muḥriqa bi-al-quṭūʿ*, India Office 1270, fol. 20^{v}	

No.	Treatise of al-Ḥasan ibn al-Ḥasan ibn al-Haytham	Variants of his name
38	***Fī istikhrāj khaṭṭ niṣf al-nahār ʿalā ghāyat al-taḥqīq*** — *On the Determination of the Meridian with the Greatest Precision* Berlin, Oct. 2970/5, fols 46^{v}–59^{r} Istanbul, ʿĀṭif 1714/3, fols 13^{v}–26^{r}	 Ibn al-Haytham al-Ḥasan ibn al-Ḥasan ibn al-Haytham
39	***Fī istikhrāj khaṭṭ niṣf al-nahār bi-ẓill wāhid*** — *On the Determination of the Meridian by means of One Shadow* Berlin, Oct. 2970/4, fols 44^{r}–46^{r} Istanbul, ʿĀṭif 1714/2, fols 11^{r}–13^{r} Tehran, Malik 3086/4, fols 59^{v}–62^{r}	 Ibn al-Haytham al-Ḥasan ibn al-Ḥasan ibn al-Haytham *Abū ʿAlī* al-Ḥasan ibn al-Ḥasan ibn al-Haytham
40	***Fī istikhrāj mas'ala ʿadadiyya*** *On the Solution of an Arithmetical Problem* London, India Office 1270/20, fols 121^{r-v} Tehran, Malik 3086/5, fols 62^{v}–66^{r}	 al-Ḥasan ibn al-Ḥasan ibn al-Haytham *Abū ʿAlī* al-Ḥasan ibn al-Ḥasan ibn al-Haytham
41	***Fī istikhrāj samt al-qibla*** *On the Determination of the Azimuth of the Qibla* St. Petersburg, B 1030, fols 111^{r}–121^{v} Oxford, Seld. A.32, fols 107^{r}–115^{r}	 al-Ḥasan ibn al-Ḥasan ibn al-Haytham al-Ḥasan ibn al-Ḥasan ibn al-Haytham
42	***Fī jamʿ*** (or ***jamīʿ***) ***al-ajzā'*** *On the Sum (or all) of the Parts*	
43	***Fī al-juz' alladhī lā yatajazza'*** *On the Indivisible Part*	
44	***Fī al-kawākib al-ḥāditha fī al-jaww*** (or ***Fī al-kawākib al-munqaḍḍa***) *On the Shooting Stars*	

I	II	III	Own references	References by other scholars
29	31	29	**quotes** : *Fī al-tanbīh 'alā mawāḍi' al-ghalaṭ* 'Āṭif 1714, fol. 13v	
42	44	44		
11	92			
	59	56	**quotes** : *Fī samt al-qibla bi-al-ḥisāb* Oxford, Seld. A 32, fol. 107r	
45	32	30		
32	65	62		
	5	4		

No.	Treatise of al-Ḥasan ibn al-Ḥasan ibn al-Haytham	Variants of his name
45	***Fī kayfiyyat al-aẓlāl*** *On the Formation of Shadows* Berlin 5668, fols 14^{r}–27^{r} Isfahān, Dānishka 17435, fol. 61^{r}–81^{r} Istanbul, Askari Müze 3025, 14 fols Istanbul, ʿĀṭif 1714/5, fols 31^{r}–46^{r} Istanbul, Fātiḥ 3439, fols 124r–130^{v} St. Petersburg 600 (= Kuibychev), fols 297^{v}–302^{v} Tehran, Majlis Shūrā 2998, fols 100–130	 al-Ḥasan ibn *al-Ḥusayn* ibn al-Haytham al-Ḥasan ibn *al-Ḥusayn* ibn al-Haytham al-Ḥasan ibn *al-Ḥusayn* ibn al-Haytham al-Ḥasan ibn *al-Ḥusayn* ibn al-Haytham al-Ḥasan ibn *al-Ḥusayn* ibn al-Haytham *al-Ḥusayn* ibn al-Haytham al-Ḥasan ibn *al-Ḥusayn* ibn al-Haytham
46	***Fī kayfiyyat al-arṣād*** — *On the Method of <Astronomical> Observations* Dublin, Ch. Beatty 4549, 19 fols Alexandria, Baladiyya 3688, fols 31^{v}–46^{r}	 *Abū ʿAlī* al-Ḥasan ibn al-Ḥasan ibn al-Haytham *Abū ʿAlī* al-Ḥasan ibn al-Ḥasan ibn al-Haytham
47	***Fī khawāṣṣ al-dawā'ir*** *On the Properties of Circles* St. Petersburg 600 (= Kuibychev), fols 421^{r}–431^{r}	 *al-Ḥusayn* ibn *al-Ḥusayn* ibn al-Haytham
48	***Fī khawāṣṣ al-muthallath min jihat al-ʿamūd*** *On the Properties of the Triangle relatively to the Perpendicular* Patna, Khudabakhsh 2468, fols 189^{r}–191^{r}	 Ibn al-Haytham
49	***Fī khawāṣṣ al-quṭūʿ*** *On the Properties of <Conic> Sections* *Fī khawāṣṣ al-qiṭʿ al-mukāfi'* *Fī khawāṣṣ al-qiṭʿ al-zā'id*	
50	***Fī khuṭūṭ al-sāʿāt*** *On the Lines of Hours* Istanbul, Askari Müze 3025, fols 1^{v}–19^{v} Istanbul, ʿĀṭif 1714/7, fols 57^{r}–76^{v}	 al-Ḥasan ibn al-Ḥasan ibn al-Haytham al-Ḥasan ibn al-Ḥasan ibn al-Haytham

I	II	III	Own references	References by other scholars
64	36	31	**quotes**: *Fī al-manāẓir* ʿĀṭif 1714, fol. 32^{v}	**quoted by** al-Fārisī, *Kitāb tanqīḥ al-manāẓir*, vol. II, p. 358
34	4	3		**quoted by** al-Qalqashandī, *Ṣubḥ al-aʿshā*, vol. I, p. 477
	72			
19	71			
	33 34	27		
27	66		**quoted in**: *Fī al-kura al-muḥriqa* Berlin, Oct. 2970, fol. 75^{r}	**quoted by** al-Fārisī, *Kitāb tanqīḥ al-manāẓir*, Leiden 201, fol. 277^{r}

No.	Treatise of al-Ḥasan ibn al-Ḥasan ibn al-Haytham	Variants of his name
51	***Fī al-kura al-muḥriqa*** *On the Burning Sphere* Istanbul, ʿĀṭif 1714/10, fols 91^v–100^v Berlin, Oct. 2970/8, fols 74^r–83^r	 al-Ḥasan ibn al-Ḥasan ibn al-Haytham Ibn al-Haytham
52	***Fī al-kura al-mutaḥarrika ʿalā al-saṭḥ*** — *On the Sphere Moving on a Plane*	
53	***Fī māhiyyat al-athar alladhī fī wajh al-qamar*** — *On the Quiddity of the Marks on the Face of the Lune* Alexandria, Baladiyya 2096 Cairo, Taymūr 78, 15 fols	 *Ibn ʿAlī* al-Ḥasan ibn al-Ḥasan ibn al-Haytham *Abū ʿAlī* al-Ḥasan ibn al-Ḥasan ibn al-Haytham
54 *a* *b* *c*	***Fī al-majarra*** — *On the Milky Way* ***Fī māhiyyat al-majarra*** ***Jawāb ʿan su'āl sā'il ʿan al-majarra hal hiyya fī al-hawā' aw fī jism al-samā'*** Leiden, Or 184/10, fols 87^r–88^v Edirne, Selimiye 713/11 Tehran, Dānishka 15, fols 37^v–38^r	 *Abū ʿAlī* al-Ḥasan ibn al-Ḥasan ibn al-Haytham *Non vidi* *Non vidi*
55	***Fī al-makān*** — *On Place* Cairo 3823, fols 1^v–5^v Hyderabad, Salar Jung Mus. 2196, fol. 19^v–22^r Istanbul, Fātiḥ 3439, fols 136^v–138^r London, India Office 1270, fols 25^v–27^v Tehran, Majlis Shūrā 2998, fols 166–174	 al-Ḥasan ibn al-Ḥasan ibn al-Haytham al-Ḥasan ibn al-Ḥasan ibn al-Haytham (*Ḥusayn* ibn al-Haytham in the *explicit*) al-Ḥasan ibn al-Ḥasan ibn al-Haytham al-Ḥasan ibn al-Ḥasan ibn al-Haytham *Abū ʿAlī* al-Ḥasan ibn *al-Ḥusayn* ibn al-Haytham
56	***Fī al-maʿlūmāt*** — *On the Known Things* Paris, BN 2458/5, fols 11^v–26^r St. Petersburg 600 (= Kuibychev), fols 303^r–315^v	 al-Ḥasan ibn al-Ḥasan ibn al-Haytham *al-Ḥusayn* ibn al-Ḥasan ibn al-Haytham

I	II	III	Own references	References by other scholars
30	77		**quotes**: *Fī al-manāẓir*, Berlin 2970, fol. 74^{v}, 83^{r} *Fī khuṭūṭ al-sā'āt*, Berlin 2970, 75^{r}	**quoted by** al-Fārisī, *Kitāb tanqīḥ al-manāẓir*, Leiden 201, fol. 277^{r}
	52	48		
67	49	54	**quotes**: *Fī ḍaw' al-qamar*, Cairo, fol. 9, 10 *Fī al-manāẓir*, fol. 11 *Fī adwā' al-kawākib*, fol. 18	
37	46	39		
38				
39	62	59		54*c* **quoted by** Ibn Riḍwān, *Kitāb Ibn Riḍwān fī masā'il jarrat baynahu wa-bayna Ibn al-Haytham fī al-majarra wa-al-makān* (List of his works, cited by Ibn Abī Uṣaybi'a)
58	68		**quotes**: *Fī anna al-kura a'ẓam* India Office 1270, fol. 26^{r}	**quoted by**: - al-Baghdādī, *Fī al-radd 'alā Ibn al-Haytham fī al-makān*, Bursa, Celebi 323, fol. 23^{v}... - Fakhr al-Dīn al-Rāzī, *al-Mulakhkhaṣ*, Majlis Shūrā 827, fols 92-93
	54	50	**quoted in:** *Fī al-taḥlīl wa-al-tarkīb* Dublin 3652, fol. 71^{v} **quotes**: *Fī (uṣūl) al-misāḥa* Paris 2458, fol. 16^{v}	**quoted by**: Ibn Hūd, *al-Istikmāl*, Leiden 123/1, fol. 60^{v}-64^{v}, Copenhagen Or. 82, fol. 65^{v}-67^{r}

No.	**Treatise of al-Ḥasan ibn al-Ḥasan ibn al-Haytham**	**Variants of his name**
57	***Fī al-manāẓir*** — *On Optics* *(7 maqāla)*	
	Istanbul, Ahmet III 1899, fols 1^{v}–249^{r}	*Abū ʿAlī* al-Ḥasan ibn al-Ḥasan ibn al-Haytham
	Istanbul, Ahmet III 3339, fols 1^{v}–125^{r}	*Abū ʿAlī* al-Ḥasan ibn al-Ḥasan ibn al-Haytham
	Istanbul, Aya Sofya 2448, 678 p.	*Abū ʿAlī* al-Ḥasan ibn al-Ḥasan ibn al-Haytham
	Istanbul, Fātiḥ 3212, fols 1^{v}–141^{r}	*Abū ʿAlī* al-Ḥasan ibn al-Ḥasan ibn al-Haytham
	Istanbul, Fātiḥ 3215, fols 138^{r}–331^{v}	al-Ḥasan ibn al-Ḥasan ibn al-Haytham
	Istanbul, Fātiḥ 3216, fols 1^{v}-138^{v}	al-Ḥasan ibn al-Ḥasan ibn al-Haytham
	Istanbul, Köprülü 952, fols 1^{r}–135^{v}	al-Ḥasan ibn al-Ḥasan ibn al-Haytham
58	***Fī al-manāẓir ʿalā ṭarīqat Baṭlamiyūs*** *On Optics according to the Method by Ptolemy*	
59	***Fī marākiz al-athqāl*** *On the Centres of Gravity*	
60	***Fī al-marāyā al-muḥriqa bi-al-dawā'ir*** *On Spherical Burning Mirrors*	
	Aligarh, ʿAbd al-Ḥayy 678, fol. 44^{r-v}	ms. truncated
	Hyderabad, S.J.M. 2196, fols 12^{v}–19^{r}	al-Ḥasan ibn al-Haytham
	Istanbul, ʿĀṭif 1714/9, fols 83^{r}–91^{r}	Ibn al-Haytham
	London, India Office 1270, fols 21^{v}–25^{r}	al-Ḥasan ibn al-Ḥasan ibn al-Haytham
	Berlin, Oct. 2970/7, fols 66^{r}–73^{v}	Ibn al-Haytham
61	***Fī al-marāyā al-muḥriqa bi-al-quṭūʿ*** *On Parabolic Burning Mirrors*	
	Aligarh, ʿAbd al-Ḥayy 678, fols 28^{r}–29^{r}	al-Ḥasan ibn al-Ḥasan ibn al-Haytham
	Hyderabad, S.J.M. 2196, fols 5^{v}–11^{v}	al-Ḥasan ibn al-Ḥasan ibn al-Haytham
	Leiden, Or. 161/3, fols 43–60	Ibn al-Haytham
	London, India Office 1270, fols 18^{r}–21^{r}	al-Ḥasan ibn al-Ḥasan ibn al-Haytham
	Florence, Laurenziana Or. 152, fol. 90^{v}–97^{v}	—

I	II	III	Own references	References by other scholars
2	3		**quoted in**: - *Fī ḥall shukūk fī kitāb al-Majisṭī*, Aligarh, ʿAbd al-Ḥayy 678, fol. 21ʳ - *Fī al-kura al-muḥriqa*, Berlin 2970, fol. 74ᵛ, 83ʳ - *Fī kayfiyyat al-aẓlāl*, ʿĀṭif 1714, fol. 32ᵛ - *Fī ṣurat al-kusūf*, Oxford, A 32, fol. 82ʳ - *Fī al-ḍaw'*, India Office 1270, 13ʳ - *Fī māhiyyat al-athar*, Cairo, Taymur 78, fol. 11 - Marsh 720, fol. 195ʳ **quotes**: *Fī 'ilm al-manāẓir*, Fātih 3212, fol. 4ᵛ (= no. 58 ?)	**quoted by**: - Ibn Hūd, *al-Istikmāl*, Copenhagen Or. 82, fol.105ʳ-107ᵛ - commented by Kamāl al-Dīn al-Fārisī, *Kitāb tanqīḥ al-manāẓir*. - Fatḥ Allāh al-Shirwānī, Teheran, Millī 799, fols 2ʳ, 4ᵛ, 5ᵛ... - al-Qalqashandī, *Ṣubḥ al-a'shā*, vol. I, p. 476 - Qāḍī Zādeh, *Fī al-hāla wa-qaws quzah*, Feyzullah 2179, fol. 89ʳ, 90ᵛ, 98ʳ
	27	26		
	14	12		**quoted by**: - al-Khāzinī, *Kitāb mīzān al-ḥikma*, ed. Hyderabad, p. 16 - al-Qalqashandī, *Ṣubḥ al-a'shā*, vol. I, p. 476
50	18	16	**quotes**: *Istikhrāj al-dawā'ir al-'iẓam* India Office 1270, fol. 24ᵛ	**quoted by**: al-Qalqashandī, *Ṣubḥ al-a'shā*, vol. I, p. 476
	19	17	**quotes**: *Fī istikhrāj jamī' al-quṭū' bi-ṭarīq al-āla* India Office 1270, fol. 20ᵛ *Fī 'amal al-quṭū'* India Office 1270, fol. 21ʳ	

No.	**Treatise of al-Ḥasan ibn al-Ḥasan ibn al-Haytham**	**Variants of his name**
62	***Fī ma'rifat irtifā' al-ashkhāṣ al-qā'ima wa-a'midat al-jibāl wa-irtifā' al-ghuyūm*** — *On Knowing the Height of Upright Objects, on the Altitude of Mountains and the Height of Clouds* Leiden Or. 14/8, fols 236–237 New York, Smith Or. 45/12, fols 243^{v}–244^{r} Tehran, Majlis 2773/2, fols 19–20 Tehran, Malik 3433, fols 1^{v}–2^{r}	 *Abū 'Alī* ibn al-Haytham *Abū 'Alī* ibn al-Haytham *Abū 'Alī* ibn al-Haytham *Abū 'Alī* ibn al-Haytham
63	***Fī mas'ala 'adadiyya*** *On a Numerical Problem*	
64	***Fī mas'ala 'adadiyya mujassama*** *On a Solid Problem of Numbers* London, India Office 1270, fols 118^{v}–119^{r}	 al-Ḥasan ibn al-Ḥasan ibn al-Haytham
65	***Fī mas'ala fī al-misāḥa*** *On a Problem of Mensuration*	
66	***Fī mas'ala handasiyya*** *On a Geometrical Problem* St. Petersburg, B 1030, fols 102^{r}–110^{v} Oxford, Seld. A32, fols 115^{v}–120^{r}	 al-Ḥasan ibn al-Ḥasan ibn al-Haytham al-Ḥasan ibn al-Ḥasan ibn al-Haytham
67	***Fī al-masā'il al-talāqī*** *On Problems of Talāqī* St. Petersburg, B. 1030, fols 90^{r}–101^{v}	 al-Ḥasan ibn al-Ḥasan ibn al-Haytham
68	***Fī misāḥat al-dā'ira*** *On the Measurement of the Circle*	This title takes part of *Fī (uṣūl) al-misāḥa* no. 96; it is not an independent treatise.
69	***Fī misāḥat al-kura*** *On the Measurement of the Sphere* Algiers, BN 1446, fols 113^{r}–119^{v} Aligarh, 'Abd al-Ḥayy 678, fol. 1^{v}–4^{v}, 13^{v}–14^{r} Berlin, Oct. 2970/13, fols 145^{r}–152^{r} Istanbul, 'Āṭif 1714/20, fols 211^{r}–218^{r} St. Petersburg, B. 1030, fols 73^{r}–77^{r}	 *Abū 'Alī* ibn al-Haytham al-Ḥasan ibn al-Ḥasan ibn al-Haytham Ibn al-Haytham al-Ḥasan ibn al-Ḥasan ibn al-Haytham ms. truncated

I	II	III	Own references	References by other scholars
				quoted by: Ibn Aḥmad al-Ḥusaynī Muḥammad al-Lāḥjānī, Majlis Shūrā 2773/1
	50	45		
8	78			
18	58	52		
40	79			
	83			
33	16	14	**quotes**: *Fī misāḥat al-mujassam al-mukāfi'* Berlin 2970, fol. 145v **quoted in**: *Fī (uṣūl) al-misāḥa* India Office 1270, fol. 28v	

No.	Treatise of al-Ḥasan ibn al-Ḥasan ibn al-Haytham	Variants of his name
70	***Fī misāḥat al-mujassam al-mukāfi'*** *On the Measurement of the Paraboloid* London, India Office 1270, fols 56ᵛ–69ᵛ	 al-Ḥasan ibn al-Ḥasan ibn al-Haytham
71	***Fī muqaddimat ḍil' al-musabba'*** *On the Lemma for the Side of the Heptagon* Aligarh, ʿAbd al-Ḥayy 678, fragment (27ʳ⁻ᵛ) London, India Office 1270, fols 122ʳ–123ᵛ Oxford, Marsh 720, fols 259ʳ–260ᵛ Oxford, Thurston 3, fols 131ʳ⁻ᵛ	 — al-Ḥasan ibn al-Ḥasan ibn al-Haytham Ibn al-Haytham Ibn al-Haytham
72	***Fī nisab al-qussī al-zamāniyya ilā irtifāʿātihā*** — *On the Ratios of Seasonal Hours to their Heights*	
73	***Fī al-qarasṭūn*** — *On the Qarasṭūn*	
74	***Fī qismat al-khaṭṭ alladhī istaʿmalahu Arshimīdis fī al-maqāla al-thāniyya min kitābihi fī al-kura wa-al-usṭuwāna*** *On the Division of the Line Used by Archimedes in his Book on the Sphere and the Cylinder* Istanbul, Ahmet III 3453/16, fols 179ᵛ Istanbul, Ahmet III 3456/18, fols 81ᵛ–82ʳ Istanbul, ʿĀṭif 1712, fol. 147ʳ⁻ᵛ Istanbul, Beshiraga 440/18, fol. 275ʳ⁻ᵛ Istanbul, Carullah 1502, fols 222ᵛ–223ʳ Istanbul, Selimaga 743, fols 135ᵛ–136ᵛ Leiden, Or. 14/26, fols 498–499 London, India Office 1270, fols 119ᵛ	 *Abū* al-Ḥasan ibn al-Ḥasan ibn al-Haytham al-Miṣrī *Abū* al-Ḥasan ibn al-Ḥasan ibn al-Haytham Ibn al-Haytham *Abū* al-Ḥasan ibn al-Ḥasan ibn al-Haytham *Abū* al-Ḥasan ibn al-Ḥasan ibn al-Haytham *Abū* al-Ḥasan ibn al-Ḥasan ibn al-Haytham *Abū* al-Ḥasan ibn al-Ḥasan ibn al-Haytham al-Ḥasan ibn al-Haytham
75	***Fī qismat al-miqdārayn al-mukhtalifayn al-madhkūrayn fī al-shakl al-awwal min al-maqāla al-ʿāshira min Kitāb Uqlīdis*** — *On the Division of Two Different Magnitudes Mentioned in the First Proposition of Book X of Euclid's Elements* St. Petersburg, B 1030, fols 78ʳ–81ʳ	 al-Ḥasan ibn al-Ḥasan ibn al-Haytham

I	II	III	Own references	References by other scholars
5	17	20	**quoted in**: *Fī misāḥat al-kura* Berlin 2970, fol. 145^v	
12	42	38	**quoted in**: *Fī 'amal al-musabba' fī al-dā'ira* 'Āṭif 1714, fol. 200^v	**quoted by** Istambulī, Cairo
	35	36		
	67			
9	43	42		
46	40	41	**quoted in**: *Fī ḥall shukūk kitāb Uqlīdis*	Ibn al-Sarī, Aya Sofya 4845, fol. 30^v See Supplementary Note [4]

No.	**Treatise of al-Ḥasan ibn al-Ḥasan ibn al-Haytham**	**Variants of his name**
76	***Fī qismat al-munḥarif al-kullī*** *On the Division of a General Trapezium*	
77	***Fī al-rukhāmāt al-ufuqiyya*** *On the Horizontal Sundials* Berlin, Oct. 2970/14, fols 153^{r}–161^{r} Istanbul, ʿĀṭif 1714/6, fols 47^{r}–55^{v} Tehran, Tungābunī 110/1, fols 1–19	Ibn al-Haytham *Abū ʿAlī* al-Ḥasan ibn al-Ḥasan ibn al-Haytham *Abū ʿAlī* al-Ḥasan ibn al-Ḥasan ibn al-Haytham
78	***Fī ru'ya al-kawākib*** *On the Visibility of the Stars* Lahore, fols 36^{v}-42^{v} Tehran, Danishka 493, fols 19^{v}–23^{r} Tehran, Milli 799, fols 20^{v}–24^{r}	al-Ḥasan ibn al-Ḥasan ibn al-Haytham Ibn al-Haytham Ibn al-Haytham
79	***Fī samt*** *On the Azimuth*	
80	***Fī samt al-qibla bi-al-ḥisāb*** *On <the Determination> of the Azimuth of the Qibla by Calculation* Berlin, Oct. 2970/1, fols 4^{r}–11^{v} Cairo, Dār al-kutub 3823, fols 14^{v}–18^{v} Istanbul, ʿĀṭif 1714/1, fols 1^{v}–9^{v} Fātiḥ 3439, fols 155^{r}–157^{v} Tehran, Tugābunī 110/2, fols 19–35	Ibn al-Haytham al-Ḥasan ibn al-Haytham al-Ḥasan ibn al-Haytham al-Ḥasan ibn al-Ḥasan ibn al-Haytham al-Ḥasan ibn al-Haytham
81	***Fī shakl Banū Mūsā*** *On the Proposition of the Banū Mūsā* Aligarh, University no.1, fols 28–38 Istanbul, ʿĀṭif 1714/16, fols 149^{r}–157^{r} Istanbul, Askari Müze 3025, 8 fols London, India Office 1270, fols 28^{r-v} London, Br. Mus. Add. 14 332/2, fols 42–61	— al-Ḥasan ibn al-Ḥasan ibn al-Haytham al-Ḥasan ibn al-Ḥasan ibn al-Haytham al-Ḥasan ibn al-Ḥasan ibn al-Haytham al-Ḥasan ibn al-Ḥasan ibn al-Haytham
82	***Fī sharḥ* al-Arithmāṭīqī *ʿalā ṭarīq al-taḥqīq*** — *On the Commentary of the* Arithmāṭīqī *by a Rigorous Method*	

I	II	III	Own references	References by other scholars
	87			
65	11	9	**quotes**: *Fī ālat al-aẓlāl* ʿĀṭif 1714, fol. 55r See Supplementary Note [3]	
13	12	10		
60	24	19		
61	7	6	**quoted in:** *Fī istikhrāj samt al-qibla (maqāla mukhtaṣara)* Oxford, Seld. A. 32, fol. 107r	**quoted by**: Ibn al-Sarrāj, Dublin, Chester Beatty 4833, fol.
49	73			
	84			

No.	Treatise of al-Ḥasan ibn al-Ḥasan ibn al-Haytham	Variants of his name
83	***Fī sharḥ muṣādarāt Kitāb Uqlīdis*** *On the Commentary on the Postulates of Euclid's Book* Algiers 1446/1, fols 1^{v}–51^{r} Bursa, Haraççi 1172/I, fols 1^{r}–81^{v} Istanbul, Ahmet III 3454/2 (fragment) Feyzullah 1359/2, fols 150–237^{r} Kasan, KGU, Arab 104, fols 151^{r}–221^{r} Oxford, Bodl. Hunt 237, fols 1^{r}–76^{r} Rampur 3657, fol. 1–223 Tunis, Aḥmad 5482/1, fols 1^{v}–61^{v}	 *Abū ʿAlī* al-Ḥasan ibn al-Ḥasan ibn al-Haytham *Abū ʿAlī* al-Ḥasan ibn al-Ḥasan ibn al-Haytham (fol. 151^{v} Ibn al-Haytham) *Abū ʿAlī* al-Ḥasan ibn al-Ḥasan ibn al-Haytham — al-Ḥasan ibn al-Ḥasan ibn al-Haytham *Abū ʿAlī al-Ḥusayn* ibn al-Haytham *Abū ʿAlī* al-Ḥasan ibn al-Ḥasan ibn al-Haytham
84	***Fī sharḥ qānūn Uqlīdis*** *On the Commentary of Euclid's Qānūn*	
85	***Fī al-shukūk ʿalā Baṭlāmiyūs*** *On the Doubts on Ptolemy* Oxford, Seld. A. 32, fols 162^{v}–184^{v} Alexandria, Baladiyya 2057, fols 1-18	 al-Ḥasan ibn al-Ḥasan ibn al-Haytham *Abū ʿAlī* al-Ḥasan ibn al-Ḥasan ibn al-Haytham
86	***Fī al-siyāsa*** *(5 maqāla) — On Politics*	
87	***Fī ṣūrat al-kusūf*** *On the Shape of the Eclipses* Istanbul, Fātiḥ 3439/3, fols 117^{r}–123^{v} St. Petersburg, B 1030, fols 21^{r}–49^{v} London, India Office 461/2, fols 8^{v}–34^{r} India Office 1270, fols 79^{r}–86^{v} Oxford, Bodl., Seld. A32, fols 81^{v}–100^{v}	 al-Ḥasan ibn al-Ḥasan ibn al-Haytham al-Ḥasan ibn al-Ḥasan ibn al-Haytham al-Ḥasan ibn al-Ḥasan ibn al-Haytham al-Ḥasan ibn al-Ḥasan ibn al-Haytham al-Ḥasan ibn al-Ḥasan ibn al-Haytham
88	***Fī tahdhīb al-Majisṭī*** *On the Correction of the Almagest*	

I	II	III	Own references	References by other scholars
3	2	2	**quoted in**: *Fī ḥall shukūk kitāb Uqlīdis fī al-Uṣūl* Istanbul, University 800, fol. 3v, 13r *Fī taḥlīl wa-tarkīb* Dublin, 3652, fol. 71v	**quoted by**: - al-Fārisī, *al-Zāwiya* - al-Anṭākī, Hyderabad, Osmāniyya 992, fol. 63v, 297v - Naṣīr al-Dīn al-Ṭūsī, *al-Risāla al-shāfiyya*, Ahmet III, 3342, fol. 258v - *Fī al-fawā'id wa-al-mustanbaṭāt min sharḥ muṣādarāt Uqlīdis*,Teheran, Majlis Shūrā 138, fol. 204
41	85			
54	64	61	**quoted in**: *Fī ḥall shukūk ḥarakat al-iltifāf* 'Āṭif 1714, fol. 139v	**quoted by**: - al-'Urḍī, *Kitāb al-hay'a*, Oxford, Marsh 621, fol. 156v - Ibn Bājja, *Min kalāmihi mā ba'atha bihi li-Ibn Ja'far Yūsuf ibn Hasdāy*, Oxford, Pococke 206, fol. 118v
	90			
7	80		**quotes**: *Fī al-manāẓir* Oxford, Seld. A.32, fol. 82r	**quoted by**: al-Fārisī, *Kitāb tanqīḥ al-manāẓir*, vol. I, p. 381
1				

No.	**Treatise of al-Ḥasan ibn al-Ḥasan ibn al-Haytham**	**Variants of his name**
89	***Fī al-taḥlīl wa-al-tarkīb*** *On Analysis and Synthesis* Dublin 3652/12, fols 69^{v}–86^{r} Cairo, Taymūr 323, fols 1–68 Istanbul, Reshit 1191/1, fols 1^{v}–30^{v} St. Petersburg 600 (= Kuibychev), fols 316^{r}–336^{r}	 al-Ḥasan ibn al-Ḥasan ibn al-Haytham Ibn al-Haytham al-Ḥasan ibn al-Ḥasan ibn al-Haytham al-Ḥasan ibn *al-Ḥusayn* ibn al-Haytham
90	***Fī ta'līq fī al-jabr*** *On the Commentary on Algebra*	
91	***Fī tamām kitāb al-makhrūṭāt li-Apollonius*** — *On the Completion of the Work on Conics by Apollonius* Manisa, Genel 1706, fols 1^{v}–25^{r}	 al-Ḥasan ibn *al-Ḥusayn* ibn al-Haytham
92	***Fī al-tanbīh 'alā mawāḍi' al-ghalaṭ fī kayfiyyat al-raṣd*** *On Errors in the Method of Making <Astronomical> Observations* Alexandria, Baladiyya 2099, fols 1^{v}–12^{r}	 *Abū 'Alī* al-Ḥasan ibn al-Ḥasan ibn al-Haytham
93	***Fī tarbī' al-dā'ira*** *On the Quadrature of the Circle* Aligarh 678, fols 10^{r}–11^{v}, 30^{v}–30^{r} Berlin, fol. 258 and quart 559 Cairo, Taymūr 140, fols 136–137 Istanbul, Aya Sofya 4832, fols 39^{v}–41^{r} Istanbul, Beshir Aga 440, fol. 151^{r} Istanbul, Carullah 1502/15, fols 124^{v}–126^{r} Meshhed 5395/1, fols 1^{v}–3^{r} Patna, Khudabakhsh 3692, 3 folios Rome, Vatican 320, fols 1^{v}–6^{v} Tehran, Danishkā 1063, fols 7^{r}–9^{v} Tehran, Majlis Shūrā 205/3, fols 93–101 Tehran, Majlis Shūrā 2998, ms. incomplete Tehran, Malik 3179, fols 107^{v}–110^{r} Tehran, Sepahsālār 559, fols 84^{v}–85^{r}	 al-Ḥasan ibn al-Ḥasan ibn al-Haytham Ibn al-Haytham Abū al-Haytham Ibn al-Haytham Ibn al-Haytham Ibn al-Haytham Ibn al-Haytham Ibn al-Haytham *Abū 'Alī al-Ḥusayn* ibn *al-Ḥusayn* ibn al-Haytham Ibn al-Haytham Ibn al-Haytham Ibn al-Haytham Ibn al-Haytham Ibn al-Haytham

I	II	III	Own references	References by other scholars
47	53	49	**quotes** : *Fī sharḥ muṣādarāt Uqlīdis* Dublin, 3652/1, fol. 71^{v} *Fī al-ma'lūmāt* Paris 2458, fol. 71^{v}	
68	91			
14	25	24	**quoted in** : *Fī istikhrāj khaṭṭ niṣf al-nahār 'alā ghāyat al-taḥqīq* 'Āṭif 1714, fol. 13^{v} Marsh 720, fol. 194^{v}	**quoted by** : Fakhr al-Dīn al-Rāzī, *al-Maṭālib*, t. III, pp. 155–6
15	30	23	**quotes :** *Fī al-hilāliyyāt* Aligarh 678, fol. 10^{v} **quoted in** : *Fī ḥall shukūk Kitāb Uqlīdis* Istanbul, University 800, fol. 167^{r}	**quoted by** : al-Shīrāzī, *Fī ḥarakat al-daḥraja*, Gotha 158, fol. 78^{v}

No.	**Treatise of al-Ḥasan ibn al-Ḥasan ibn al-Haytham**	**Variants of his name**
94	***Fī taṣḥīḥ al-aʿmāl al-nujūmiyya*** *(2 maqāla) — On the Corrections of Astrological Operations* Oxford, Bodl., Seld. A32, fols 132^{v}–162^{r}	—
95	***Fī thamara al-ḥikma*** — *On the Benefit of Wisdom* (probably apocryphal) Istanbul, Köprülü 1604, fol. 41^{v}	*Abū 'Alī* al-Ḥasan al-Ḥasan ibn al-Haytham
96	***Fī uṣūl al-misāḥa*** *On the Principles of Measurement* London, India Office 1270, fol. 28^{v}–32^{v} Istanbul, Fātiḥ 3439, fols 103^{v}–104^{v} St. Petersburg B 2139/2, fols 100^{r}–139^{v} St. Petersburg, National Library 143, fols 13^{v}–15^{v}	— (ms. truncated) *Abū 'Alī* ibn al-Ḥasan ibn al-Ḥasan ibn al-Haytham (fragment) al-Ḥasan ibn al-Ḥasan ibn al-Haytham — (fragment)

[no. 3] Quoted in (I) under the title: *Uṣūl al-kawākib*, and in (III) under the title: *Maqāla fī ḍaw' al-kawākib.*

[no. 10] Quoted in (I) under the title: *Fī anna al-kura awsa' al-ashkāl al-mujassama*, and in (III) under the title: *Maqāla fī al-ukar wa-sharḥ al-mujassamāt*. Cf. Introduction, pp. 36–7.

[no. 11] Quoted in (I) under the title: *Mā yurā min al-samā' 'aẓam min niṣfihā*, and in (III): *Maqāla fī anna mā yurā min al-samā' aktharu min niṣfihā.*

[no. 12] See Introduction, pp. 32–3.

[no. 14] We may read in the *explicit* of ms. Leiden Or. 133: *Tammat al-maqāla li-Baṭlamiyūs al-Thānī al-Shaykh 'Alī al-Ḥasan ibn al-Ḥasan ibn al-Haytham.*

[no. 16] Quoted by Fatḥ Allāh al-Shirwānī with the name: Ibn al-Haytham, ms. Tehran, Milli 799 (folios without numbering = 4^{v}, 5^{v}...)

[no. 18] Quoted by al-Fārisī under the title *Fī al-atharayn* and with the name: Abū ʿAlī al-Ḥasan ibn al-Ḥasan ibn al-Haytham.

[no. 20] Quoted in (I) under the title: *Ḥall shukūk al-Majisṭī*; in (II): *Fī ḥall shukūk fī al-maqāla al-ūlā min kitāb...*, and in (III): *Maqāla fī ḥall shukūk fī al-Majisṭī* (III). In this treatise, Ibn al-Haytham sets out some doubts relating to the difficulties encountered in Ptolemy's *Almagest*. He then gives his own solutions, each of which is introduced by the word *al-jawāb* (the answer).

I	II	III	Own references	References by other scholars
	28	28	**quotes**: *Fī anna mā yurā min al-samā' huwa akthar min niṣfihā*, fol. 163^{v} He quotes the first treatise, fol. 132^{v}	
16	15	13	**quotes**: *Fī misāḥat al-kura* India Office 1270, fol. 28^{v} **quoted in**: *Fī ḥall shukūk kitāb Uqlīdis fī al-Uṣūl*, Istanbul, University 800, fol. 87^{r-v} *Fī al-ma'lūmāt*, Paris 2458, fol. 16^{v}	

F. Sezgin ascribes to Ibn al-Haytham an anonymous text, which is extant in the manuscripts Thurston 3, fols 100^{r}-101^{r} and Marsh 720, fols 194^{r}–198^{r}, and identifies it with a work entitled *al-Masā'il <wa-al-ajwiba>* (Problems and solutions). In fact, the author quotes two treatises by Ibn al-Haytham, namely the *Optics* and *On Errors in the Method of Making <Astronomical> Observations*. However, a close examination shows that the anonymous author has borrowed some 'answers', sometimes *in verbis*, from Ibn al-Haytham's treatise *On the Resolution of Doubts on the Book of the* Almagest.

[no. 21] This title *a* is confirmed by the successors and the critics of Ibn al-Haytham as well as the extant manuscripts. It is quoted in (I) under the title *al-shukūk 'alā <Uqlīdis>*. The title *b* is given by al-Khayyām who does not mention that it is an independent title or a title of a fragment of *a*. We have no means to know if the titles *c* and *d* mentioned only by Ibn Abī Uṣaybi'a's predecessor and the author of the Lahore list (*Fī ḥall shakk fī al-mujassamāt ghayr al-ūlā* and *Fī ḥall shakk fī al-mujassamāt*), as well as the titles of *e* and *f* mentioned only by al-Qifṭī designate different chapters of *a*, or independent treatises. Moreover, we do not know is some of the titles designate more or less the same text.

The treatise 21*a* is quoted by Ibn al-Sarī in *Jawāb li-Aḥmad ibn Muḥammad ibn Sarī 'an burhān mas'ala muḍāfa ilā al-maqāla al-sābi'a min kitāb Uqlīdis fī al-Uṣūl*, under the name: Abū 'Alī ibn al-Haytham (ms. Aya Sofya 4830, fols 139^{v}–140^{r}, 142^{v}, 143^{v}, 144^{v}), Abū 'Alī (fols 140^{v}, 143^{v}, 145^{r}) or Ibn al-Haytham (143^{v}–145^{r}); in *Fī bayān mā wahama fīhi Abū 'Alī ibn al-Haytham fī kitābihi fī al-shukūk 'alā Uqlīdis,* under the name: Abū 'Alī ibn al-Haytham (fols 146^{r-v}, 147^{v}, 148^{r}, 149^{v}–151^{v}), Abū 'Alī (fols 146^{v}, 148^{v}); in *Fī īḍāḥ ghalāṭ Abī 'Alī ibn al-Haytham fī al-shakl al-awwal min al-maqāla al-'āshira min kitāb Uqlīdis fī al-Uṣūl*, under the name: Abū 'Alī ibn al-Haytham (fol. 149^{r}), Abū 'Alī (fols 150^{r}–151^{r}) or Ibn al-Haytham (fols 150^{v}, 151^{r}); and in *Fī kashf li-al-shubha allatī 'araḍat li-jamā'a mimman yansib nafsahu ilā 'ulūm al-ta'ālīm 'alā Uqlīdis fī al-shakl al-rābi' 'ashar min maqāla al-thāniya 'ashar min kitāb Uqlīdis*, under the name: Abū 'Alī ibn al-Haytham (fol. 152^{r-v}), Abū 'Alī (fols 150^r–151^r) or Ibn al-Haytham (153^v).

The treatise 21*a* is also quoted by al-Fārisī under the name: Ibn al-Haytham; and by Naṣīr al-Dīn al-Ṭūsī under the name: Abū 'Alī ibn al-Haytham (fol. 258^r) and Ibn al-Haytham (fol. 248^v).

[no. 22] Quoted in (II) under the title: *Maqāla fī ḥarakat al-iltifāf;* and in (III): *Maqāla fī ajrām al-iltifāf*. Naṣīr al-Dīn al-Ṭūsī and Ibn al-Shāṭir refer to this treatise with no precise indication of the title and with the name: Ibn al-Haytham.

[no. 26] See Introduction, pp. 29–30.

[no. 28] Quoted in (II) under the title: *Fī ḥisāb al-mu'āmalāt*. The treatise *Fī al-mu'āmalāt fī al-ḥisāb* refers to this treatise under the name Abū al-Ḥasan ibn al-Haytham (fol. 76^v), with no precise indication of the title. See Supplementary Note [1].

[no. 29] Quoted in (I) under the title: *Irtifā'āt al-kawākib*, and in (III): *Maqāla fī irtifā'āt al-kawākib*.

[no. 30] Quoted in (II) under the title: *Mas'ala fī ikhtilāf al-naẓar*.

[no. 32] See Introduction, p. 37.

[no. 33] Contrary to what has frequently been stated, this treatise is different from *Fī ma'rifat irtifā' al-ashkhāṣ al-qā'ima wa-a'midat al jibāl wa-irtifā' al-ghuyūm* [no. 62].

[no. 34] Quoted in (II) under the title: *Maqāla fī istikhrāj arba'a khuṭūṭ bayna khaṭṭayn*, and in (III): *Maqāla fī wujūd arba'a khuṭūṭ bayna khaṭṭayn*. Al-Khayyām refers to this treatise with no precise indication of the title, and with the name Abū 'Alī ibn al-Haytham (fol. 55^r), Ibn al-Haytham (fol. 55^v).

[no. 35] See Introduction, p. 37.

[no. 36] Quoted in (I) under the title: *Irtifā' al-quṭr* (I).

[no. 38] Quoted in (I) under the title: *Khaṭṭ niṣf al-nahar*, and in (III): *Maqāla fī istikhrāj niṣf al-nahar*.

[no. 41] Quoted in (II, III) under the title: *Maqāla mukhtaṣara fī samt al-qibla*. This title may be confirmed from the introduction of Ibn al-Haytham's treatise. He writes: 'We have already composed a treatise on the determination of the azimuth of the *Qibla* in all northerly and southerly localities on the earth, by an arithmetical method and geometrical proofs. We were able after that to determine the azimuth of the *Qibla* for all northerly localities of the planet, by a brief method which does not require any arithmetics. We have then composed this treatise' (Oxford, Seld. A. 32, fol. 107^r).

[no. 44] Quoted in (III) under the title: *Maqāla fī al-kawākib al-munqaḍḍa*.
It seems that the two titles, quoted by Ibn Abī Uṣaybi'a and in the list of Lahore, designate the same treatise. The order of the titles on the two lists (Number 4 and Number 5 respectively) confirms this hypothesis. The word *al-munqaḍḍa* has already been used by Ḥunayn ibn Isḥāq in his translation of Ps.-Olympiodorus' commentary on the *Meteorologica*. We find the expression '*fī al-kawākib al-munqaḍḍa* (shooting stars)' (*Commentaires sur Aristote perdus en grec et autres épîtres*, published and annotated by A. Badawi, Beirut, 1986, p. 95), which is the phenomenon designated by Aristotle as

διᾴσσοντες ἀστέρες (*Meteorologica*, I, 4, 341 b). We still need to know why this same treatise is quoted under two different titles. We suggest that one of the two lists quotes the title of the treatise, whereas the other quotes the sentence which immediately follows the title.

[no. 45] Quoted in (III) under the title: *Fī al-aẓlāl.*

[no. 48] Quoted in (I, II) under the title: *Fī aʿmidat al-muthallathāt*; this corresponds to the *explicit*: *Tammat al-maqāla fī aʿmidat al-muthallath.*

[no. 50] See Supplementary Note [3]: Ibn Sinān and Ibn al-Haytham on the subject of 'lines of hours'.

[no. 53] Quoted in (I, II, III) under the title: *Maqāla fī al-athar alladhī fī al-qamar.*

[no. 54] The treatise 54*c* is quoted in (I) under the title: *Fī jawāb man khālafa fī al-majarra,* in (II) *Maqāla fī al-radd ʿalā man khālafahu fī ma'iya al-majarra,* and in (III) *Maqāla fī al-radd ʿalā man khālafahu fī al-majarra.* This treatise has probably been composed in response to Ibn Riḍwān, as indicated in the title of a book written by Ibn Riḍwān, mentioned by Ibn Abī Uṣaybiʿa.

[no. 56] In his book, *al-Istikmāl*, al-Mu'taman ibn Hūd borrows some propositions from Ibn al-Haytham, without referring explicitly to his treatise on the *Knowns*. We may compare, for example, Proposition 13 (ms. Leiden 123, fol. 60^v–61^r) and Proposition 14 of the second chapter of the *Knowns* (pp. 244–5 of our edition in 'La philosophie des mathématiques d'Ibn al-Haytham. II: *Les connus*', *MIDEO*, 21, 1993, pp. 87–275). He has been inspired by some other propositions in Ibn al-Haytham's treatise on the *Knowns*. Cf. Jan P. Hogendijk, 'The geometrical parts of the *Istikmāl* of Yūsuf al-Mu'taman ibn Hūd (11th century). An analytical table of contents', *Archives internationales d'histoire des sciences*, 41, 127, 1991, pp. 207–81, especially pp. 249–53.

[no. 57] Quoted by al-Qalqashandī under the title: *al-mabsūṭa*. Quoted by al-Shirwānī under the name: Abū ʿAlī al-Ḥasan ibn al-Ḥusayn ibn al-Haytham (fol. 2^r), Ibn al-Haytham (fols 4^v, 5^v…).

[no. 58] Quoted in (III) under the title: *Maqāla fī al-manāẓir*. However, considering the similarity of the order of titles on the two lists, it is likely that this title designates the same treatise.

[no. 59] Quoted by al-Khāzinī under the name: Ibn al-Haytham al-Miṣrī.

[no. 60] al-Qifṭī quotes only: *Fī al-marāyā al-muḥriqa*, and does not distinguish this treatise from no. 61. Similarly, al-Qalqashandī quotes *Fī al-marāyā al-muḥriqa*, without doing any distinction between the two treatises, and under the name: Ibn al-Haytham

[no. 62] cf. treatise no. 33.

[no. 64] Quoted in (I) under the title: *al-adad wa-al-mujassam.*

[no. 65] Quoted in (II) under the title: *Qawl fī jawāb mas'ala fī al-misāha*, and in (III): *Maqāla fī mas'ala misāḥiyya.* This treatise is perhaps the one he evokes in *Uṣūl al-misāḥa.*

[no. 69] See Introduction, p. 34–5.

[no. 70] See Introduction, pp. 33–4. Some bibliographers have mentioned that there is another manuscript of this treatise at Zanjān (Iran). We have examined the catalogue of the Zanjānī collection without success.

[no. 71] Quoted in (II) under the title: *Qawl fī istikhrāj muqaddimat ḍil' al-musabba'.*

[no. 72] *irtifā'ihā* in (II).

[no. 75] Quoted in (I) under the title: *Fī qismat al-miqdārayn*, dans (III): *Qawl fī qismat al-miqdārayn al-mukhtalifayn.* See Introduction, pp. 35–6 and Supplementary Note [4].

[no. 77] Quoted in (II, III) under the title: *Maqāla fī al-rukhāma al-ufuqiyya.*

[no. 78] This treatise is reproduced by Fatḥ Allāh al-Shirwānī in his book (ms. Tehran, Milli 799, fol. 20^{v} and Danishka 493, fols 19^{v}–23^{r}).

[no. 80] Quoted in (III) under the title: *Maqāla fī samt al-qibla.*

[no. 82] *'alā ṭarīq al-ta'līq* (II). We also find *Fī sharḥ al-rumūnīqī (?) 'alā ṭarīq al-ta'līq* (86 in Ibn Abī Uṣaybi'a's list).

[no. 83] Quoted by al-Anṭākī under the name: Abū ibn al-Haytham (fol. 297^{v}), but with no precise indication of the title.

[no. 84] Quoted in (II) under the title: *Maqāla fī sharḥ al-qanūn 'alā ṭarīq al-ta'līq.*

[no. 85] Quoted by al-'Urḍī under the name Abū 'Alī ibn al-Haytham (fol. 156^{v}) and Ibn al-Haytham (fol. 196^{r}), with no precise indication of the title.
Quoted by Ibn Bājja under the name: Ibn al-Haytham, cf. Jamal al-Dīn al-'Alawī, *al-Matn al-Rushdī,* Rabat, 1986.

[no. 89] See al-Samaw'al, *al-Bāhir*, p. 148: *Qāla Abū 'Alī ibn al-Haytham nurīd an nubayyn kayfa na'mal muthallathan qā'im al-zāwiya….* The statement of this problem is equivalent to problem 8 of Ibn al-Haytham's *Analysis and Synthesis* (cf. *MIDEO*, 20, p. 104). Al-Samaw'al himself wrote a treatise on analysis and synthesis, which is lost. This treatise by al-Samaw'al might have been a source of information about the destiny of al-Ḥasan ibn al-Haytham's treatise.

[no. 90] Quoted in (II) under the title: *Ta'līq 'allaqahu Isḥāq ibn Yūnus al-Mutaṭabbib bi-Miṣr 'an Ibn al-Haytham fī Kitāb Diyūfanṭus fī masā'il al-jabr (5 maqāla).*

[no. 92] Quoted in (I) under the title: *al-Tanbih 'alā mā fī al-raṣd min al-ghalaṭ*; and in (III): *Maqāla fī al-mawāḍi' al-ghalaṭ fī al-raṣd.*

[no. 93] Ibn al-Haytham informs that he will write an independent treatise in which he would prove how to find a square equal to a circle (p. 103). See also Introduction pp. 30–1 and p. 103, n. 6.

[no. 95] This treatise explicitly attributed to al-Ḥasan ibn al-Haytham deals with the classification of sciences. However, it contradicts in many occurrences what is written in *The Analysis and Synthesis* and *The Knowns*, two treatises which do not raise any problem of authenticity. Moreover, when the author of the classification of sciences mentions the optics, he discusses only reflection, what is in a way contradictory with al-Ḥasan ibn al-Haytham's own works. Finally, al-Ḥasan ibn al-Haytham did not use metaphorical titles, such as *thamara* (fruit), for his writings.

[no. 96] Quoted in (I) under the title: *Fī uṣūl al-misāḥa wa-dhikrihā bi-al-barāhīn.*

Let us mention before concluding the apocryphal titles attributed to al-Ḥasan ibn al-Haytham – those of Muḥammad which became al-Ḥasan's own writings, because of the confusion of the names. We have proved in the Introduction, the Supplementary Notes and the previous List of Works that the titles are evidently apocryphal. However, the list of al-Ḥasan ibn al-Haytham's books will not be completed before the final examination of the titles inscribed in the previous list. Some of them raise difficult questions insofar as we ignore their true content, whereas their titles indicate strange themes in Ibn al-Haytham's own research – such as (1), (5), (86).

1. Treatises of doubtful authenticity:

Fī thamara al-ḥikma (see comment, no. 95).

Fī 'uqūd al-abniya — On Architecture. Mentioned by later authors: al-Qalqashandī, *Ṣubḥ al-a'shā*, vol. I, p. 476 and Tashkupri-Zadeh, *Miftāḥ al-sa'āda*, vol. I, p. 375. This treatise is not extant. However, al-Bayhaqī had suggested that al-Ḥasan ibn al-Haytham wrote a book on 'ingenious procedures'. Nothing confirms that the ancient biographer and the later authors were speaking about the same book. We have not either any assurance that the later authors were not confusing al-Ḥasan and Muḥammad, as far as the latter wrote, according to Ibn Abī Uṣaybi'a: *Maqāla fī ijārāt al-ḥufūr wa-al-abniya bi-jamī' al-ashkāl al-handasiyya.*

2. Treatises by Muḥammad attributed to al-Ḥasan:

Fī sharḥ al-Majistī, ms. Ahmet III 3329/2, fols 38^v–158^r.

Maqāla Manālaūs fī ta'arruf aqdār al-jawāhir al-mukhtalifa, ms. Lahore.

3. Treatises probably by Muḥammad, attributed to al-Ḥasan:

Fī hay'at al-'ālam (see List, no. 25 and Supplementary Note [2]).

Fī wujūd khaṭṭayn yaqrabāni wa-lā yaltaqiyāni, ms. Cairo 4528, fols 15^v–20^r.

4. Apocryphal treatises:

Tuḥfat al-ṭullāb bi-'amal al-asṭurlāb (Abū al-Ḥasan 'Alī ibn al-Ḥusayn ibn al-Haytham), ms. Istanbul, Köprülü 1177, fols 15^r–23^r.

Sharḥ qaṣīda Ibn al-Haytham fī tarḥīl al-shams fī al-manāzil, attributed by Abū 'Abd Allāh Muḥammad ibn Aḥmad ibn Hishām al-Lakhmī (ms. Cairo, Dār al-Kutub, Miqāt 1051: *Amma qā'il hādhihi al-qaṣīda fa-al-Shaykh Abū 'Alī al-Ḥasan ibn al-Ḥasan ibn al-Haytham* ...).

Fī tawṭi'at muqaddamāt li-'amal al-quṭū' 'alā saṭḥ mā bi-ṭarīq ṣinā'ī, Florence, Laurenziana Or. 152, fols 97^v–100^r.

Fī al-mu'āmalāt fī al-ḥisāb, ms. Istanbul, Nur Osmaniye, fols 39–125, Istanbul, Feyzullah 1365, fols 73^v–164^r. See Supplementary Note [1].

De Crepusculis.

BIBLIOGRAPHY

1.1. MANUSCRIPTS

Ibn al-Haytham

Qawl fī al-hilāliyyāt
Aligarh, ʿAbd al-Ḥayy 678, fols 14^v–16^v.

Qawl fī tarbīʿ al-dāʾira
Aligarh, ʿAbd al-Ḥayy 678, fols 10^r–11^v, 30^v–30^r.
Cairo, Dār al-Kutub, Taymūr - Riyāḍa 140, fols 136–137.
Istanbul, Aya Sofya 4832 II/21, fols 39^v–41^r.
Istanbul, Beshir Aga 440, fol. 151^r.
Istanbul, Carullah 1502/15, fols 124^v–126^r.
Meshhed 5395/1, fols 1^v–3^r.
Patna, Khudabakhsh 3692, without numbering, 3 fols.
Rome, Vatican 320, fols 1^v–6^v.
Tehran, Danishgāh 1063, fols 7^r–9^v.
Tehran, Majlis Shūrā 205/3, fols 93–101.
Tehran, Majlis Shūrā 2998, 1 folio without numbering.
Tehran, Malik 3179, fols 107^v–110^r.
Tehran, Sepahsālār 559, fols 84^v–85^r.

Maqāla mustaqṣāt fī al-ashkāl al-hilāliyya
Berlin, Staatsbibliothek, Oct. 2970, fols 24^r–43^v.
Istanbul, ʿĀṭif 1714/17, fols 158^r–177^v.
Istanbul, Süleymaniye, Fātiḥ 3439, fols 115^r–117^r.
London, India Office 1270/12, Loth 734, fols 70^r–78^v.
St. Petersburg, Oriental Institut 89, coll. B 1030, fols 50^r–72^v, 133^v–144^r.

Maqāla fī misāḥat al-mujassam al-mukāfiʾ
London, India Office 1270/11, Loth 734, fols 56^v–69^v.

Qawl fī misāḥat al-kura
Algiers, National Library 1446, fols 113^r–119^v.
Aligarh, ʿAbd al-Ḥayy 678, fols 1^v–5^v, 13^v–14^v.
Berlin, Staatsbibliothek, Oct. 2970/13, fols 145^r–152^r.
Istanbul, ʿĀṭif 1714/20, fols 211^r–218^r.
St. Petersburg, Oriental Institut 89, coll. B 1030, fols 73^r–77^r.

Qawl fī qismat al-miqdārayn al-mukhtalifayn al-madhkūrayn fī al-shakl al-awwal min al-maqāla al-ʿāshira min kitāb Uqlīdis
St. Petersburg, Oriental Institut 89, coll. B 1030, fols 78^v–81^r.

Qawl fī anna al-kura awsaʿ al-ashkāl al-mujassama allatī iḥāṭātuhā mutasāwiya wa-anna al-dāʾira awsāʿ al-ashkāl al-musaṭṭaḥa allatī iḥāṭātuhā mutasāwiya
Berlin, Staatsbibliothek, Oct. 2970/9, fols 84ʳ–105ʳ.
Istanbul, ʿĀṭif 1714/18, fols 178ʳ–199ᵛ.
Tehran, Majlis Shūrā, Tugābunī 110, fols 462–502.

Maqāla fī ʿillat al-jadhr wa-iḍʿāfihi wa-naqlihi
Aligarh, ʿAbd al-Ḥayy 678, fols 17ʳ–19ʳ.

Qawl fī istikhrāj ḍilʿ al-mukaʿʿab
Kuibychev, fols 401ᵛ–402ʳ.

1.2. OTHER MANUSCRIPTS CONSULTED FOR THE ANALYSIS AND THE SUPPLEMENTARY NOTES

Anonymous treatise
Fī ujūd khaṭṭayn yaqrabāni wa-lā yaltaqiyāni
Cairo 4528, fols 15ᵛ–20ʳ.

Apocryphal treatises (attributed al-Ḥasan ibn al-Haytham)
Fī hayʾat al-ālam
London, India Office, Loth 734, fols 101–116.
Kastamonu (Turkey) 2298, fols 1–43.
Rabat, Bibliothèque Royale, no. 8691, fols 29ʳ–48ʳ.

Tuḥfa al-ṭullāb bi-ʿamal al-asṭurlāb (Abū al-Ḥasan ʿAlī ibn al-Ḥusayn ibn al-Haytham)
Bursa, Haraççi 1177, fols 15ʳ–23ʳ.

Sharḥ qaṣīda Ibn al-Haytham fī tarḥīl al-shams fī al-manāzil, attributed by Abū ʿAbd Allāh Muḥammad ibn Aḥmad ibn Hishām al-Lakhmī
Cairo, Dār al-Kutub, Miqāt 1051.

Fī tawṭiʾat muqaddamāt li-ʿamal al-quṭūʿ ʿalā saṭḥ mā bi-ṭarīq ṣināʿī
Florence, Laurenziana Or. 152, fols 97ᵛ–100ʳ.

Fī al-muʿāmalāt fī al-ḥisāb
Istanbul, Nur Osmaniye 2978, fols 39–125.
Istanbul, Feyzullah 1365, fols 73ᵛ–164ʳ.

Archimedes
The Sphere and the Cylinder
Istanbul, Süleymaniye, Fātiḥ 3414.

Euclid
Elements (Isḥāq-Thābit's version)
Tehran, Malik 3433.

Ibn al-Haytham, al-Ḥasan[1]

Fī birkār al-dawā'ir al-'iẓām
London, India Office 1270, Loth 734, fols 116ᵛ–118ʳ.

Fī ḥall shukūk kitāb Uqlīdis fī al-Uṣūl
Istanbul, University 800, 181 fols.

Fī ḥall shukūk fī kitāb al-Majisṭī yushakkiku fīhā ba'ḍ ahl al-'ilm
Aligarh, 'Abd Ḥayy 21.
Istanbul, Beyazit 2304, fols 1^v–20^v.
Istanbul, Fātiḥ 3439, fols 142^r–154^v.

Ibn al-Haytham, Muḥammad

Fī sharḥ al-Majisṭī
Istanbul, Topkapi Saray, Ahmet III 3329, 124 fols.

Maqāla Menelaus fī ta'arruf aqdār al-jawāhir al-mukhtalifa
Lahore, fols 44–51 and Nabī Khān M81.

Ibn Hūd (al-Mu'taman)

Al-Istikmāl
Leiden, Universiteitsbibliotheek 123/1, fols 1^r–80^v.
Copenhagen, The Royal Library, Or. 82, fols 1^r–128^r.

Ibn al-Sarī

Jawāb li-Aḥmad ibn Muḥammad ibn Sarī 'an burhān mas'ala muḍāfa ilā al-maqāla al-sābi'a min kitāb Uqlīdis fī al-Uṣūl
Istanbul, Aya Sofya 4830, fols 139^r–145^v.

Fī bayān mā wahama fīhi Abū 'Alī ibn al-Haytham fī kitābihi fī al-shukūk 'alā Uqlīdis
Istanbul, Aya Sofya 4830, fols 146^v–149^v.

Fī kashf li-l-shubha allatī 'araḍat li-jamā'a mimman yansib nafsahu ilā 'ulūm al-ta'ālīm 'alā Uqlīdis fī al-shakl al-rābi' 'ashar min maqāla al-thāniya 'ashar min kitāb Uqlīdis
Istanbul, Aya Sofya 4830, fols 151^v–154^v.

Al-Khiraqī

Kitāb muntahā al-idrāk fī taqāsīm al-aflāk, Paris, BNF 2499.

2. BOOKS AND ARTICLES

Abū Kāmil: Algèbre et analyse diophantienne, édition, traduction et commentaire par R. Rashed, Berlin/New York, Walter de Gruyter, 2012.

[1] See also the manuscripts mentioned in the List of Works by Ibn al-Haytham.

A. Akhmedov, 'Kniga ob izvletcheni rebra kouba', *Matematika i astronomia v troudakh outchionikh srednebekovovo vostoka izdatel'stvo 'fan'*, Tashkent, 1977, pp. 113–17.

Al-ʿAlawī, *al-Matn al-Rushdī*, Rabat, 1986.

Apollonius, *Les Coniques (I–VII),* commentaire historique et mathématique, édition et traduction du texte arabe par R. Rashed, Berlin/New York, Walter de Gruyter, 2008–2010.

R. C. Archibald, *Euclid's Book on Division of Figures*, Cambridge, 1915.

Archimède, *De la sphère et du cylindre*, French transl. Ch. Mügler, Collection des Universités de France, Paris, Les Belles Lettres, 1970.

Aristote, *Météorologiques*, ed. and French transl. by P. Louis, Paris, Les Belles Lettres, 1982.

Al-Bayhaqī, *Tārīkh ḥukamā' al-Islām,* ed. Kurd ʿAlī, 1st ed., Damascus, Arab Language Academy of Damascus, 1946; 2nd ed., 1976.

O. Becker, *Grundlagen der Mathematik*, 2nd ed., Munich, 1964.

C. Brockelmann, *Geschichte der arabischen Literatur*, 2nd ed., I, Leiden, 1943; Suppl. I, Leiden, 1937; Suppl. III, Leiden, 1942.

J. al-Dabbāgh, 'Infinitesimal Methods of Ibn al-Haitham', *Bulletin of the College of Science*, Baghdad, University of Baghdad, vol. 11, 1970, pp. 8–17.

Al-Dhahabī, *Siyar a'lām al-nubalā'*, ed. S. al-Aranu'ūṭ *et al.,* Beirut, 1984, vol. XVIII.

P. Duhem, *Le système du monde*, vol. 2: *Histoire des doctrines cosmologiques de Platon à Copernic*, Paris, 1965.

Euclid, *The Thirteen Books of Euclid's Elements*, translation and commentary by Th. Heath, 3 vols, 2nd ed., Cambridge, 1926.

Al-Fārisī, Kamāl al-Dīn, *Kitāb tanqīḥ al-manāẓir li-dhawī al-abṣār wa-al-baṣā'ir*, 2 vols, Hyderabad, Osmānia Oriental Publications Bureau, 1347–48/1928–30.

G. Graf, *Geschichte der christlichen arabischen Literatur*, Rome, 1947.

T. L. Heath, *A History of Greek Mathematics*, 2 vols, Oxford, 1921; repr. Oxford, 1965.

A. Heinen, 'Ibn al-Haiṯams Autobiographie in einer Handschrift aus dem Jahr 556 H/1161 A. D.', in U. Haarmann and P. Bachmann (eds), *Die islamische Welt zwischen Mittelalter und Neuzeit, Festschrift für Hans Robert Roemer zum 65. Geburtstag*, Beiruter Texte und Studien 22, Beirut, 1979, pp. 254–77.

M. J. Hermosilla, 'Aproximación a la *Taṭimmat ṣiwān al-ḥikma* de Al-Bayhakī', in *Actas de las II Jornadas de Cultura Arabe e Islámica*, Instituto Hispano-Arabe de Cultura, Madrid, 1980, pp. 263–72.

Ḥijāb, Muḥammad ʿAlī, 'Qā'ima bi-al-mawjūd min kutub Ibn al-Haytham wa-makān ūjūdihi', *Publications of the Egyptian Society for the History of Science*, 2, Cairo, 1948, pp. 139–43.

Jan P. Hogendijk, 'The geometrical parts of the *Istikmāl* of Yūsuf al-Mu'taman ibn Hūd (11th century). An analytical table of contents', *Archives internationales d'histoire des sciences*, 41, 127, 1991, pp. 207–81.

Ibn Abī Uṣaybiʿa, *'Uyūn al-anbā' fī ṭabaqāt al-aṭibbā'*, ed. A. Müller, 2 vols, Cairo/Königsberg, 1882–1884; ed. N. Riḍā, Beirut, 1965.

Ibn 'Asākir, *Tārīkh Madīnat Dimashq,* vol. 43, ed. Sakīna al-Shīrabī, Damascus, 1993.

Ibn Bashkuwāl, *Kitāb al-Ṣila*, ed. Sayyid ʿIzzat al-ʿAṭṭār al-Ḥusaynī, Cairo, 1955.

Ibn Duqmāq, *Kitāb al-Intiṣār li-wāsiṭat 'aqd al-amṣār*, ed. Būlāq, Cairo, n.d.

Ibn al-Haytham

Majmūʿ al-rasā'il, Hyderabad, Osmānia Oriental Publications Bureau, 1938–1939.

Maqāla fī al-shukūk 'alā Baṭlamiyūs ('Dubitationes in Ptolemaeum'), ed. A. I. Sabra and N. Shehaby, Cairo, 1971.

The Optics of Ibn al-Haytham, Books I–III, On Direct Vision, 2 vols, London, 1989.

On the Configuration of the World, edition, translation and commentary by Y. Tzvi Langermann, New York/London, 1990.

Maqālah 'an thamrat al-Ḥikmah. A Treatise on the Fruit (Benefit) of Wisdom, ed. M. ʿAbd al-Hādī Abou-Rīdah, Cairo, 1991.

Ibn al-ʿIbrī, Abū al-Faraj, *Tārīkh mukhtaṣar al-duwal,* ed. O.P. A. Ṣāliḥānī, 1st ed., Beirut, 1890; repr. 1958.

Ibn al-ʿImād, *Shadharāt al-dhahab fī akhbār man dhahab*, Beirut, n.d., vol. II.

Ibn Taghrī Birdī, Abū al-Maḥāsin, *al-Nujūm al-zāhira fī mulūk Miṣr wa-al-Qāhira*, 4 vols, Cairo, 1933; repr. 12 vols, Beirut, 1992.

Al-Khāzinī, *Mīzān al-ḥikma,* Hyderabad, Osmānia Oriental Publications Bureau, 1940–1941.

P. Kunitzsch, *Ibn aṣ-Ṣalāḥ. Zur Kritik der Koordinatenüberlieferung im Sternkatalog des* Almagest, Abhandlungen der Akademie der Wissenschaften in Göttingen, Philologisch-Historische Klasse, series 3, no. 94, Göttingen, 1975.

Kūshyār ibn al-Labbān, *Principles of Hindu Reckoning*, A translation with Introduction and Notes by Martin Levey and Marvin Petruck of the *Kitāb fī uṣūl ḥisāb al-hind*, The University of Wisconsin Press, Publications in Medieval Science, Madison/Milwaukee, 1965; ed. of the Arabic text by A. S Saidan in *Majallat Ma'had al-Makhṭūṭāt al-'Arabiyya*, 13, 1967, pp. 55–83.

Al-Maqrīzī, *Kitāb al-mawā'iẓ wa-al-i'tibār bi-dhikr al-khiṭaṭ wa-al-āthār*, ed. Būlāq, 2 vols, Cairo, n.d.

R. Morelon: *see* Thābit ibn Qurra.

A. Müller, 'Über das sogenannte *Ta'rīkh al-ḥukamā'* des al-Qifṭī', in *Actes du VIII*[e] *Congrès International des Orientalistes tenu en 1889 à Stockholm et à Christiana*, Sect. I, Leiden, 1891, pp. 15–36.

M. Munk, 'Notice sur Joseph ben-Iehouda ou Aboul'hadjâdj Yousouf ben-Ya'hya al-Sabti al-Maghrebi, disciple de Maïmonide', *Journal Asiatique*, 3rd series, 14, 1842, pp. 5–70.

Al-Nadīm, *Kitāb al-fihrist*, ed. R. Tajaddud, Tehran, 1971.

C. Nallino, *Arabian Astronomy, its History during the Medieval Times*, conferences pronounced in Arabic at the University of Cairo, Rome, 1911, pp. 50–64.

M. Naẓīf, *Al-Ḥasan ibn al-Haytham, buḥūthuhu wa-kushūfuhu al-baṣariyya*, 2 vols, Cairo, 1942–1943.

G. Nebbia, 'Ibn al-Haytham, nel millesimo anniversario della nascita', *Physis* IX, 2, 1967, pp. 165–214.

Al-Nu'aymī, *al-Dāris fī ta'rīkh al-madāris*, ed. Ja'far al-Ḥasanī, Damascus, 1951, vol. II.

Olympiodorus, *Commentaires sur Aristote et d'autres épîtres*, ed. 'A. Badawī, Beirut, 1986.

Al-Qalānsī, *Dhayl Tārīkh Dimashq*, Beirut, 1908.

Al-Qifṭī, *Ta'rīkh al-ḥukamā'*, ed. J. Lippert, Leipzig, 1903.

R. Rashed

'La construction de l'heptagone régulier par Ibn al-Haytham', *Journal for the History of Arabic Science*, III, 2, 1979, pp. 309–87.

'Ibn al-Haytham et la mesure du paraboloïde', *Journal for the History of Arabic Sciences*, 5, 1982, pp. 191–262.

'Al-Sijzī et Maïmonide: Commentaire mathématique et philosophique de la proposition II.14 des *Coniques* d'Apollonius', *Archives internationales d'histoire des sciences* 37, 119, 1987, pp. 263–96; repr. in *Optique et mathématiques: recherches sur l'histoire de la pensée scientifique en arabe*, Variorum Reprints, London, 1992, XIII; English transl. in *Fundamenta Scientiæ* 8, 3–4, 1987, pp. 241–56.

'L'analyse et la synthèse selon Ibn al-Haytham', in R. Rashed (ed.), *Mathématiques et philosophie de l'antiquité à l'âge classique*, Paris, 1991, pp. 131–62; repr. in *Optique et mathématiques: recherches sur l'histoire de la pensée scientifique en arabe*, Variorum Reprints, London, 1992, XIV.

'La philosophie des mathématiques d'Ibn al-Haytham. I: *L'analyse et la synthèse*', *Mélanges de l'Institut Dominicain d'Études Orientales du Caire (MIDEO)*, 20, 1991, pp. 31–231.

'Fūthīṭos (?) et al-Kindī sur "l'illusion lunaire"', in M.-O. Goulet-Cazé, G. Madec, D. O'Brien (eds), *ΣΟΦΙΗΣ ΜΑΙΗΤΟΡΕΣ. Hommage à Jean Pépin*, Paris, Institut d'Études Augustiniennes, 1992, pp. 533–59.

Géométrie et dioptrique au Xe siècle: Ibn Sahl, al-Qūhī et Ibn al-Haytham, Paris, Les Belles Lettres, 1993.

'La philosophie des mathématiques d'Ibn al-Haytham. II: *Les connus*', *MIDEO*, 21, 1993, pp. 87–275.

Les Mathématiques infinitésimales du IXe au XIe siècle, vol. I: *Fondateurs et commentateurs: Banū Mūsā, Thābit ibn Qurra, Ibn Sinān, al-Khāzin, al-Qūhī, Ibn al-Samḥ, Ibn Hūd*, London, al-Furqān, 1996; English translation: *Founding Figures and Commentators in Arabic Mathematics*. A history of Arabic sciences and mathematics, Culture and Civilization in the Middle East, London, Centre for Arab Unity Studies/Routledge, 2012.

Les Mathématiques infinitésimales du IXe au XIe siècle, vol. II: *Ibn al-Haytham*, London, al-Furqān, 1993.

Les Mathématiques infinitésimales du IXe au XIe siècle, vol. III: *Ibn al-Haytham. Théorie des coniques, constructions géométriques et géométrie pratique*, London, al-Furqān, 2000.

Œuvre mathématique d'al-Sijzī, vol. I: *Géométrie des coniques et théorie des nombres au Xe siècle*, Les Cahiers du Mideo 3, Louvain/Paris, Peeters, 2004.

Les Mathématiques infinitésimales du IXe au XIe siècle, vol. IV: *Méthodes géométriques, transformations ponctuelles et philosophie des mathématiques*, London, al-Furqān, 2002.

Geometry and Dioptrics in Classical Islam, London, al-Furqān, 2005.

Les Mathématiques infinitésimales du IXe au XIe siècle, vol. V: *Ibn al-Haytham: Astronomie, géométrie sphérique et trigonométrie*, London, al-Furqān, 2006.

'*The Configuration of the Universe*: a book by al-Ḥasan ibn al-Haytham?', *Revue d'histoire des sciences,* 60, no. 1, janvier–juin 2007, pp. 47–63.

'Le pseudo-al-Ḥasan ibn al-Haytham: sur l'asymptote', in R. Fontaine, R. Glasner, R. Leicht and G. Veltri (eds), *Studies in the History of Culture and Science. A Tribute to Gad Freudenthal*, Leiden/Boston, 2011, pp. 7–41.

See also al-Samaw'al and al-Ṭūsī.

R. Rashed and H. Bellosta, *Ibrāhīm ibn Sinān. Logique et géométrie au* x^e^ *siècle*, Leiden, Brill, 2000.

N. Rescher, *Galen and the Syllogisms*, Pittsburgh, 1966.

F. Rosenthal, 'Die arabische Autobiographie', *Studia Arabica: Analecta Orientalia*, 14, 1937, pp. 3–40.

B. A. Rozenfeld, 'The list of physico-mathematical works of Ibn al-Haytham written by himself', *Historia Mathematica*, 3, 1976, pp. 75–6.

A. I. Sabra

'A twelfth-century defence of the figure of the syllogism', *Journal of the Warburg and Courtauld Institutes*, XXVIII, 1965, pp. 14–28.

'The authorship of the *Liber de crepusculis*', *Isis*, 58, 1967, pp. 77–85.

'Ibn al-Haytham', *Dictionary of Scientific Biography,* ed. C. Gillispie, vol. VI, New York, 1972, pp. 204–8.

See also Ibn al-Haytham.

Ṣā'id al-Andalūsī, *Ṭabaqāt al-umam,* ed. H. Bū'alwān, Beirut, 1985; French transl. by R. Blachère, *Livre des Catégories des Nations*, Paris, 1935.

Al-Samaw'al, *al-Bāhir en Algèbre d'as-Samaw'al,* annoted edition and introduction by S. Ahmad and R. Rashed, Damascus, 1972.

J. Schacht and M. Meyerhof, *The Medico-Philosophical Controversy between Ibn Butlan of Baghdad and Ibn Ridwan of Cairo. A Contribution to the History of Greek Learning Among the Arabs,* Faculty of Arts no. 13, Cairo, 1937.

M. Schramm, *Ibn al-Hay<u>th</u>ams Weg zur Physik,* Wiesbaden, 1963.

C. J. Scriba, 'Welche Kreismonde sind elementar quadrierbar? Die 2400 jährige Geschichte eines Problems bis zur endgültigen Lösung in den Jahren 1933/1947', *Mitteilungen der mathematischen Gesellschaft in Hamburg*, XI, 5, 1988, pp. 517–34.

F. Sezgin, *Geschichte des arabischen Schrifttums*, V: *Mathematik*, Leiden, 1974, VI: *Astronomie*, Leiden, 1978.

Al-Shahrazūrī, *Nuzhat al-arwāḥ wa-rawaḍat al-afrāḥ fī tārīkh al-ḥukamā' wa-al-falāsifa*, Hyderabad, Osmānia Oriental Publications Bureau, 1976.

S. M. Stern, 'Ibn al-Samḥ', *Journal of the Royal Asiatic Society*, 1956; repr. in S. M. Stern, *Medieval Arabic and Hebrew Thought*, ed. F. W. Zimmermann, London, 1983.

H. Suter

'Die Kreisquadratur des Ibn el-Haiṯam', *Zeitschrift für Mathematik und Physik*, 44, 1899, pp. 33–47.

Die Mathematiker und Astronomen der Araber und ihre Werke, Leipzig, 1900; Johnson Reprint, New York, 1972, pp. 91–5.

'Corrigenda et addenda', *Bibliotheca Mathematica*, 3rd series, 4, 1903, pp. 295–6.

'Die Abhandlung über die Ausmessung des Paraboloides von el-Ḥasan b. el-Ḥasan b. el-Ḥaiṯam', *Bibliotheca Mathematica,* 3rd series, 12, 1911–1912, pp. 289–332.

Tashkupri-Zadeh, *Miftāḥ al-Sa'āda*, *Miftāḥ al-Sa'āda*, ed. Kamil Bakry and Abdel-Wahhab Abū' L-Nur, Cairo, 1968.

Thābit ibn Qurra, *Œuvres d'astronomie*, ed., transl. and com. by Régis Morelon, Paris, Les Belles Lettres, 1987.

Türker Küyel, Mubahat

'Les critiques d'Ibn al-Ṣalāḥ sur le *De Caelo* d'Aristote et sur ses commentaires', *Araştirma,* II, 1964, pp. 19–30 and 52–79.

'Ibn uṣ-Salâḥ comme exemple à la rencontre des cultures', *Araştirma,* VIII, 1972 (published in 1973); as well as his edition and translation in Turkish: 'Aristoteles' in Burhân Kitabi'nin ikinci makalesi'nin sonundaki kismin şerhi ve oradaki yanlişin düzeltilmesi hakkinda', *Araştirma,* VIII, 1972 (published in 1973).

Al-Ṭūsī, Sharaf al-Dīn, *Œuvres mathématiques. Algèbre et géométrie au XIIe siècle*, Texte établi et traduit par R. Rashed, 2 vols, Paris, Les Belles Lettres, 1986.

E. Wiedemann, 'Ibn al-Haiṯam, ein arabischer Gelehrter', *Festschrift J. Rosenthal... gewidmet*, Leipzig, 1906, pp. 149–78.

Yāqūt, *Mu'jam al-buldān*, Beirut, n.d., vol. III.

M. A. Youschkevitch, *Les Mathématiques arabes*, Paris, 1976.

INDEX OF NAMES

SUBJECT INDEX

INDEX OF WORKS

Milton Keynes UK
Ingram Content Group UK Ltd.
UKHW031138141024
449569UK00024B/1239

9 781138 109131

CULTURE AND CIVILIZATION IN THE MIDDLE EAST

General Editor: Ian Richard Netton
Professor of Islamic Studies, University of Exeter

This series studies the Middle East through the twin foci of its diverse cultures and civilizations. Comprising original monographs as well as scholarly surveys, it covers topics in the fields of Middle Eastern literature, archaeology, law, history, philosophy, science, folklore, art, architecture and language. While there is a plurality of views, the series presents serious scholarship in a lucid and stimulating fashion.

PREVIOUSLY PUBLISHED BY CURZON

THE ORIGINS OF ISLAMIC LAW
The Qur'an, the *Muwatta'* and Madinan *Amal*
Yasin Dutton

A JEWISH ARCHIVE FROM OLD CAIRO
The history of Cambridge University's Genizah collection
Stefan Reif

THE FORMATIVE PERIOD OF TWELVER SHI'ISM
Hadith as discourse between Qum and Baghdad
Andrew J. Newman

QUR'AN TRANSLATION
Discourse, texture and exegesis
Hussein Abdul-Raof

CHRISTIANS IN AL-ANDALUS 711–1000
Ann Rosemary Christys

FOLKLORE AND FOLKLIFE IN THE UNITED ARAB EMIRATES
Sayyid Hamid Hurriez